Wireless Edge Caching

Understand both uncoded and coded caching techniques in future wireless network design. Expert authors present new techniques that will help you improve backhaul; minimize load; reduce deployment cost; and increase security, energy efficiency, and the quality of the user experience. Covering topics from high-level architectures to specific requirement-oriented caching design and analysis, including big data–enabled caching, caching in cloud-assisted 5G networks, and security, this is an essential resource for academic researchers, postgraduate students, and engineers working in wireless communications.

Thang X. Vu is a Research Associate at the Interdisciplinary Centre for Security, Reliability and Trust, University of Luxembourg.

Ejder Baştuğ is a member of technical staff at Nokia Bell Labs, France.

Symeon Chatzinotas is Full Professor/Chief Scientist at the University of Luxembourg and a visiting professor at the University of Parma, Italy.

Tony Q.S. Quek is a Professor at Singapore University of Technology and Design. He is a Distinguished Lecturer of the IEEE Communications Society and a Fellow of IEEE.

Wireless Edge Caching

Modeling, Analysis, and Optimization

Edited by

THANG X. VU
University of Luxembourg

EJDER BAŞTUĞ
Nokia Bell Labs, France

SYMEON CHATZINOTAS
University of Luxembourg

TONY Q.S. QUEK
Singapore University of Technology and Design

CAMBRIDGE
UNIVERSITY PRESS

University Printing House, Cambridge CB2 8BS, United Kingdom

One Liberty Plaza, 20th Floor, New York, NY 10006, USA

477 Williamstown Road, Port Melbourne, VIC 3207, Australia

314–321, 3rd Floor, Plot 3, Splendor Forum, Jasola District Centre, New Delhi – 110025, India

79 Anson Road, #06–04/06, Singapore 079906

Cambridge University Press is part of the University of Cambridge.

It furthers the University's mission by disseminating knowledge in the pursuit of education, learning, and research at the highest international levels of excellence.

www.cambridge.org
Information on this title: www.cambridge.org/9781108480833
DOI: 10.1017/9781108691277

First published 2021

Printed in the United Kingdom by TJ Books Limited, Padstow Cornwall

A catalogue record for this publication is available from the British Library.

ISBN 978-1-108-48083-3 Hardback

Contents

Contributors

Mehrnaz Afshang
Bradley Department of Electrical and Computer Engineering,
Virginia Tech, Blacksburg, Virgina, USA

Italo Atzeni
Communication Systems Department, EURECOM, Sophia Antipolis,
France

Ejder Baştuğ
Nokia Bell Labs, Paris-Saclay, France

Mehdi Bennis
Center for Wireless Communications, University of Oulu, Oulu, Finland

Youlong Cao
Department of Electronic Engineering, Shanghai Jiao Tong University, Shanghai,
China

Jacob Chakareski
College of Engineering, University of Alabama, Tuscaloosa, USA

Symeon Chatzinotas
Interdisciplinary Center for Security, Reliability and Trust, University of
Luxembourg, Luxembourg City, Luxembourg

Erkai Chen
Department of Electronic Engineering, Shanghai Jiao Tong University, Shanghai,
China

Zheng Chen
Department of Electrical Engineering (ISY), Linköping University, Linköping, Sweden

Mingzhe Chen
Beijing University of Posts and Telecommunications, Beijing, China

Mérouane Debbah
CentraleSupélec, Gif-sur-Yvette, France

Harpreet S. Dhillon
Bradley Department of Electrical and Computer Engineering,
Virginia Tech, Blacksburg, Virgina, USA

Suhas Diggavi
Electrical and Computer Engineering Department, University of California,
Los Angeles, USA

Khai Nguyen Doan
Information Systems Technology and Design Pillar, Singapore
University of Technology and Design, Singapore

Eryk Dutkiewicz
School of Electrical and Data Engineering, University of Technology Sydney,
Sydney, Australia

Jad Hachem
Software Engineer, Google, Sunnyvale, CA, USA

Dinh Thai Hoang
School of Electrical and Data Engineering, University of Technology Sydney,
Sydney, Australia

George Iosifidis
School of Computer Science & Statistics, Trinity College Dublin, Ireland

Mingyue Ji
Department of Electrical and Computer Engineering, University of Utah,
Salt Lake City, USA

Nikhil Karamchandani
Department of Electrical Engineering, Indian Institute of Technology, Bombay,
India

Marios Kountouris
Communication Systems Department, EURECOM, Sophia Antipolis,
France

Shankar Krishnan
Bradley Department of Electrical and Computer Engineering,
Virginia Tech, Blacksburg, Virgina, USA

Jeongho Kwak
Daegu University, Gyeongsan, South Korea

Kangqi Liu
Department of Electronic Engineering, Shanghai Jiao Tong University, Shanghai, China

Ya-Feng Liu
Institute of Computational Mathematics and Scientific/Engineering Computing, Academy of Mathematics and Systems Science, Chinese Academy of Sciences, Beijing, China

Marco Maso
Nokia Bell Labs, Paris-Saclay, France

Sharayu Moharir
Department of Electrical Engineering, Indian Institute of Technology Bombay, India

Derrick W. K. Ng
School of Electrical Engineering and Telecommunications, University of New South Wales, Sydney, Australia

Diep Nguyen
School of Electrical and Data Engineering, University of Technology Sydney, Sydney, Australia

Thang Van Nguyen
Information Systems Technology and Design Pillar, Singapore University of Technology and Design, Singapore

Dusit Niyato
School of Computer Science and Engineering, Nanyang Technological University, Singapore

Björn Ottersten
Interdisciplinary Center for Security, Reliability and Trust, University of Luxembourg, Luxembourg City, Luxembourg

Nikolaos Pappas
Department of Science and Technology (ITN), Linköping University, Linköping, Sweden

Georgios Paschos
France Research Center, Huawei Technologies, Boulogne-Billancourt, France

Dario Pompili
Department of Electrical and Computer Engineering, Rutgers University,
New Brunswick, NJ, USA

Tony Q.S. Quek
Information Systems Technology and Design Pillar, Singapore University of Technology and Design, Singapore

Walid Saad
Department of Electrical and Computer Engineering, Virginia Tech, Blacksburg, Virginia, USA

Yuris Mulya Saputra
School of Electrical and Data Engineering, University of Technology Sydney, Sydney, Australia

Robert Schober
Institute for Digital Communications, Friedrich-Alexander-Universität Erlangen-Nürnberg, Erlangen, Germany

Omid Semiari
Department of Electrical and Computer Engineering, University of Colorado Colorado Springs, Colorado, USA

Meixia Tao
Department of Electronic Engineering, Shanghai Jiao Tong University, Shanghai, China

Tuyen X. Tran
Department of Electrical and Computer Engineering, Rutgers University,
New Brusnswick, NJ, USA

Thang X. Vu
Interdisciplinary Center for Security, Reliability and Trust, University of Luxembourg, Luxembourg City, Luxembourg

Vincent W. S. Wong
Department of Electrical and Computer Engineering, University of British Columbia, Vancouver, British Columbia, Canada

Lin Xiang
Interdisciplinary Centre for Security, Reliability and Trust, University of Luxembourg, Luxembourg City, Luxembourg

Fan Xu
Department of Electronic Engineering, Shanghai Jiao Tong University, Shanghai, China

Changchuan Yin
Beijing University of Posts and Telecommunications, Beijing, China

Wei Yu
Department of Electrical and Computer Engineering, University of Toronto, Toronto, ON, Canada

Guosen Yue
FutureWei Technologies (Huawei R&D), Bridgewater, NJ, USA

Engin Zeydan
Centre Tecnològic de Telecomunicacions de Catalunya (CTTC), Barcelona, Spain

Preface

Wireless edge caching refers to a novel distributed architecture for storing contents at the wireless edge networks, such as at base stations and user terminals, to efficiently accommodate the proliferation of mobile devices and new data-hungry applications. Efficient deployment of edge caching is facing various challenges and requires a cross-layer design method when applied to wireless networks. One major challenge is how to jointly design caching (at the network layer) and signal transmission (at the physical layer) in an effective manner. Another problem is to deal with dynamic content popularity and network topology. Since content popularity is time varying, caching operations can be optimized only if a fresh view of the system is maintained. This requires massive data collection and processing and statistical inference from these data, which by itself is a complex task to handle. The successful implementation of wireless edge caching therefore heavily depends on joint research developments in different scientific domains, such as networking, information theory, machine learning, and wireless communications.

This book is intended to give a broad overview of the current literature as well as the latest research results for both *uncoded* and *coded* caching techniques for future wireless network design. The book covers the range from high-level architectures to specific requirement-oriented caching design and analysis. A number of new techniques are also presented to improve the edge caching systems in terms of backhaul load minimization, deployment cost reduction, security, energy efficiency, and user quality of experience.

It is our hope that this book will serve as a useful reference for academic researchers, postgraduate students, and engineers and that it will motivate more researchers to contribute to the wireless edge caching research, which will be fruitful for the next generation of networks.

1 Introduction

Ejder Baştuğ, Thang X. Vu, Symeon Chatzinotas, and Tony Q.S. Quek

1.1 History of Caching

The idea of caching can be traced back to the sixties in context of fast memory access in operating systems. According to [1], the most effective online cache replacement strategy is earliest deadline first (EDF) policy that replaces contents in memory blocks that are not going to be requested in the nearest future, provided that the future content demand profile is perfectly available. Other examples include analysis of least recently used (LRU) cache placement policy under stationary demand, either using a Markovian model [2] or an approximate cache placement method [3]. Implementation of these methods in the systems of that era are largely based on LRU, least frequently used (LFU), hybrid, and/or randomized approaches, while maintaining low-complexity design. The main objective therein is to maximize cache hits subject to ultra-low memory capacity constraints, thus achieving the goal of fast content access in such a localized setting.

The nineties witnessed an explosion of interconnected devices and systems over the internet, with the rise of world wide web contributing to creation of millions of websites, startups, and projects. This growth together with the *dot-com bubble* introduced a huge congestion over the web infrastructure, even leading its inventor, Sir Tim Berners-Lee, to mention the network congestion and the conventional client–server connectivity model as the main bottlenecks for the scalability of the web-based internet. Regarding the client–server model by which a web page is downloaded from the same centralized server by every internet client/user (often) multiple times, a workaround solution to alleviate the bottleneck was to replicate contents across several proxy servers that are geographically close to users, mostly supporting heuristic caching placements of unencrypted traffic and static contents. This allowed websites and content providers to minimize their global network bandwidth, provide rapid content access, and offload their servers.

The automation of this technical process and new business models ultimately led to creating content delivery networks (CDNs) around the late 1990s (see [4] for a brief literature review), and over the time, this evolution yielded complex infrastructures supporting various features, like the caching of large video streaming, social network data, and high-traffic websites, mostly over secure connections and distributed around the planet. Several cache placement strategies for CDN have been proposed [9] and implemented in this regard, allowing content providers to minimize access delays to the requested contents. The questions that have been targeted in this scope are (1) where to deploy the cache servers, (2) how much cache capacity is required for each node, (3) which content to cache, and (4) how and when to redirect/route the contents to end users. Concepts like content centric networking (CCN) and information centric networking (ICN) emerged during this period (see [5, 6] for examples), aiming to fundamentally shift the way that content is stored and accessed on the internet and looking to pave the way of successful implementations. Most of CDN-like caching technologies are now crucial elements of the world networking infrastructure.

Nowadays, all these networking-level caching challenges are being revisited in the wireless domain, mostly driven by the steep increase of mobile data traffic and billions of devices/users connected to the mobile networks, where telecom operators and organizations are looking for innovative ways to design and deploy cellular networks of the future. The aforementioned technical challenges of caching are not only being revisited but also taking a twisted step, with majority of traditional caching problems taking into account limited backhaul capabilities in dense cellular deployments, base station cooperation and coordination, coded/uncoded techniques, learning in large scale, mobility, economics, ultra-performance demanding new applications, level of content placement at the wireless edge (namely at base stations and user devices), and others. The research community is growing and aims to bring the wireless caching into reality (see [7, 8, 10] for examples), also with industrial and startup activities taking place. The aim of our book is to cover most of these technical aspects of wireless edge caching, with the help of experts and researchers active in the domain.

1.2 Summary of the Book

The book presents a collection of invited chapters on a wide range of issues and open challenges related to edge caching applied to future wireless networks, which are coherently presented in four parts:

- Part I: Optimal Cache Placement and Delivery
- Part II: Proactive Caching
- Part III: Cache-Aided Interference and Physical Layer Management
- Part IV: Energy-Efficiency, Security, Economics, and Deployment

Part I provides a comprehensive view on optimal cache design for both placement and delivery phases. The five chapters in this part cover most advanced techniques in coded caching, cache-aided device-to-device communications, and cooperative caching. More specifically, Chapter 2 provides the comprehensive performance analysis of coded caching in heterogeneous wireless networks via a joint design of storage and delivery and the optimal trade-off between the cache memory size and the broadcast delivery rate. Chapter 3 investigates the performance of cache-aided device-to-device networks under both uncoded and coded caching strategies. Chapter 4 proposes a cooperative hierarchical caching framework in cloud radio access networks (C-RAN) and explores the synergies of the in-network computing and storage resources. Chapter 5 proposes the concept of stochastic caching in large wireless networks and analyzes the three main performance metrics: cache-hit probability, successful delivery probability, and content delivery latency. Chapter 6 studies the edge caching via a joint design of caching, routing, and channel assignment for video delivery over coordinated small-cell cellular systems.

Part II provides key aspects in designing proactive caching algorithms with users' behavior prediction and learning techniques. In particular, Chapter 7 proposes a novel popularity-predicting-based caching procedure based on raw video data to determine an optimal cache placement policy, which deals with both published and unpublished videos. Chapter 8 studies wireless edge caching paradigms for mobile social networks to improve reliable and low-latency communication services for mobile users on social networking. In Chapter 9, a big data analytic-based framework is proposed for content popularity estimations and proactive caching at base stations. Chapter 10 investigates the impact of mobility on edge caching and proposes the optimal cache in both static and mobile user scenarios.

Part III consists of four chapters that provide the cross-layer cache-aided design for interference and physical layer resources management under both coded caching and uncoded methods. More precisely, Chapter 11 studies the caching effects on multicast-enabled access downlinks and proposes cache-aware joint designs of the content-centric base station clustering and multicast beamforming. Chapter 12 analyzes the impact of caching in the interference networks under both fully and partially connected topologies. Chapter 13 studies the performance enhancement brought by the caching capabilities in full-duplex radios in the context of ultra-dense networks. Chapter 14 investigates the impact of edge caching in mobile millimeter wave systems via a mobility management framework that exploits broadband millimeter wave connectivity to cache the contents of interest.

Part IV highlights the edge caching in future wireless networks from various aspects such as energy efficiency, security, and economics as well as the edge caching deployment in unmanned aerial vehicle (UAV) and virtual reality systems. Chapter 15 investigates the energy-efficiency and delivery time performance of the wireless edge caching systems via both uncoded and coded caching strategies. Chapter 16 studies the cache-enabled UAVs in C-RAN to reduce the content delivery latency and improve the users' quality of experience. Chapter 17 investigates the application of edge caching to enhance the physical layer security of cellular networks via proactive content sharing policy across a subset of base stations. Chapter 18 proposes a framework for the delivery of 360°-navigable videos to 5G virtual reality wireless clients in future cooperative multi-cellular systems. Finally, Chapter 19 investigates the elastic wireless edge caching that reveals the economic interactions of different stake holders in the network and provides the key differences between in-network and edge caching.

References

[1] L. A. Belady, "A study of replacement algorithms for a virtual-storage computer," *IBM Systems Journal*, vol. 5, no. 2, pp. 78–101, 1966.

[2] A. J. Smith, "Analysis of the optimal, look-ahead demand paging algorithms," *SIAM Journal on Computing*, vol. 5, no. 4, pp. 743–757, 1976.
[3] R. Fagin, "Asymptotic miss ratios over independent references," *Journal of Computer and System Sciences*, vol. 14, no. 2, pp. 222–250, 1977.
[4] J. Wang, "A survey of web caching schemes for the internet," *ACM SIGCOMM Computer Communication Review*, vol. 29, no. 5, pp. 36–46, 1999.
[5] A. Araldo, M. Mangili, F. Martignon, and D. Rossi, "Cost-aware caching: optimizing cache provisioning and object placement in ICN," arXiv:1406.5935, 2014.
[6] B. Ahlgren, C. Dannewitz, C. Imbrenda, D. Kutscher, and B. Ohlman, "A survey of information-centric networking," *IEEE Communications Magazine*, vol. 50, no. 7, pp. 26–36, 2012.
[7] G. S. Paschos, G. Iosifidis, M. Tao, D. Towsley, and G. Caire, "The role of caching in future communication systems and networks," *IEEE Journal on Selected Areas in Communications*, vol. 36, no. 6, pp. 1111–1125, 2018.
[8] G. S. Paschos, G. Iosifidis, M. Tao, D. Towsley, and G. Caire, "Guest editorial caching for communication systems and networks—part II," *IEEE Journal on Selected Areas in Communications*, vol. 36, no. 8, pp. 1663–1665, Aug. 2018.
[9] S. Borst, V. Gupta, and A. Walid, "Distributed caching algorithms for content distribution networks," in *INFOCOM, 2010 Proceedings IEEE*, IEEE, 2010, pp. 1–9.
[10] G. Paschos, E. Bastug, I. Land, G. Caire, and M. Debbah, "Wireless caching: technical misconceptions and business barriers," *IEEE Communications Magazine*, vol. 54, no. 8, pp. 16–22, 2016.

Part I

Optimal Cache Placement and Delivery

2 Coded Caching for Heterogeneous Wireless Networks

Nikhil Karamchandani, Jad Hachem, Suhas Diggavi, and Sharayu Moharir

2.1 Introduction

Broadband data consumption has witnessed a tremendous growth over the past few years, due in large part to multi-media applications such as video-on-demand. Increasing data demand has been managed in the wired internet via content distribution networks (CDNs) that mirror data in various locations and effectively push content closer to end users. CDNs help reduce host server load by serving user requests locally via content cached locally. This solution works best when neither local storage nor data rates are bottlenecks [1]. Neither of these is true in cellular networks; the last-hop wireless link has low throughput (improvements in cellular data rates do not sufficiently compensate for exploding demand) and there is virtually no storage at base stations. To address the throughput issue, a heterogeneous wireless network (HetNet) architecture has been proposed for 5G systems [2, 3]. HetNets consist of a dense deployment of very small cells (pico/femto) with high data rates, combined with a sparse deployment of larger macro-cellular base stations (BSs) of comparatively lower data rates; WiFi access points (APs) can be a typical small cell. However, this architecture is "incomplete" because the APs are connected to the backbone via best-effort backhaul, which is a bottleneck [4]. Even joint management of APs and BSs cannot provide enough improvement to deal with projected demand growth [2, 5]. This leads us to argue that the traditional CDN approach in an enhanced wireless system design is an incomplete solution: the CDNs optimize content placement without accounting for characteristics of wireless communications, and wireless system design focuses only on increasing delivery rates agnostic to content. To fully enable content-centric wireless networks, we need a joint design of content placement, access, and delivery. Broadly, the proposal is to provide storage capabilities at the network nodes (base stations and WiFi access points) and create a large-scale distributed cache. Users will be served by connecting them to one or more nodes hosting their requested content. Delivery protocols will use algorithms that are aware of attributes of wireless networks like the broadcast medium and interference. Figure 2.1 illustrates this.

In this chapter we describe a problem based on an architecture where content is stored at multiple APs without a priori knowing the user requests, and the base station broadcast is used judiciously to complement the local caching, *after* the user requests

Nikhil Karamchandani would like to acknowledge support from SERB, Govt. of India, in the form of a grant titled "Content Caching and Delivery over Wireless Networks." Suhas Diggavi's work was supported in part by NSF grant 1514531, UC-NL grant LFR-18-548554 and ARL cooperative grant W911NF-17-2-0196.

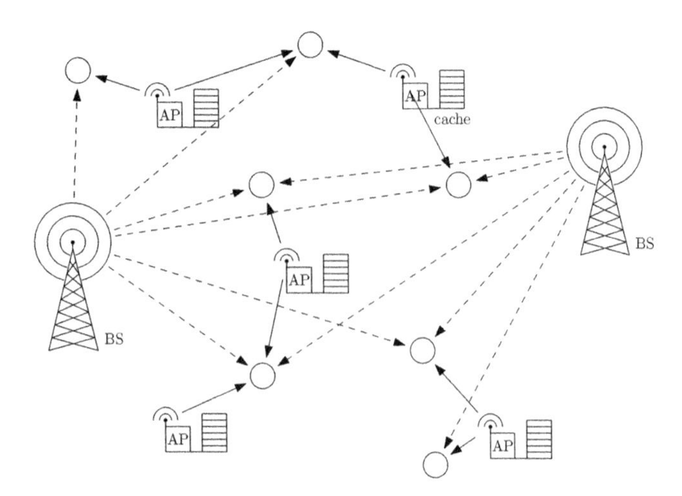

Figure 2.1 Caching in a wireless heterogeneous network (HetNet).

are known. This is motivated by the new approach initiated in the seminal works [6, 7], where it has been shown that joint design of storage and delivery (also known as "coded caching") can significantly improve content delivery-rate requirements. This was enabled by content placement that creates (network-coded) multicast opportunities among users with access to different storage units, even when they have different (and a priori unknown) requests. This enables an examination of the optimal trade-off between the cache memory size and the broadcast delivery rate.

We will begin by discussing the setup studied in [6, 7], which introduced the idea of coded caching, and describing their main results. These works considered the case where all files in the catalog have the same popularity. However, it is well understood that content demand is non-uniform in practice, with some files being more popular than others. Motivated by this, we then describe models that take this non-uniform popularity into account and discuss how it impacts the results and proposed caching and delivery schemes. While all these focus on a setup where each user has (fixed) access to a single unique cache and hears the common message broadcast by the base station, several generalizations to the network structure have been studied recently. In particular, we then discuss in detail the case where, based on the user requests, each user can be adaptively matched to one cache (possibly among a subset of caches). Finally, we end the chapter by considering a setting where multiple base stations simultaneously communicate with users over an interference network. We will argue for a separation-based architecture for such a setting where reliable (physical-layer) wireless communication is appropriately matched to coded caching.

2.2 Overview of Coded Caching

Coded caching was first introduced in 2012 by Maddah-Ali and Niesen [6] as a solution to the content distribution problem in a wireless setting. In order to focus on this new technique, the setup ignored variations in content popularity and limited user-to-cache access to exactly one user connecting to one cache. The authors showed that conventional caching techniques are inefficient in such a setup. Instead, one can leverage the broadcast capabilities inherent to wireless communications in order to send a small network-coded message that can serve a large number of users at once. The setup and ideas became fundamental to much of the following literature on the subject, and so in this chapter we give an overview of the results and insights from [6].

2.2.1 Setup and Notation

We begin by describing the setup studied in the seminal work of Maddah-Ali and Niesen [6], illustrated in Figure 2.2. Consider a server hosting a content library with N files, labeled $W_1, \ldots, W_N$, of size F bits each. There are K users in the network, each of which is equipped with a local cache of size MF bits. The server is connected to the users via an error-free broadcast link.

The system operates in two phases. We start with a *placement phase* in which all the user caches are populated with content related to the N files. No restrictions are posed on this placement phase aside from the cache memory constraint. In particular, the caches are not restricted to holding just files or parts of files; any function (deterministic or not) of the files can be used, and the caches do not necessarily have to hold the same information. Crucially, this is done before the user requests are revealed to the system. Next, we move to the *delivery phase*, which starts with each user making a request for one file from the content library. Based on the user requests and the content stored in the caches, the server transmits a message of size RF bits across the error-free broadcast link, intended to serve the requests of all the users. Each user then combines this message with the contents of its own cache in order to recover the file that it requested.

Our resources here are the *cache memory* M and the *broadcast rate* R. Clearly there is a trade-off between them: the larger the cache, the smaller the size of the broadcast message needed to serve the users. The goal is to characterize the optimal trade-off.

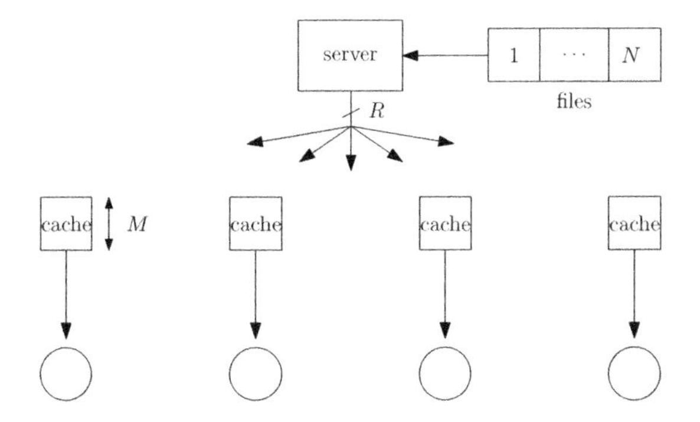

Figure 2.2 The basic coded caching problem, with N files, K caches, and one user at every cache. There is an error-free broadcast link from the server to the users.

Formally, we say that a pair (R, M) is achievable if there exists a caching scheme $\mathcal{S}_F$ for every file size F such that

- $\mathcal{S}_F$ uses a cache memory at each user of capacity at most MF bits and a broadcast rate from the server of size at most RF bits; and
- for any collection of user requests, the probability that each user recovers its requested file without error goes to one as $F \to \infty$.

Then, our goal is to find, for every $M \geq 0$, the information-theoretically optimal rate defined as

$$R^\star(M) = \inf\{R\colon (R, M) \text{ is achievable}\}. \tag{2.1}$$

Note that the infimum in (2.1) is over all possible caching and delivery schemes, without any restrictions.

We discuss a small example in the next section to illustrate the setup as well as some representative caching and delivery schemes.

2.2.2 A Small Illustrative Example

Consider a special case of the system described earlier with $N = 2$ files, $K = 2$ users, and $M = 1$ memory at each user. Denote the two files by A and B. Any caching and delivery scheme has to specify what to store in the caches during the placement phase and what the server should transmit during the delivery phase so that the user requests can be served.

For instance, a natural cache placement strategy is to split file A into two equal parts A_1, A_2 and similarly file B into B_1, B_2. Each cache stores one-half of each file. In a conventional caching and delivery system, each cache would store (A_1, B_1), and the server would handle each user request via a separate transmission. For example, if one user wants file A and the other wants file B, the server would transmit A_2 to the first user and B_2 to the second user, and thus the broadcast message size is equivalent to the size of one file. This scheme leverages the local presence of a cache at every user: each user has access to the half file present in its cache, which reduces the message size by that amount per user. However, using network coding techniques, we can design the cache contents in such a way that each user benefits from the contents of both its cache *and the other user's cache*.

To do so, we consider an alternate placement and delivery strategy whose main idea is to store different file parts in each user's cache in a way that enables sending linear combinations of file parts that are simultaneously useful to both users. First, in the placement phase the first user's cache stores (A_1, B_1) while the second user's cache stores (A_2, B_2). Second, instead of treating the two user requests separately during the delivery phase, we consider them jointly. For example, if the first user requests file A and the second user requests file B, then the server sends a linear combination $A_2 \oplus B_1$ on the shared broadcast link, where $\oplus$ denotes the bitwise-XOR operation. The first user has A_1 available in its local cache and can combine B_1 with $A_2 \oplus B_1$ in order to recover

A_2; the second user can similarly obtain both B_1 and B_2. Thus the users' requests were both served with a broadcast message size of only half a file.

Note that this scheme reduces the server transmission rate by a factor of two over a conventional scheme. The main idea is to carefully design the cache contents so as to maximize the number of coded multicasting opportunities during server transmission, enabling the server to send a single message satisfying multiple users, possibly requesting different files, simultaneously.

2.2.3 Achievable Rate

These ideas were generalized in [6], which proposed a new caching and delivery scheme for the general setup described in Section 2.2.1 and also characterized its achievable rate, as shown in the following result.

THEOREM 2.1 *For the system described in Section 2.2.1 with* $M \in \frac{N}{K} \cdot \{0, 1, \ldots, K\}$, *there exists a placement and delivery scheme that achieves the following server transmission rate:*

$$R(M) = K \cdot \left(1 - \frac{M}{N}\right) \cdot \frac{1}{1 + KM/N}.$$

For $M \in [0, N]$, *the lower convex envelope of these points can be achieved.*

A rate of $N - M$ can also be achieved (without coded caching) and is useful when the number of files is small, but in the more relevant case where $N \geq K$, the rate in Theorem 2.1 is smaller, and we will henceforth focus on it.

The scheme achieving this rate is a generalization of the ideas described in Section 2.2.2, and we describe it later in the chapter. Before we do that, we can gain some insights about the achievable rate in Theorem 2.1 by factoring it into three terms. The first term, K, is the total number of users and represents the rate needed without caching, since in the worst case the server might be required to transmit K distinct files. The second term, $1 - M/N$, is referred to as the *local caching gain*. It is the gain obtained by the fact that each user already has a fraction M/N of its requested file stored locally. The third term, $1/(1 + KM/N)$, is referred to as the *global caching gain*. It is specifically achieved by the fact that the server sends coded multicast messages that are useful to many users at once (more precisely, each bit is used by exactly $1 + KM/N$ users). In effect, the coded multicast allows each user to benefit from the caches of all the other users as well, hence the appearance of the total system memory KM in the expression. This is the gain that the proposed scheme derives over a conventional caching and delivery scheme that serves the user requests through separate unicast server messages.

The significance of the global caching gain can be captured by noticing that the achievable rate in Theorem 2.1 can be upper-bounded by

$$R(M) \leq \min\left\{K, \frac{N}{M}\right\}\left(1 - \frac{M}{N}\right). \tag{2.2}$$

Consequently, as long as the total memory in the network is large enough to hold the entire library (i.e., $KM \geq N$), the achievable rate is at most $N/M - 1$, and is thus *independent of the number of users*!

Proof of Theorem 2.1 We now describe a placement and delivery scheme for the general system described in Section 2.2.1.

Placement phase: Denote the files by $W_1, W_2, \ldots, W_N$. Let $t \triangleq MK/N$. From the statement of the theorem, note that t is an integer between 0 and K. Divide each file W_i into $\binom{K}{t}$ equal parts and index them as follows:

$$W_i = \left(W_i^S : S \subseteq \{1, 2, \ldots, K\}, |S| = t\right).$$

Note that each subfile is of size $F/\binom{K}{t}$. For any user i, its cache stores $\binom{K-1}{t-1}$ pieces of each file W_j given by

$$\left(W_j^S : i \in S, S \subseteq \{1, 2, \ldots, K\}, |S| = t\right).$$

The total amount of storage each cache needs to store these pieces is given by

$$N \cdot \binom{K-1}{t-1} \cdot \frac{NF}{\binom{K}{t}} = \frac{Ft}{K} = MF,$$

where the last equality follows from the definition of t. Thus, the storage constraint is satisfied at each cache.

Delivery phase: For $t = K$, the memory at each cache is $M = N$ and is sufficient to store the entire file catalog. Thus, the required server transmission rate is zero. Later we consider $t \in \{0, 1, 2, \ldots, K - 1\}$. Denote by d_i the index of the file requested by user i, i.e., user i requests file W_{d_i}. Consider a subset $S \subseteq \{1, 2, \ldots, K\}$ of size $t + 1$. Note that for each $j \in S$, there is a subfile of its requested file W_{d_j}, given by $W_{d_j}^{S \setminus \{j\}}$, which is stored in the caches of all the other users in S. Corresponding to this subset S, the server transmits the message

$$\oplus_{j \in S} W_{d_j}^{S \setminus \{j\}}. \tag{2.3}$$

We repeat this procedure for each of the $\binom{K}{t+1}$ subsets of $\{1, 2, \ldots, K\}$ of size $t + 1$.

We now show that this placement and delivery scheme allows each user i to recover its requested file W_{d_i}. Consider a subset $S \subseteq \{1, 2, \ldots, K\}$ of size $t + 1$ such that $i \in S$. Note from (2.3) that among the $t + 1$ subfiles involved in the transmitted message corresponding to S, all except its desired subfile $W_{d_i}^{S \setminus \{i\}}$ is already available in the cache of user i. Thus, the user is able to recover its desired subfile.

Repeating this argument, user i is able to recover all the subfiles of the form

$$\left(W_{d_i}^T : T \subseteq \{1, 2, \ldots, K\} \setminus \{i\}, |T| = t\right).$$

Furthermore, all the other subfiles of the requested file W_{d_i} are already available in the cache of user i. Thus, the described placement and delivery strategy represents a feasible scheme for the general system.

To complete the proof of the theorem, we have to evaluate the server transmission rate for the proposed scheme. From (2.3), the size of the transmission corresponding to any subset S of size $t+1$ is $F/\binom{K}{t}$. Since there is one such transmission corresponding to each such subset, the total transmission size is given by

$$\binom{K}{t+1} \cdot \frac{F}{\binom{K}{t}} = \frac{K-t}{t+1} = K \cdot \left(1 - \frac{M}{N}\right) \cdot \frac{1}{1+KM/N}. \qquad \square$$

Decentralized scheme Note that the described scheme carefully orchestrates the placement phase to create simultaneous coded multicasting opportunities during the delivery phase. In particular, the number of users and their identities are required to ensure that each cache is populated with the right file pieces. Since this kind of information might not always be available in practice, the authors develop a decentralized placement (and corresponding delivery scheme) scheme in [7], where each user randomly samples MF bits from the NF bits in the content library. The achievable rate of this scheme is characterized in [7] and is presented in the following:

THEOREM 2.2 *Consider the system described in Section 2.2.1. For $M \in [0, N]$, there exists a decentralized placement scheme and a corresponding delivery scheme that achieves a server transmission rate arbitrarily close to*

$$R_D(M) = K \cdot \left(1 - \frac{M}{N}\right) \cdot \min\left\{\frac{N}{KM}\left(1 - (1 - M/N)^K\right), \frac{N}{K}\right\},$$

for a large enough file size F.

Although the decentralized scheme cannot control the placement as precisely as the centralized scheme, the authors nevertheless show that the decentralized placement creates almost as many coding opportunities as the centralized placement with very high probability. In fact, the resulting achievable rates are within a constant multiplicative factor of each other.

2.2.4 Approximate Optimality

Next, we examine how the performance of the centralized placement and delivery scheme proposed in Section 2.2.3 compares to the optimal scheme for this setup with no restrictions on the placement and delivery phases. The next theorem [6] states the approximate optimality of the achievable rate $R(M)$ in Theorem 2.1 with respect to the optimal rate $R^\star(M)$ as defined in (2.1).

THEOREM 2.3 *The rate achieved in Theorem 2.1 is within a constant multiplicative factor of the optimal rate. Specifically, for all values of N, K, and M,*

$$1 \leq \frac{R(M)}{R^\star(M)} \leq 12.$$

Note that the bound is independent of the problem parameters: the achievable rate is within a factor of 12 of the optimum, even if N and K are arbitrarily large. Proving Theorem 2.3 requires deriving information-theoretically lower bounds on the optimal rate using cut-set-based arguments [8]. We will not cover this here and instead point the interested reader to [6] for details. Finally, while the constant gap factor of 12 is indeed quite large, there have been significant improvements in terms of both the achievable rates [9–11] and the lower-bound arguments [12–14], which can be used to tighten the gap significantly. In fact, under the restriction of uncoded placement, the rate proposed in Theorem 2.1 is shown to be exactly optimal in [15, 16].

2.3 Non-uniform Content Popularity

In general, the popularity of a file can be thought of as the likelihood that a given user will request this file. Under a stochastic popularity model, this translates to a probability distribution over the files such that each user requests one file based on this probability. Note that, since the number of files is typically large compared to the number of users, we cannot reliably predict the number of users requesting each file from prior requests, especially for the less popular files. However, if the files are partitioned into a small number of levels by grouping together contents of similar popularity, we can more reliably estimate the cumulative popularity across these levels. If the number of users is large compared to the number of levels, the number of users *per level* under the stochastic popularity model will concentrate around the average value. The multi-level popularity model, introduced in [17], captures this aspect by making the number of users requesting files from each level fixed, deterministic, and known a priori; the worst-case rate (under this restriction on the demand) is then analyzed in a similar vein as in Section 2.2.1. We discuss this multi-level popularity model in this section.

More formally, in the multi-level popularity model, the files in the content library are partitioned into a certain number of groups called popularity levels. Each level $i \in \{1, \ldots, L\}$ consists of N_i files, and there are a total of K_i users requesting files from this level. Each of these K_i users can request any file belonging to level i. It is useful to think of the popularity of each file in level i as being proportional to the number of users per file of the level, K_i/N_i. For simplicity, we restrict the discussion here to the case where there are more files than users for every level, i.e., $N_i \geq K_i$ for all $i \in \{1, \ldots, L\}$. Note that the setup studied in Section 2.2 is a special case, with $L = 1$ level, $N_1 = N$, and $K_1 = K$.

The multi-level popularity model turns out to be useful in studying how the total number of users in the network, as compared to the number of caches, affects the system under non-uniform popularity. We will look at two extremes: one in which each cache has exactly one associated user (the single-user setup) and one in which each cache has a large number of associated users (the multi-user setup). In the single-user setup, only one level is represented at each cache since each user requests one file from one popularity level, as shown in Figure 2.3a. In the multi-user setup, the number of users is large enough for every level to be represented at every cache by at least one user,

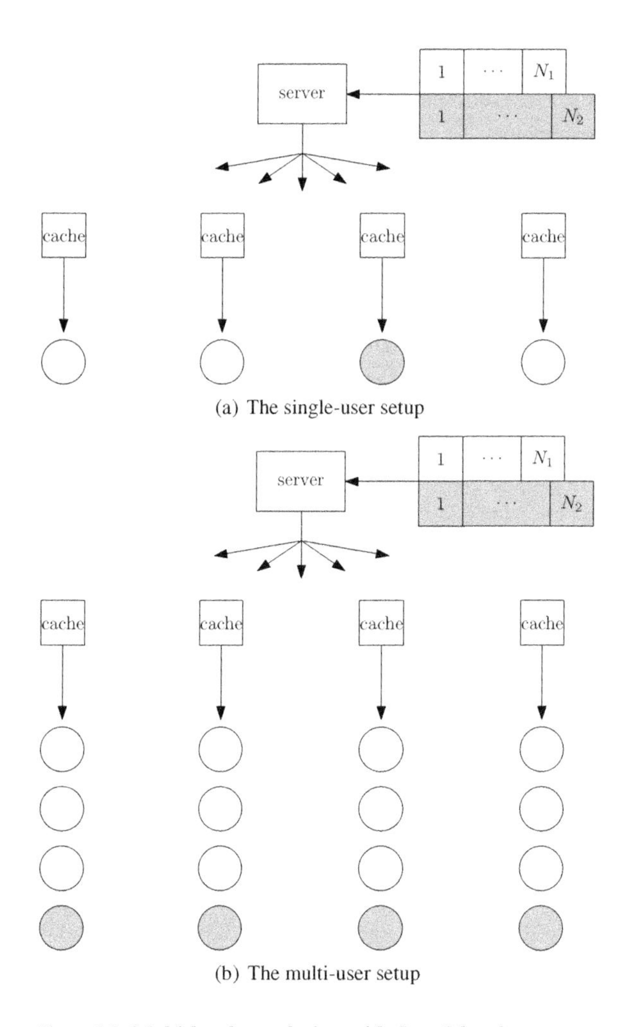

Figure 2.3 Multi-level popularity, with $L = 2$ levels.

as shown in Figure 2.3b. Interestingly, it turns out that the strategies required for these two setups are quite different: a level-merging approach works for the single-user setup, while a level-separation approach works for the multi-level setup.

In order to understand the major difference between these two setups, it is useful to first reflect on what enables the coding gains in the original setup in Section 2.2. Recall that a coded message from the server consists of a linear combination of parts of files requested by a subset of users, for example see (2.3). Each such user has access to *different* side information through its distinct caches. It is precisely this difference in side information that allows the same linear combination to be beneficial to multiple users possibly requesting distinct files. If two users' cache contents were identical, then their side information is identical and no coding gains can be achieved among them.

Moreover, because of the symmetry of the delivery message in (2.3), the procedure is most efficient when the involved subfiles are of the same size or equivalently when the files involved are stored equally in each cache. When adapting this to the non-uniform popularities setup, this creates a conflict: since the more popular files are more likely to be requested, we would want to give them a larger portion of the cache memory than the less popular files. However, as mentioned before, this would negatively impact the efficiency of coded messages involving requests for files with significantly different popularities. Hence there is a potential dilemma between giving the more popular files a larger memory share on the one hand and obtaining more efficient coding opportunities on the other hand.

The dilemma is naturally resolved for the multi-user setup. Notice from Figure 2.3b that the users can be partitioned into "rows" of K users each, such that each user in the row connects to one cache and requests a file from the same popularity level. A natural strategy here is to ensure that each linear combination the server sends is intended for only a subset of users belonging to the same row. Since all involved requests in a row will be for files belonging to the same level, they will have the same popularity and hence the same allocated cache memory. Furthermore, grouping an additional user (requesting a distinct file) with a row of K users in a single coded-multicast transmission cannot be beneficial since this user will necessarily share a cache with another user in the considered row. These two users will thus have access to the same side information, and hence, as discussed before, no coding gains can be obtained between them. As we will see later, the approximately optimal strategy here is to partition each cache among the various levels during the placement phase and then address the demands in each row of users separately during the delivery phase using coded-multicast transmissions, as discussed in Section 2.2.

On the other hand, the dilemma is not so easily resolved in the single-user setup. Notice from Figure 2.3a that in this case there is only one "row" of users in which all the file popularity levels are represented. This is unlike the multi-user setup where all users in a row requested files from the same popularity level, and hence if we allow all linear combinations in the server transmission, we might have to combine requests for files with very different popularities. However, if we restrict server transmissions to combine only requests belonging to the same popularity level, that will limit the coded-multicasting opportunities severely and increase the required server transmission rate. As we will see later, it turns out that the approximately optimal strategy here is to "merge" a subset of the higher popularity levels so that all the files belonging to them are given the same amount of memory and so that the requests belonging to these levels can be efficiently combined in the coded-multicast messages.

Thus, the schemes corresponding to the multi-user and single-user setups have different philosophies, and this difference marks the dichotomy between the two setups. Next, we study each of these setups in more detail.

2.3.1 The Single-User Setup

In the single-user setup, there is exactly one user connected to each of the K caches. As mentioned before, K_i users in the system request a file from level i and $K_1 + \cdots + K_L = K$. Importantly, while placing content in the caches, we know exactly how many users will request a file from each level, but we do not know *which* users will request from which level.

As discussed before, the idea in this setup is to strike a balance between two opposing principles: creating coding opportunities across popularity levels on the one hand and allocating more of the cache memory to the more popular files on the other hand. The balance that turns out to be approximately optimal is to partition the levels into two groups, which we will call H and I. The files belonging to levels in the set I will all be treated as if they were of the same popularity and were all allocated the same amount of memory; effectively the levels in I are merged into one superlevel with $\sum_{i \in I} N_i$ files and $\sum_{i \in I} K_i$ users. All the cache memory will be given to the set I, while the files in set H will not be stored at all. A coded caching scheme is then used on the set I as described in Section 2.2.3, and all requests for files from set H are handled by direct unicast transmissions from the server.

We can therefore apply Theorem 2.1 on each of H and I separately, which using (2.2) yields an achievable rate upper bounded by

$$R_{\mathrm{SU}}(M) \leq \max\left\{\frac{\sum_{i \in I} N_i}{M} - 1, 0\right\} + \sum_{h \in H} K_h. \tag{2.4}$$

The maximization with zero is necessary since, depending on the choice of I, the memory M could be larger than $\sum_{i \in I} N_i$.

Example 1 Consider an example multi-level single-user setup with $L = 3$ file popularity levels; $N_1 = 100, N_2 = 500$, and $N_3 = 1{,}000$ files; and $K_1 = 100, K_2 = 50$, and $K_3 = 5$ users. Consider the memory per cache to be $M = N_1 = 100$. We evaluate the rate of this proposed strategy for different choices of H, I:

1. *Store most popular only*: In this case, we set $I = \{1\}$ and $H = \{2, 3\}$ and thus store the files of only the most popular level, level 1, in the caches. From (2.4), the rate of the scheme for this choice is $\max\{N_1/N_1 - 1, 0\} + K_2 + K_3 = 55$.
2. *Treat all levels as uniform*: In this case, we set $I = \{1, 2, 3\}$ and $H = \phi$ and thus allocate equal memory to all the files. From (2.4), the rate of the scheme for this choice is $\max\{(N_1 + N_2 + N_3)/N_1 - 1, 0\} = 15$.
3. *Merge subset of levels*: Let us set $I = \{1, 2\}$ and $H = \{3\}$ and thus allocate equal memory to all the files belonging to the more popular levels, levels 1 and 2. From (2.4), the rate of the scheme for this choice is $\max\{(N_1 + N_2)/N_1 - 1, 0\} + K_3 = 10$.

Thus this example suggests that the optimal choice of H and I is nontrivial and greatly impacts the rate of the proposed scheme.

To understand how to, in general, choose the sets H and I optimally, consider the following back-of-the-envelope calculation. Suppose that all levels except one (call it level ℓ) have been partitioned into two sets H' and I'. If we put ℓ with H', we get the achievable rate

$$R_1(M) \approx \frac{\sum_{i \in I'} N_i}{M} + \sum_{h \in H'} K_h + K_\ell,$$

whereas if we combine it with I' we get

$$R_2(M) \approx \frac{\sum_{i \in I'} N_i + N_\ell}{M} + \sum_{h \in H'} K_h.$$

Then, $R_1(M) \leq R_2(M)$ if and only if $K_\ell / N_\ell \leq 1/M$, in which case the better choice is to group level ℓ with H'. Following this intuition, we choose the sets H and I as

$$H = \{h \in \{1, \ldots, L\} : K_h/N_h < 1/M\}; \quad I = \{1, \ldots, L\} \setminus H, \tag{2.5}$$

where, as mentioned before, the cache memory is divided equally among only the files belonging to levels in I in accordance with the scheme described in Section 2.2.3. Recall that for our setup, we can think of the popularity of each file in level i as being proportional to the number of users per file of the level, K_i/N_i. Thus, the decision rule suggests $1/M$ as a popularity threshold: all files with popularity above this threshold are assigned equal memory and all files with lower popularity are not allocated any memory during the placement phase.

The scheme leads to the following achievable rate for the single-user multi-level caching setup.

THEOREM 2.4 *In the single-user setup, the following rate is achievable for all L, K, $\{N_i, K_i\}$, and M:*

$$R_{\text{SU}}(M) \leq \max\left\{\frac{\sum_{i \in I} N_i}{M} - 1, 0\right\} + \sum_{h \in H} K_h,$$

where H and I are as defined in (2.5).

As we did for the single-level setup in Section 2.2.4, next we examine how the performance of the proposed scheme compares to that of the optimal scheme. The next result [17] states the approximate optimality of the achievable rate $R_{\text{SU}}(M)$ in Theorem 2.4 with respect to the optimal rate $R^\star_{\text{SU}}(M)$ for this setup.

THEOREM 2.5 *The rate $R_{\text{SU}}(M)$ achieved in Theorem 2.4 for the system with multi-level popularity and a single user per cache is within a constant multiplicative factor of the information-thoeretically optimal rate $R^\star_{\text{SU}}(M)$. Specifically, for all values of L, K, $\{N_i, K_i\}$ with $N_i \geq K_i$, and M,*

$$1 \leq \frac{R_{\text{SU}}(M)}{R^\star_{\text{SU}}(M)} \leq 72.$$

Note that the bound is independent of the problem parameters. As before, the proof derives information-theoretically lower bounds on the optimal rate using cut-set-based arguments; details are available in [17]. Finally, the focus of this result is on proving constant factor optimality (irrespective of system parameters), and while the factor of 72 is very large, this can be vastly improved using the aforementioned progress made on designing better achievable strategies and lower-bound arguments.

2.3.2 Multi-user Setup

We begin with some notation; see Figure 2.3b for an illustration of a multi-level multi-user setup. For each popularity level i, each cache has exactly U_i users requesting files from level i, which implies that the total number of users demanding files from level i is $K_i = KU_i$. As mentioned earlier, we assume that $N_i \geq K_i = KU_i$ for each level i.

As compared to the single-user setup described in the previous section, the biggest difference in the multi-user setup is that every level is represented at each cache, equally across the caches. In other words, every cache has the same user profile, where a user profile is an indicator of the number of users requesting a file from each given level. This allows separating the popularity levels and restricting all coding opportunities to be among users requesting files from a single level and not across levels.

More precisely, the idea is to partition the memory M among the popularity levels, giving level $i \in \{1, \ldots, L\}$ a memory of $\alpha_i M$ for some $\alpha_i \in [0, 1]$, and then apply the single-level coded caching scheme from Section 2.2 on each row of users separately, with each row consisting of users requesting files from a single level. Under this strategy, we can derive the achievable rate using Theorem 2.1 and (2.2) to be

$$R_{\mathrm{MU}}(M) \leq \sum_{i=1}^{L} U_i \cdot \min\left\{K, \max\left\{\frac{N_i}{\alpha_i M} - 1, 0\right\}\right\}. \tag{2.6}$$

The factor U_i appears because there are exactly U_i rows of users for level i.

By optimizing the overall rate over the memory-sharing parameters $\alpha_1, \ldots, \alpha_L$, we establish a memory allocation that we will show achieves a rate that is information-theoretically order optimal. At a high level, this allocation is done by partitioning the popularity levels into three sets: H, I, and J. The levels in H have such a small popularity that they will get no cache memory. Thus, for all levels $h \in H$, we will assign $\alpha_h M = 0$. On the opposite end of the spectrum, the most popular levels are assigned to J and are given enough cache memory to completely store all their files in every cache. Thus, for every level $j \in J$, we have $\alpha_j M = N_j$, since that is the amount of memory needed to completely store all files of level j in each cache. Finally, the rest of the levels, in the set I, will share the remaining memory among themselves, obtaining some non-zero amount of memory per cache but not enough to completely store all of their files in every cache. The more popular files should get more memory, and as discussed before we can think of KU_i/N_i as representing the popularity of a level i. For the order-optimal strategy we propose, we choose to give level i a memory per cache of roughly $\alpha_i M \propto N_i \cdot \sqrt{U_i/N_i}$ (hence the memory *per file* is proportional to $\sqrt{U_i/N_i}$).[1]

The above assignment will represent a valid choice for the memory-sharing parameters as long as the partition (H, I, J) is selected so that each $\alpha_i \in [0, 1]$. When we plug the this choice of the memory-sharing parameters into (2.6), we get the following result.

[1] The square root comes from minimizing the rate expression in (2.6), which has an inverse function of $\{\alpha_i\}$.

THEOREM 2.6 *Given a multi-user caching setup with K caches; L levels; and, for each level i, N_i files and U_i users per cache; and a cache memory of M, the following rate*[2] *is achievable:*

$$R_{\mathrm{MU}}(M) \approx \sum_{h\in H} KU_h + \frac{\left(\sum_{i\in I}\sqrt{N_i U_i}\right)^2}{M - \sum_{j\in J} N_j} - \sum_{i\in I} U_i, \tag{2.7}$$

where (H, I, J) is the unique *partition of the set of popularity levels that satisfies:*

$$\forall h \in H, \qquad \tilde{M} < \frac{1}{K}\sqrt{\frac{N_h}{U_h}};$$

$$\forall i \in I, \qquad \frac{1}{K}\sqrt{\frac{N_i}{U_i}} \le \tilde{M} \le \left(1 + \frac{1}{K}\right)\sqrt{\frac{N_i}{U_i}};$$

$$\forall j \in J, \qquad \left(1 + \frac{1}{K}\right)\sqrt{\frac{N_j}{U_j}} < \tilde{M},$$

where $\tilde{M} \approx (M - \sum_{j\in J} N_j)/\sum_{i\in I}\sqrt{N_i U_i}$.

The proof of the this result is rather involved, and we point the reader to [17] for details. Intuitively, since a level $h \in H$ receives no cache memory, all requests from its KU_h users must be handled directly from the broadcast. Since we have $N_i \ge K_i = KU_i$ for all levels i, then in the worst case a total of KU_h distinct files must be completely transmitted for the users requesting files from level h. This contributes the term $\sum_{h\in H} KU_h$ in the expression of the achievable rate (2.7). The users in set J require no transmission as the files are completely stored in all the caches; however, it does affect the rate through the memory available for levels in I. This is apparent in the expression $M - \sum_{j\in J} N_j$ in (2.7). Finally, the levels in I, having received some memory, require a rate that is inversely proportional to the effective memory and that depends on the level-specific parameters N_i and U_i.

Notice in the theorem statement that in the inequalities defining the chosen partition (H, I, J) the different sets are largely determined by the quantity $\sqrt{N_i/U_i}$ for each level i, which is a function of the file popularities. Moreover, the inequalities satisfy the natural choice that the most popular levels (i.e., those with the *smallest* N_i/U_i) will be in J, while the least popular levels (those with the *largest* N_i/U_i) will go to the set H.

Example 2 Consider an example multi-level multi-user setup with $K = 10$ caches; $L = 3$ file popularity levels; $N_1 = 100$, $N_2 = 200$, $N_3 = 300$ files; and $U_1 = 10$, $U_2 = 5$, $U_3 = 1$ users/cache. Consider the memory per cache to be $M = N_1 = 100$. We evaluate the rate of the proposed strategy for different choices of H, I, J:

1. *Store most popular only*: In this case, we set $J = \{1\}$, $I = \phi$, and $H = \{2, 3\}$ and thus, store the files of only the most popular level, level 1, in the caches. From (2.7), the rate of the scheme for this choice is $KU_2 + KU_3 = 60$.

[2] This expression of the rate is a slight approximation, which we use here for simplicity as it is more intuitive. An exact and complete description of the achievable rate can be found in [17].

2. *Share memory among all levels*: In this case, we set $I = \{1,2,3\}$ and thus allocate memory to each level in proportion to the square root of its popularity. From (2.7), the rate of the scheme for this choice is approximately $\frac{(\sqrt{N_1U_1}+\sqrt{N_2U_2}+\sqrt{N_3U_3})^2}{N_1} - (U_1+U_2+U_3) \approx 65 - 16 = 49$.
3. *Share memory among a subset of levels*: In this case, we set $I = \{1,2\}$ and $H = \{3\}$, and thus allocate memory to only the more popular level in proportion to the square root of its popularity. From (2.7), the rate of the scheme for this choice is approximately $KU_3 + \frac{(\sqrt{N_1U_1}+\sqrt{N_2U_2})^2}{N_1} - (U_1+U_2) = 50 - 15 = 35$.

Thus, we say that the optimal choice of H, I, J is non-trivial and greatly impacts the rate of the proposed scheme.

The next result [17] states the approximate optimality of the achievable rate $R_{\text{MU}}(M)$ in Theorem 2.1 with respect to the optimal rate $R^{\star}_{\text{MU}}(M)$ for this setup.

THEOREM 2.7 *The rate $R_{\text{MU}}(M)$ achieved in Theorem 2.6 for the system with multi-level popularity and multiple users per cache is within a constant multiplicative factor of the information-thoeretically optimal rate $R^{\star}_{\text{MU}}(M)$. Specifically, for all values of L, K, M, $\{N_i, U_i\}$ with $N_i \geq K_i$ and satisfying regularity condition[3] $\sqrt{\frac{U_i/N_i}{U_j/N_j}} \geq \frac{1}{\beta}$,*

$$1 \leq \frac{R_{\text{MU}}(M)}{R^{\star}_{\text{MU}}(M)} \leq c,$$

where $\beta = 198$ and $c = 9909$ are constants (independent of all problem parameters).

Unlike the approximate optimality results presented before, the proof of this theorem requires the use of *non cut-set*-based arguments to derive information-theoretically lower bounds on the optimal rate; details are available in [17]. As before, the constants involved can all potentially be improved greatly.

2.4 Multiple Cache Access

So far, we have considered situations only in which each user accesses exactly one cache, with no flexibility. However, in a wireless heterogeneous network such as the one in Figure 2.4, the density of access points that have caches could be high enough for each user to potentially access a large number of caches. This enables some interesting capabilities that can be harnessed to achieve a lower broadcast rate R for the same cache memory M. For instance, each user could have access to the contents of multiple caches at once, effectively increasing the memory available to it. Alternatively, we could allow the system to adaptively assign to each user one cache out of a set of nearby caches, based on the file that it requested. In this section, we explore the latter approach in detail as studied in [18], and leave the former as a short discussion at the end.

[3] The reasoning behind this condition is that, if it did not hold for some levels i and j, then we can think of them as essentially one level with $N_i + N_j$ files and $U_i + U_j$ users per cache. The resulting popularity $\frac{U_i+U_j}{N_i+N_j}$ would be close to both U_i/N_i and U_j/N_j.

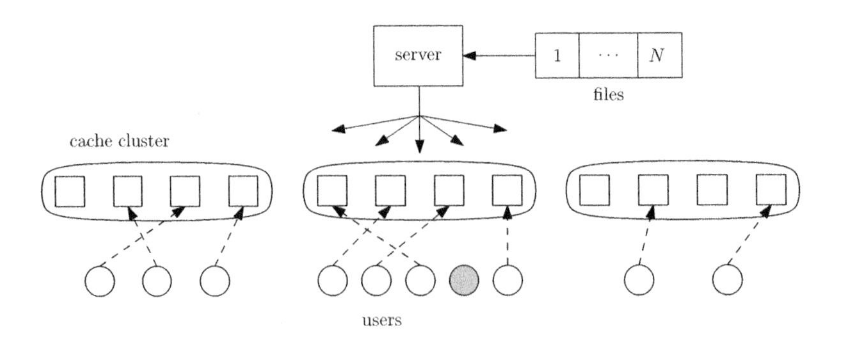

Figure 2.4 Partial adaptive matching setup. The caches are partitioned into clusters, and each user can be matched to one cache in its cluster, with a load constraint on the caches. Excess users in a cluster cannot be matched, such as the user colored in gray.

2.4.1 Overview of Adaptive User-to-Cache Matching

In the adaptive matching setup, we keep the restriction of each user accessing the contents of exactly one cache but allow the flexibility of choosing which cache (possibly among some subset of caches) the user should access based on its requested file. An additional restriction is a load constraint on the caches: each cache can serve at most only one user. Such a problem was studied in [19, 20], in the extreme case where all users are able to access any cache. The surprising insight in both papers is that, contrary to the "static matching case" where each user is preattached to a unique cache (the setting described in Section 2.2), an approximately optimal scheme is to replicate complete files across multiple caches in proportion to their popularity in the placement phase, and then during the delivery phase, match as many users as possible to a cache that holds their requested file.

We thus observe a dichotomy between two extremes: In the static matching case (when each user is restricted to one cache), appropriate splitting of files and careful placement of subfiles to enable coded-multicast transmissions during delivery, as described in Section 2.2, is approximately optimal, while simple file replication is not. On the other hand, in the fully "adaptive matching case," where each user can be matched to any cache during the delivery phase, appropriate file replication coupled with maximum matching during delivery is approximately optimal, while a static pairing of users and caches along with the coded caching approach of Section 2.2 is suboptimal. The natural next question is then: What happens when each user can be matched adaptively to one of a *subset* of caches? This problem was studied in [18], and we will discuss the main results here.

2.4.2 System Model

Suppose there is a content library of N files, called $W_1, \dots, W_N$. There are K caches, partitioned into K/d mutually exclusive *clusters* of d caches each (assume d divides K). At each cluster c, there is a stochastic number of users $u_n(c)$ that request file W_n, where $u_n(c)$ is a Poisson random variable of parameter $\rho d/N$, with $\rho \in (0, 1/2)$ a constant.

Thus at every cluster (whose size is d caches), the expected number of users is ρd. We will refer to $\mathbf{u} = \{u_n(c)\}_{n,c}$ as the *user profile*.

In addition to the usual placement and delivery phases, there is an intermediate *matching phase*, which occurs after the users have made their requests. In this phase, we assign each user in a cluster to one cache in the same cluster, subject to a load constraint of no more than one user per cache. Thus if there are more users than caches in a cluster ($\sum_n u_n(c) > d$ for some c), there will necessarily be some unmatched users who will have access to the contents of no cache. Note that the placement phase occurs without knowing the user profile, while both the matching phase and the delivery phase have knowledge of the user profile.

Let $R_{\mathbf{u}}$ denote the broadcast rate given a specific user profile $\mathbf{u}$. We are interested in the expected rate $\bar{R} = \mathbb{E}_{\mathbf{u}}[R_{\mathbf{u}}]$, and more specifically the optimal expected rate $\bar{R}^\star(M)$ for every memory M over all possible placement, matching, and delivery strategies.

The choice of a Poisson number of users is useful as it not only more closely models real-world user requests but also simplifies the analysis in this problem. There is also little difference, fundamentally, between the Poisson model and the model with a fixed number of users (such as the one studied in the previous sections), as long as the cluster size d is large enough, namely $d = \Omega(\log K)$. This means that comparisons with other works in the literature are possible. Note that for smaller d, the Poisson model is less meaningful; when $d = 1$ for instance, each cluster has a constant positive probability ($e^{-\rho}$) to have more than one user, which means that with high probability, a significant fraction of the users cannot be matched to any cache, and consequently a high server transmission rate is necessary irrespective of the cache memory size.

As mentioned earlier, the Poisson model makes sense only for $d = \Omega(\log K)$, and so we adopt this regularity condition in this section. More precisely, we assume that

$$d \geq \frac{2(1+t_0)}{\alpha} \log K, \tag{2.8}$$

where $\alpha = -\log(2\rho e^{1-2\rho}) > 0$, and $t_0 > 0$ is some constant. Finally, we restrict our attention to the case when $N \geq K$.

2.4.3 Balancing Two Extremes

The decribed model, known as the partial adaptive matching setup, is a generalization of the two extremes (Figure 2.4). When $d = 1$, we have a static matching setup as in [6] (while the Poisson model is not meaningful for $d = 1$, insights can still be gained). When $d = K$, we have the full adaptive matching setup as in [19, 20].

As discussed, there is a dichotomy between these two extremes: the former favors a coded delivery scheme, while the other favors an uncoded replication scheme. In what follows, we examine how each scheme performs if adapted to the partial adaptive matching setup. Specifically, we look at

- pure coded delivery (PCD), which ignores any potential adaptive matching benefits by arbitrarily assigning users to caches and applying a standard Maddah-Ali–Niesen scheme, as discussed in Section 2.2; and

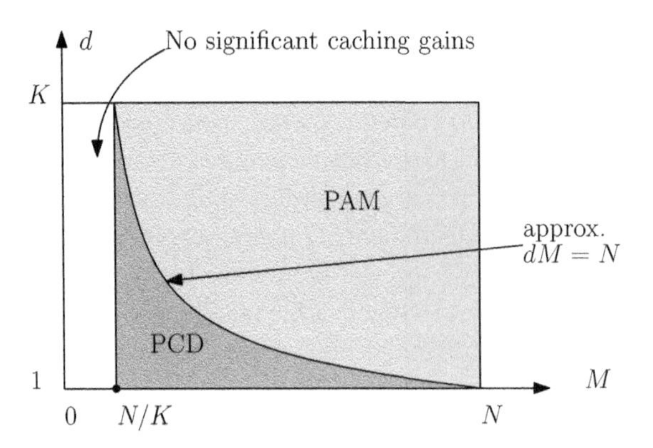

Figure 2.5 An approximate visualization of the regimes in which each scheme is more favorable. The boundary between the PCD- and PAM-dominated regions is blurry; the regime $\Omega(N) < dM < O(N \log N)$ is still not very well understood.

- pure adaptive matching (PAM), which ignores any potential coding gains and focuses only on file replication within a cluster and on adaptively matching users to caches within a cluster.

As we will see, in the general case we observe two regimes, and each scheme will be preferred in one regime. These regimes are roughly defined by a threshold on the *total cluster memory* dM: when $dM \ll N$ then PCD is favorable, and when $dM \gg N$ then PAM is favorable. Furthermore, in each regime, the favorable scheme is approximately optimal for almost all values of the cache memory. This is illustrated in Figure 2.5. Notice that in the special case $d = 1$ (respectively, $d = K$), Figure 2.5 shows that PCD (respectively, PAM) is always preferred, as expected from the previous results.

Each of the two schemes focuses on one idea: PCD ignores adaptive matching in favor of coding gains, and PAM ignores coded delivery in favor of adaptive matching gains. A hybrid coding and matching (HCM) scheme is introduced in [18] that performs better than both schemes in most memory regimes.

2.4.4 The Pure Coded Delivery (PCD) Scheme

The PCD scheme is a straightforward adaptation of the Maddah-Ali–Niesen scheme described in Section 2.2; the placement phase is identical to the one described there. During the matching phase, we pick any valid user-to-cache matching and provide each user with access to the corresponding cache; this is sufficient since the placement is completely symmetric with respect to the caches and the files. The delivery phase is conducted in two parts:

1. For the subset of users that were matched to caches, delivery proceeds in the same fashion as in Section 2.2 by creating coded-multicast transmissions.
2. Any users that were not matched (because there were more users than caches in their cluster) will simply be served directly by the server.

Note that for the model described in Section 2.4.2, the expected number of such unmatched users is very small. In fact it can be shown that

$$\mathbb{E}[U^0] \le K^{-t_0}/\sqrt{2\pi},$$

where U^0 is the total number of excess users across all clusters and $t_0 > 0$ is from (2.8). Note that the expected number of excess users goes to zero as K increases. All the other users (i.e., those that are matched to some cache) will be served by the basic Maddah-Ali–Niesen scheme, and so PCD can achieve the rate in the following theorem.

THEOREM 2.8 *For the partial adaptive matching model described in Section 2.4.2, the expected rate achieved by PCD is*

$$\bar{R}^{\mathrm{PCD}}(M) \le \min\left\{\rho K, \left[\frac{N}{M} - 1\right]^+ + \frac{K^{-t_0}}{\sqrt{2\pi}}\right\}.$$

Notice that this is not much different from the rate achieved in the static matching setup. This is expected since we are not making any intelligent use of the adaptive matching feature at all in PCD. However, this turns out to be approximately optimal when the total memory in any cluster is not enough to hold the entire library, as stated next.

THEOREM 2.9 *When $M \le (1 - e^{-1}/2)N/2d$, the expected rate achieved by PCD is approximately optimal in the sense that*

$$\bar{R}^{\mathrm{PCD}}(M) \le C \cdot \bar{R}^\star(M) + o(1),$$

where $\bar{R}^\star(M)$ is the information-theoretically optimal rate, C is a constant independent of the problem parameters, and the $o(\cdot)$ notation is to be understood with respect to the growth of K.

We skip the proof of these results here and instead point the interested reader to [18], which has all the details and discusses more general scenarios with non-uniform content popularity.

2.4.5 The Pure Adaptive Matching (PAM) Scheme

As previously mentioned, the PAM scheme takes the opposite approach to PCD. It ignores all possible coding in favor of a more intelligent matching of users to caches. The idea is to store only replicas of files in every cluster and rely as much as possible on the matching phase to connect each user to a cache that contains the file that it requested.

More precisely, the three phases work as follows. In the placement phase, the total cluster memory is dM, and we store a complete copy of every file in $\lfloor dM/N \rfloor$ caches in every cluster. In the matching phase, we find the best matching of users to caches so that the number of users matched to a cache containing their requested file is maximized. In the delivery phase, any users that could not be successfully matched to a suitable cache are served directly from the server.

Notice that the scheme really takes off only once $dM \geq N$: for smaller memory values, there is a significant fraction of users whose requests cannot be satisfied locally and have to be served directly by the server. What's more interesting is that after this threshold of $dM \geq N$, the achieved expected rate decays exponentially with the cluster memory! The precise rate expression is given in the following theorem.

THEOREM 2.10 *For the partial adaptive matching model described in Section 2.4.2, the expected rate achieved by PAM is*

$$\bar{R}^{\mathrm{PAM}}(M) \leq \begin{cases} \rho K & \text{if } M < N/d; \\ KMe^{-\rho h d M/N} & \text{if } M \geq N/d, \end{cases}$$

where $h = (1/\rho)\log(1/\rho) + 1 - 1/\rho$.

The proof follows along similar lines as [19], which focuses on the fully adaptive matching case and generalizes the results to the partially adaptive matching case.

Notice that $\bar{R}^{\mathrm{PAM}}(M) = o(1)$ when $dM > \Omega(N \log N)$. Thus, once the total cluster memory is slightly larger than the total catalog size, the PAM scheme requires negligible server transmission rate. This also trivially implies that PAM is approximately information-theoretically optimal in that regime. Combining with Theorem 2.9, we have that PCD is approximately optimal when $dM < O(N)$, and PAM is approximately optimal when $dM > \Omega(N \log N)$, as illustrated in Figure 2.5 (ignoring the $\log N$ factor for simplicity).

2.4.6 The Hybrid Coding and Matching (HCM) Scheme

So far, we have looked at two schemes that each focuses on a single gain: either a coding gain or an adaptive matching gain. The schemes are approximately optimal in complementary regimes, as illustrated in Figure 2.5. This section explores a hybrid scheme that unifies coded delivery and adaptive matching by incorporating ideas of both PCD and PAM. This HCM scheme, first introduced in [18], turns out to perform better than both PCD and PAM in most memory regimes.

The hybrid scheme combines ideas from both PCD and PAM by introducing a coloring scheme at both the cache level and the file level. First, we choose a certain number of colors $\chi \in \{1, \ldots, d\}$. The exact value is not important for now; the ideas work for any χ. We then partition the caches in every cluster into χ subsets of (almost) equal size, and color each subset with a unique color. Similarly, we partition the files in the content library into χ subsets of (almost) equal size, and color each subset with one color. Finally, we apply the coded delivery ideas *within each color*, while applying the adaptive matching ideas *across colors*.

More precisely, the three phases proceed as follows. In the placement phase, for every color x we perform a Maddah-Ali–Niesen placement of the files *of color* x in only the caches of the same color. The placement phase is agnostic to the cluster to which a cache belongs. In the matching phase, every user can be matched to an arbitrary cache in its cluster whose *color* matches the file that the user requested; the user is matched to the color, but the choice of cache within that color is arbitrary. In the delivery phase, the Maddah-Ali–Niesen coded delivery is performed for every color x separately, and unmatched users are served directly. Like the placement phase, the delivery phase ignores clusters and serves all users of the same color in the same broadcast message.

Since we have χ subsystems of N/χ files, each running a separate Maddah-Ali–Niesen scheme, using Theorem 2.8 we can show that the hybrid scheme can achieve a rate of

$$\bar{R}(M) \approx \min\left\{\rho K, \chi \cdot \left(\frac{N/\chi}{M} - 1\right) + \bar{U}^0(\chi)\right\},$$

where $\bar{U}^0(\chi)$ is the expected number of unmatched users when choosing χ colors. What is left is therefore to choose the right value of χ.

If the number of colors is too small, then there is little benefit in adaptive matching since the number of choices is reduced. Conversely, if the number of colors is too large, then the number of caches in each color becomes small, and it becomes likely that a significant number of colors have fewer caches than there are users requesting a file from them; the number of unmatched users thus becomes too large. The balance is struck when $\chi \approx d/\log K$ colors, as stated more precisely in the next theorem.

THEOREM 2.11 *For any $t \in [0, t_0]$, the HCM scheme can achieve an expected rate of*

$$\bar{R}^{\mathrm{HCM}}(M) \le \begin{cases} \min\left\{\rho K, \frac{N}{M} - \chi + \frac{K^{-t}}{\sqrt{2\pi}}\right\} & \text{if } M \le \lfloor N/\chi \rfloor; \\ \frac{K^{-t}}{\sqrt{2\pi}} & \text{if } M \ge \lceil N/\chi \rceil, \end{cases}$$

where $\chi = \lfloor \alpha d/(2(1+t)\log K) \rfloor$ and $t_0 > 0$ is a positive constant.

The rate expression can be approximately written as

$$\bar{R}^{\mathrm{HCM}}(M) \approx \min\left\{\rho K, \left[\frac{N}{M} - \Theta\left(\frac{d}{\log K}\right)\right]^+ + o(1)\right\}.$$

Comparing the performances of PCD, PAM, and HCM, we find that HCM performs better than both of them in most memory regimes. In fact, HCM is a unified scheme that is approximately optimal for almost all memory regimes. Specifically, we have

- for all $M \ge 0$, HCM performs better than PCD;
- when $dM \le O(N)$, both HCM and PCD are approximately optimal, while PAM is not;
- when $dM \ge \Omega(N \log N)$, both HCM and PAM achieve a rate of $o(1)$ and are trivially approximately optimal, while PCD is not;[4] and
- the intermediate regime $\Omega(N) \le dM \le O(N \log N)$ is not very well understood, and the exact relationship between the different rates, as well as their approximate optimality, is not known.

More details can be found in [18].

[4] In fact, HCM achieves a rate of $o(1)$, even for $dM \ge \Omega(N \log K)$. If N grows polynomially with K, then $N \log N = \Theta(N \log K)$.

2.4.7 Simultaneous Cache Multi-access

In Section 2.4.6, we studied the adaptive matching setup where each user has many nearby caches but is matched to only one among them, based on its file request. In this section, we briefly discuss an alternative setting where each user is allowed access to the information stored in *all* the neighboring caches. This problem was introduced in [17] where it was analyzed within the larger multi-level popularity setting. In this section, we restrict the discussion to a uniform popularities setup in order to focus on the simultaneous multi-access aspect of the problem.

The first question we must ask ourselves is: What sort of multi-access model should we adopt here? A setting with caches divided into clusters like in Section 2.4.2 is not very interesting in this scenario. Indeed, suppose like in Section 2.4.2 that the caches are partitioned into clusters of d caches each, and that every user could access all the caches in its cluster. Thus, any two users in the same cluster will have access to the same subset of caches. Then the problem is effectively reduced to the basic setup seen in Section 2.2.1, but with K/d caches of memory dM each and multiple users accessing each cache. This is a special case of the multi-level multi-user setup described in Section 2.3, restricted to a single popularity level. Hence a cache-cluster model with simultaneous cache access does not introduce any new concepts.

A more interesting scenario is when the sets of caches that users can access have nontrivial intersections. In other words, two users can have a few caches in common but also a few caches that are exclusive to one or the other. One way to model this is using a "sliding window" approach, where user k accesses caches $k, k+1, \ldots, k+d-1$ for some $d \in \{1, 2, \ldots, K\}$, using a cyclic wraparound to preserve symmetry. Specifically, if we label the caches as Z_1 through Z_K, then user $k \in \{1, \ldots, K\}$ has access to the d caches

$$Z_k, Z_{\langle k+1 \rangle}, \ldots, Z_{\langle k+d-1 \rangle},$$

where $\langle m \rangle = m$ if $m \leq K$ and $\langle m \rangle = m - K$ if $m > K$. We call d the *access degree*. Thus if $K = 4$ and $d = 2$, then user 1 has access to caches Z_1 and Z_2, while user 4 accesses caches Z_4 and Z_1.

This problem setup can be motivated by a scenario in which caches are arranged linearly and users access the d nearest caches to them. While this linearity assumption is simplistic, the problem can be easily extended to a more realistic scenario in which the caches are arranged in a 2-dimensional lattice and, as before, every user accesses the d nearest caches.

At this point, it is interesting to think about how the local and global caching gains would be different in this scenario compared with the basic setup in Section 2.2.1. Recall that the global caching gain is caused by the total memory in the system, KM. In this scenario, the total memory is still KM, so we might not expect the global caching gain to be different. However, also recall that the local caching gain is caused by the cache memory available for each user, which in the basic setup was M. But a key difference in this simultaneous multi-access problem is that every user actually has access to a memory of dM. Thus one might expect that in the simultaneous multi-access problem we can achieve a rate of

$$R_{\text{SM}}(M) \approx K \cdot \left(1 - \frac{dM}{N}\right) \cdot \frac{1}{1 + KM/N}, \tag{2.9}$$

which is similar to the expression in Theorem 2.1 except for the dM term in the factor that represents the local caching gain. Note that it is not immediate whether the above rate expression is achievable for the setup being considered here, since no two users share the same cache-access structure.

We will now analyze (2.9) in order to get insights into schemes that can achieve this rate. Notice that the effect of multi-access appears only when dM is larger than some fraction of N, e.g., $dM > N/2$. Below this threshold, the global caching gain, which is not affected by multi-access, dominates. This inspires the following simple scheme:

- When $dM \leq N/2$, we ignore multi-access and assume that user k accesses only cache Z_k. We apply the Maddah-Ali–Niesen scheme under this assumption, achieving a rate of $R(M) \leq \min\{K, N/M\}$.
- When $dM = N$, we apply a (K,d)-erasure-correcting code on each file, creating K coded messages of size F/d bits each, such that any d of them can re-create the entire file. We store each such coded message in one unique cache. Thus every user, by accessing the d caches in its neighborhood, can recover any file by retrieving the corresponding d coded messages in those caches. This achieves a rate of zero.
- When $N/2 < dM < N$, we use memory sharing between the two schemes at $dM = N/2$ and $dM = N$, respectively, to achieve a linear combination of the two rates.

The described scheme achieves the rate expression in the theorem that follows.

THEOREM 2.12 *In the simultaneous multi-access problem with N files, K caches and users, and a cyclic cache-access structure with a per-user access degree of d, we can achieve a rate of*

$$R_{\text{SA}}(M) \leq 4 \cdot \min\left\{K, \frac{N}{M}\right\}\left(1 - \frac{dM}{N}\right),$$

for all cache memory $M \in [0, N/d]$. A rate of zero is achieved for $M \geq N/d$. Furthermore, the gap of the achievable rate to the information-theoretically optimal rate $R^{\star}_{\text{SA}}(M)$ is given by

$$1 \leq \frac{R_{\text{SA}}(M)}{R^{\star}_{\text{SA}}(M)} \leq c \cdot d,$$

where c is some constant.

The proof of Theorem 2.12 and further details can be found in [17].

2.5 Wireless Interference Networks: A Separation Architecture

In the previous sections, we studied a simple network structure in which the server communicates directly with the users via an error-free broadcast link. Such a structure was useful for deriving the coded multicasting gains of caching. However, in practice we observe more complex network structures such as interference networks, device-to-device networks, and hierarchical structures. In this section we mostly focus on caching in interference networks and argue that an efficient way to operate content delivery in cache-aided wireless networks is to appropriately *separate* the role of physical-layer communication from the role of the messages it delivers, using coded caching.

2.5.1 Caching in Interference Networks

Wireless networks, by their very nature, exhibit both the *broadcast* of signals as well as their *superposition*. The previous sections discussed the former and established the caching gains of broadcast (or, more precisely, multicast). This however is limited to networks with a single base station. An interesting problem is to consider multiple base stations, each of which has access to the content library one way or another. These base stations will then interfere with each other, and the question becomes how to manage this interference for the purpose of content distribution. We will therefore study cache-aided interference networks, which incorporate both the broadcast and superposition properties.

Moreover, our discussion has so far ignored the physical layer, instead treating all channels as error-free bit pipes. In broadcast networks, this is not a big issue since separating what to send from how to transmit it is natural in a broadcast setting. However, this is no longer the case when several base stations are present, and in this section we also take the physical layer into consideration when studying the interference networks.

The problem of caching in interference networks was first studied in [21], which had an interference channel with three transmitters that were equipped with caches and three receivers that were requesting content. This was later extended in [22] to consider an arbitrary number of transmitters and receivers. Problems with caches both at the transmitters and at the receivers were then studied, but with restrictions on the schemes: [23] was limited to one-shot linear schemes while [24] prohibited coding across files during the placement phase. Furthermore, both [24, 25] looked at a limited number of transmitters and receivers (no more than three).

The first general result was published in [26], which found an approximate characterization of the information-theoretically optimal rate-memory trade-off in the high signal-to-noise ratio (SNR) regime. Three key insights into the problem are derived. First, it is shown that a separation of the physical and network layers is approximately optimal: a physical-layer scheme focuses on transmitting some message set across the interference network by generalizing a technique known as interference alignment [27], and the network-layer scheme uses this message set as error-free bit pipes to implement a coded caching scheme. Second, it is shown that, as long as the transmitters can collectively store exactly the entire content library, then increasing the transmitter memory has no effect on the optimal rate beyond a constant multiplicative factor. A consequence of this is that it is not necessary for the transmitters to share information: they can all store distinct parts of the content library for most gains to be obtained. Third, there is a trade-off between the receiver memory and the number of transmitters needed to approximately achieve maximal system performance: as the receiver memory increases, fewer transmitters are required. In particular, when each receiver can hold a fraction of the library, then a constant number of transmitters is sufficient to achieve most benefits. We discuss these results in more detail later in the chapter.

At the other extreme, the low-SNR regime was studied in [28], where a similar separation of the network and physical layers is proposed. This separation architecture is shown to be approximately optimal in some cases, namely the single-receiver and the single-transmitter cases. Contrary to the high-SNR regime, it is shown in [28] that transmitter cooperation, by storing shared content in their caches, is crucial in the low-SNR regime.

2.5.2 The Separation Architecture

Although the separation architecture was studied specifically for a Gaussian interference network, we first describe it in a very general context as the same ideas hold. We then show how it applies to the Gaussian network.

2.5.2.1 High-Level Overview of the Separation Architecture

We will first describe a general cache-aided interference network. There is a content library containing N files of size F bits. The library is separated from the users by an interference channel. The interference channel has K_t transmitters and K_r receivers, that act as the users. Each transmitter has a cache of memory $M_t F$ bits, and each receiver has a cache of memory $M_r F$ bits.

During the placement phase, we place information about the files in every transmitter and receiver cache. During the delivery phase, each transmitter $\ell \in \{1, \ldots, K_t\}$ sends a code word $\mathbf{x}_\ell = (x_\ell(1), \ldots, x_\ell(T))$ over a block length of T through the interference network. Importantly, the code word $\mathbf{x}_\ell$ can depend only on the file requests and the contents of transmitter ℓ's cache. Each receiver $k \in \{1, \ldots, K_r\}$ then receives a signal $\mathbf{y}_k = (y_k(1), \ldots, y_k(T))$, which is some (noisy) function of the input code words. Finally, each receiver k uses the received signal $\mathbf{y}_k$ in combination with the contents of its cache to recover the requested file. The goal is to find the largest possible transmission rate defined as $R = F/T$ over all possible strategies.[5]

A key aspect of this problem that has not been discussed in previous sections is that it combines the content delivery problem with that of physical-layer transmission. One of the main questions that arise is whether a joint design of the network layer and

[5] The "rate" defined in this section is not the same as the one in previous sections. In this section it represents the rate of communication as is common in the literature on interference networks, whereas in the previous sections it denoted the *size* of the message sent by the server, which is sometimes called the "normalized delivery time" (NDT). The two terms are inversely proportional to each other.

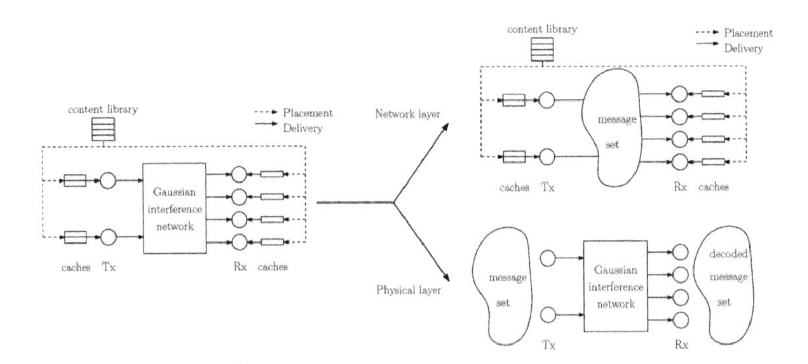

Figure 2.6 Illustration of the separation architecture. The cache-aided interference channel is split into a physical layer and a network layer; the two layers interface using a message set.

the physical layer is necessary. The question is answered in [26] in the negative by proposing the separation architecture and showing that it is approximately optimal.

The key idea of the separation architecture is to separate the caching aspect (what to store in the caches and what to send from transmitters to receivers) from the physical-layer aspect (how to send it through the interference network). Thus the system is split into an overlay network layer and a physical layer. The two layers interface through a set of messages from (subsets of) transmitters to (subsets of) receivers. This is illustrated in Figure 2.6.

In the most general sense, we define a set of n messages

$$\mathscr{V} = \left\{ V_{\mathcal{K}_1\mathcal{L}_1}, V_{\mathcal{K}_2\mathcal{L}_2}, \ldots, V_{\mathcal{K}_n\mathcal{L}_n} \right\},$$

where $\mathcal{K}_i \subseteq \{1, \ldots, K_r\}$ and $\mathcal{L}_i \subseteq \{1, \ldots, K_t\}$, and $V_{\mathcal{K}_i\mathcal{L}_i}$ is a message from the transmitters in $\mathcal{L}_i$ to the receivers in $\mathcal{K}_i$. At the physical layer, we have the sub-problem of transmitting this message set across the interference network reliably. It is assumed that all the transmitters in $\mathcal{L}_i$ have access to the message $V_{\mathcal{K}_i\mathcal{L}_i}$. At the network layer, we can use these messages as orthogonal, noninteracting bit pipes through which we can pass information from the transmitters to the receivers. The constraint is that every transmitter in $\mathcal{L}_i$ must be able to cosntruct $V_{\mathcal{K}_i\mathcal{L}_i}$ from the contents of its cache.

Suppose the physical-layer scheme can transmit every message in the message set at a rate of at least R', over a block length of T. Suppose also that the network-layer scheme can deliver the requested files using at most vF bits through each bit pipe represented by the messages in the message set. Then the message set can be supported by both the network layer and the physical layer as long as $R'T \geq vF$. Therefore, by choosing the smallest supported block length $T = vF/R'$, we can reliably deliver the files to all the users at a rate of $R = F/T = R'/v$.

2.5.2.2 The High-SNR Gaussian Interference Network

In the particular case of the memoryless Gaussian interference network, the inputs and outputs to the channel are real valued. At each time step τ, every output symbol $y_k(\tau)$ is a linear combination of all the input symbols $x_k(\tau)$, plus a Gaussian unit-variance random noise variable. Furthermore, a power limit of P is imposed on the input code words, $\|\mathbf{x}_\ell\|^2 \leq PT$.

We are interested in the high-SNR regime, i.e., the regime in which P is large. More specifically, we look at the degrees of freedom (DoF) of the system, which is the behavior of the rate as a scaling of the capacity of a point-to-point Gaussian channel. Specifically, if we write the information-theoretically optimal rate for a specific power P as $R^\star(P)$, then the degrees of freedom is defined as

$$\mathsf{DoF} = \lim_{P\to\infty} \frac{R^\star(P)}{\frac{1}{2}\log P}.$$

We next describe the approximately optimal strategy within the context of the high-SNR cache-aided Gaussian interference network. This strategy makes use of the separation architecture in the following way.

1. The message set $\mathcal{V}$ that is chosen is a set of messages from every single transmitter ℓ to every subset $\mathcal{K}$ of receivers of a fixed size $|\mathcal{K}| = \kappa + 1$, where $\kappa \approx K_r M_r/N$. Notice that this is exactly the same as the multicast size in the broadcast setup in Section 2.2. Thus the message set represents a set of single-transmitter multicast channels.
2. At the network layer, we partition every file in the content library into K_t parts and store each part at one transmitter. Thus each transmiter has a content sublibrary consisting of part of every file. A standard centralized Maddah-Ali–Niesen scheme is performed on each sublibrary, and each multicast message (intended for $\kappa + 1$ receivers) is sent through the corresponding bit pipe.
3. At the physical layer, we apply a technique known as interference alignment in order to transmit the message set as efficiently as possible.[6]

The following theorem gives the approximate degrees of freedom of the network, as determined in [26].

THEOREM 2.13 *The degrees of freedom of the cache-aided Gaussian interference network is approximately given by*

$$\mathsf{DoF} \approx \frac{K_t K_r}{K_t + K_r - 1} \cdot \frac{1}{1 - M_r/N} \cdot \frac{K_r M_r/N + 1}{\frac{M_r}{N}(\frac{1}{K_r} + \frac{1}{K_t - 1})^{-1} + 1},$$

for all N, K_t, K_r, $M_r \in [0, N]$, and $M_t \geq N/K_t$. The approximation is within a constant multiplicative factor.

Notice that the degrees of freedom can be written as the product of three gains, in a similar way to the rate expression in Section 2.2. The first term is the interference alignment gain and represents the DoF when no receiver caches are present. The second term is the local caching gain, as with the broadcast case. The third term is the global caching gain.

[6] In fact, interference alignment can exactly achieve the degrees of freedom of the communication problem that arises at the physical layer. Note that this is the degrees of freedom associated with the rate R' described earlier; it is not the degrees of freedom of the entire cache-aided interference network.

Finally, some powerful insights can be gained from Theorem 2.13. The first is that a separation of the network and physical layers is approximately optimal. Second, one can see that the DoF expression does not involve the transmitter memory M_t, except through the constraint $K_t M_t \geq N$. This implies that there is no benefit (no more than a constant factor) in increasing the transmitter memory. In particular, it shows that transmitter cooperation is unnecessary: the DoF can be approximately achieved even if the transmitters share no information at all. Third, as the receiver memory increases, the number of transmitters needed for approximately achieving the DoF decreases. These insights together show that the cache-aided interference network problem can be solved with a system that is layered and simple to design and implement.

Before we conclude the discussion on the interference networks, some thoughts on the low-SNR regime. We can apply a very similar separation architecture in the low-SNR regime. However, some crucial differences arise. First, the message set used for the high-SNR regime is insufficient: it is necessary for each message to be from *subsets* of transmitters to subsets of receivers. Second, the physical-layer scheme consists in the transmitters co-operating and beam-forming signals to multiple receivers, yielding a beamforming gain (in lieu of the interference alignment gain in the high-SNR regime). Third, there is a trade-off between the beam-forming gain and the global caching gain; consequently it is sometimes better to ignore any coded caching gains in favor of an increase in the beam-forming gain. We point the interested reader to [28] for a more detailed analysis of the low-SNR regime.

2.5.3 Other Network Topologies

There have been several works in the literature that study other complex topologies for the cache network. We briefly describe two such models:

1. *Hierarchical networks*: In this model, the server is connected to the users via a tree network, with caches of possibly different sizes at each level of the tree and where each cache at level i communicates with its children at level $i + 1$ via an error-free broadcast link. Note that the server is the root of this hierarchical caching network, and the users are the leaves. One study [29] looked at the special case of a two-level hierarchical tree caching network with N files of size F bits each at the server, communicating via an error-free broadcast link with K_1 *mirrors* each with a cache of memory size $M_1 F$ bits, at level 1. Each of these mirror nodes is connected to K_2 *users* each with a cache of size $M_2 F$ bits. The system operates as before in two phases: a placement phase when all caches are populated with content and then, after the user requests are revealed, a delivery phase where the server sends a common message of size $R_1 F$ bits to the mirrors and each mirror sends a message of size $R_2 F$ bits to its connected users. [29] proposed a scheme and showed that for any M_1, M_2, the required rates R_1, R_2 for the proposed scheme are within a constant factor (independent of all problem parameters) of the information-theoretically optimal rates. A desired feature of the proposed scheme is that the delivery phase only uses messages that involve coding across a single layer of storage at a time. Details can be found in [29].

2. *Device-to-device networks*: In this model, there is no designated server in the network that hosts the entire catalog of N files. The system consists of K co-located users, each with a cache of size MF bits, which can communicate with each other over an error-free broadcast link. This setup was studied in [30, 31], where the authors proposed a caching and delivery scheme for this setup and analyzed the total required transmission size on the shared link, as well as compared it to information-theoretically lower bounds.

References

[1] M. R. Korupolu, C. G. Plaxton, and R. Rajaraman, "Placement algorithms for hierarchical cooperative caching," in *Proceedings of the ACM-SIAM Symposium on Discrete Algorithms (SODA)*, pp. 586–595, 1999.

[2] "The 1000x challenge," Qualcomm, 2013, www.qualcomm.com/solutions/wireless-networks/technologies/1000x-data/small-cells.

[3] "Intel heterogeneous network solution brief," Intel, 2013, www.intel.com/content/dam/www/public/us/en/documents/solution-briefs/communications-heterogeneous-network-brief.pdf.

[4] "Rethinking the small cell business model," Intel, 2012, www.intel.com/content/dam/www/public/us/en/documents/white-papers/communications-small-cell-study.pdf.

[5] "Visual networking index (VNI) global mobile data traffic forecast update," Cisco, 2017, www.cisco.com/c/en/us/solutions/collateral/service-provider/visual-networking-index-vni/mobile-white-paper-c11-520862.pdf.

[6] M. A. Maddah-Ali and U. Niesen, "Fundamental limits of caching," *IEEE Transactions on Information Theory*, vol. 60, no. 5, pp. 2856–2867, May 2014.

[7] M. A. Maddah-Ali and U. Niesen, "Decentralized coded caching attains order-optimal memory-rate tradeoff," *IEEE/ACM Transactions on Networking*, vol. 23, no. 4, pp. 1029–1040, 2015.

[8] T. M. Cover and J. A. Thomas, *Elements of Information Theory* Hoboken, NJ: John Wiley & Sons, 2012.

[9] M. M. Amiri and D. Gündüz, "Fundamental limits of coded caching: improved delivery rate-cache capacity tradeoff," *IEEE Transactions on Communications*, vol. 65, no. 2, pp. 806–815, 2017.

[10] K. Zhang and C. Tian, "Fundamental limits of coded caching: from uncoded prefetching to coded prefetching," *IEEE Journal on Selected Areas in Communications*, vol. 36, no. 6, pp. 1153–1164, June 2018.

[11] Y.-P. Wei and S. Ulukus, "Novel decentralized coded caching through coded prefetching," in *IEEE Information Theory Workshop*, 2017, pp. 1–5.

[12] C.-Y. Wang, S. H. Lim, and M. Gastpar, "A new converse bound for coded caching," in *Information Theory and Applications Workshop*, 2016, pp. 1–6.

[13] C.-Y. Wang, S. S. Bidokhti, and M. Wigger, "Improved converses and gap-results for coded caching," in *IEEE International Symposium on Information Theory*, 2017, pp. 2428–2432.

[14] H. Ghasemi and A. Ramamoorthy, "Improved lower bounds for coded caching," *IEEE Transactions on Information Theory*, vol. 63, no. 7, pp. 4388–4413, May 2017.

[15] Q. Yu, M. A. Maddah-Ali, and A. S. Avestimehr, "The exact rate-memory tradeoff for caching with uncoded prefetching," *IEEE Transactions on Information Theory*, vol. 64, no. 2, pp. 1281–1296, 2018.

[16] K. Wan, D. Tuninetti, and P. Piantanida, "On the optimality of uncoded cache placement," in *IEEE Information Theory Workshop*, 2016, pp. 161–165.

[17] J. Hachem, N. Karamchandani, and S. N. Diggavi, "Coded caching for multi-level popularity and access," *IEEE Transactions on Information Theory*, vol. 63, no. 5, pp. 3108–3141, May 2017.

[18] J. J. Hachem, N. Karamchandani, S. Moharir and S. N. Diggavi, "Caching With Partial Adaptive Matching," in IEEE Journal on Selected Areas in Communications, vol. 36, no. 8, pp. 1831–1842, Aug. 2018.

[19] M. Leconte, M. Lelarge, and L. Massoulié, "Bipartite graph structures for efficient balancing of heterogeneous loads," *SIGMETRICS Performance Evaluation Review*, vol. 40, no. 1, pp. 41–52, Jun. 2012.

[20] S. Moharir and N. Karamchandani, "Content replication in large distributed caches," in *2017 9th International Conference on Communication Systems and Networks (COMSNETS)*, 2017, pp. 128–135.

[21] M. A. Maddah-Ali and U. Niesen, "Cache-aided interference channels," in *2015 IEEE International Symposium on Information Theory (ISIT)*, June 2015, pp. 809–813.

[22] R. Tandon and O. Simeone, "Cloud-aided wireless networks with edge caching: fundamental latency trade-offs in fog radio access networks," in *2016 IEEE International Symposium on Information Theory (ISIT)*, July 2016, pp. 2029–2033.

[23] N. Naderializadeh, M. A. Maddah-Ali, and A. S. Avestimehr, "Fundamental limits of cache-aided interference management," in *2016 IEEE International Symposium on Information Theory (ISIT)*, July 2016, pp. 2044–2048.

[24] F. Xu, K. Liu, and M. Tao, "Cooperative Tx/Rx caching in interference channels: a storage-latency tradeoff study," in *2016 IEEE International Symposium on Information Theory (ISIT)*, July 2016, pp. 2034–2038.

[25] J. Hachem, U. Niesen, and S. Diggavi, "A layered caching architecture for the interference channel," in *2016 IEEE International Symposium on Information Theory (ISIT)*, July 2016, pp. 415–419.

[26] J. Hachem, U. Niesen, and S. N. Diggavi, "Degrees of freedom of cache-aided wireless interference networks," *IEEE Transactions on Information Theory*, vol. 64, no. 7, pp. 5359–5380, July 2018.

[27] V. R. Cadambe and S. A. Jafar, "Interference alignment and the degrees of freedom of wireless *x* networks," *IEEE Transactions on Information Theory*, vol. 55, no. 9, pp. 3893–3908, Sept 2009.

[28] J. Hachem, U. Niesen, and S. Diggavi, "Energy-efficiency gains of caching for interference channels," *IEEE Communications Letters*, vol. 22, no. 7, pp. 1434–1437, July 2018.

[29] N. Karamchandani, U. Niesen, M. A. Maddah-Ali, and S. N. Diggavi, "Hierarchical coded caching," *IEEE Transactions on Information Theory*, vol. 62, no. 6, pp. 3212–3229, June 2016.

[30] M. Ji, G. Caire, and A. F. Molisch, "Wireless device-to-device caching networks: basic principles and system performance," *IEEE Journal on Selected Areas in Communications*, vol. 1, no. 34, pp. 176–189, 2016.

[31] M. Ji, G. Caire and A. F. Molisch, "Fundamental limits of caching in wireless D2D networks," *IEEE Transactions on Information Theory*, vol. 62, no. 2, pp. 849–869 2016.

3 Wireless Device-to-Device Caching Networks

Mingyue Ji

3.1 Overview

Internet traffic has been increasing significantly in the past few years, mainly due to on-demand video streaming. While users prefer to use wireless to connect to the internet, seamless and cost-effective on-demand video streaming service could not be supported by today's cellular wireless technologies. For instance, a *single* streaming session of a standard definition (SD) movie, from a service such as iTunes, Netflix, Hulu, or Amazon Prime (duration of 1 hour and 30 minutes with size of about 2 GB), could completely consume most of the viewer's monthly cellular data plan. It is evident that a dramatic technological paradigm shift will be required in order to fill the gap between users' expectations and the provided services' limitations. Recently, in this perspective, *caching at wireless edge*, i.e., caching part of or entire content library directly in the wireless nodes such as femtocell base stations and/or user devices, has been recognized for its potential to solve this problem by providing per-node throughput scaling much higher than that of conventional unicast transmissions in a variety of scenarios.

One important feature of on-demand video streaming services is that user demands are highly redundant over space and time. For example, consider a university campus where $n \approx 10,000$ users (distributed over an area of approximately 1 km^2) stream movies from a library of approximately 100 files, such as the weekly top-of-the chart titles of iTunes, Netflix, Hulu, or Amazon Prime. In this scenario, each user's demand can be satisfied via short-range local communications from a cache, instead of arranging thousands of unicast sessions from a cellular base station and without requiring deployment of a large number of small cell access points or femtocell/pico base stations, each of which requires a costly high-throughput backhaul. Intuitively, caching can effectively take advantage of the inherent redundancy of user demands, although, differently from live streaming; users in on-demand streaming do not request the same content at the exact same time. This type of redundancy is referred to in [1–3] as *asynchronous content reuse*. It has been shown recently that the throughput for delivery of wireless video files can be greatly enhanced by wireless device-to-device (D2D) communications in conjunction with caching on the devices. In particular, systems have been proposed in which each device is equipped with storage capacity for caching carefully designed parts of library content either randomly or deterministically. When a user requests a file not already in its own cache, it can obtain it from one of its neighbors through a spectrally and cost-efficient, short-range D2D link. As user density increases, the aggregate storage capacity of the D2D network increases linearly with the user size, while the average communication distance decreases (and hence the spatial reuse increases). For these reasons, D2D caching networks are scalable, such that both demand and throughput increase linearly with the user density. An overview of D2D for video is given by [11]. A series of papers about this topic includes [2, 3, 5–13].

In this chapter, we first describe the general network model used in wireless D2D caching networks in Section 3.2. Under the simple and deterministic protocol channel model, in Section 3.3, we focus on the case of uncoded D2D caching networks, and we relax the uncoded constraint and study the coded D2D caching network in Section 3.4. In Section 3.5, we relax the constraint of the protocol channel model and study the D2D caching over a noisy physical layer channel. Finally, we study caching in D2D networks with user mobility patterns in Section 3.6.

3.2 General Network Model

In this chapter, we consider a network formed by wireless user nodes $\mathcal{U} = \{1,\ldots,n\}$, which could be placed deterministically on a grid (e.g., Section 3.3.1) or randomly according to either a uniform distribution (e.g., Section 3.3.2) or a point Poisson process (PPP) (e.g., Section 3.5.2). The area of the network is assumed to be $A(n)$, which is a function of the number of users n.[1] Each user $u \in \mathcal{U}$ makes an arbitrary request or a random request for a file $f \in \mathcal{F} = \{1,\ldots,m\}$ in an independent and identically distributed (I.I.d.) manner, according to a probability mass function $P_r(f)$, which is assumed to be a Zipf distribution as follows:[2]

$$P_r(f) = \frac{\frac{1}{f^\gamma}}{\sum_{j=1}^{m} \frac{1}{j^\gamma}}, \quad 1 \le f \le m, \tag{3.1}$$

where γ is the parameter of the Zipf distribution. We assume that each user caches M files, each of which consists of F bits uniformly generated over $\{1,\ldots,2^F\}$. In general, the caching networks consists of a *cache placement phase* and a *delivery phase*. Specifically, delivery phase includes *coded delivery* (generates and decodes the [coded] messages) and *transmission policy* (transmits the [coded] messages)[3]. Therefore, we have the following definitions.

DEFINITION 3.1 **(Cache Placement)** *The cache placement is a map of the file library $\{W_f : f \in \mathcal{F}\}$ onto the cache of the users in $\mathcal{U}$. Each cache has size MF bits (i.e., M files). For each $u \in \mathcal{U}$, the function $\phi_u : \mathbb{F}_2^{mF} \to \mathbb{F}_2^{MF}$ generates the cache content $Z_u \triangleq \phi_u(W_f : f \in \mathcal{F})$. The cache messages Z_u are stored in the user caches at the beginning of time and kept fixed through the subsequent network operations.* ◇

[1] Note that $A(n)$ can also be a constant.

[2] In practice, the realistic demand distribution may follow a MZipf distribution, which is a modified version of Zipf distribution. The scheme based on the MZipf distribution may have similar performance [14].

[3] Note that the uncoded message is a special case of the coded message.

DEFINITION 3.2 **(Coded Delivery)** *The delivery phase is defined by two sets of functions: the node encoding functions, denoted by* $\{\psi_u : u \in \mathcal{U}\}$, *and the node decoding functions, denoted by* $\{\lambda_u : u \in \mathcal{U}\}$. *Let* R_u^{T} *denote the number of* coded bits *transmitted by node u to satisfy the request vector* f. *The rate of node u is defined by* $R_u = \frac{R_u^{\mathrm{T}}}{F}$. *The function* $\psi_u : \mathbb{F}_2^{MF} \times \mathcal{F}^n \to \mathbb{F}_2^{FR_u}$ *generates the transmitted message* $X_{u,\mathsf{f}} \triangleq \psi_u(Z_u, \mathsf{f})$ *of node u as a function of its cache content* Z_u *and of the demand vector* f.

Let $\mathcal{D}_u$ *denote the set of users whose transmit messages are received by user u (according to some transmission policy that will be given in Definition 3.3). The function* $\lambda_u : \mathbb{F}_2^{F \sum_{v \in \mathcal{D}_u} R_v} \times \mathbb{F}_2^{MF} \times \mathcal{F}^n \to \mathbb{F}_2^F$ *decodes the request of user u from the received messages and its own cache, i.e., we have*

$$\hat{W}_{u,\mathsf{f}} \triangleq \lambda_u(\{X_{v,\mathsf{f}} : v \in \mathcal{D}_u\}, Z_u, \mathsf{f}). \tag{3.2}$$

$\diamondsuit$

As anticipated in Section 3.1, in video on-demand services, the probability that two users wish to stream a file at the exact *same time* is essentially zero, although there is a large redundancy in the demands when $n \gg m$. A rigorous mathematical model for asynchronous content reuse is introduced in [3]. This involves the request of random segments, each formed by K' packets of F bits each, from files of K packets, and then letting $K \to \infty$ while keeping K' fixed to some arbitrary constant. Here, for the sake of brevity, we simply *forbid* the possibility of naive multicasting.

With this in mind, we move on and define the worst-case error probability as

$$P_e = \max_{\mathsf{f} \in \mathcal{F}^n} \max_{u \in \mathcal{U}} P\left(\hat{W}_{u,\mathsf{f}} \neq W_{f_u}\right). \tag{3.3}$$

For given number of users n and library size m, by letting $R = \sum_{u \in \mathcal{U}} R_u$, the cache-rate pair (M, R) is achievable if $\forall\ \varepsilon > 0$ there exists a sequence indexed by the file size $F \to \infty$ of cache encoding functions $\{\phi_u\}$, delivery functions $\{\psi_u\}$, and decoding functions $\{\lambda_u\}$, with rate $R^{(F)}$ and probability of error $P_e^{(F)}$ such that $\limsup_{F \to \infty} R^{(F)} \leq R$ and $\limsup_{F \to \infty} P_e^{(F)} \leq \varepsilon$. The optimal achievable transmission rate is given by

$$R^*(M) \triangleq \inf\{R : (M, R) \text{ is achievable}\}. \tag{3.4}$$

The definition of transmission policy is given by the following definition.

DEFINITION 3.3 **(Transmission Policy)** *The transmission policy* Π *is a rule to activate the D2D links in the network at different time slots. Let* $\mathcal{L}$ *denote the set of all directed D2D links. Let* $\mathcal{A} \subseteq 2^{\mathcal{L}}$ *be the set of all possible feasible subsets of links (this is a subset of the power set of* $\mathcal{L}$, *formed by all sets of links forming independent sets in the network interference graph induced by the protocol model). Let* $\mathsf{A} \subset \mathcal{A}$ *denote a feasible set of simultaneously active links. Then,* Π *is a conditional probability mass function over* $\mathcal{A}$ *given* f *(requests) and the caching functions, assigning probability* $\Pi(\mathsf{A})$ *to* $\mathsf{A} \in \mathcal{A}$. $\diamondsuit$

Note that the deterministic transmission policy is a special case of Definition 3.3.

3.3 Uncoded D2D Caching Networks Based on the Protocol Channel Model

In this section, we study the D2D caching networks when the channel model follows a seminal deterministic and noiseless *protocol model* [15]. The protocol channel model is defined as follows. When node i transmits a packet to node j, the transmission is successful if and only if (1) the distance between i and j is less than r and (2) any other node k transmitting simultaneously is at distance $d(k, j) \geq (1 + \Delta)r$ from the receiver j, where $r, \Delta > 0$ are parameters of the protocol model. In practice, nodes often send data at some constant rate C_r bit/s/Hz, where C_r is a nonincreasing function of the transmission range r. In this section, we focus on the uncoded D2D cache placement and delivery, which means that only packets or the entire files will be cached and transmitted directly without coding among them. The request distribution is assumed to be a Zipf distribution given by (3.1). In the following, first, we focus on the case where only single-hop transmissions are allowed. Second, we consider the case where the multi-hop transmission opportunities are exploited.

3.3.1 Throughput-Outage Trade-off in Single-Hop D2D Caching Networks

When a grid node location is considered (see Figure 3.1a) and only single hop transmissions are allowed, under the uncoded constraint, we design an independent and randomized cache placement of the entire files and unicast delivery.

3.3.1.1 Random Cache Placement and Unicast Delivery

When a random and independent caching placement scheme is considered, we divide the network into clusters of equal size and independent of the users' demands and cache placement realizations (see Figure 3.1b). The D2D communications can take place only inside the corresponding cluster, given the cardinality of the number of users in each cluster as g_c.[4] In [3], it is shown that the optimal caching distribution P_c^* for maximizing the probability that any user finds its demanded file inside its own cluster is given (for a node deployment on a grid as described earlier) by

$$P_c^*(f) = \left[1 - \frac{\nu}{z_f}\right]^+, \quad f = 1, \dots, m, \tag{3.5}$$

where $\nu = \frac{m^*-1}{\sum_{f=1}^{m^*} \frac{1}{z_f}}$, $z_f = P_r(f)^{\frac{1}{M(g_c-1)-1}}$, $m^* = \min\left\{\frac{M}{\gamma_r} g_c, m\right\}$ and $[\Lambda]^+ = \max[\Lambda, 0]$.

The delivery scheme is as follows. Inside a cluster, all the potential links (the links established by one source-destination pair) are served in a round robin manner. Different clusters will be scheduled in a time division multiple access (TDMA) manner. It can be seen that it is possible that users may not find the requested files in their own cluster such that they cannot be served by the delivery scheme. We call this event an *outage*, and the probability of outage is p_o. We define the random variable T_u as the number of useful

[4] We also use $g_c(m)$ when we need to emphasize the dependence of g_c on m.

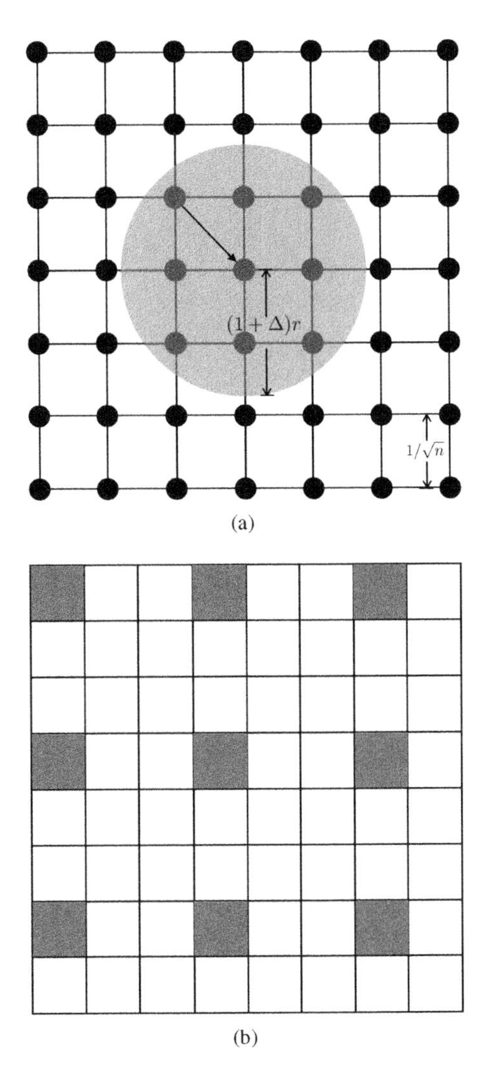

Figure 3.1 (a) An example of a grid D2D network with $n = 49$ users (black points), minimum separation pf $\frac{r}{\sqrt{2}} = \frac{1}{\sqrt{n}}$, and the protocol channel model. In this figure, the gray area is the disk where the protocol model does not allow other concurrent transmissions, r is the common worst-case transmission range, and Δ is the interference parameter. (b) An example of single-cell deployment and the interference avoidance (e.g., TDMA) scheme. In this figure, each square represents a cluster and the gray squares represent the simultaneous transmitting clusters. In this example, the TDMA parameter is $K = 9$, which means each cluster can be activated every 9 scheduling slots.

received information bits per slot unit time by user u and the average throughput for user u is $\overline{T} = E[T_u]$. We focus on max–min fairness and express the throughput-outage trade-off in terms of minimum average user throughput defined as $\overline{T}_{\min} = \min_{u \in \mathcal{U}} \overline{T}_u$. Hence, we will focus on solving the following optimization problem:

$$\text{maximize} \quad \overline{T}_{\min}, \quad \text{s.t.} \quad p_o \leq p, \tag{3.6}$$

where the maximization is with respect to the cache placement and transmission policies Π_c, Π_t. Hence, it is immediately clear that $T^*(p)$ is nondecreasing in p, which is the requirement of the outage probability.

Let $\alpha_\gamma = \frac{1-\gamma}{2-\gamma}$. It can be shown that the only practically interesting regime, meaning vanishing outage probability, is the case when $\lim_{n\to\infty} \frac{m^{\alpha_\gamma}}{n} = 0$. Based on this fact, the optimal achievable throughput-outage trade-off is given by the following theorem.

THEOREM 3.4 *Assume* $\lim_{n\to\infty} \frac{m^{\alpha_\gamma}}{n} = 0$ *and* M *is a constant. Then, the throughput-outage trade-off* $T(p)$ *achievable by random caching and clustering behaves as*

$$T(p) = \begin{cases} \frac{C}{K}\frac{M}{\rho_1 m} + o(1/m), & p = (1-\gamma)e^{\gamma-\rho_1} \\ \frac{CA}{K}\frac{M}{m(1-p)^{\frac{1}{1-\gamma}}} + o\left(\frac{1}{m(1-p)^{\frac{1}{1-\gamma}}}\right), & p = 1 - a\left(\frac{g_c(m)}{m}\right)^{1-\gamma} \\ \frac{CB}{K} m^{-\alpha} + o\left(m^{-\alpha_\gamma}\right), 1 - a\rho_2^{1-\gamma} m^{-\alpha_\gamma} & \le p \le 1 - ab^{1-\gamma} m^{-\alpha_\gamma} \\ \frac{CD}{K} m^{-\alpha_\gamma} + o\left(m^{-\alpha_\gamma}\right), & 1 - ab^{1-\gamma} m^{-\alpha_\gamma} \le p \le 1, \end{cases} \tag{3.7}$$

where we define $a = \gamma^\gamma M^{1-\gamma}$, $b = \left(\frac{1-\gamma}{a}\right)^{\frac{1}{2-\gamma}}$, $A \triangleq \gamma^{\frac{\gamma}{1-\gamma}}$, $B \triangleq \frac{a\rho_2^{1-\gamma}}{1+a\rho_2^{2-\gamma}}$, *and* $D \triangleq \frac{ab^{1-\gamma}}{1+ab^{2-\gamma}}$ *and where* ρ_1 *and* ρ_2 *are positive parameters satisfying* $\rho_1 \ge \gamma$ *and* $\rho_2 \ge b$. *The cluster size* $g_c(m)$ *is any function of* m *satisfying* $g_c(m) = \omega(m^\alpha)$ *and* $g_c(m) \le \gamma m/M$. □

Based on Theorem 3.4, it can be seen that when $Mn \ge m$ (the entire library can be cached by the nodes in the network), for an arbitrarily small outage probability, in [3], it is shown that the per-user throughput scales as $\Theta\left(\frac{M}{m}\right)$. This means that the per-user throughput is independent of and increases linearly with the number of users. Moreover, the throughput also grows linearly with the per-user memory M. This can be very attractive since, for example, in order to double the per-user throughput, instead of increasing the bandwidth and/or power, we can just double the (relatively cheap) storage capacity per user.

Interestingly, this result of throughput scaling coincides with using the subpacketized cache placement and coded multicasting delivery algorithms by [11, 15]. However, the practical performance in realistic channels can be quite different. Under more realistic channel models, D2D caching is with conventional unicasting, harmonic broadcasting and coded multicasting. Consider a network of size 600 m $\times$ 600 m, and let $n = 10{,}000$ users distributed uniformly. The file library has size $m = 300$ (e.g., 300 popular films and TV shows to be refreshed on a daily basis at off-peak hours by the conventional cellular network). The per-user storage capacity is assumed to be $M = 20$, and the Zipf distribution parameter is $\gamma_r = 0.4$. The channel model is a mixture of models from [16]; for more details see [2]. The simulation results of the throughput-outage trade-off for different schemes are given in Figure 3.2. We can see that in this realistic propagation scenario (not the protocol model) the D2D single-hop caching network with simple transmission scheme can provide both large throughput, sufficient for streaming video at standard definition quality, and low outage probability. Also, the D2D caching scheme

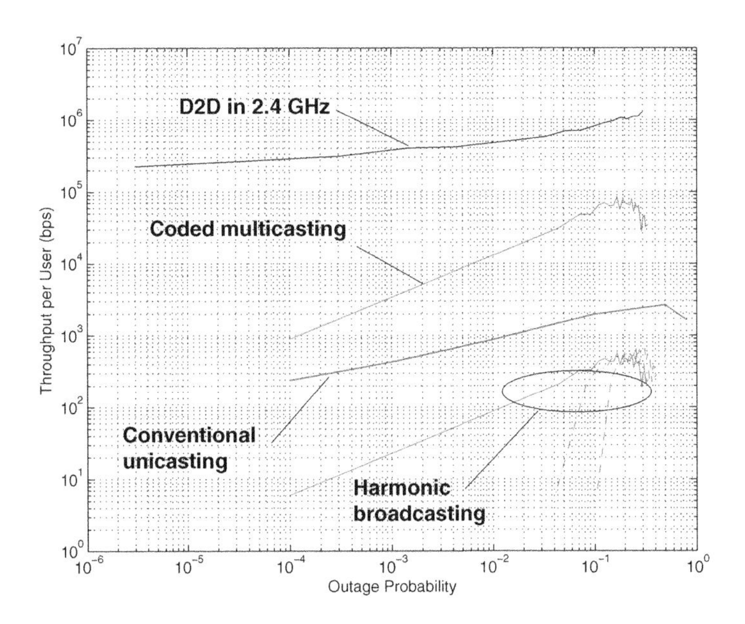

Figure 3.2 Simulation results for the throughput-outage trade-off for different schemes under the realistic indoor/outdoor propagation environment, where $n = 10,000$, $m = 300$, $M = 20$, and $\gamma = 0.4$. For harmonic broadcasting with the m' most popular files, the solid, dash-dot, and dashed lines means are $m' = 300$, $m' = 280$, and $m' = 250$, respectively.

significantly outperforms the other schemes for all the regimes of outage probability.[5] This performance gain is particularly notable with respect to the current technology, such as conventional unicasting and harmonic broadcasting from the base station. We also observe the outstanding performance advantages compared to the coded multicasting, despite the fact that the two schemes have the same throughput scaling laws. The main reason for this observation is that the capacity of multicasting is limited by the "weakest link" between the base station and the various mobile devices (see [17]), while for the D2D scheme, short-range transmission (which usually has a high signal-to-noise ratio [SNR], shallow fading, and thus high capacity) determines the overall performance.

It is also worth noticing that the scheduling scheme used in the simulations is based on the clustering structure and the interference avoidance (TDMA). An advanced interference management scheme such as FlashLinQ by [18] or ITLinQ by [19] may provide additional gains for the D2D caching networks (see Section 3.5.1 for more details).

3.3.2 Uncoded Multi-hop D2D Caching

In this section, we present a natural extension of the single-hop D2D network by allowing multi-hop transmissions as shown in [20]. The multi-hop D2D caching networks under the protocol model have been studied in the literature as in [21]. The main

[5] When the demand distribution is MZipf, using a similar cache placement and delivery scheme, similar performance can also be observed, as shown in [14].

objective of [21] is to minimize the expected number of flows passing through each node. Such expected number of flows is proportional to the reciprocal of the expected per-node throughput; on the other hand, [20] adopts the throughput metric guaranteed for any demand realization with high probability as the number of nodes n increases, which is a natural extension from the conventional ad hoc network model, e.g., see [15, 23–26]. It directly derives the lower and upper bounds of throughput by using only appropriate inequalities (holding with high probability in the limit of large n) without any approximation. Moreover, in order to provide a comprehensive understanding on throughput scaling laws, it assumes a general scaling environment, i.e., n, m, and M scale with a general relationship instead of a fixed M considered in [3, 21]. Finally, a centralized and deterministic caching placement was proposed in [21] based on the popularity distribution; in contrast, in [20], it presents a completely decentralized random caching placement according to a uniform distribution over the entire file library when $m = o(nM)$, which is "universal" in the sense that it is independent of the specific popularity distribution. Remarkably, while the placement and the achievability scheme of [21] would break under the node mobility, such that one should reallocate the cache content when the nodes are in the presence of node mobility, our scheme is robust since any random movement of the nodes would generate the same caching distribution and, therefore, yields the same throughput scaling with high probability.

The proposed caching placement and delivery schemes in [20] yield a per-user throughput scaling of $\Theta(\sqrt{M/m})$, which achieves order optimality when the popularity distribution has the "heavy tail" property defined as follows.

DEFINITION 3.5 (**Heavy-Tailed Popularity Distributions**) *Define a class of popularity distributions such that, for any $0 < c_1 < a_1$, there exists $c_2 > 0$ satisfying that*

$$\lim_{n\to\infty} \sum_{i=1}^{c_1 n^\alpha} P_r(i) \le 1 - c_2, \tag{3.8}$$

where c_1 and c_2 are some constants and independent of n. ◊

For example, a Zipf distribution (3.1) with an exponent less than one satisfies the heavy tail property.[6] This result shows that multi-hop achieves a much better per-node throughput scaling than that of single-hop D2D caching networks, which is $\Theta(M/m)$. Furthermore, it is shown that for other popularity distributions, where the heavy tail property is not satisfied or the user demands strongly concentrate, a further improvement of the per-node throughput scaling law beyond $\Theta(\sqrt{M/m})$ is achievable, similar to the case of single-hop D2D caching networks in [21] and the case of shared-link caching networks in [27].

In order to provide a comprehensive understanding on capacity scaling laws of the wireless multi-hop D2D caching network defined earlier, we consider a general relationship between the parameters n, m, and M as follows:

$$m = a_1 n^\alpha \text{ and } M = a_2 n^\beta, \tag{3.9}$$

[6] Throughout this section and Section 3.5.3.2, an "order-optimal" scheme means that it achieves the optimal scaling law of the per-node throughput within a multiplicative factor of n^ϵ for any $\epsilon > 0$.

where $\alpha, a_1, a_2 > 0$, and $\beta \in [0, \alpha]$. Notice that the delivery phase becomes trivial if $\alpha < \beta$ (each node is able to store the entire library $\mathcal{F}$ for this case). For the same reason, we assume that $a_1 > a_2$ if $\alpha = \beta$. It can be observed that the regimes where $m = \omega(M)$ and $m = o(nM)$ might be the most important and interesting regimes in practice, since each node can cache only a subset of the library but there are some caching redundancies in the entire network so that a nontrivial caching gain is expected in these regimes.

In the following, we partition the entire parameter space into five regimes given by

- Regime I: $\alpha - \beta > 1$.
- Regime II: $\alpha - \beta = 1$ and $a_1 > a_2$.
- Regime III: $\alpha - \beta = 1$ and $a_1 \leq a_2$.
- Regime IV: $\alpha - \beta \in (0, 1)$.
- Regime V: $\alpha - \beta = 0$ and $a_1 > a_2$.

Notice that the shifting from Regimes I to V tends to increase the relative storage capability at each node, compared to the library size (recall the relation between m and M in (3.9)).

The distribution of nodes in this network model is assumed to be a random uniform distribution in contrast to the deterministic grid distribution as in Section 3.3.1. For a given feasible delivery strategy, we let T_n denote the corresponding per-node symmetric throughput, i.e., the rate (in bit/s/Hz) at which the request of any node in the network can be served. Note that the definition of the outage event in this network model is more strict compared to that in Section 3.3.1. In this case, if there exists an unserved user $u \in \mathcal{U}$, we say that the network is in outage. In this case, conventionally, we let $T_n = 0$.

3.3.2.1 Random Caching and Unicast Multi-hop Delivery

We consider the regimes of interest, Regimes IV and V, in this section.[7]

Decentralized Cache Placement: Each node u caches M distinct files in its own memory, chosen uniformly at random from the library $\mathcal{F}$, independently of other nodes.

Local Multi-Hop Protocol: We first explain how each node finds its source node caching the demanded file (source node selection):

- The entire network is divided into square *traffic cells* of area $a_c = n^{-\eta}$ for some $\eta \in [0, 1)$, where η will be determined later.
- Each node chooses one of the nodes that caches the requested file in the same traffic cell as its *source* node. If there are multiple candidates, it picks one of them uniformly at random.

For the ease of illustration, we refer to the pair formed by a node and its source node as a *source–destination (SD) pair*. Note that in our model, each SD pair is located in the

[7] For Regimes I and II, the outage probability cannot go to zero as $n \to \infty$. For Regime III, which corresponds to the case $nM = \Theta(m)$, a *centralized* file placement and a *globally* multi-hop protocol can be designed as shown in [20].

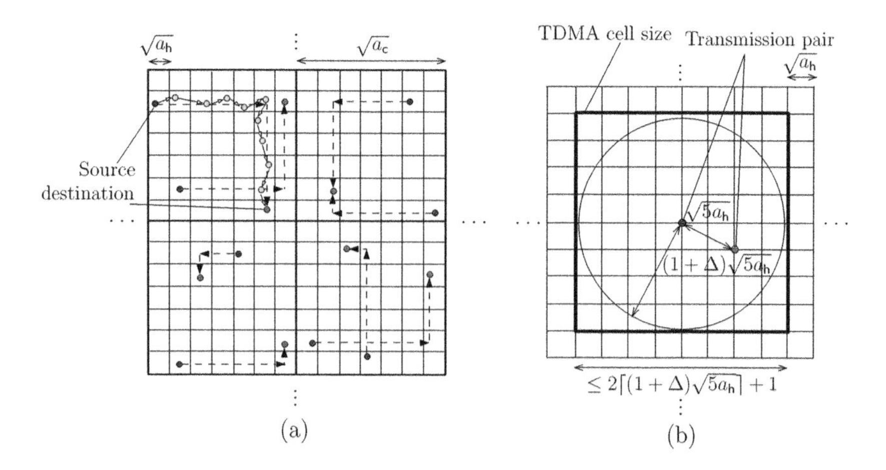

Figure 3.3 (a) The proposed multi-hop routing protocol for file delivery after the source node selection. (b) TDMA cell size based on the protocol model.

same traffic cell while in the classical wireless ad hoc network, SD pairs are randomly located over the entire network. It can be seen that caching can reduce the distance of each SD pair, which means that all nodes are able to find their source nodes within their traffic cells with high probability.

Next, we illustrate the proposed multi-hop transmission scheme for the file delivery between n SD pairs, see also Figure 3.3a (multi-hop transmissions):

- Each traffic cell is divided into square *hopping cells* of area $a_h = \frac{2\log n}{n}$.
- We define the horizontal data path (HDP) and the vertical data path (VDP) of a SD pair as the horizontal line and the vertical line connecting a source node to its destination, respectively. Each source node transmits the demanded file to its destination by first hopping to the adjacent hopping cells on its HDP and then on its VDP.[8]
- The TDMA scheme is used with the reuse factor K for which each hopping cell is activated only once out of K time slot durations (see Figure 3.3b).
- A transmitter node in each active hopping cell sends a file (or a fragment of a file) to the receiver node in an adjacent hopping cell. For simplicity, round-robin scheduling is used for all transmitters in the same hopping cell.

In this scheme, each hopping cell should contain at least one node for relaying, which is satisfied with high probability due to the fact that $a_h = \frac{2\log n}{n}$.

3.3.2.2 The Characterization of Throughput and Discussions

The following throughput scaling laws hold *universally* for any popularity distribution.

[8] If a source node and its destination node are in the same hopping cell, then the source node directly transmits the requested file to its destination.

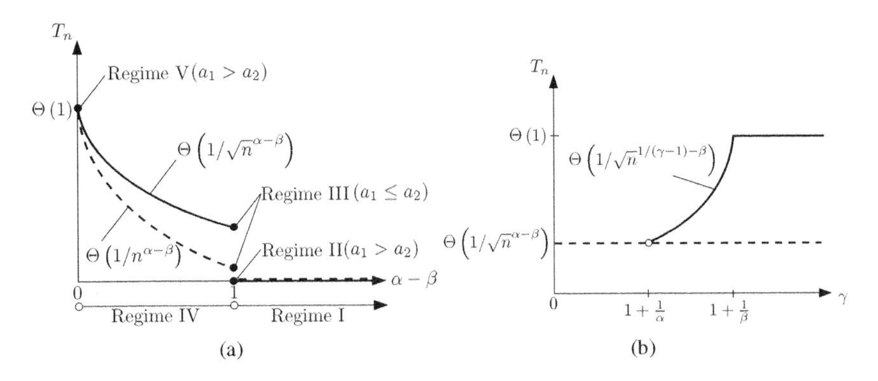

Figure 3.4 (a) Achievable throughput scaling laws, where the solid curve is for caching wireless multi-hop D2D networks and the dashed curve is for caching wireless single-hop D2D networks. (b) Achievable throughput scaling laws in (3.11) (solid curve) and (3.10) (dashed curve) with respect to γ for Regime IV.

THEOREM 3.6 *For the caching wireless D2D network defined in this section, under heavy-tail demand distribution, the order optimal achievable throughput satisfies the following scaling laws with high probability:*

$$T_n = \begin{cases} 0 & \text{for Regimes I and II,} \\ \Omega(n^{-\frac{1}{2}-\epsilon}) & \text{for Regime III,} \\ \Omega(n^{-\frac{\alpha-\beta}{2}-\epsilon}) & \text{for Regime IV,} \\ \Omega(n^{-\epsilon}) & \text{for Regime V,} \end{cases} \tag{3.10}$$

where $\epsilon > 0$ is arbitrarily small.

Figure 3.4a compares the achievable throughput scaling laws of the wireless D2D caching networks between multi-hop and single-hop file deliveries,[9] where the term $n^{-\epsilon}$ is omitted for simplicity. Regimes I and II correspond to the case where the aggregate storage capacity in the entire network is strictly less than the library size, i.e., $Mn < m$. Hence, an outage is inevitable even if a centralized caching scheme is used, which leads to $T_n = 0$. As the aggregate storage capacity increases, i.e., $\alpha - \beta$ decreases, each node can find its demanded file in the network and thus a non-zero T_n is achievable for Regimes III and IV. As will be clear from the achievable schemes, the geometric interpretation of this behavior is as follows: as $\alpha - \beta$ decreases (i.e., the storage capacity M increases), the file delivery distance decreases, such that an increased spatial reused can be achieved (multiple links can be activated at the same time under the constraint of the protocol model). As a result, T_n increases when $\alpha - \beta$ decreases for the throughputs of both multi-hop and single-hop D2D communications. Finally, when $\alpha = \beta$ (i.e., Regime V), each node can find its requested file from its nearest neighbors. Therefore, the delivery distance is $O(1/\sqrt{n})$ and $T_n = \Theta(1)$ is achievable.

[9] The throughput of the single-hop D2D caching network in this in this case can be obtained similarly as in Section 3.3.1.

One important fact is that single-hop D2D delivery is order optimal only for Regime V. For almost all parameter regimes of interest (Regimes III and IV), multi-hop D2D delivery significantly improves the scaling of per-user throughput by a factor $\sqrt{\frac{m}{M}}$. Intuitively, spatial reuse is much more effective with multi-hop transmissions since more concurrent short-rage transmissions can be exploited; meanwhile, the cost of duplicated transmissions by multi-hop is not comparable with the gains obtained by the simultaneously active links. It is worth mentioning that, for a Zipf demand distribution (3.1) with $\gamma < 1$, this throughput scaling law matches with that in [21], although a decentralized random caching is used in this work rather than a centralized and deterministic caching scheme introduced in [21]. Furthermore, due to the universality of the proposed scheme (random and independent caching), the proposed scheme is robust to random user mobility since random movement of each user does not change the overall distribution of node locations when n is large and the proposed random caching placement is affected by only the overall node distribution, not by each user's location.

As the demand distribution becomes more skewed (e.g., γ increases in a Zipf distribution), the condition in Definition 3.5 may not hold. In this case, it can be expected that the throughput scaling law can be improved by a more refined cache placement strategy, biased toward the files demanded with higher probabilities. In fact, we show that caching only an appropriately optimized subset of the most popular files can guarantee that the aggregate "tail" probability of the least popular files vanishes, such that no outage event can be obtained. In the following, we demonstrate this strategy for a Zipf demand distribution with $\gamma > 1 + \frac{1}{\alpha}$.

THEOREM 3.7 *Consider the wireless D2D caching networks defined in this section and assume that the requests follow a Zipf demand distribution with exponent $\gamma > 1 + \frac{1}{\alpha}$. Then the achievable throughput satisfies whp the scaling law:*

$$T_n = \Omega\left(n^{-\frac{1-\min(1,\beta+1-1/(\gamma-1))}{2}-\epsilon}\right) \quad \textit{for Regime IV}, \tag{3.11}$$

where $\epsilon > 0$ is arbitrarily small.

In Figure 3.4b, we compare the improved scaling laws in (3.11) and the scaling laws in (3.10) for Regime IV, where we omit the term $n^{-\epsilon}$. When the demands follow a Zipf distribution, the improved throughput scaling law $\Theta(1/\sqrt{n}^{1/(\gamma-1)-\beta})$ is achievable instead of $\Theta(1/\sqrt{n}^{\alpha-\beta})$ in (3.10) when $\gamma > 1 + \frac{1}{\alpha}$, and eventually $\Theta(1)$ is achievable when $\gamma \geq 1 + \frac{1}{\beta}$ (see Figure 3.4b). Moreover, the improved throughput scaling laws in Theorem 3.7 can still be achieved by a fully decentralized random cache placement over an appropriately reduced *effective* library size, i.e., *decentralized random caching uniformly across a subset of popular files*. In this regime, we can ignore some files, whose probability is small enough such that an outage will not occur with high probability as $n \to \infty$.

3.4 Coded D2D Caching under the Protocol Model

One important property or constraint of the described scheme is that both the caching placement and the delivery schemes exploit an uncoded approach. The gain of the throughput is mainly obtained by spatial reuse (TDMA) and multi-hop if applicable. At this point, a natural and reasonable question to ask is whether coded multicasting for D2D transmissions can provide an additional gain or whether the coding gain and the spatial reuse gain can accumulate. In [11], the authors design a (deterministic or random) subpacketized caching and a network-coded delivery scheme for the single-hop D2D caching networks. The scheme is explained by the example shown in Figure 3.5, where it is assumed that no spatial reuse can be used or only one transmission per time-frequency slot is allowed but the transmission range can be arbitrary (e.g., covering the entire network). This scheme can be generalized to arbitrary n, m, and M. Without using spatial reuse, the achievable normalized number of transmissions (transmission rate) without outage is given by $R = \frac{m}{M}\left(1 - \frac{M}{m}\right)$, which is surprisingly almost the same as the result shown in [15], where instead of D2D communications, one central server (base station) with the access to the entire library multicasts coded packets.

Suppose that (M, R) is achievable and that there exists a transmission policy that can deliver to each user the coded packets necessary to decode its requested file in no more than D channel uses for the worst-case demands. Then, the per-user throughput, defined as the useful information bits per channel use, is given by $T = \frac{F}{D}$. Thus, we can see that the per-user throughput $\Theta\left(\frac{M}{m}\right)$ has the same scaling law as the throughput by using

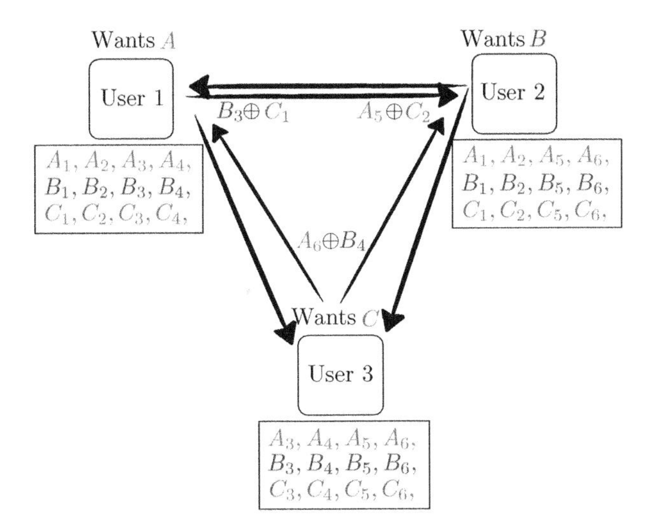

Figure 3.5 An example of 3 users, 3 files, $M = 2$, and achieving half normalized file transmissions. Each file is divided into 6 packets (e.g., A is divided into $A_1, \ldots, A_6$.) User 1 demands A; user 2 demands B, and user 3 demands C. The cached packets are shown in the rectangles under each user. For the delivery phase, user 1 sends $B_3 \oplus C_1$, user 2 sends $A_5 \oplus C_2$, and user 3 sends $A_6 \oplus B_4$. The normalized number of file transmissions is $3 \cdot \frac{1}{6} = \frac{1}{2}$, which also achieves the information theoretically optimal rate for this network.

the decentralized random cache placement and uncoded delivery scheme described in Section 3.3.1. In addition, it can be shown that if spatial reuse is also exploited, no further order gain is achievable. In other words, *the gains from spatial reuse and coding cannot accumulate*. Intuitively, if spatial reuse is not allowed, a complex caching scheme can be designed such that one transmission can be useful for as many users as possible. If the transmission range is reduced and coding is exploited in each cluster, then the number of nodes benefited by one coded transmission is reduced but the D2D transmissions can operate concurrently at a higher rate. Moreover, the complexity of file subpacketizations and coding can also be reduced. Hence, the benefit of coding depends on the actual physical layer link throughput (bits/s/Hz) and the coding complexity rather than throughput scaling laws.[10]

3.4.1 Discussions

3.4.1.1 Extension to Multi-hop Coded D2D Caching Networks

In Section 3.3, under an uncoded cache placement and delivery constraint, we consider edd both single-hop and multi-hop D2D transmission schemes. In this section, we allow coding for both cache placement phase and delivery and transmission phase but focus on only single-hop communications. It is natural to consider the extension that relaxes all the constraints on the cache placement phanse and delivery and transmission phase under the protocol channel model. It means that we can allow coding on both cache placement and coded delivery phases and also allow multi-hop communications. In [29], the authors consider this case. They propose a coded caching placement strategy based on deterministic assignment of minimum distance separable (MDS)-coded packets of the files, a coded multicast delivery strategy where the users send linearly coded messages to each other in order to collectively satisfy their demands, and a randomized Euclidean minimum spanning tree–based routing strategy for transmission. Under the worst-case demands, it shows that this approach actually achieves the throughput scaling law of $\Theta\left(\sqrt{\frac{M}{m}}\right)$, which is slightly improved over the throughput scaling, as shown in Section 3.3.2.2 for the case of uncoded cache placement and delivery when multi-hop transmission is allowed. This throughput scaling law in [29] also achieves information theoretically outer bound within a multiplicative constant factor in practical parameter regimes.

3.4.1.2 Consideration of Lower Subpacketizations

One of the biggest practical constraints for coded D2D caching network is the large number of packets that each file needs to be partitioned into. For example, in the work of [11], each file needs to be partitioned into $\binom{n}{t}t$, where $t = \frac{Mn}{m}$. When $\frac{M}{m}$ is fixed, $\binom{n}{t}t$ grows exponentially with n. To address this concern, in [30–32], the authors propose several approaches to design coded caching networks with reduced subpacketizations. In particular, the authors propose two combinatorial designs for centralized D2D caching networks that have reduced subpacketization compared to [11]. The first approach uses a "hypercube" to define the cache placement and demonstrates how the geometry of

[10] An extensive analysis of the performance for D2D multicast is given by [28].

this hypercube relates to coded multicasting opportunities for delivery. The hypercube approach is optimized specifically for D2D caching networks, as opposed to adapting an already studied shared link scheme. The number of required packets (or bits) per file is reduced to $\left(\frac{m}{M}\right)^t$ from $\binom{n}{t}t$, as in [11], while the transmission rate is $\frac{m}{M}$, which is almost the same as $\frac{m}{M}-1$, as in [11]. In addition, by adopting the idea recently proposed in [33], this scheme can also be extended to a decentralized coded D2D caching scheme, which allows a much more flexible design for given network parameters. Meanwhile, the advantage of the reduced subpacketization of the hypercube approach still remains in the decentralized D2D caching networks.

The second approach is based on an application of the Ruzsa–Szeméredi graph [34, 35], which was first used for the design in shared link caching networks in [36]. [31] extends the use of the Ruzsa–Szeméredi graph to D2D caching networks and shows that the requirement of file subpacketization is at most subquadratic in terms of the number of users if no spatial reuse is allowed while the per-user throughput scales as $\Theta(n^{-\delta})$ for some arbitrarily small δ under some parameter regimes. Both D2D combinatorial designs sustain the significant throughput gain compared to conventional uncoded unicast [2], and the required packetizations are reduced exponentially compared to [11], with respect to the number of users n while keeping the library size m and memory size M fixed. Finally, the impact of enabling spatial reuse in these caching network designs has also been studied, showing this can further reduce the required packetizations, while also improving the per-user throughput significantly for some parameter regimes, in contrast with the case in [11].

3.5 Physical Layer Caching in D2D Networks

In this section, we relax the constraint of the deterministic protocol channel model considered in Section 3.3 and consider the realistic physical layer channel model. In particular, we consider the following scenarios. First, we discuss the direct extension of Section 3.3.1 by using the optimal rule of treating interference as noise in addition to interference avoidance (e.g., TDMA) [37]. Second, besides treating interference as noise, we consider the case when the D2D nodes are distributed according to a point Poisson process. Finally, we relax the constraints of treating interference as noise and interference avoidance and allow full cooperation among the D2D users.

In this section, the channel coefficient between a transmitter node j and a receiver node i is

$$h_{i,j} = h_0 e^{\sqrt{-1}\theta_{i,j}} r_{i,j}^{-\frac{\eta}{2}}, \tag{3.12}$$

where h_0 is a (random) small-scale fading coefficient, $\theta_{i,j}$ is the random phase with uniform distribution on $(0, 2\pi]$, $r_{i,j}$ is the distance between i and j, and $\eta \geq 2$ is the pathloss exponent. Let the transmit power be P_o, the signal-to-interference-plus-noise-ratio (SINR) of the user i can be expressed as

$$\mathrm{SINR_i} = \frac{P_o h_{i,j} r_{i,j}^{-\eta}}{\sigma^2 + I_r}, \tag{3.13}$$

where $I_r = \sum_{i \neq j} P_o |h_{i,j}|^2$ is the cumulative interference and σ^2 is the variance of the random noise. The link rate from node j to node i is given by $\log(1 + \mathrm{SINR_i})$.

3.5.1 D2D Caching with the Optimal Rule of Treating Interference by Noise

In this section, we extend the study of Section 3.3.1 to the case of using the approximately optimal condition of treating interference as noise. This scheme is based on the information-theoretically link scheduling (ITLinQ) introduced in [38]. There are three main advantages of using ITLinQ-based caching schemes. First, it can achieve a high D2D link throughput by exploiting the approximately optimal condition of treating interference as noise. Second, unlike the protocol channel model, where all the transmission ranges are the same, by using ITLinQ, we do not impose this constraint. Third, it can relax the “clustering” scheme by allowing each user to be served by any node in the network satisfying the ITLinQ conditions. In the following, we first introduce the ITLinQ and then we describe the ITLinQ-based delivery in D2D caching networks.

3.5.1.1 Information-Theoretically Link Scheduling

Let SNR_i denote the signal-to-noise ratio at user i, and for two different users i and j, INR_{ij} denotes the interference-to-noise ratio of source j at destination i; then the approximately optimal condition of treating interference as noise (TIN) is that for each user i,

$$\mathrm{SNR}_i \geq \max_{i \neq i} \mathrm{INR}_{ij} \cdot \max_{k \neq i} \mathrm{INR}_{ki}, \tag{3.14}$$

where SNR_i denotes the signal-to-noise ratio of the receiver i and INR_{ij} denotes the interference-to-noise ratio from user j to user i. The condition (3.14) means that if for each user, the desired channel strength is at least the sum of the strengths of the strongest interference from this user and the strongest interference to this user (on a dB scale), then treating interference as noise achieves the entire information-theoretically capacity region of the network within a constant additive gap. Based on condition (3.14), we define the *information-theoretically independent set (ITIS)*, which is a subset of users satisfying the condition (3.14) and leads to the spectrum sharing scheme of ITLinQ as follows [38].

DEFINITION 3.8 *At each time slot, the information-theoretically link scheduling spectrum sharing scheme schedules the sources in an information-theoretic, independent set to transmit simultaneously. In addition, all the destinations will treat their incoming interference as noise.*

3.5.1.2 The Greedy Closest-Source Policy Based on ITLinQ

Based on a slightly (modified) optimal cache placement scheme as described in Section 3.3.1,[11] [37] proposes an algorithm that can achieve superior performance compared to the original caching scheme. The main challenge is to find the user

[11] The modified cache placement scheme is obtained by enlarging $g_c(m)$.

association that leads to the best performance. [37] presents a greedy user association algorithm based on ITLinQ as follows.

1. We assign a random ordering to users.
2. Based on the assigned order, we associate each user with the closest potential source that has not been associated with a previous user.

We call this user associate policy the *greedy closest-source policy*. Let n^{α_r} users request distinct files at any given time, where $\alpha_r \in (0,1)$ is a constant; using this approach, each user has a distinct source node and the delivery phase is over an n^{α_r} user interference channel. Afterward, it uses an ITLinQ scheme in this resulting interference channel in order to schedule the deliveries from sources to destinations, which is referred to as the *distributed ITLinQ algorithm* proposed in [38]. This algorithm was demonstrated through numerical simulations that it provides throughput gains over the cluster based state-of-the-art delivery scheme proposed in [2] under the same channel model. In particular, the gain of the ITLinQ-based delivery scheme can achieve a $265\%\times$ to $488\%\times$ gain, depending on the library size.

3.5.2 D2D Caching Networks with Poisson Point Processes

In this section, with the TIN condition, we let the D2D node locations be distributed as a homogeneous PPP instead of either the grid distribution or random uniform distribution considered in the previous sections. PPP modeling is considered to be more practical because unlike the cluster-based model in [13], where only a pair of users are allowed to communication in a square region, this scheme requires no constraint on the link distance and allows a random number of simultaneous D2D transmissions. In [39], the authors propose a new file caching strategy exploiting stochastic geometry and introduce the concept of the *density of successful reception (DSR)*, which is closely related to the outage probability and can be obtained through the scaling of the coverage, i.e., the complement of the outage probability, with the number of receivers per unit area.

The mathematical description of the system model is as follows. We consider a D2D network where users are spatially distributed as a homogenous PPP Φ of density λ, where a randomly selected user can transmit or receive information. At any time slot, only a fraction of the D2D users can be scheduled. Differently from the previous sections, we let any user transmit with probability γ_1 and receive with probability $\gamma_2 = 1 - \gamma_1$ independently of other users. Each user has a cache with storage size of 1 file. If it is selected as a receiver at a time slot, it draws a sample from the request distribution $P_r(\cdot)$, which is assumed to be Zipf distributed. If it is selected as transmitter at a time slot, it draws a sample from the caching distribution $P_c(\cdot)$. At any time slot, each receiver is scheduled based on the closest transmitter association. Without any loss of generality, we assume that the mobile user under consideration is located at the origin. A user is in coverage when its SINR from its nearest transmitter is larger than some threshold T_{r} and it is in outage when SINR is below T_{r}. The definition of DSR is given by [39].

DEFINITION 3.9 **(Density of Successful Receptions)** *The performance of a randomly chosen receiver is determined by its SINR coverage. For the homogeneous PPP Φ with density γ, let γ_1 fraction of all users be the transmitter process Φ_t, and γ_2 fraction of users be the receiver process Φ_r, where $0 \leq \gamma_1, \gamma_2 < 1$. The coverage probability of a randomly chosen receiver is $p_{cov}(T_r, \lambda\gamma_1, \eta)$, which is the same for all receivers, and the total average number of receiver is proportional to the density $\lambda\gamma_2$. Hence, the DSR, which denotes the mean number of successful receptions per unit area, is given by*

$$\begin{aligned} \text{DSR} &= \lambda\gamma_2 p_{cov}(T_r, \lambda\gamma_1, \eta) \\ &= \lambda\gamma_2 \left(\pi\lambda\gamma_1 \int_0^\infty e^{-\pi\lambda\gamma_1 r \beta(T_r,\eta) - \mu T_r \sigma^2 r^{\eta/2}} dr \right), \end{aligned} \tag{3.15}$$

where $p_{cov}(T_r, \lambda\gamma_1, \eta) = \pi\lambda \int_0^\infty e^{-\pi\lambda v \beta(T_r,\eta) - \mu T_r \sigma^2 r^{\eta/2}} dr$ with the thinning property of the PPP (Φ_t), which is obtained from the thinning of Φ and is a homogeneous PPP with density $\lambda\gamma_1$; $\beta(T_r, \eta) = \frac{2(\mu T_r)^{\frac{2}{\eta}}}{\eta} E\left[g^{\frac{2}{g}} \left(\Gamma\left(-2\eta, \mu T_r g\right) - \gamma\left(-2/\eta\right) \right) \right]$, and the expectation is with respect to the interferer's channel distribution denoted by g. Also, $\Gamma(a,x) = \int_x^\infty t^{a-1} e^{-t} dt$ denotes the incomplete gamma function, and $\Gamma(x) = \int_0^\infty t^{x-1} e^{-t} dt$ the standard gamma function.

3.5.2.1 DSR for a Single File

In this case, we assume that there is a single file in the network. The single-file case is the baseline model for the more general multi-file model. The receivers demanding only one file from the nearest transmitter while all other transmitters are interferers, and each transmitter can serve multiple receivers. Since the total density of receivers is $\lambda\gamma_2$, and each receiver is successfully covered with probability $p_{\text{cov}}(T_r, \lambda\gamma_1, \eta)$, the DSR, is given by their product as in (3.15). In the single-file scenario, since only one file is transmitted in the network, there is no need to design cache placement. Hence, the objective is to determine the optimal fractions of transmitter γ_1 and receivers γ_2 in the PPP network that maximize the DSR. This will be used for characterizing the optimal random caching PMF later. The optimization problem to determine γ_1 and γ_2 is given by

$$\begin{aligned} \text{DSR}^* = &\max_{\gamma_1 > 0, \gamma_2 > 0} \lambda\gamma_2 p_{\text{cov}}(T, \lambda\gamma_1, \eta) \\ &s.t. \quad \gamma_1 + \gamma_2 = a, \quad 0 < a \leq 1, \end{aligned} \tag{3.16}$$

where a is the total fraction of transmitters and receivers in the D2D network.[12]

3.5.2.2 Sequential Serving Model with Multiple Files

In this section, we aim to determine the optimal caching distribution for the transmitters to maximize the DSR. In particular, we focus on the sequential serving-based strategy due to its analytical tractability. Under a Zipf demand distribution (3.1), in this model, only the set of transmitters having a common file transmits simultaneously, which is the special case in which only one file is transmitted in the network at a given time. In

[12] For the readers who are interested in the computation of the optimization problem (3.16), please refer to [39].

particular, if a user is selected as a receiver at a time slot, it makes a random request according to the known P_r. If any user is randomly selected as the transmitter at a time slot with probability γ_1, it randomly caches a file according to the caching distribution $P_c(\cdot)$, which is not known yet. At any time slot, each receiver is scheduled based on closest transmitter association. Since file i is available at each transmitter with $P_c(i)$, using the thinning property of the PPP, the probability of coverage file i is

$$p_{\text{cov}}(T, \lambda_t P_c(i), \eta) = \pi \lambda_t P_c(i) \int_0^{\infty} e^{-\pi \lambda_t P_c(i) r \beta(T,\eta) - \mu T \sigma^2 r^{\alpha/2}} dr, \tag{3.17}$$

where $\lambda_t = \lambda \gamma_1$ is the total density of the transmitters. Our objective is to maximize the DSR of users for the sequential serving-based model, denoted by DSR_S, for a PPP model with density λ:

$$\begin{aligned} \max_{P_c} \quad & \text{DSR}_S \\ \text{s.t.} \quad & \sum_{i=1}^{m} P_c(i) = 1, \quad P_r(i) = \frac{i^{-\gamma}}{\sum_{j=1}^{m} j^{-\gamma}}, i = 1, \ldots, m, \end{aligned} \tag{3.18}$$

where $\text{DSR}_S = \lambda \gamma_2 \sum_{i=1}^{M} P_r(i) p_{\text{cov}}(T, \lambda \gamma_2 P_c(i), \eta)$. The optimal values of γ_1 can be found by taking the derivative of (3.18) with respect to γ_1. Note that the optimal value of γ_1 and P_c are coupled. Hence, we can first solve (3.18) by optimizing P_c and then determine γ_1. We illustrate the optimal solution of (3.18) by considering the case in which the noise is small but non-zero and $\eta > 2$ in the following theorem.[13]

THEOREM 3.10 *The optimal caching distribution is* $P_c(i) = \frac{i^{-\gamma_c}}{\sum_{j=1}^{m} j^{-\gamma_c}}$, $i = 1, \ldots, m$, *which is also a Zipf distribution, where* $\gamma_c = \frac{\gamma_c}{\eta/2+1}$ *is the Zipf exponent for the caching probability mass function (PMF).*

Assuming $\eta > 2$, the caching PMF exponent satisfies $\gamma_c < \frac{\gamma_r}{2}$, which implies that the optimal caching PMF that maximizes the DSR has a more uniform distribution, which exhibits less locality compared to the demand distribution that is more skewed toward the most popular files.

3.5.3 D2D Caching Networks with Cooperations

In Section 3.5.1 and Section 3.5.2, we focused on the case under the TIN condition. In this section, we relax this constraint by allowing the cooperation among the D2D nodes. It turns out that caching can be used to mitigate interference and enable cooperative transmissions in the physical layer. In particular, we consider two schemes, that are proposed to extended networks, and dense networks respectively.[14] In the physical layer, both schemes are based on the hierarchical multiple-input multiple output (MIMO)

[13] Please refer to [39] for the case of arbitrary noise with $\eta > 2$.

[14] The area of extended networks can grow linearly with the number of users n and the nearest distance between two nodes are above a constant, while the area of dense networks is a constant.

cooperation scheme proposed in [25]. In both schemes, no *coded caching enabled multicasting* is allowed. Throughout this subsection, we let h_0 in (3.12) be 1, which means that only random phase fading will be considered.

3.5.3.1 Hierarchical Cooperation in Dense Network Model

In [40], the authors propose a scheme based on the hierarchical cooperation and focus on the dense network model. The network is divided into square clusters of area A_c with N_c nodes and $A_c = n^\nu$ for some $\nu > 0$. The user requests follow a uniform distribution (Zipf distribution (3.1) with $\gamma = 0$). We describe the cache placement phase and the transmission policy in the following.

Cache Placement: Each file in the library is divided into B equal sized packets. In the cache placement phase, each user randomly picks MB packets from the total mB packets in the library and stores them in its own cache based on Π_c such that the packets stored at one node are different from the packets of other nodes with high probability.

Transmission Policy: The transmission policy has two stages. In the *first stage*, it designs the transmission policy Π_t based on the fact that each user caches different packets from distinct files with high probability. For each user, in order to receive all B packets of its requested file, it finds B source nodes, which store the B distinct packets in their caches, respectively, and are not in the neighboring clusters of the sink node. Then, to serve the destination user, a virtual MIMO transmission is formed from these B source nodes to the destination cluster $\mathcal{D}$.[15] After finding all the source nodes, we choose all the N_c nodes in $\mathcal{D}$ as receivers. All the packets of the requested file are simultaneously transmitted from the B source nodes to the N_c receivers via the virtual MIMO transmission, which can be formed if the received signals at all receivers can be jointly processed in the second stage. Note that to form the virtual MIMO transmissions, all the source nodes are required to be synchronized for their transmissions. For serving all nodes in the network, the same transmission policy is applied in a TDMA manner. In the *second stage*, we apply the *hierarchical cooperations*. The goal of this stage is to collect and jointly process the received signals from the virtual MIMO transmissions from the first stage at each destination node. To this end, we apply the *hierarchical cooperative MIMO* similar to that in [25, 41]. Within a destination cluster, each node quantizes the received signal and transmits it to the destination node, which then jointly processes the N_c copies of superimposed signals received from previous virtual MIMO transmissions. Hence, an actual MIMO transmission from the source nodes to the destination node has been formed through the two-stage transmissions.[16] The achievable throughput scaling law is given by the following theorem.

[15] If there are multiple candidates of source nodes, we randomly pick up any B candidates that meet the criteria.

[16] Note that the quantization does not change the linear scaling of MIMO capacity, which is shown in [25].

THEOREM 3.11 *For the considered wireless D2D caching network with a library of m files and local caches of M files, if* $nM > m$, *as* $n \to \infty$, *the following per-user throughput scaling law is achievable with high probability:*

$$T_n = \Theta\left(n^{\frac{\tau}{\tau+1}}\right), \tag{3.19}$$

where $\tau, \tau \geq 1$, *is an integer constant independent of* n.

From Theorem 3.11, it can be seen that the throughput scaling law by using hierarchical cooperation based D2D caching scheme has a order gain compared to both single-hop ($T_n = M/m$) and multi-hop ($T_n = \sqrt{M/m}$) based caching schemes in the corresponding parameter regimes discussed in Section 3.3.1 and Section 3.3.2.

3.5.3.2 Hierarchical Cooperation in Extended Network Model

In [42], the authors propose a different hierarchical cooperation–based scheme by considering an extended network model with area of $A(n)$, which consists of $n = 4^L$ nodes, where L is some positive integer.[17] All nodes are distributed on a grid (see Figure 3.6a). In this work, the capacity scaling laws in the extended D2D caching networks under the physical model as in (3.12) with $h_0 = 1$ is characterized. The proposed scheme is called *cache-induced hierarchical MIMO cooperation*.

The Hierarchical Cache Content Placement: This phase decides how to distribute files into caches of different nodes in the network. In detail, nodes are grouped into clus-

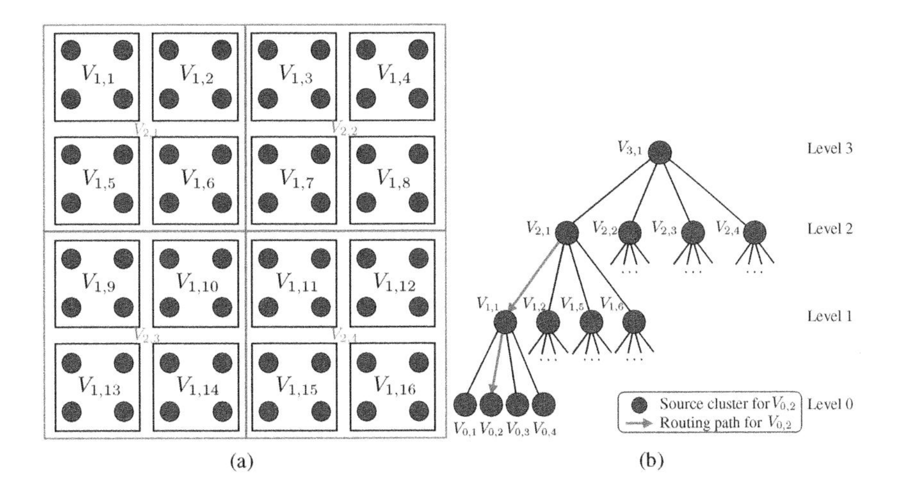

Figure 3.6 (a) An illustration of clusters at different levels for the D2D caching network with $n = 64$ nodes. (b) An illustration of the tree-graph-based content delivery for the D2D caching network on the left, with $n = 64$ nodes. Suppose node $V_{0,2}$ requests a file cached in the second level. Then the set of source nodes is $V_{2,1}$ and the routing path from the source cluster $V_{2,1}$ to the destination $V_{0,2}$ is demonstrated with the arrows.

[17] It could be shown that the result still holds when $n \neq 4^L$.

ters of different levels. In the ℓth level, $A(n)$ is divided into $4^{L-\ell}$ equally sized squares with 4^{ℓ} nodes each. These squares form a cluster in the ℓth level. Let $V_{\ell,i} \subset V(n)$ be the i-th cluster in the ℓth level where $i \in \{1,\ldots,4^{L-\ell}\}$. In this cache placement, each node caches $q_f F$ bits of file f, where $q_f \in \left\{0, \frac{1}{4^L}, \frac{1}{4^{L-1}}, \ldots, \frac{1}{4}, 1\right\}$. If $q_f = \frac{1}{4^\ell}$, the file f is equally distributed over the nodes in $V_{\ell,i}$ for any $i \in \{1,\ldots,4^{L-\ell}\}$. In other words, each node in $V_{m,i}$ caches a portion of the $4^{-\ell}F$ bits of file f such that file f can be reconstructed by collecting all portions from the nodes in $V_{m,i}$. Let $\mathbf{x} = [x_0, x_1, \ldots, x_L] \in Z_+^{L+1}$, where $x_\ell = \sum_{f=1}^{m} 1(q_f = 4^{-\ell})$ is the number of files cached at the ℓth level. In addition, we also have $\sum_{\ell=0}^{M} x_\ell = m$ and $\sum_{\ell=0}^{L} x_\ell 4^{-\ell} \leq M$ in order to satisfy the cardinality of library and the memory constraint.

The Tree-Graph-Based Content Delivery: This phase exploits each node's cached content to serve users' demands, and it has of four layers, which are (1) the source determination layer, ((2) routing layer, (3) cooperation layer, and (4) physical layer. In this phase, a tree graph (see Figure 3.6b) is used to abstract the original graph. Based on this tree graph, we can determine the source and design the routing scheme. Specifically, in the tree graph, each user is a leaf node, and an internal node is a set of source nodes for a leaf node. The routing layer routes packets between the source nodes and the sink node. The cooperation layer designs this tree abstraction to the routing layer by appropriate concentrating traffic over the D2D network. Finally, the physical layer implement of this concentration of traffic in the wireless network is based on two physical layer transmission modes, namely, the hierarchical cooperation mode [25][18] and multi-hop mode. Using this scheme, given cache content placement parameter $\mathbf{x}$, we can formulate the following optimization problem:

$$\max_{\mathbf{x} \in Z_+^{\ell}} \quad T_n$$

$$\text{s.t.} \quad \sum_{f=\sum_{i=0}^{\ell-1} x_i + 1}^{m} P_r(f) R \leq C_\ell 4^{-(\ell-1)}, \quad \sum_{\ell=0}^{L} x_\ell 4^{-\ell} \leq M, \quad \sum_{\ell=0}^{L} x_\ell = m, \tag{3.20}$$

where $C_\ell = \frac{4^\ell R_u(4^\ell)}{3L_b}$, $\forall \ell \in \{L_b+1,\ldots,L\}$; $L_b = \max\{\ell\colon x_\ell > 0\}$, and $R_u(g_c)$ is the per-node throughput for a grid network under per-cluster uniform permutation traffic with cluster size g_c. The optimization problem (3.20) is an integer optimization problem, and a closed form solution is difficult to obtain. However, it is possible to obtain the order-optimal scaling laws for the optimal value by allowing some relaxations on C_m.

Throughput Scaling Laws under Zipf Popularity Distribution: We assume the same parameters scaling as in Section 3.3.2.2. In particular, we focus on Regimes III, IV, and V. Then we have the following theorem for the order-optimal achievable scaling laws.[19]

[18] This is similar to the stage two of the transmission policy in the previous scheme for dense networks.

[19] This order optimality is subject to an independent delivery mechanism, which means that messages of different users are treated as independent messages. It is not guaranteed that this scheme is order optimal without any constraint. Nevertheless, it can be proved in some parameter regimes, such as $\gamma > \frac{\min\{3,\eta\}}{2}$, this scheme can still achieve order optimal throughput. Moreover, the order optimality also means that the ratio between the optimal throughput and the achievable throughput is upper bounded by n^{ε}, where ε is arbitrarily small.

THEOREM 3.12 *(Achievable Scaling Law in Extended Networks): For extended networks with $A(n) = n$, in Regimes III and IV, the order-optimal achievable throughput T_n^* obtained by the cache-induced hierarchical cooperation is given by*

$$T_n^* = \begin{cases} \Omega\left(n^{\beta-\alpha}(\gamma_\eta - 1) - \varepsilon_\eta\right), & \gamma \in [0,1] \\ \Omega\left(n^{\alpha(\gamma-\gamma_\eta)+\beta(\gamma_\eta-1)-\varepsilon_\eta}\right), & \gamma \in (1, \gamma_\eta] \\ \Omega\left(n^{\beta(\gamma-1)-\varepsilon_\eta}\right), & \gamma > \gamma_\eta, \end{cases} \tag{3.21}$$

where $\gamma_\eta = \frac{\min\{3,\eta\}}{2}$, $\varepsilon_\eta = \Theta\left(\frac{1}{\log n}\right)$ for $\eta \in (2,3)$ and $\varepsilon_\eta = 0$ for $\eta \geq 3$.

In Regime V, we have

$$T_n^* = \begin{cases} \Omega(1), & \gamma \in [0,1] \\ \Omega\left(n^{\beta(\gamma-1)}\right), & \gamma > 1. \end{cases} \tag{3.22}$$

From Theorems 3.12, we can see the order improvement by using the order-optimal physical layer schemes compared to the case when only a multi-hop scheme is allowed. In particular, when $\gamma \in [0,1]$, $\eta < 3$ and in Regimes III and IV, the improvement is from $\Theta\left(n^{-\frac{\alpha-\beta}{2}}\right)$ (see Section 3.3.2.2) to $\Theta\left(n^{-(\alpha-\beta)(1-\frac{\eta}{2})}\right)$. Depending on the value of η and α, when $\gamma > 1$ and in Regime IV, the improvement of the throughput can be from $\Omega\left(n^{-\frac{1-\min(1,\beta+1-1/(\gamma-1))}{2}}\right)$ (see Section 3.3.2.2) to $\Theta\left(n^{\alpha(\gamma-\gamma_\eta)+\beta(\gamma_\eta-1)}\right)$ or $\Theta\left(n^{\beta(\gamma-1)}\right)$.

3.6 Mobile D2D Caching

In previous sections, we assume all the D2D users are static or do not assume an explicit model for their mobility. In this section, we present how mobility can effect the behavior and the performance of D2D caching networks. The effect of user mobility on D2D caching was investigated by simulations in [13], which shows that user mobility does not have a significant impact on a random caching scheme. Nevertheless, such a caching strategy may not take advantage of the specific user mobility pattern. In [43], the authors show that user mobility has positive effect on D2D caching network and in [44], the authors consider the case where mobile users can update cache placement based on the demand and user mobility. However, it is assumed that one complete file can be transmitted via any D2D link when two users contact, which may not be practical. In this section, we consider two explicit user mobility models, which are based on the distribution of contact and intercontact time [45, 46] and random walk model [47], respectively.

3.6.1 Mobility-Aware D2D Caching Based on Contact and Intercontact Time

3.6.1.1 User Mobility Model

In the user mobility model, when mobile users are within the transmission range, they may contact each other. Hence, the *contact time* for two mobile users is defined as the time that they can transmit to or serve each other. Then, the *intercontact time* for two mobile users is defined as the time between two consecutive contact times. In particular, the locations of contact times in the timeline for any two users i and j are modeled as a Poisson process with intensity $\lambda_{i,j}$. For simplicity, the timelines for different user pairs are independent. $\lambda_{i,j}$ is defined as the pairwise contact rate between users i and j, which is the average number of contacts per unit time slot.

3.6.1.2 Cache Placement and File Delivery

Coded cache placement within each file is considered. Each file is partitioned and encoded into a large number of distinct segments, and it can be retrieved by collecting enough encoded segments. Accordingly, it can be guaranteed that no repetitive encoded segment is in the network. Specifically, it assumes that each file is encoded into multiple segments, each within size s bits, and the file f can be recovered by collecting K_f encoded packet. Notice that K_f depends on the size of file f. The number of encoded packets of file f cached in user i is denoted as $y_{i,f}$. With a little bit of abuse of the notation, we let each user cache M packets. Similar to before, each user requests file f with probability $P_r(f)$ and it will start to download coded packets of file f from the encountered mobile nodes, and it will also check its own cache. The duration of each contact of user i and j and the link rate from user j to user i are assumed to be $t^c_{i,j}$ seconds and $r_{i,j}$ bits/second, respectively. Therefore, it can be computed that $B_{i,j} = \left\lfloor \frac{t^c_{i,j} r_{i,j}}{s} \right\rfloor$ packets can be sent within one contact from user j to user i. Moreover, a delay constraint, denoted as T^d, is assumed, which means that if each user cannot collect a least K_f packets of the requested file within T^d, it will be served via the base station for the remaining requested packets.

3.6.1.3 Problem Formulation and Discussions

The data offloading ratio of user i is

$$E_i = \sum_{f \in \mathcal{F}} \frac{P_r(f)}{K_f} \left(E \left[\min \left(\sum_{j \in \mathcal{N}} \min(B_{i,j} M_{i,j}, y_{j,f}), K_f \right) \right] \right),$$

where $M_{i,j}$ denotes the number of contact times from user i and j within time T^d, and it follows a Poisson distribution with mean $\lambda_{i,j} T^d$. Hence, we obtain the following optimization problem:

$$\max_{y_{j,f}} \quad \frac{1}{n} \sum_{j \in \mathcal{U}} E_i, \quad \text{s.t.} \sum_{f \in \mathcal{F}} y_{j,f} \leq M, y_{j,f} \in N, y_{j,f} \leq K_f, \forall j \in \mathcal{U}, f \in \mathcal{F}. \quad (3.23)$$

This optimization problem is a mixed integer nonlinear program, and it is NP-hard. In [45], a divide and conquer algorithm is designed to evaluate the objective function, and a dynamic programing algorithm is proposed to solve the optimization problem. However, the complexity of the dynamic programing algorithm increases exponentially with the number of mobile users. To solve this problem, a suboptimal algorithm can be designed by reformulating the optimization problem (3.23) into one that maximizes a monotone submodular function over a matroid constraint [48] that has been widely studied; effective algorithms with provable approximation ratios have been proposed.

Simulation results show that this well-designed mobility-aware caching scheme provides significant performance gains over commonly used caching strategies, including popular caching and random caching placement. In addition, it can be observed that users with very low speed tend to rarely contact others, and hence it is more advantageous for them to cache the most popular files to meet their own demands. On the other hand, the users with high velocity can contact others more frequently, and they also need to cache the most popular files in order to serve other mobile users to download the demanded files.

A recent work [46] by the same authors studies the case that relaxes the assumption of constant contact durations and provides analytical performance evaluations. In particular, it considers users' contact and intercontact durations via an alternating renewal process. Then a tractable expression of the data offloading ratio can be derived and maximized, which was proved to be increasing with the user moving speed. In addition, it was shown that the variation of contact durations is important while designing cache placement, especially when the average contact duration is relatively short or comparable to the intercontact duration.

3.6.2 Mobility-Aware Centralized D2D Caching Based on Random Walks

In [47], the authors consider a different D2D mobility model based on random walks. The user demand distribution considered here varies over time. For any time t, let $p_{u,t}^{f}$ be the probability that user u requests file f in time slot t. Let the requests be independent among time slots and users. Moreover, the demand profiles of each user follows a *cyclo-stationary* pattern that repeats itself in a period of T time slots. That is, $p_{u,t}^{f} = p_{u,t+kT}^{f}$ for nonnegative integer k. We assume that single-hop D2D communication is allowed and can be used to transfer data items between users. A fixed link rate between all users is assumed. We capture the cost of caching each byte by a parameter $r_{\mathrm{D2D}} > 0$.

3.6.2.1 User Mobility Model

We assume that the service provider is interested in P popular locations $\mathcal{P} = \{1, 2, \ldots, P\}$ where high demand can be related with mobility of users. Moreover, a service provider can track, learn, and predict the mobility of each user over time and hence constructs a mobility profile for user u denoted by $\theta_{u,t}^{\rho}$, which is the probability that user u will be present at location ρ in time slot t, where $\sum_{\rho=1}^{P} \theta_{u,t}^{\rho} = 1$, $\forall u, t$. It assumes that users stay in the same location within a time slot and may move to another location at the beginning of each time slot. Let $\lambda_{u,t}^{l,k}$ be the transition probability that user u moves from location l to location k in time slot t, where $\sum_{k=1}^{P} \lambda_{u,t}^{l,k} = 1$, $\forall u, t, l$. These transition probabilities may change from one time slot to another to capture the mobility of each user. In addition, the probability of being at a certain location in a time slot t depends on the location in the previous time slot $t-1$ only, i.e., $\theta_{u,t}^{\rho} = \sum_{k=1}^{P} \theta_{u,t-1}^{k} \lambda_{u,t}^{k,\rho}$, where $\theta_{u,1}^{\rho} = \lambda_{u,1}^{\rho,\rho}$. We assume that each user randomly takes a trajectory every day, starting from one location and moving to other locations. However, the mobility profile of each user follows a *cyclo-stationary* pattern that repeats itself in a period of T time slots.

3.6.2.2 Cost Function

Let user u cache an amount x_u^f of data item f, where $0 \leq x_u^f \leq S_f$. The service provider replaces the data stored in users devices when they are expired at the end of time slot T_e. Hence, the total network load is given by

$$L_t^{\mathrm{P}} = \sum_{f=1}^{m} \sum_{u=2}^{n-1} \sum_{\alpha_k \in \mathcal{A}_k} \left(S_f - \sum_{u \in \alpha_k} x_u^f \right)^+ \sum_{u \in \alpha_k} p_{u,t}^f \sum_{l=1}^{P} \Pi_{u \in \alpha_k} \theta_{u,t}^l \Pi_{j \notin \alpha_n} \left(1 - \theta_{j,t}^l\right)$$
$$+ \sum_{f=1}^{m} \left(S_f - \sum_{u=1}^{n} x_u^f \right)^+ \sum_{u=1}^{n} p_{u,t}^f \sum_{l=1}^{P} \Pi_{u=1}^{n} \theta_{u,t}^l$$
$$+ \sum_{f=1}^{m} \sum_{u=1}^{n} (S_f - x_u^f) p_{u,t}^f \left(1 - \sum_{l=1}^{P} \sum_{k=2}^{n} \sum_{\alpha_k \in \mathcal{A}_k} \Pi_{j \in \alpha_k} \theta_{j,t}^l \Pi_{i \notin \alpha_k (1 - \theta_{i,t}^l)} \right), \quad (3.24)$$

where $\mathcal{A}_k = \{\alpha_k = \{a_1, \ldots, a_k, a_j \in \{1, \ldots, n\}, \forall j\}\}$ and $|\mathcal{A}_k| = \binom{n}{k}$. In (3.24), the first term is the case when some users get together, the second term represents the case when all users get together, and the third term is the case when each user is moving alone. We assume users share cached data items when they meet each other and get the remaining portion from the service provider. In addition, the shared items between users are free. Hence, the corresponding objective cost under the proactive operation is given by $C^{\mathrm{P}} = \limsup_{T \to \infty} \frac{1}{T} \sum_{t=1}^{T} E\left[C(L_t^{\mathrm{P}})\right] + r_{\mathrm{D2D}} \sum_{u=1}^{n} \sum_{f=1}^{m} x_u^f$.

3.6.2.3 Problem Formulations and Discussions

Our objective is to find an optimal caching policy $\{x_u^{f*}\}$ that minimize C^{P}. The problem is defined as

$$\min \quad C^{\mathrm{P}}, \quad \text{s.t.} \quad 0 \leq x_u^f \leq S_f. \quad (3.25)$$

The optimization problem (3.25) depends mainly on the cost function C, which may be linear, quadratic, or a polynomial of higher order. The exact solution of (3.25) for nonlinear cost functions can be obtained using convex optimization techniques. However, this case may not provide enough insights on the effect of a user's mobility, and finding an optimal caching policy will be nontractable in the sense that the complexity of the optimal policy grows exponentially with the number of users n.

We will not discuss how to solve the optimization problem (3.25).[20] However, we can get some insights from the optimal solution. First, the service provider's optimal caching decision depends on the exact value of the caching cost r_{D2D}. In particular, a smaller value of r_{D2D} yields more caching and vice versa. Second, the possibility of over-caching reduces when the probability of meeting increases. Third, higher demand values yield more caching.

[20] For the interested readers, please see [47].

References

[1] K. Shanmugam, N. Golrezaei, A. G. Dimakis, A. F. Molisch, and G. Caire, "Femtocaching: wireless content delivery through distributed caching helpers," *IEEE Transactions on Information Theory*, vol. 59, no. 12, pp. 8402–8413, 2013.

[2] M. Ji, G. Caire, and A. F. Molisch, "Wireless device-to-device caching networks: basic principles and system performance," *IEEE Journal on Selected Areas in Communications*, vol. 34, no. 1, pp. 176–189, 2016.

[3] M. Ji and G. Caire and A. F. Molisch, "The throughput-outage tradeoff of wireless one-hop caching networks," *IEEE Transactions on Information Theory*, vol. 61, no. 12, pp. 6833–6859, 2015.

[4] A. F. Molisch, G. Caire, D. Ott, J. Foerster, D. Bethanabhotla, and M. Ji, "Caching eliminates the wireless bottleneck in video-aware wireless networks," arXiv:1405.5864, 2014.

[5] N. Golrezaei, A. Dimakis, and A. Molisch, "Device-to-device collaboration through distributed storage," in *Global Communications Conference (GLOBECOM), 2012 IEEE*, 2012, pp. 2397–2402.

[6] N. Golrezaei, A. G. Dimakis, and A. F. Molisch, "Wireless device-to-device communications with distributed caching," in *Information Theory Proceedings (ISIT), 2012 IEEE International Symposium*, 2012, pp. 2781–2785.

[7] J. Kim, F. Meng, P. Chen, H. E. Egilmez, D. Bethanabhotla, A. F. Molisch, M. J. Neely, G. Caire, and A. Ortega, "Adaptive video streaming for device-to-device mobile platforms," in *Proceedings of the 19th Annual International Conference on Mobile Computing & Networking*. 2013, pp. 127–130.

[8] N. Golrezaei, A. F. Molisch, and A. G. Dimakis, "Base-station assisted device-to-device communications for high-throughput wireless video networks," in *Communications (ICC), 2012 IEEE International Conference*, 2012, pp. 7077–7081.

[9] M. Ji, G. Caire, and A. F. Molisch, "Optimal throughput-outage trade-off in wireless one-hop caching networks," in *Information Theory Proceedings (ISIT), 2013 IEEE International Symposium*, 2013, pp. 1461–1465.

[10] M. Ji, G. Caire, and A. Molisch, "Fundamental limits of distributed caching in D2D wireless networks," in *Information Theory Workshop (ITW), 2013 IEEE*, 2013, pp. 1–5.

[11] M. Ji, G. Caire, and A. F. Molisch, "Fundamental limits of caching in wireless D2D networks," *IEEE Transactions on Information Theory*, vol. 62, no. 2, pp. 849–869, 2016.

[12] J. Kim, G. Caire, and A. F. Molisch, "Quality-aware streaming and scheduling for device-to-device video delivery," *IEEE/ACM Transactions on Networking*, vol. 24, no. 4, pp. 2319–2331, 2016.

[13] N. Golrezaei, P. Mansourifard, A. F. Molisch, and A. G. Dimakis, "Base-station assisted device-to-device communications for high-throughput wireless video networks," *IEEE Transactions on Wireless Communications*, vol. 13, no. 7, pp. 3665–3676, 2014.

[14] M.-C. Lee, M. Ji, A. F. Molisch, and N. Sastry, "Performance of caching-based D2D video distribution with measured popularity distributions," arXiv:1806.05380, 2018.

[15] M. A. Maddah-Ali and U. Niesen, "Fundamental limits of caching," *IEEE Transactions on Information Theory*, vol. 60, no. 5, pp. 2856–2867, 2014.

[16] P. Kyosti et al., "WINNER II channel models. Part I: channel models," Information Society Technologies, Technical Report IST-4-027756, WINNER II, D1.1.2 V1.1, Public, Sept. 2007.

[17] M. Ji and R.-R. Chen, "Caching and coded multicasting in slow fading environment," in *Wireless Communications and Networking Conference (WCNC), 2017 IEEE*, 2017, pp. 1–6.

[18] X. Wu, S. Tavildar, S. Shakkottai, T. Richardson, J. Li, R. Laroia, and A. Jovicic, "Flashlinq: a synchronous distributed scheduler for peer-to-peer ad hoc networks," in *Communication, Control, and Computing (Allerton), 2010 48th Annual Allerton Conference*, 2010, pp. 514–521.

[19] N. Naderializadeh and A. S. Avestimehr, "Itlinq: a new approach for spectrum sharing in device-to-device communication systems," *IEEE Journal on Selected Areas in Communications*, vol. 32, no. 6, pp. 1139–1151, 2014.

[20] S. Jeon, S. Hong, M. Ji, G. Caire, and A. F. Molisch, "Wireless multihop device-to-device caching networks," *IEEE Transactions on Information Theory*, vol. 63, no. 3, pp. 1662–1676, 2017.

[21] S. Gitzenis, G. Paschos, and L. Tassiulas, "Asymptotic laws for joint content replication and delivery in wireless networks," *IEEE Transactions on Information Theory*, vol. 59, no. 5, pp. 2760–2776, 2013.

[22] P. Gupta and P. Kumar, "The capacity of wireless networks," *IEEE Transactions on Information Theory*, vol. 46, no. 2, pp. 388–404, 2000.

[23] M. Franceschetti, O. Dousse, D. Tse, and P. Thiran, "Closing the gap in the capacity of wireless networks via percolation theory," *IEEE Transactions on Information Theory*, vol. 53, no. 3, pp. 1009–1018, 2007.

[24] U. Niesen, P. Gupta, and D. Shah, "The balanced unicast and multicast capacity regions of large wireless networks," *IEEE Transactions on Information Theory*, vol. 56, no. 5, pp. 2249–2271, 2010.

[25] A. Ozgur, O. Leveque, and D. N. C. Tse, "Hierarchical cooperation achieves optimal capacity scaling in ad hoc networks," *IEEE Transactions on Information Theory*, vol. 53, no. 10, pp. 3549–3572, 2007.

[26] A. Ozgur, R. Johari, D. N. C. Tse, and O. Leveque, "Information-theoretic operating regimes of large wireless networks," *IEEE Transactions on Information Theory*, vol. 56, no. 1, pp. 427–437, 2010.

[27] M. Ji, A. M. Tulino, J. Llorca, and G. Caire, "Order-optimal rate of caching and coded multicasting with random demands," *IEEE Transactions on Information Theory*, vol. 63, no. 6, pp. 3923–3949, 2017.

[28] X. Lin, R. Ratasuk, A. Ghosh, and J. G. Andrews, "Modeling, analysis and optimization of multicast device-to-device transmissions," *IEEE Transactions on Wireless Communications*, vol. 13, no. 8, pp. 4346-4359, Aug. 2014.

[29] M. Ji, R. Chen, G. Caire, and A. F. Molisch, "Fundamental limits of distributed caching in multihop D2D wireless networks," in *2017 IEEE International Symposium on Information Theory (ISIT)*, 2017, pp. 2950–2954.

[30] N. Woolsey, R.-R. Chen, and M. Ji, "Towards practical file packetizations in wireless device-to-device caching networks," arXiv:1712.07221, 2017.

[31] N. Woolsey, R. Chen, and M. Ji, "Device-to-device caching networks with subquadratic subpacketizations," in *GLOBECOM 2017–2017 IEEE Global Communications Conference*, 2017, pp. 1–6.

[32] N. Woolsey, R. Chen, and M. Ji, "Coded caching in wireless device-to-device networks using a hypercube approach," in *2018 IEEE International Conference on Communications Workshops (ICC Workshops)*, 2018, pp. 1–6.

[33] S. Jin, Y. Cui, H. Liu, and G. Caire, "New order-optimal decentralized coded caching schemes with good performance in the finite file size regime," arXiv:1604.07648, 2016.

[34] I. Ruzsa and E. Szemerédi, "Triple systems with no six points carrying three triangles," *Combinatorics (Keszthely, 1976), Colloquia Mathematica Societatis János Bolyai*, vol. 18, pp. 939–945, 1978.

[35] N. Alon, A. Moitra, and B. Sudakov, "Nearly complete graphs decomposable into large induced matchings and their applications," in *Proceedings of the Forty-Fourth Annual ACM Symposium on Theory of Computing*, 2012, pp. 1079–1090.

[36] K. Shanmugam, A. M. Tulino, and A. G. Dimakis, "Coded caching with linear subpacketization is possible using ruzsa-szem\'eredi graphs," arXiv:1701.07115, 2017.

[37] N. Naderializadeh, D. T. H. Kao, and A. S. Avestimehr, "How to utilize caching to improve spectral efficiency in device-to-device wireless networks," in *2014 52nd Annual Allerton Conference on Communication, Control, and Computing (Allerton)*, 2014, pp. 415–422.

[38] N. Naderializadeh and A. S. Avestimehr, "Itlinq: a new approach for spectrum sharing in device-to-device communication systems," *IEEE Journal on Selected Areas in Communications*, vol. 32, no. 6, pp. 1139–1151, 2014.

[39] D. Malak, M. Al-Shalash, and J. G. Andrews, "Optimizing content caching to maximize the density of successful receptions in device-to-device networking," *IEEE Transactions on Communications*, vol. 64, no. 10, pp. 4365–4380, 2016.

[40] J. Guo, J. Yuan, and J. Zhang, "An achievable throughput scaling law of wireless device-to-device caching networks with distributed mimo and hierarchical cooperations," *IEEE Transactions on Wireless Communications*, vol. 17, no. 1, pp. 492–505, 2018.

[41] S. Hong and G. Caire, "Beyond scaling laws: on the rate performance of dense device-to-device wireless networks," *IEEE Transactions on Information Theory*, vol. 61, no. 9, pp. 4735–4750, 2015.

[42] A. Liu, V. K. N. Lau, and G. Caire, "Cache-induced hierarchical cooperation in wireless device-to-device caching networks," *IEEE Transactions on Information Theory*, vol. 64, no. 6, pp. 4629–4652, 2018.

[43] S. Krishnan and H. S. Dhillon, "Effect of user mobility on the performance of device-to-device networks with distributed caching," *IEEE Wireless Communications Letters*, vol. 6, no. 2, pp. 194–197, 2017.

[44] R. Lan, W. Wang, A. Huang, and H. Shan, "Device-to-device offloading with proactive caching in mobile cellular networks," in *2015 IEEE Global Communications Conference (GLOBECOM)*, 2015, pp. 1–6.

[45] R. Wang, J. Zhang, S. H. Song, and K. B. Letaief, "Mobility-aware caching in d2d networks," *IEEE Transactions on Wireless Communications*, vol. 16, no. 8, pp. 5001–5015, 2017.

[46] R. Wang, J. Zhang, S. H. Song, and K. B. Letaief, "Exploiting mobility in cache-assisted d2d networks: performance analysis and optimization," *IEEE Transactions on Wireless Communications*, vol. 17, no. 8, pp. 5592–5605, 2018.

[47] S. Hosny, A. Eryilmaz, A. A. Abouzeid, and H. E. Gamal, "Mobility-aware centralized d2d caching networks," in *2016 54th Annual Allerton Conference on Communication, Control, and Computing (Allerton)*, 2016, pp. 725–732.

[48] G. L. Nemhauser, L. A. Wolsey, and M. L. Fisher, "An analysis of approximations for maximizing submodular set functions," *Mathematical Programming*, vol. 14, no. 1, pp. 265–294, 1978.

4 Cooperative Caching in Cloud-Assisted 5G Wireless Networks

Tuyen X. Tran, Guosen Yue, and Dario Pompili

Cloud-assisted wireless networks are emerging solutions that unite wireless networks and cloud-computing to provide cloud services at the edge of the network in order to support the foreseen massive demands from data and computation hungry mobile users. In this chapter, we first provide an overview of the two emerging cloud-assisted wireless network paradigms, namely, cloud radio access network (C-RAN), where the functionalities at the base stations (BSs) are centralized, and mobile-edge computing (MEC), recently renamed multi-access edge computing, which aims at providing the RAN with computing and storage resources. We then leverage the C-RAN and MEC paradigms to design novel cooperative caching frameworks that explore the synergies of the in-network computing and storage resources. Specifically, a novel cooperative hierarchical caching framework is designed in C-RAN, where caching is performed both at the distributed BSs and at the center processing unit (CPU) that bridges the gap between the traditional edge-based and core-based caching schemes. Furthermore, a joint cooperative caching and processing framework is designed in an MEC network, where the MEC servers perform both cache storage and video transcoding to support adaptive bitrate (ABR) video streaming. Numerical simulations are performed using real-world video requests on YouTube and synthetic content requests. It is shown that important gains can be achieved in terms of content access delay, cache-hit ratio, and backhaul traffic load using the envisioned cooperative caching frameworks.

4.1 Cloud-Assisted Wireless Networks

The next generation of mobile wireless systems, e.g., 5G, will imply major changes in the implementation and deployment of networking infrastructure, based on software-defined networking (SDN) and network functions virtualization (NFV). Network operations and services are becoming cloud enabled in almost every industry, and it creates an apparent opportunity to generate value for the telecommunication industry from exploiting distributed storage and cloud computing toward specific clients and services. To overcome the limitations of current connection-centric RANs, cloud-assisted wireless networks are promising solutions that unite wireless networks and cloud computing to deliver cloud services directly from the network edges. The two emerging paradigms for cloud-assisted wireless networks are C-RAN, which leverages virtualization technology to consolidate the BS functionalities in a centralized cloud, and MEC, which aims at equipping storage, computing, and networking resources at the edge of the mobile RAN.

4.1.1 Cloud Radio Access Network (C-RAN)

C-RAN was introduced as a revolutionary redesign of the cellular architecture to address the increase in data traffic and to reduce the capital expenditure (CAPEX) and operating expenditure (OPEX) [1]. The idea of C-RAN is to decouple the computational functionalities from the distributed BS (a.k.a. eNodeB in long-term evolution [LTE]) and to consolidate them in a centralized processing center. Its main characteristics are (1) centralized and flexible management of spectrum and computing resources, (2) coordinated radio control algorithms, and (3) softwarization of BS functionalities on generic computing platforms.

A typical C-RAN is composed of (1) lightweight, distributed radio remote heads (RRHs), which are deployed at the remote site and are controlled by a centralized virtual base station pool; (2) the central processing unit (CPU) (also known as the baseband unit [BBU] pool), which utilizes high-speed programmable processors and real-time virtualization technology to carry out the digital processing tasks; and (3) low-latency high-bandwidth optical fibers, which connect the RRHs to the CPU. The communication functionalities of the CPU are implemented on virtual machines (VMs) hosted over general-purpose computing platforms, which are housed in one or more racks of a small cloud data center. As a precautionary measure and to be on the safe side, the optical fiber transmission latency is limited to less than 1% of the physical layer (PHY) processing latency. Hence, the range of CPU is limited by latency constraints of wireless systems.

4.1.2 Mobile-Edge Computing (MEC)

In the past decade, we have witnessed cloud computing playing a significant role for control, massive data storage, and computation offloading. However, severe demands have been posed by the rapid proliferation of mobile applications and the internet of things (IoT) over the last few years on cloud infrastructure and wireless access networks. On the other hand, fast developments and deployments of the 5G cellular specifications open opportunities for many vertical services. The flexible frame structure and design in 5G not only provides significant enhancement on the mobile broadband (MBB) services but also enables massive machine-type communication (MTC) and ultra-reliable low-latency communication (URLLC). Stringent requirements such as ultra-low latency, user experience continuity, and high reliability in the new generation of cellular network necessitate the support of localized services that are closer to the end users. To this end, MEC can play a key role in assisting wireless networks with context-aware and low-latency services that are deployed directly at the network edge.

Different from conventional cloud computing systems that utilize public clouds, the MEC system relies on the deployment of the MEC servers as commodity servers at the edge of the wireless network. Depending on the deployment topology of the BSs in the C-RAN and the functional splitting therein, a MEC server can be deployed at each BS or at an aggregation point serving several BSs. With this position, MEC enables application execution and data transfer in relatively short distances from the end users, resulting in reduced end-to-end (E2E) latency and reduced backhaul traffic load [2, 3]. Additionally, there are various benefits for MEC to empower the network potentially, including (1) hosting compute-intensive services at the network edge for optimizing the utilization of mobile resources, (2) reducing backhaul traffic by preprocessing raw data before forwarding them (or some extracted features) to the cloud, and (3) context-aware applications with the help of the RAN state information such as radio resource utilization, cell load, and user locations.

4.1.3 Co-deployment of C-RAN and MEC

While C-RAN and MEC propose to shift computing capabilities in a different direction (to the cloud instead of to the edge), they are highly complementary technologies, and their co-location will help make the economics of each significantly more attractive [4]. On the one hand, the centralized nature of C-RAN can be exploited to address the problem of capacity fluctuation and to improve spectrum and energy efficiency in mobile networks. The full centralization principle of C-RAN and the densification of cellular BSs, however, entail heavy exchanges of raw physical-layer data between the radio heads and CPU, which impose stringent requirements in terms of throughput and latency to the fronthaul connections [5]. On the other hand, the main benefits of the MEC paradigm are in improving localized user experience and reducing service latency. However, the computing and storage capacity at the MEC server would be much lower than that of the centralized cloud in C-RAN, making the resource provisioning and allocation problem a critical challenge in MEC networks.

The rest of this chapter is organized as follows. We give a brief overview of the state of the art in cooperative caching in Section 4.2. We then present details of the innovative solutions for cooperative caching, including a cooperative hierarchical caching for C-RANs, in Section 4.3 and a joint cooperative caching and transcoding for adaptive bitrate video streaming in MEC networks in Section 4.4. Finally, conclusions are discussed in Section 4.5.

4.2 State of the Art in Cooperative Caching

In wireless caching systems, the limited cache capacity at each BS often results in moderate cache-hit ratio. To overcome this limitation, cooperative caching has been proposed to utilize the collective cache storage capacities at different BSs. In small-cell networks, cooperative caching has been studied among the BSs in [6–9], and among the network operators in [10]. In this direction, the works in [7, 11] proposed online cooperative caching algorithms that minimize the total cost incurred to content providers without knowing in advance the content popularity. Recently, a collaborative joint caching and processing strategy for multi-bitrate video streaming in MEC networks has been proposed [12]. In this scheme, each MEC server is envisioned to act as both a cache server and a transcoding server. Additionally, Poularakis et al. [13] proposed a method that combines multicast and caching in a 5G network with massive demand for delay-tolerant content to improve energy efficiency. Considering user mobility, Tran et al. [14] proposed a coded caching algorithm that aims at minimizing the system energy consumption.

Existing cooperative caching schemes have focused on two categories: (1) *horizontal cooperation*, which exploits the coordination between caches at the BSs, and (2) *vertical cooperation* (hierarchical caching), which exploits the coordination between caches at the BSs and at the core network (CN). While these cooperative caching schemes offer great potential to improve the performance over the non-cooperative approaches in terms of cache-hit ratio; there are several challenges that fundamentally limit their effectiveness. First, in hierarchical caching, considerable delay is incurred by fetching contents from the CN's cache to the BSs, which is usually many-fold higher than the delay of transferring content among the caches at the BSs [6, 7]. Second, direct interconnections between the BSs are relied on for current cooperative caching techniques. Those interconnections usually have very limited capacity and thus are not suitable for handling huge amounts of content sharing. In the current 4G LTE cellular network, the BSs communicate with each other via the X2 interface, which is designed for exchanging control messages or users' data buffer during handover [15]. Therefore, it is not practical to use such interface for transferring of data between edge caches at BSs and to realize the benefits of cooperative caching.

4.3 Cooperative Hierarchical Caching in C-RANs

Leveraging the ample storage and computing resources, the CPU in C-RAN can provide a central port for content management and traffic offloading to help address the rapidly growing multimedia traffic from mobile users. As shown in Figure 4.1, by taking advantage of the C-RAN architecture, a *cooperative hierarchical caching* scheme is proposed utilizing both the *cloud cache* at the CPU and distributed *edge caches* at the BSs. The deployments of cloud cache and edge caches are complementary and interoperable. Different from the existing solutions, the cloud cache serves as an additional layer in the in-network cache hierarchy, consisting of the core-based caching schemes (high access latency, large cache capacity) and the edge based caching (low access latency, small cache capacity) to reduce the average content access latency. Specifically, we leverage the low-latency, high-bandwidth fronthaul interconnections (e.g., optical fiber) among the BSs [1] for transferring the contents between the edge caches. Hence, each BS can retrieve cache contents available at the neighboring BSs through a "U-turn" path (BS–CPU–BS) with lower latency than fetching the contents from the original remote sources in the content delivery network (CDN) via backhaul links [6, 7]. The proposed scheme exploits the high flexibility provided by C-RAN to utilize the collective storage capacity in order to improve cache-hit performance and to reduce backhaul traffic due to content requests going to the higher-level network units.

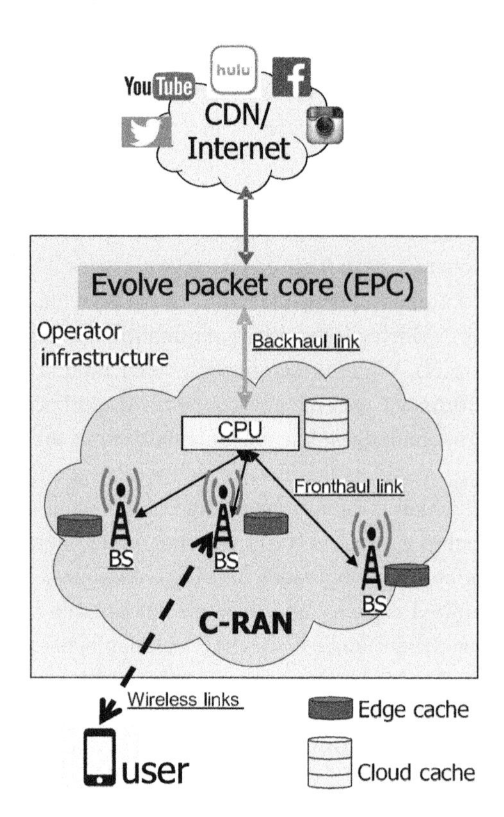

Figure 4.1 C-RAN caching system.

In order to design an efficient cooperative hierarchical caching in C-RAN, it is important to determine the allocations of contents among the cache nodes subject to their capacity constraints such that the content access delay is minimized. The cache placement problem becomes difficult due to the presence of multiple cache layers, which are coupled with the geographical variation of content demand at different cell sites. In the following, we formulate the optimization problem of cache placement, which is then shown to be NP-hard. We then propose the design of a heuristic cache management strategy with low complexity, which achieves a constant approximation ratio compared to the optimal solution.

4.3.1 System Model

As shown in Figure 4.1, we consider a C-RAN system that consists of K BSs that are distributed at different cell sites and a common CPU that is connected to all BSs via low-latency and high-capacity fronthaul links. The backhaul link connects the CPU with the evolve package core (EPC), which is further connected to the CDN in the internet. We assume that there is a central cache manager (CCM) that is co-located with the CPU and is responsible for monitoring content requests from users and for making cache allocation decisions.

Denote $\mathcal{U} = \{1,2,\ldots,U\}$ and $\mathcal{K} = \{1,2,\ldots,K\}$ as the set of active users and the set of BSs in the considered C-RAN system, respectively. Denote $\mathcal{V} = \{1,2,\ldots,V\}$ as the set of indices for all content files that the users can download. Additionally, we let $k = 0$ represent the CPU. For convenience, we assume that all files have an equal size [MB]; however this assumption could be easily lifted by considering a finer packetization to break a longer file into small file packets of the equal length. Define the popularity distribution of the files at the kth cell served by the kth BS as $\mathcal{P} = \{p_k^1, p_k^2, \ldots, p_k^V\}$, where $p_k^v \in [0,1]$ is the probability that a user in cell k requests for file v. Hence, $\sum_{v\in\mathcal{V}} p_k^v = 1, \forall k \in \mathcal{K}$. In this C-RAN system, it is assumed that there is one *edge-cache* co-located at each BS that provides certain capabilities for caching, such as content storage and look up besides the radio frequency (RF) front-end integration. Similarly there is a cloud cache co-located with the CPU. We assume that the edge cache at the BS $k \in \mathcal{K}$ has a normalized storage capacity of M_k [files], and the cloud cache at the CPU has a normalized storage capacity of M_0 [files]. One valid assumption on M_0 is $M_0 \gg M_k, k = 1,\ldots,K$. Denote $\mathcal{C}_k$, $k \in \mathcal{K}$ as the set of files stored at BS k, which is referred to as the cache at BS k. Additionally, we denote $\mathcal{C}_0$ as the set of files stored at the cloud cache. To formulate the cache placement decision, which determines the set of files being stored at each cache, we define the ground set for cache placement as, $\sum_{v\in\mathcal{V}} p_k^v = 1, \forall k \in \mathcal{K}$:

$$\mathcal{G} = \{f_{10}, f_{20}, \ldots, f_{V0}, \ldots, f_{1K}, f_{2K}, \ldots, f_{VK}\}, \tag{4.1}$$

where f_{vk} denotes the presence of file v in cache $\mathcal{C}_k, k \in \mathcal{K} \cup \{0\}$. In the sequel, we will refer to file v and to its presence indicator f_{vk} interchangeably, unless stated otherwise. The ground set $\mathcal{G}$ can be divided into $K+1$ non-overlapping sets, $\mathcal{G}_0, \mathcal{G}_1, \ldots, \mathcal{G}_K$, where $\mathcal{G}_k = \{f_{1k}, f_{2k}, \ldots, f_{Vk}\}$ is the collection of all files that can be stored in the cache $\mathcal{C}_k$, and thus $\mathcal{C}_k \subseteq \mathcal{G}_k$. For a cache placement decision to be feasible, the following storage capacity constraints must be satisfied

$$|\mathcal{C}_k| \leq M_k, \forall k \in \mathcal{K} \cup \{0\}. \tag{4.2}$$

When a request is initiated from a user to BS k for file v that is available in $\mathcal{C}_k$, the user can directly download file v from $\mathcal{C}_k$ without incurring traffic on the fronthaul and backhaul links. Otherwise, the request is deferred to the CCM. Once the CCM receives a request for file v originated from BS k, it will first search for v in the cloud cache $\mathcal{C}_0$, and then in the edge caches of neighboring BS k, i.e., $\mathcal{C}_j$'s, $j \in \mathcal{K} \setminus \{k\}$. If file v is found, the CCM will direct the user to download the file from the cache of a neighboring BS (with lowest cost) via the fronthaul link; otherwise the CCM will direct the user to retrieve the file from the original content server in the CDN. For a given cache placement decision $\mathcal{C}$, we define the following binary variables ($\forall v \in \mathcal{V}, j \in \mathcal{K}, k \in \{0\} \cup \mathcal{K}$),

$$c_k^v = \begin{cases} 1 & f_{vk} \in \mathcal{C}_k, \\ 0 & \text{otherwise}, \end{cases} \tag{4.3}$$

$$z_{k,j}^v = \begin{cases} 1 & \text{request of file } v \text{ from BS } k \text{ is retrieved from } \mathcal{C}_j, \\ 0 & \text{otherwise}, \end{cases} \tag{4.4}$$

$$z_{k,K+1}^{v} = \begin{cases} 1 & \text{request of file } v \text{ from BS } k \text{ is retrieved from the CDN,} \\ 0 & \text{otherwise.} \end{cases} \quad (4.5)$$

We introduce the following constraint to ensure that each request is served by one source,

$$\sum_{j=0}^{K+1} z_{k,j}^{v} = 1, \forall k \in \mathcal{K}, v \in \mathcal{V}. \quad (4.6)$$

Content Access Delay Cost: Although the cost for storage (e.g., hask disk) is getting much lower nowdays, it is still not cost-efficient and sometimes infeasible to store all available files in the caches. When a file is requested by a user and is not present in the serving BS's edge cache, the file has to be retrieved from other places, which incurs higher access delay. In this case, the content access delay is defined as the time period between the time when a user initiates the request for content to the time when the user receives the content data. Denote d_k as the delay cost incurred when retrieving a file from the cloud cache to the edge cache $\mathcal{C}_k$, which is assumed to be the same as the delay cost of the reverse path of the file transferring, i.e., from edge cache $\mathcal{C}_k$ to the CPU. Thus, the delay cost of transferring a file from cache of BS j to BS k is $d_{kj} = d_j + d_k$. Denote d_0 as the delay cost incurred when retrieving a file from the CDN to the CPU. In practice, d_0 is usually much larger than d_j and d_{kj} [6] and thus it is most effective to retrieve content from the edge caches in the RAN whenever possible. The delay cost when a user downloads a file directly from the cache of its serving BS will be mostly determined by the delay of the wireless access channel, which is independent of the caching decision. Hence, without loss of generality when studying cache placement policy, we assume that such last-mile delay is zero.

Denote the set of users being served by BS k as $\mathcal{U}_k \subseteq \mathcal{U}$. For a given cache-placement decision $\mathcal{C}$ and a given content popularity distribution $\mathcal{P}$, the average delay cost of user $u \in \mathcal{U}_k$ can be calculated as

$$\bar{D}_{u,k} = \sum_{v \in \mathcal{V}} p_k^v \left(z_{k,0}^v d_k + z_{k,K+1}^v (d_k + d_0) + \sum_{j \in \mathcal{K} \setminus \{k\}} z_{k,j}^v d_{kj} \right). \quad (4.7)$$

This delay cost reflects the expected content access delay that the users have to experience before having access to the requested contents. Since $d_0 \gg d_j$ and d_{jk}, reducing the average access delay cost will decrease the usage in backhaul network, i.e., the amount of data traffic going through the backhaul links, consequently, reduce the network resource consumption.

4.3.2 Cache Management Algorithms

4.3.2.1 Problem Formulation

We consider the design of a cache-management scheme that proactively distributes content files in the caches and dynamically updates these caches based on cache-hit/miss statistics. Note that the same file can be placed at different caches. The strategy design can be formulated as the optimization problem below,

$$\min_{\mathcal{C}, z_{k,j}^v} \sum_{k \in \mathcal{K}} \sum_{u \in \mathcal{U}_k} \bar{D}_{u,k}, \tag{4.8a}$$

$$\text{s.t.} \sum\nolimits_{v \in \mathcal{V}} c_k^v \leq M_k, \ \forall k \in \{0\} \cup \mathcal{K}, \tag{4.8b}$$

$$\sum\nolimits_{j=0}^{K+1} z_{k,j}^v = 1, \ \forall k \in \mathcal{K}, \tag{4.8c}$$

$$z_{k,j}^v \leq c_j^v, \ \forall k \in \mathcal{K}, j \in \{0\} \cup \mathcal{K}, v \in \mathcal{V}, \tag{4.8d}$$

$$z_{k,j}^v \in \{0,1\}, z_{k,K+1}^v \in \{0,1\}, c_j^v \in \{0,1\}, \forall k \in \mathcal{K}, j \in \{0\} \cup \mathcal{K}, v \in \mathcal{V}, \tag{4.8e}$$

where $\bar{D}_{u,k}$ is given in (4.7). The objective function (4.8a) is the total average delay cost incurred by serving all users' requests. The constraint in (4.8b) imposes the cache capacities at the BSs and at the CPU and (4.8d) represents the cache availability constraint, i.e., it is possible to retrieve a content file from a cache only if such cache stores the request file. From (4.8c), by substituting $z_{k,K+1}^v$ by $1 - \sum_{j=0}^{K} z_{k,j}^v$ into (4.7), we get,

$$\bar{D}_{u,k} = \sum\nolimits_{v \in \mathcal{V}} p_k^v \left(d_0 + d_k - S_k^v\right), \tag{4.9}$$

where

$$S_k^v = z_{k,0}^v d_0 + z_{k,k}^v (d_0 + d_k) + \sum_{j \in \mathcal{K} \setminus \{k\}} z_{k,j}^v \left(d_0 - d_j\right). \tag{4.10}$$

Observe that S_k^v can be viewed as the reduction of delay cost when there is a request for file v by user u at BS k, and that only the term S_k^v in (4.9) is dependent on the optimization variables. Therefore, problem (4.8) can be reformulated as an optimization problem of maximizing the average delay cost saving, given by,

$$\max_{\mathcal{C}, z_{k,j}^v} \sum_{k \in \mathcal{K}} \sum_{u \in \mathcal{U}_k} \sum_{v \in \mathcal{V}} p_k^v S_k^v, \tag{4.11a}$$

$$\text{s.t.} \sum\nolimits_{v \in \mathcal{V}} c_k^v \leq M_k, \ \forall k \in \{0\} \cup \mathcal{K}, \tag{4.11b}$$

$$\sum\nolimits_{j=0}^{K} z_{k,j}^v \leq 1, \ \forall k \in \mathcal{K}, v \in \mathcal{V}, \tag{4.11c}$$

$$z_{k,j}^v \leq c_j^v, \ \forall k \in \mathcal{K}, j \in \{0\} \cup \mathcal{K}, v \in \mathcal{V}, \tag{4.11d}$$

$$z_{k,j}^v \in \{0,1\}, z_{k,K+1}^j \in \{0,1\}, c_j^v \in \{0,1\}, \forall k \in \mathcal{K}, j \in \{0\} \cup \mathcal{K}, v \in \mathcal{V}. \tag{4.11e}$$

The objective function (4.11a) adds up the *utility value* corresponding to each BS, and the goal here is to maximize the sum utility value for all BSs. Problem (4.11) can be shown to be NP-complete [16], and thus it is very difficult to derive a global optimal solution with practical complexity.

This motivates us to design a suboptimal strategy with low complexity, as will be detailed in the following. In particular, we exploit the special structure of problem (4.11) and reformulate it as a submodular function maximization problem subject to matroid constraints [17]. Specifically, we show that the objective function in (4.11) can be expressed as a monotone submodular function and the constraints can be represented as the independent sets of a matroid.[1]

4.3.2.2 Cache Placement Design via Submodular Optimization

First, according to the definition of the ground set $\mathcal{G}$ in (4.1), we have that $\mathcal{C}_k \subseteq \mathcal{G}$ and $\mathcal{C}_k = \mathcal{C} \cap \mathcal{G}_k, \forall k = 0, 1, \ldots, K$. As such, the cache-capacity constraint in (4.16b) can be expressed as $\mathcal{C} \subseteq \mathcal{I}$, where,

$$\mathcal{I} = \{\mathcal{C} \subseteq \mathcal{G}\colon\ |\mathcal{C} \cap \mathcal{G}_k| \leq M_k, \forall k = 0, 1, \ldots, K\}. \tag{4.12}$$

It can be seen that the constraint in (4.12) is in the form of a partition matroid $\mathcal{M} = (\mathcal{G}, \mathcal{I})$. Additionally, according to (4.3), the set $\left\{c_k^v\colon v \in \mathcal{V}\right\}$ can be considered as the Boolean representation of $\mathcal{C}_k$. By defining the objective function in (4.11a) as a set function of the cache placement set $\mathcal{C}$, i.e., $g(\mathcal{C}) = \sum_{k\in\mathcal{K}} \sum_{u\in\mathcal{U}_k} \sum_{v\in\mathcal{V}} p_k^v S_k^v$, the following lemma has been proved in [18].

LEMMA 4.1 *$g(\mathcal{C})$ is a monotone submodular function.*

The result in Lemma 4.1 provides a valuable foundation that allows us to exploit the techniques developed for the monotone submodular function maximization problem with a matroid constraint. Specifically, we extend the greedy algorithm [17] to solve our problem in (4.11). The cache-management solution we proposed contains two stages: first, the content files are allocated to the caches using the proposed proactive caching algorithm; then, every time a requested file is missing on the caches and is retrieved from the remote server, the CCM determines whether to swap this file with an existing one in the caches via reactive caching.

4.3.2.3 Proactive Cache Distribution

For the initial cache placement phase, we propose the proactive cache distribution (PCD) algorithm that starts with an empty cache placement set and distributes content files one by one to the caches in a greedy manner. In each step, a new file is added to the cache placement set such that the marginal value of the objective function when adding this file is the highest compared to adding all other files. The process stops when all caches are filled up. Due to the submodularity of the objective function, the marginal value at each step will decrease as the cache placement set extends. We outline the procedure for the proposed greedy PCD algorithm as in Algorithm 1. Since problem (4.16) maximizes a monotone submodular objective function, it can be shown that Algorithm 1 achieves $\frac{1}{2}$-approximation of the optimal solution [17].

In Algorithm 1 step 1, all the caches are initialized as empty sets, and in step 2 iteration process begins. In each iteration, given the cache placement set $\mathcal{C}$ from the previous iteration, step 3 searches from all hypotheses and obtains the best solution, i.e., placement of file v' in cache $\mathcal{C}_{k'}$, represented by $f_{v'k'}$, that provides the highest marginal value to the objective function. Hence, among all the files $\{f_{vk} \in \mathcal{G}\backslash\mathcal{C}\}$ that

[1] Refer to [17] for definition of submodular set functions and their properties.

Algorithm 1 Proactive Cache Distribution (PCD)

1: Initialize: $\mathcal{G}_k = \{f_{1k}, f_{2k}, \dots, f_{Vk}\}, \mathcal{C}_k = \emptyset, k = 0, 1, \dots, K,$
$\mathcal{G} = (\mathcal{G}_0, \mathcal{G}_1, \dots, \mathcal{G}_K), \mathcal{C} = (\mathcal{C}_0, \mathcal{C}_1, \dots, \mathcal{C}_K).$
2: **repeat**
3: $\quad f_{v'k'} = \arg\max_{f_{vk} \in \mathcal{G} \backslash \mathcal{C}} \left[g\left(\mathcal{C} + f_{vk}\right) - g\left(\mathcal{C}\right)\right]$
4: $\quad \mathcal{C} \leftarrow \mathcal{C} + f_{v'k'}$
5: $\quad$ **if** $|\mathcal{C}_{k'}| = M_{k'}$ **then** $\mathcal{G} \leftarrow \mathcal{G} \backslash \mathcal{G}_{k'}$
6: $\quad$ **end if**
7: **until** $\mathcal{G} = \emptyset$
8: Output: $\mathcal{C}$

have not been allocated to any cache, $f_{v'k'}$ is the best candidate and is added to the current cache placement set in step 4. Once a cache has reached its full capacity, the corresponding candidate file set for this cache will be excluded from the super candidate set $\mathcal{G}$ in step 5. The process will terminate in step 7 when all caches are filled up, or equivalently, the candidate set is empty. We can see from Algorithm 1 that it takes total $\sum_{k=0}^{K} M_k$ iterations to fill up the caches. In each iteration, the algorithm evaluates the marginal values of at most $(K + 1)V$ files in the candidate set. The complexity of evaluating each marginal value is $\mathcal{O}(U)$. Therefore, the total complexity order is $\mathcal{O}\left((K + 1)\, VU \sum_{k=0}^{K} M_k\right)$.

4.3.2.4 Reactive Cache Replacement

In Algorithm 1 above we describe the solution for the initial cache distribution or the thorough cache redistribution, which is often performed during off-peak traffic hours (e.g., nighttime) when the backhaul bandwidth is highly unutilized. During high-traffic time, upon a cache miss, the requested file will be fetched from the CDN server. As the missing file, which is new to the current cache-placement set, is downloaded to the BS, the CCM can reevaluate the marginal value of the objective function. Specifically, the CCM can decide to replace the new file with an existing file in the cache if this operation will result in greater objective function. Described in Algorithm 2, the proposed reactive cache replacement (RCR) algorithm will adapt the cache placement set based on actual content requests.

Step 1 in Algorithm 2 is trigged whenever there is a cache miss that leads to downloading of a new file v^*. Step 2 initiates the evaluation for $K + 1$ caches to determine whether one file in each of these cache sets can be replaced by v^* so to improve the utility objective but also to satisfy the cache capacity constraints. For each cache, step 3 identifies a file with the smallest utility value to be the candidate for removal. If replacing the file of the least utility value with the new file can result in a higher objective value, as verified in step 4, the replacement operation will take place in step 5. The complexity of step 3 is $\mathcal{O}\left(U \sum_{k=0}^{K} M_k\right)$. Since step 3 is performed $K+1$ times, the overall complexity of Algorithm 2 is then $\mathcal{O}\left((K + 1)\, U \sum_{k=0}^{K} M_k\right)$.

Algorithm 2 Reactive Cache Replacement (RCR)

1: Upon a request of a file v^* such that $f_{v^*k} \notin \mathcal{C}, \forall k = 0, 1, \ldots, K$
2: **for** $k = 0 : K$ **do**
3: $\quad f_{v'k'} = \arg\min_{f_{vk} \in \mathcal{C}} \left[g(\mathcal{C}) - g(\mathcal{C} - f_{vk}) \right]$
4: $\quad$ **if** $g(\mathcal{C} - f_{v'k'} + f_{v^*k'}) > g(\mathcal{C})$ **then**
5: $\quad\quad \mathcal{C} \leftarrow \mathcal{C} - f_{v'k'} + f_{v^*k'}$
6: $\quad$ **else** Break
7: $\quad$ **end if**
8: **end for**
9: Output: $\mathcal{C}$

4.3.3 Performance Evaluation

In this section, we evaluate the effectiveness of our proposed algorithms through simulation results with content requests extracted from YouTube trace as well as generated using statistical models. In particular, we evaluate the impact of the proposed cooperative hierarchical caching scheme compared to the existing approaches. We assume that there are seven hexagonal cells in the C-RAN system. The content requests are generated one by one with equal probability of arriving at each BS. The latency of downloading a file from the CDN's server to the CPU, i.e., d_0, is set as 80 ms.[2] The latency of transferring a video between the CPU and the BSs, d_k's, are randomly generated from a uniform distribution in the range [10–30 ms] [6]. All the video files are assumed to have the same size of 20 MB. For brevity of presentation, we consider an example cache setting where for a given total cache capacity in the system, 10% of this amount is allocated to each edge cache and the remaining 30% cache capacity is allocated to the cloud cache. It is straightforward to extend the cache setting to any arbitrary cache capacity allocation. Two key performance metrics are considered: (1) *cache-hit ratio*: the fraction of video requests that can be served by retrieving files from one of the caches, and (2) *average access delay* (ms): average latency that the users have to wait before receiving the requested content. The following caching schemes are compared.

- *Cooperative hierarchical caching (CHC)*: our proposed scheme comprising of the PCD and RCR algorithms.
- *Exclusive most popular caching (ExMPC)*: the cache placement decision is made independently at each edge cache that stores the most popular files according to local popularity. Excluding the files that have been cached at the edge caches, the cloud cache stores the most popular files from the rest files in the candidate file set based on the global popularity. This scheme realizes the greedy cache placement algorithm in [19] for interlevel cache cooperation.
- *Femtocaching extension (FemtoX)*: this scheme is an extension of the Femtocaching scheme [20] to a hierarchical caching system in C-RAN.

[2] Refer to 3GPP TS 23.203, V13.5.1, September 2015, available at www.3gpp.org.

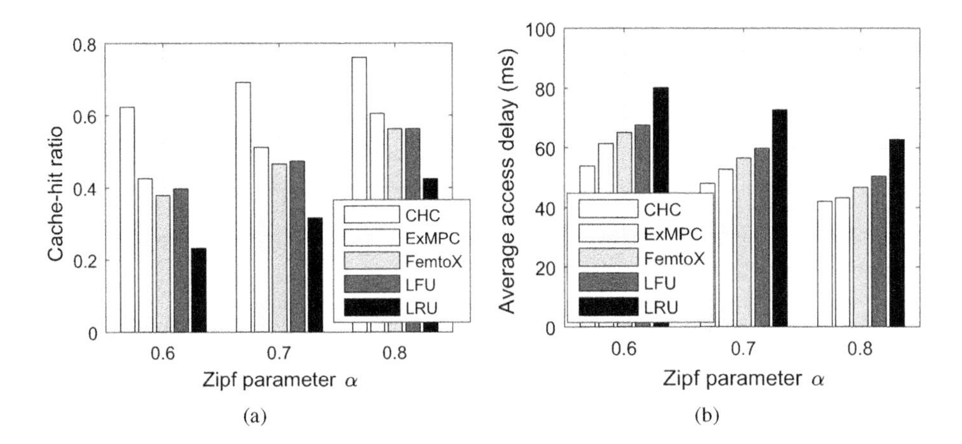

Figure 4.2 Comparison of different caching schemes when increasing the relative total cache capacity (as the fraction of the total content library size).

- *Least frequently used (LFU)*: the LFU scheme is introduced in [21]. We now apply it to the hierarchical caching system. Upon a cache miss, the missing file will be retrieved from the remote server and stored in the cache of the serving BS. If the cache is already full, the least frequently used file will be evicted.
- *Least recently used (LRU)*: this scheme is similar to the LFU scheme, except that in the replacement phase, the least recently used file will be evicted.

4.3.3.1 Simulations with Trace-Based Requests

We first carry out simulations based on the record of video requests to YouTube originating from the University of Massachusetts Amherst campus during the day of March 12, 2008 [22]. In this record, there were 122,280 requests made for 77,414 different videos. We calculated the video popularity based on the frequency each video was requested in the trace. To vary the content popularity at different BSs, we randomly shuffled the global popularity distribution calculated from the trace. In Figure 4.2a and b, we show the performance comparisons of the proposed CHC caching policy with the four baselines listed earlier. It is seen that CHC provides the best performance in terms of both cache-hit ratio and average access delay.

4.3.3.2 Simulations with Synthetic Requests

To evaluate our proposed scheme in a more generic scenario, we consider a content-request model in which the popularity of the files follows a Zipf distribution [23]. Specifically, the probability that the ith most popular content at BS k is expressed as $p_{ik} = \frac{1/i^{\alpha}}{\sum_{v=1}^{V} 1/v^{\alpha}}, \forall k \in \mathcal{K}$, where V is the set of all contents and α determines the skewness of the distribution. The estimated value of α may vary from different measurements. According to the measurements in [24], the estimated value of α ranges from 0.64 to 0.83. We consider a library of 10,000 files, each having a size of 20 MB and randomly generate 100,000 requests following the Zipf-based popularity distribution

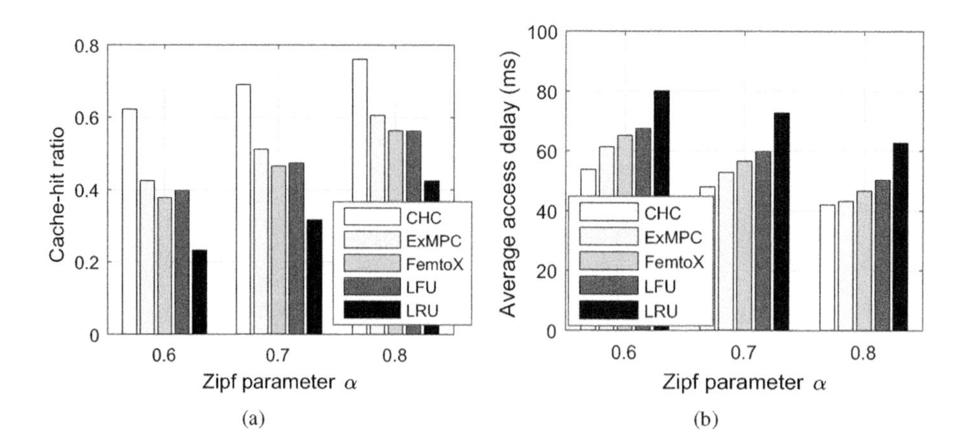

Figure 4.3 Performance of different caching schemes with synthetic content requests generated using the Zipf-based popularity distribution.

with $\alpha \in [0.6, 0.7, 0.8]$. Figure 4.3a and b illustrates the performance of the CHC caching scheme as well as the baseline schemes with different α settings for the content popularity distributions when the relative total cache capacity is set to 30% total library size. We observe from Figure 4.3 that as α increases, the performance of all schemes on both cache-hit ratio and average content access delay significantly improves. For all cases, CHC always performs better than other schemes although the performance gaps between CHC and the baselines become smaller as α increases.

4.4 Cooperative Caching and Video Transcoding in MEC Networks

Given the dynamic nature of the wireless network connections and the variety of user devices' configuration, the users might request different quality variants of the same video. For example, users with less capable devices and slow network conditions might prefer low-resolution videos so they can watch the video without stalls; on the other hand, users with highly capable devices and strong connections may request high-quality videos to maximize their quality of experience (QoE). Adaptive bitrate (ABR) video streaming [25] is widely adopted as an effective technique that maximizes the users QoE by generating and transmitting different variants of a video with different bitrates based on the channel condition and capabilities of the requesting users. The dependency between different video variants of the same video (e.g., one can be transcoded from one another [26]) is often not considered in traditional video caching solutions, in which each video variant is treated as an independent stream.

We present in this section a framework that jointly considers both video transcoding and caching capabilities at the MEC servers in order to facilitate ABR video streaming. In particular, each MEC server is envisioned to perform both a caching function and transcoding function. We consider that a higher bitrate variant can be transcoded into a low bitrate variant [27]. For instance, a video at bitrate of 10 Mbps (1080p) can be transcoded into the same video with bitrate of 5 Mbps. In addition, we consider a new dimension to the cooperative caching scheme, where the MEC servers can utilize the help from each other not only by retrieving video from neighboring servers but also by having the requested video being transcoded at those servers. For example, when a MEC server asks for a video variant from its neighboring server, it can also ask for this neighbor to transcode the video before transferring it through the backhaul link. That way, one can better improve the transcoding load balancing among the servers. This strategy offers several advantages: (1) it is not necessary to generate and store all video variants at the content server, (2) different users can get video qualities that are optimized for their network conditions and processing capabilities, and (3) the overall cache-hit ratio and load balancing can be improved via collaboration among the MEC servers.

There are multiple challenges associated with the envisioned cooperative caching framework. First, there is high storage overhead when caching multiple variants of each video. Hence, in order to design an effective caching strategy, it is important to consider the dependency (transcodability) among different video variants, which in turn increase the complexity of the conventional cache placement problem where different video variants are cached independently. Second, transcoding videos in real time is a computation-intensive task and thus only a moderate number of videos can be transcoded at the same time given the limited computing capacity at the MEC servers. These challenges necessitate the design of a joint caching and request scheduling scheme, which effectively utilizes the limited cache storage and processing capacities. To design such scheme, we formulate a cooperative joint caching and processing problem that aims at minimizing the average access delay cost of delivering all requested videos, subject to the cache storage and processing capacity constraints. This is cast as an integer linear program (ILP). Due to the NP-completeness and the prohibitive complexities of the existing offline approaches, we employ the online cache placement policy based on an LRU eviction rule and propose a novel online video transcoding scheduling scheme that makes decision upon arrival of each new request. One main advantage of our solution is that it does not assume a priori knowledge of content popularity as well as request arrivals as commonly considered.

4.4.1 System Model

Our considered wireless network consists of multiple BSs, each of which is equipped with a MEC server that can act as both a caching server and a transcoding server. Each MEC server can retrieve videos from the remote content server in the internet as well as from other neighboring MEC servers. In our caching system, when a user requests a video, one of the following events will occur (as shown in Figure 4.4: (a) the requested video is available in the cache of the serving BS; (b) only a higher bitrate variant of the requested video is available at the cache of the home BS; (c) the requested video variant is available in the cache of a neighboring BS or in the remote server; (d) a higher bitrate variant of the requested video is available in the cache of the neighboring BS, which will also do the transcoding; and (e) same as (d) but the video is transferred to the serving BS before being transcoded. We refer to the events in (a,c) as *exact hit* and to those in (b,d,e) as *soft hit*.

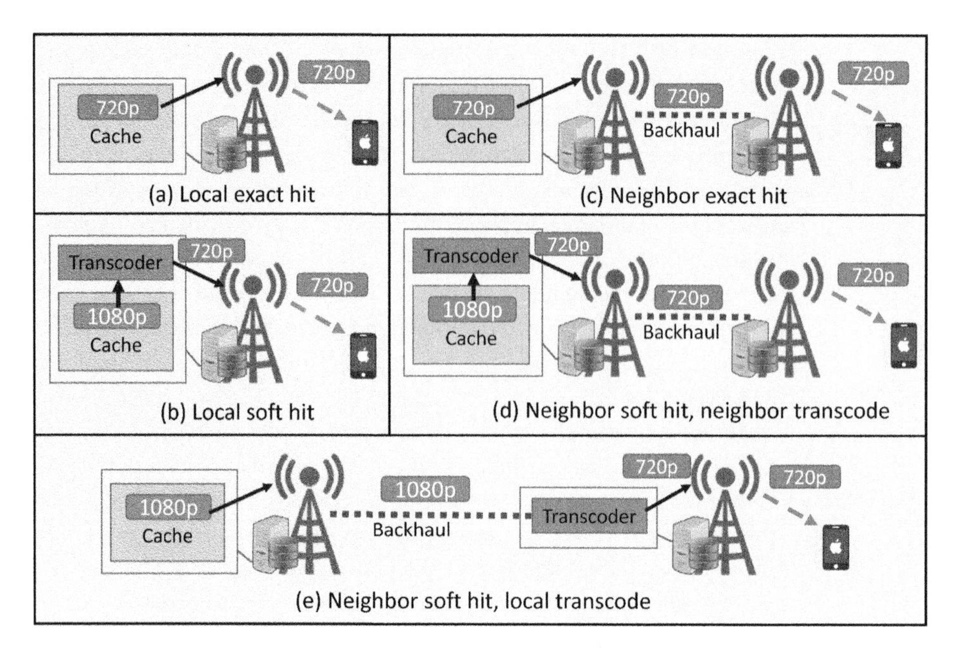

Figure 4.4 Possible events that occur when there is a video request to the cache system.

Let $\mathcal{K} = \{1,2,\ldots,K\}$ be the set of K MEC servers that are connected to each other through a backhaul mesh network. The remote content server is denoted as $k = 0$. The indexes of all videos that users can request is denoted by $\mathcal{T} = \{1,2,\ldots,V\}$. Without loss of generality, it is assumed that the duration of all videos is the same, and each video has L different bitrate variants represented by the list $\{v_1,v_2,\ldots,v_L\}$, which is sorted by the ascending order bitrate. The size of each video variant l, denoted as r_l $\left[\text{bytes}\right]$, is therefore proportional to its bitrate. The collection of all video variants that a user can download can be written as $\mathcal{V} = \{v_l\,|v \in \mathcal{T}, l = 1,\ldots,L\}$. For simplicity, hereafter we refer to video and video variant interchangeably. It is assumed that variant v_l can be transcoded from the higher bitrate variant v_h, $l \leq h$ and that the computation cost of transcoding v_l from v_h is given by b_{hl}, $\forall v \in \mathcal{T}$ and $l,h = 1,\ldots,L$. Similar to [27], we consider that b_{hl} is dependent only on r_l, i.e., $b_{hl} = b_l$. Our proposed scheme can also be generalized to the scenario where b_{hl} is a function of both r_l and r_h.

We consider that each user is served by only one BS, which is usually the BS with the strongest channel condition. We define the cache placement variables as $c_k^{v_l} \in \{0,1\}$, $k \in \mathcal{K}, v_l \in \mathcal{V}$, where $c_k^{v_l} = 1$ if v_l is stored at server k and $c_k^{v_l} = 0$ otherwise. Accordingly, the cache placement set is denoted by $\mathcal{C} = \left\{c_k^{v_l}\,\middle|c_k^{v_l} = 1, k \in \mathcal{K}, v_l \in \mathcal{V}\right\}$. Assuming that the storage capacity of each cache server k is M_k $\left[\text{bytes}\right]$, we can express the cache capacity constraint at each server $j \in \mathcal{K}$ as,

$$\sum\nolimits_{v_l \in \mathcal{V}} r_l c_k^{v_l} \leq M_k, \forall k \in \mathcal{K}. \tag{4.13}$$

To represent different events that occur when file v_l is being requested at server $k \in \mathcal{K}$, we define the binary variables $\left\{x_k^{v_l}, y_k^{v_l}, z_{kj}^{v_l}, t_{kj}^{v_l}, w_{kj}^{v_l}\right\} \in \{0,1\}$ as follows.

- $x_k^{v_l} = 1$ if v_l is served directly from cache of BS k (as shown in Figure 4.4a); and $x_k^{v_l} = 0$ otherwise.
- $y_k^{v_l} = 1$ if there is a soft hit at BS k, i.e., v_l is obtained from BS k by transcoding from a higher bitrate variant (as illustrated in Figure 4.4b); and $y_k^{v_l} = 0$ otherwise.
- $z_{kj}^{v_l} = 1$ if v_l is obtained from cache of BS $j \neq k, j \in \mathcal{K} \cup \{0\}$ (which could also be the remote server, as shown in Figure 4.4c); and $z_{kj}^{v_l} = 0$ otherwise.
- $t_{kj}^{v_l} = 1$ if v_l is obtained by transcoding a higher bitrate variant from cache of BS $j \neq k, j \in \mathcal{K}$ (as shown in Figure 4.4d); and $t_{kj}^{v_l} = 0$ otherwise.
- $w_{kj}^{v_l} = 1$ if v_l is obtained by retrieving a higher bitrate variant from cache of BS $j \neq k, j \in \mathcal{K}$ to BS k where transcoding is performed to render v_l (as shown in Figure 4.4e); and $w_{kj}^{v_l} = 0$ otherwise.

When a specific video variant is requested, depending on the cache availability, the system will schedule one of the described events to satisfy the request. To ensure there is one delivery for each request, we take into account the following constraint ($\forall k \in \mathcal{K}$, $v_l \in \mathcal{V}$),

$$x_k^{v_l} + y_k^{v_l} + \sum_{j \neq k,\, j \in \mathcal{K}} \left(z_{kj}^{v_l} + t_{kj}^{v_l} + w_{kj}^{v_l} \right) + z_{k0}^{v_l} = 1. \tag{4.14}$$

For content access delay cost, the delay cost incurred when cache server k retrieves a video from a neighbor cache server j and from the remote server are denoted as d_{kj} and d_{k0}, respectively. Thus, the delay cost incurred when the request for video v_l is served at BS k is calculated as ($\forall k \in \mathcal{K}, v_l \in \mathcal{V}$),

$$D_{v_l,k} = d_{k0} z_{k0}^{v_l} + \sum_{j \neq k,\, j \in \mathcal{K}} d_{kj} \left(z_{kj}^{v_l} + t_{kj}^{v_l} + w_{kj}^{v_l} \right). \tag{4.15}$$

4.4.2 Joint Cooperative Caching and Processing Algorithm

4.4.2.1 Problem Formulation

We consider an arbitrary period of time during which the set of videos being requested is $\mathcal{N}_k$. Our design objective is to minimize the total access delay cost, expressed as $\Omega = \sum_{k \in \mathcal{K}} \sum_{v_l \in \mathcal{N}_k} D_{v_l,k}$. The optimization variables consist of the *cache placement policy* $\mathcal{C}$ and the *video request scheduling policy* $\mathcal{R} \triangleq \left\{ x_k^{v_l}, y_k^{v_l}, z_{kj}^{v_l}, t_{kj}^{v_l}, w_{kj}^{v_l} \right\}$ that must conform to the cache storage and processing capacity constraints. At every time slot t, we denote the set of videos being served at BS k as $\mathcal{N}_k^t$ and thus $\bigcup_t \mathcal{N}_k^t = \mathcal{N}_k$. With this position, we can derive the optimal solutions $\{\mathcal{C}, \mathcal{R}\}$ via solving the static joint collaborative caching and processing problem $\mathcal{J}^t$ in (4.16) in every time slot t.

In problem (4.16), constraints (4.16b) and (4.16c) ensure feasibility of the exact hit; constraints (4.16d), (4.16e), and (4.16f) ensure the feasibility of the soft hit; constraint (4.16g) ensures that there is one and at most one delivery for each request. Additionally, the limited cache storage capacity and computing capacity (represented by transcoding throughput) at each cache server is enforced in constraint (4.16h) and

(4.16i), respectively. It can be seen that problem $\mathcal{J}^t$ in (4.16) is an ILP, which can be shown to be NP-complete by reduction from a multiple knapsack problem [29]. To overcome the intractability of this problem, we employ a decomposition approach to transform the original problem of solving a series of problems $\mathcal{J}^t$ into a *cache placement problem* and a *request scheduling problem* whose solutions are derived in the following:

$$(\mathcal{J}^t):\ \min_{\mathcal{C},\mathcal{R}} \sum_{k\in\mathcal{K}} \sum_{v_l\in\mathcal{N}_k^t} D_{v_l,k}, \tag{4.16a}$$

$$\text{s.t. } x_k^{v_l} \le c_k^{v_l},\ \ \forall k \in \mathcal{K}, v_l \in \mathcal{V}, \tag{4.16b}$$

$$z_{kj}^{v_l} \le c_j^{v_l},\ \ \forall j,k \in \mathcal{K},\ v_l \in \mathcal{V}, \tag{4.16c}$$

$$y_k^{v_l} \le \min\left(1, \sum\nolimits_{m=l+1}^{L} c_k^{v_m}\right),\ \ \forall k \in \mathcal{K}, v_l \in \mathcal{V}, \tag{4.16d}$$

$$t_{kj}^{v_l} \le \min\left(1, \sum\nolimits_{m=l+1}^{L} c_j^{v_m}\right),\ \ \forall k,j \in \mathcal{K}, v_l \in \mathcal{V}, \tag{4.16e}$$

$$w_{kj}^{v_l} \le \min\left(1, \sum\nolimits_{m=l+1}^{L} c_j^{v_m}\right),\ \ \forall k,j \in \mathcal{K}, v_l \in \mathcal{V}, \tag{4.16f}$$

$$x_k^{v_l} + y_k^{v_l} + \sum\nolimits_{j\neq k, j\in\mathcal{K}} \left(z_{kj}^{v_l} + t_{kj}^{v_l} + w_{kj}^{v_l}\right) + z_{k0}^{v_l} = 1,\ \ \forall k \in \mathcal{K}, \tag{4.16g}$$

$$\sum\nolimits_{v_l\in\mathcal{V}} r_l c_k^{v_l} \le M_k,\ \ \forall k \in \mathcal{K}, \tag{4.16h}$$

$$\sum_{v_l\in\mathcal{N}_k^t} b_l \left(y_k^{v_l} + \sum_{j\neq k, j\in\mathcal{K}} w_{kj}^{v_l}\right) + \sum_{j\neq k, j\in\mathcal{K}} \sum_{v_l\in\mathcal{N}_j^t} b_l t_{jk}^{v_l} \le B_k,\ \ \forall k \in \mathcal{K}, \tag{4.16i}$$

$$c_k^{v_l}, x_k^{v_l}, y_k^{v_l}, z_{kj}^{v_l}, t_{kj}^{v_l}, w_{kj}^{v_l} \in \{0,1\},\ \ \forall k \in \mathcal{K}, v_l \in \mathcal{V}. \tag{4.16j}$$

4.4.2.2 Cache Placement

Without having prior knowledge about request arrival, it is important that the reactive cache placement policy is simple and can be calculated in real time to avoid additional delay. To this end, we employ the popular LRU cache placement policy [21, 27, 28]. Following each cache miss, the requested video will be retrieved from either the neighboring caches or the remote server. This retrieved video will be placed in the cache, and if it was already full, the LRU will evict the existing files that are least recently used.

4.4.2.3 Request Scheduling

In each time slot t, let us use $\mathcal{Q}^t$ to denote the request-scheduling problem, which is reduced from problem $\mathcal{J}^t$ in (4.16) for a given cache placement decision $\mathcal{C}$ as:

$$(\mathcal{Q}^t):\ \min_{\mathcal{R}} \sum_{k\in\mathcal{K}} \sum_{v_l\in\mathcal{N}_k^t} D_j\left(v_l\right) \tag{4.17a}$$

$$\text{s.t. } (4.16\text{b}) - (4.16\text{g}), (4.16\text{i}), \tag{4.17b}$$

$$x_k^{v_l}, y_k^{v_l}, z_{kj}^{v_l}, t_{kj}^{v_l}, w_{kj}^{v_l} \in \{0,1\}, \forall k \in \mathcal{K}, v_l \in \mathcal{V}. \tag{4.17c}$$

Given the formulation of problem $\mathcal{Q}^t$, we discuss the alternatives to solve problem $\mathcal{J}^t$ as follows.

Optimal request scheduling (OptRS): The optimal request scheduling scheme solves problem $\mathcal{Q}^t$ when there is a new request arrival at time t to optimize the request scheduling specific to this time slot. In this way, the long-term solution is also optimal. Since $\mathcal{Q}^t$ is an ILP, the optimal solution can be obtained using standard optimization solver. The solution to the optimal request scheduling scheme, referred to as OptRS, can be obtained by using standard ILP solver which usually have very high complexity.

Online request scheduling (OnRS): To avoid the high complexity of the OptRS approach, we propose here a low-complexity online request scheduling algorithm, referred to as OnRS. In particular, at each time slot t, OnRS is responsible for determining how to redirect the incoming request to appropriate cache servers or remote server. For brevity of presentation, we focus on a particular time slot in the sequel and omit the index t.

The processing load (due to transcoding) at BS k when serving the set $\mathcal{N}_k$ of incoming requests to this BS can be calculated as,

$$A_k(\mathcal{N}) = \sum_{v_l \in \mathcal{N}_k} b_l \left(y_k^{v_l} + \sum_{j \in \mathcal{K} \setminus \{k\}} w_{kj}^{v_l} \right) + \sum_{j \in \mathcal{K} \setminus \{k\}} \sum_{v_l \in \mathcal{N}_j} b_l t_{jk}^{v_l}. \tag{4.18}$$

Additionally, we define the *closest* (in terms of bitrate) transcodable variant of video v_l at BS k as $T(k, v_l) = v_h$, where,

$$h = \min\left\{ m \,\middle|\, m \in \{l+1, \ldots, L\}, c_k^{v_m} = 1 \right\}. \tag{4.19}$$

For each request of file v_l at BS $k \in \mathcal{K}$, the proposed OnRS scheme is described as in Algorithm 3. Specifically, if there cannot be an exact hit (line 2) or a soft hit (lines 3, 4) at BS k, the algorithm will check whether there is an exact hit (lines 5, 6, 7) or a soft hit at a neighboring cache (line 8). If a neighboring BS j happens to have the transcodable variant of v_l, this variant can be transcoded either at BS j or BS k to render v_l. Our algorithm will pick the cache server with lower processing load to perform transcoding in order to balance the load. Finally, if there is neither an exact hit nor a soft hit in the MEC cache system, v_l will be retrieved from the remote server in the CDN (line 20).

The proposed online joint cooperative caching and processing strategy contains OnRS as the inner loop, which is trigged upon each new request arrival, and the LRU policy as the conditional routine. If line 20 is reached in an inner-loop iteration, the LRU policy will be triggered to add the new cache entry $c_k^{v_l}$ while removing some other entries.

4.4.3 Performance Evaluation

Here, we present numerical simulations to demonstrate the effectiveness of the proposed joint cooperative caching and processing algorithm, which uses LRU policy for cache placement and OnRS for request scheduling, under various cache sizes and processing capacities at the MEC servers. We compare LRU–OnRS with the competing schemes using LRU for cache placement and different request scheduling algorithms:

Algorithm 3 Online Video Request Scheduling (OnRS)

1: Upon each request for video v_l at BS $k \in \mathcal{K}$, initiate the algorithm.
2: **if** $c_k^{v_l} = 1$ **then** BS k gets v_l from its cache and transmits the file to the user
3: **else if** $T(k, v_l) \neq \emptyset$ and $B_k - A_k(\mathcal{N}) - b_l \geq 0$ **then**
4: BS k transcodes $T(k, v_l)$ to v_l and transmits it to the user
5: **else if** $\sum_{j \in \mathcal{K} \setminus \{k\}} c_j^{v_l} \geq 1$ **then**
6: Find $j^* = \arg\min_{j \in \mathcal{K} \setminus \{k\}} d_{kj}$ s.t. $c_j^{v_l} = 1$
7: BS k retrieves v_l from BS j^* and then forwards it to the user
8: **else if** $\tilde{\mathcal{K}} \triangleq \{j \in \mathcal{K} \setminus \{k\} \mid T(j, v_l) \neq \emptyset\} \neq \emptyset$ **then**
9: Evaluate the available processing capacity

$$Q_j(\mathcal{N}) = B_j - A_j(\mathcal{N}) - b_l, \forall j \in \tilde{\mathcal{K}} \cup \{k\} \tag{4.20}$$

10: Find $j^* = \arg\max_{j \in \tilde{\mathcal{K}} \cup \{k\}} Q_j(\mathcal{N})$ s.t. $Q_j(\mathcal{N}) \geq 0$
11: **if** $j^* = k$ **then**
12: BS k fetches $T(j, v_l)$ from BS $j \in \tilde{\mathcal{K}}$ with the lowest delay cost
13: BS k transcodes $T(j, v_l)$ to v_l and then sends the file to the user.
14: **else if** $j^* \neq \emptyset$ **then**
15: BS j^* transcode $T(f, v_l)$ to v_l.
16: BS k retrieves v_l from BS j^* and sends it to the user.
17: **else** Return to line 20.
18: **end if**
19: **else**
20: BS k fetches v_l from the remote server and then transmits it to the user.
21: **end if**

(1) LRU–OptRS: the optimal request scheduling algorithm; (2) LRU–CachePro: joint caching and processing scheme without collaboration among the cache servers, as proposed in [27]; and (3) LRU–CoCache: a simplified version of OnRS that does not consider transcoding.

We consider that there are $K = 3$ MEC servers in the studied network. Users can request for video variants from the library $\mathcal{V}$ in which there are $V = 1{,}000$ different videos with the same playtime duration of 10 minutes. Each video has $L = 4$ bitrate variants whose bitrates are $0.90, 1.10, 1.34$, and 1.64 Mbps, respectively. We randomly generate $10{,}000$ request arrivals to each BS following a Poisson distribution with arrival rate λ_k $[\text{reqs/min}]$. For each request, the probability that a particular video is requested is calculated based on a Zipf distribution with the skewness parameter equals to 0.6. The video transferring delay d_{kj} and d_{k0} are randomly assigned based on the uniform distribution in the rage of $[5, 10]$(ms), $[20, 50]$(ms), and $[100, 200]$(ms), respectively. In the simulation results, the storage capacity at each BS is represented by the percentage of the total video library size, and the processing capacity is regarded as the transcoding throughput [bits/s].

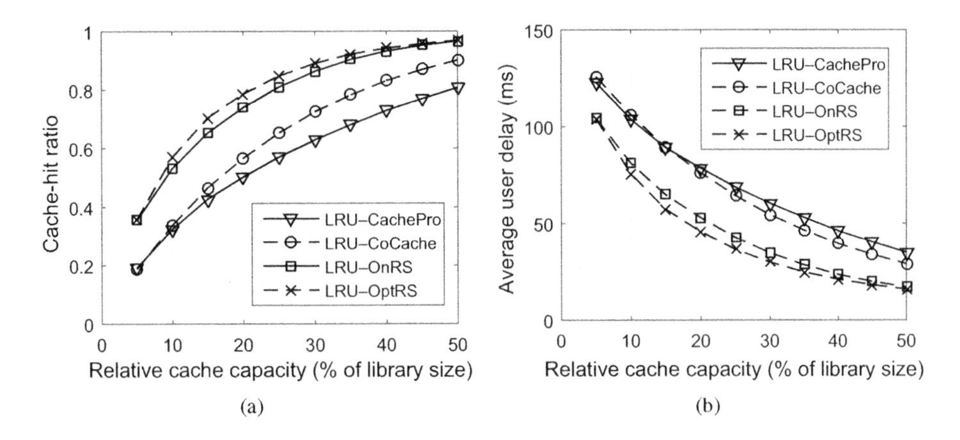

Figure 4.5 Cache-hit ratio and average user delay of different schemes when varying the relative cache capacity at each server; $B_k = 10$ Mbps, $-_k = 8$ reqs/minute, $\forall k \in \mathcal{K}$.

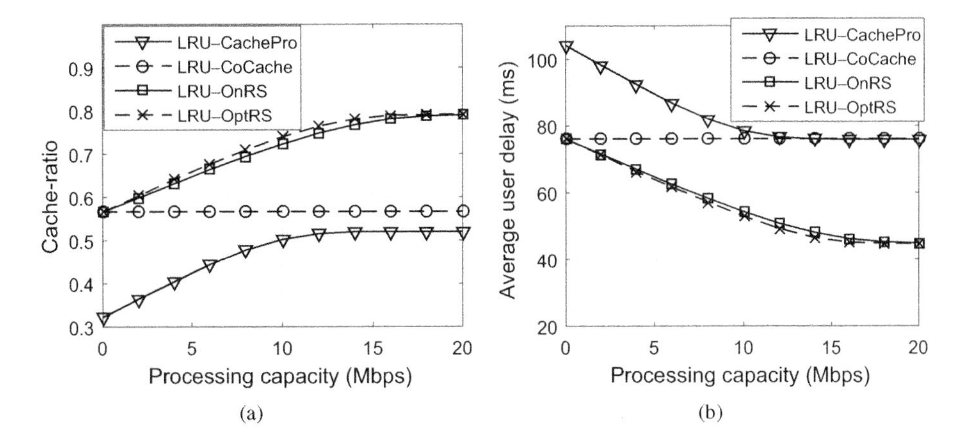

Figure 4.6 Cache-hit ratio and average user delay of different schemes when varying transcoding throughput at each server; $M_k = 20\%$ [Lib.size], $-_k = 8$ reqs/minute, $\forall k \in \mathcal{K}$.

4.4.3.1 Impact of Cache Capacity

The relative cache capacity is given by the ratio between the cache size at each server and the total size of the video library. The results shown in Figure 4.5a and b demonstrate that increasing the cache size results in higher cache-hit ratio and lower average access delay in all schemes. It can be seen that our proposed schemes LRU–OnRS always yield better performance compared to the baselines. With moderate cache size, there is a small performance gap between the LRU–OnRS and LRU–OptRS, however at higher cache size, this gap diminishes.

4.4.3.2 Impact of Processing Capacity

Figure 4.6a and b compares the performance in terms of cache-hit ratio and average user delay of different schemes when varying the processing capacity at each cache server. Notice that the LRU–CoCache scheme does not involve transcoding, hence, its performance is not dependent on the processing capacity. We observe that at low processing capacity regime, the performance of all schemes improve as the processing capacity increases. Additionally, our proposed LRU–OnRS scheme always outperforms the baseline schemes. Moreover, the performance gap between LRU–OnRS and LRU–OptRS diminishes when processing capacity is high.

4.5 Conclusions

In this chapter, we discussed the two emerging cloud-assisted wireless network paradigms—C-RAN and MEC—and proposed novel cooperative caching schemes that exploit the specific topology of C-RAN and the in-network processing capability of MEC. Specifically, a cooperative hierarchical caching in C-RAN is proposed to exploit both the vertical collaboration between the edge caches and the cloud cache and the horizontal collaboration among the edge caches to form a heterogeneous cache storage pool. We then presented a joint cooperative caching and processing framework in a MEC network where the MEC servers perform both cache storage and video transcoding in order to enhance the performance of ABR video streaming. Numerical results have shown the significant advantages of the proposed caching solutions in terms of improvement in cache-hit ratio as well as reduction in content access latency.

References

[1] "C-RAN: the road towards green RAN," white paper, China Mobile Research Institute, Sept. 2013.

[2] T. X. Tran, A. Hajisami, P. Pandey, and D. Pompili, "Collaborative mobile edge computing in 5g networks: new paradigms, scenarios, and challenges," *IEEE Commun. Mag.*, vol. 55, no. 4, pp. 54–61, 2017.

[3] T. X. Tran, M.-P. Hosseini, and D. Pompili, "Mobile edge computing: recent efforts and five key research directions," in *IEEE COMSOC MMTC Commun. Frontiers*, 2017.

[4] A. Reznik et al., "Cloud RAN and MEC: a perfect pairing," ETSI White Paper No. 23, Feb. 2018.

[5] T. X. Tran, A. Younis, and D. Pompili, "Understanding the computational requirements of virtualized baseband units using a programmable cloud radio access network testbed," in *Proc. IEEE Int. Conf. Autonomic Comput. (ICAC)*, 2017, pp. 221–226.

[6] X. Wang, M. Chen, T. Taleb, A. Ksentini, and V. Leung, "Cache in the air: exploiting content caching and delivery techniques for 5G systems," *IEEE Commun. Mag.*, vol. 52, no. 2, pp. 131–139, 2014.

[7] A. Gharaibeh, A. Khreishah, B. Ji, and M. Ayyash, "A provably efficient online collaborative caching algorithm for multicell-coordinated systems," *IEEE Trans. Mobile Comput.*, vol. 15, no. 8, pp. 1863–1876, 2016.

[8] K. Poularakis and L. Tassiulas, "On the complexity of optimal content placement in hierarchical caching networks," *IEEE Trans. Commun.*, vol. 64, no. 5, pp. 2092–2103, 2016.

[9] A. Khreishah, J. Chakareski, and A. Gharaibeh, "Joint caching, routing, and channel assignment for collaborative small-cell cellular networks," *IEEE J. Sel. Areas Commun.*, vol. 34, no. 8, pp. 2275–2284, 2016.

[10] K. Poularakis, G. Iosifidis, A. Argyriou, I. Koutsopoulos, and L. Tassiulas, "Caching and operator cooperation policies for layered video content delivery," in *Proc. IEEE Int. Conf. Comput. Commun. (INFOCOM)*, 2016, pp. 874–882.

[11] P. Ostovari, J. Wu, and A. Khreishah, "Efficient online collaborative caching in cellular networks with multiple base stations," in *Proc. IEEE Int. Conf. Mobile Ad Hoc Sensor Syst. (MASS)*, 2016, pp. 136–144.

[12] T. Tran and D. Pompili, "Adaptive bitrate video caching and processing in mobile-edge computing networks," *IEEE Trans. Mobile Comput.*, vol. 18, no. 9, pp. 1965–1978, Sept. 2018.

[13] K. Poularakis, G. Iosifidis, V. Sourlas, and L. Tassiulas, "Exploiting caching and multicast for 5G wireless networks," *IEEE Trans. Wireless Commun.*, vol. 15, no. 4, pp. 2995–3007, 2016.

[14] T. X. Tran, F. Kazemi, E. Karimi, and D. Pompili, "Mobee: mobility-aware energy-efficient coded caching in cloud radio access networks," in *Proc. IEEE Int. Conf. Mobile Ad Hoc Sensor Syst. (MASS)*, 2017, pp. 461–465.

[15] J. Robson, "Small cell backhaul requirements," NGMN white paper, pp. 1–40, 2012.

[16] T. X. Tran and D. Pompili, "Octopus: a cooperative hierarchical caching strategy for cloud radio access networks," in *Proc. IEEE Int. Conf. Mobile Ad Hoc Sensor Syst. (MASS)*, Oct. 2016, pp. 154–162.

[17] G. Calinescu, C. Chekuri, M. Pál, and J. Vondrák, "Maximizing a monotone submodular function subject to a matroid constraint," *SIAM J. Comput.*, vol. 40, no. 6, pp. 1740–1766, 2011.

[18] T. X. Tran, D. V. Le, G. Yue, and D. Pompili, "Cooperative hierarchical caching and request scheduling in a cloud radio access network," *IEEE Trans. Mobile Comput.*, vol. PP, pp. 1–15, Mar. 2018.

[19] S. Borst, V. Gupt, and A. Walid, "Distributed caching algorithms for content distribution networks," in *Proc. IEEE Int. Conf. Comput. Commun. (INFOCOM)*, 2010, pp. 1–9.

[20] N. Golrezaei, K. Shanmugam, A. G. Dimakis, A. F. Molisch, and G. Caire, "Femtocaching: wireless video content delivery through distributed caching helpers," in *Proc. IEEE Int. Conf. Comput. Commun. (INFOCOM)*, 2012, pp. 1107–1115.

[21] D. Lee, J. Choi, J. H. Kim, S. H. Noh, S. L. Min, Y. Cho, and C. S. Kim, "LRFU: A spectrum of policies that subsumes the least recently used and least frequently used policies," *IEEE Trans. Comput.*, vol. 50, no. 12, pp. 1352–1361, 2001.

[22] http://traces.cs.umass.edu/index.php/Network.

[23] L. Breslau, P. Cao, L. Fan, G. Phillips, and S. Shenker, "Web caching and Zipf-like distributions: evidence and implications," in *Proc. IEEE Int. Conf. Comput. Commun. (INFOCOM)*, 1999, pp. 126–134.

[24] H. Yu, D. Zheng, B. Y. Zhao, and W. Zheng, "Understanding user behavior in large-scale video-on-demand systems," *ACM SIGOPS Oper. Syst. Rev.*, vol. 40, no. 4, pp. 333–344, 2006.

[25] T. Stockhammer, "Dynamic adaptive streaming over http–: standards and design principles," *In Proc. Annual ACM Conf. Multimedia Syst.*, pp. 133–144, 2011.

[26] A. Vetro, C. Christopoulos, and H. Sun, "Video transcoding architectures and techniques: an overview," *IEEE Signal Process. Mag.*, vol. 20, no. 2, pp. 18–29, 2003.

[27] H. A. Pedersen and S. Dey, "Enhancing mobile video capacity and quality using rate adaptation, RAN caching and processing," *IEEE/ACM Trans. Netw.*, vol. 24, no. 2, pp. 996–1010, 2016.
[28] B. Shen, S.-J. Lee, and S. Basu, "Caching strategies in transcoding-enabled proxy systems for streaming media distribution networks," *IEEE Trans. Multimedia*, vol. 6, no. 2, pp. 375–386, 2004.
[29] M. R. Gary and D. S. Johnson, *Computers and Intractability: A Guide to the Theory of NP-completeness*, New York: W. H. Freeman, 1979.

5 Stochastic Caching Schemes in Large Wireless Networks

Zheng Chen, Nikolaos Pappas, and Marios Kountouris

In this chapter, we present the concept of stochastic caching in large wireless networks with randomly distributed nodes. Specifically, we consider a random network where user devices can directly communicate and exchange information through device-to-device (D2D) communication. The distribution of D2D-enabled devices follows a Poisson point process (PPP), and each user stores proactively the popular files based on some probabilistic caching policy. The optimal caching probabilities depend on the specific objective functions to be optimized. We investigate three different caching schemes, namely maximizing the cache-hit probability, maximizing the density of successfully served requests by local caches, and minimizing the delay to receive the requested content. By comparing the performance achieved with these schemes, we show that the success probability of physical layer (PHY) transmission plays a critical role in the throughput and delay performance of large wireless networks with stochastic caching methods.

5.1 Introduction

The mobile data traffic is under exponential growth and it is highly likely to have an eightfold increase between 2015 and 2020, and 75% of the data traffic is expected to be video [1]. One effective way to reduce the cellular data traffic is to allow mobile devices to communicate directly without passing through the base stations. This can be done by establishing short-distance links through D2D communication. As multiple D2D links can be active at the same time, sharing the same spectrum resources, the spectral efficiency per area can also be improved. Another perspective is to investigate the source of wireless data traffic. It is well understood that wireless demand for video content has become the dominant source of the wireless data traffic. An important property of video-type content is the high degree of asynchronous content reuse [2], which means that the same content can be viewed by different users at different time. In general, the time differences among the user requests are large enough that the multicast system is not an available option to handle multiple requests for the same content. Furthermore, user preferences regarding the requested contents are often spatially correlated and affected by the connections between mobile users on social networks. Therefore, if the network treats each user request independently, the same video files will be repeatedly transmitted to the users in proximity. Enabling caching capabilities at the network edge has emerged as a potential solution to exploit the content correlation of user requests. The main idea is to store proactively some popular contents at the wireless network edge nodes (e.g., small base stations, user devices) during off-peak hours. When the cached contents are requested by nearby mobile users, they can be delivered with lower latency and less energy consumption.

Wireless edge caching has been under extensive discussions for its promising advantages, such as improved spectral efficiency, energy efficiency, and reduced end-to-end latency [3–6]. The early works in this topic always assume deterministic caching strategy, that is, all the cache-equipped nodes will choose the files to cache in a deterministic way. The most commonly used strategy is to *cache the most popular content* (MPC) everywhere, which gives an optimal cache service performance in the case of isolated caches or when the caching helpers have nonoverlapping service range. In large wireless networks with randomly distributed caching helpers, one user can be within the coverage area of multiple caching helpers. In this case, stochastic caching schemes will clearly outperform the MPC strategy if the caching probabilities are properly optimized.

The concept of random/probabilistic caching was first proposed in [7], referred to as geographical caching. Considering a fixed number of base stations (BSs), each BS is assumed to have the same storage capabilities. The BSs decide to cache different files according to the same probabilistic distribution with respect to its storage constraint. The optimal caching probabilities have been obtained by maximizing the cache-hit probability, i.e., the probability that a randomly requested file can be found in the cache storage of the covering BSs. In [7], an optimization problem has been formulated by targeting at maximizing the *cache-hit probability*, which represents the probability that a randomly requested file of a user can be found in the caching helpers within the coverage range.

A probabilistic caching scheme has been further investigated in large random networks, where the distribution of the caching helpers is modeled by some stochastic point processes, e.g., PPP. In this case, the number of caching helpers within a certain distance to a user is a random variable that follows a Poisson distribution. The distribution of distances between a user to its kth nearest BS can be found in [8]. The optimal caching placement in Poisson networks has been extensively investigated in many scenarios [2, 9–13]. In [14], a hard-core placement (HCP) policy is considered, which captures the pairwise correlation between the caching nodes. When assuming that the file popularity may change over time, the optimal probabilistic caching and recaching policies has been investigated in [15].

Except the commonly considered cache-hit maximization, recently, many works started to consider the impact of PHY transmission on the performance of random caching schemes in large wireless networks. Using existing results on the coverage/outage probability in PPP networks, we can define a performance metric that is related to the successful transmission probability and formulate the caching probability optimization problem with respect to the storage capabilities of the caching helpers. In [16], the channel selection diversity gain has been considered as the performance metric to optimize. The density of successful transmission/reception or the network throughput is another popular metric to characterize the performance of cache-enabled networks [17–19]. Extending the results in single-tier random caching scheme, the analysis and optimization of probabilistic caching in N-tier heterogeneous networks is presented in [20]. In [21], a random caching with cooperative transmission design has been proposed and optimized to maximize the successful transmission probability.

In this chapter, we present the analysis and optimization of random/probabilistic caching in cache-enabled D2D networks. Compared to caching-helper-based networks where the requested file of a user can be found only in nearby caching helpers, in D2D-assisted caching networks, the requested file might be cached within the device itself. In this case, the user request is handled without any delay or energy cost.

5.2 Network Model

The network model we consider is a D2D-enabled wireless network with randomly distributed mobile user devices, whose distribution follows a homogeneous PPP Φ_{u} with intensity λ_{u}. Due to the randomness in user activity, we assume that each mobile user is actively requesting for some video content with probability $\rho \in [0, 1]$. In the meanwhile, the inactive devices can act as caching helpers and potential D2D transmitters to the active users. Because of independent thinning, the distributions of actively requesting devices and potential caching helpers also follow homogeneous PPPs $\Phi_{\mathrm{u}}^{\mathrm{r}}$ and $\Phi_{\mathrm{u}}^{\mathrm{t}}$ with intensity $\rho\lambda_{\mathrm{u}}$ and $(1 - \rho)\lambda_{\mathrm{u}}$, respectively. We assume that every device has limited caching capabilities, i.e., popular video files will be stored proactively inside the caches of the devices. Denote by M_{d} the cache memory size inside each device, and the video files have equal unit size.

We assume that the popular video files online form a finite-size content category $\mathcal{F} = \{f_1, \ldots, f_N\}$, with f_i being the ith most popular file and $N > M_{\mathrm{d}}$ being the size of the content category. The content popularity is assumed to follow a Zipf distribution, where similar assumptions can be found in many existing works in the literature [2, 11, 14, 22]. In this case, the request probability of file f_i is

$$p_i = \frac{1}{i^{\gamma} \sum\limits_{j=1}^{N} j^{-\gamma}}. \tag{5.1}$$

Here, γ is the shape parameter of the Zipf distribution, which represents the skewness of the popularity [23]. A large γ means that most of the requests are generated for very few most popular files. The request probability p_i represents the probability that a randomly generated request from a random user is for file f_i.

Our random caching policy is based on some predetermined probabilistic distribution, i.e., each user device independently caches file f_i with a certain probability q_i with respect to its storage capacity. Denote by $\mathbf{q} = [q_1, \ldots, q_N]$ the caching probabilities of the files $i \in [1, N]$, we have $\sum_{i=1}^{N} q_i \leq M_{\mathrm{d}}$ due to the storage capacity of each device.

When an active user requests a file in $\mathcal{F}$, a *cache-hit* event refers to the case that the requested file can be found within local caches, including **self-request** and **D2D cache hit**. The first case is when the requested file of a user can be found in its own cache, while the second refers to the case when the requested file is not cached in its own device, but in one of its nearby devices. We assume that there is a maximum searching distance R_{d} to establish D2D connection, due to the fact that a successful D2D transmission is usually within a short distance. If there is more than one neighbor device within the searching distance that has the requested file, the file is transmitted from the nearest one.

In case of a *cache miss*, which means that the requested file is not cached locally, the file will be retrieved from the core network to the nearest macro base station and then transmitted to the user. We consider D2D *overlaid* cellular networks, which means that orthogonal frequency bands are used for D2D and cellular communication, to avoid cross-tier interference.

5.3 Performance Metrics and Analysis

To determine the optimal caching probabilities, we need to first define the performance metrics to be optimized. The performance of a random caching policy in large wireless networks is very often characterized by the cache-hit probability, which is the probability that a randomly requested file can be found in local caches. However, finding the requested file locally does not guarantee that the request can be successfully handled. The transmission might fail in case of bad channel condition. Another important metric is the density of cache-served requests. It can be seen as a throughput-like metric since it combines the probability to find the requested file and the success probability of PHY transmission. In this chapter, we also consider the average content delivery delay as the third metric, which has been rarely investigated in the literature of random caching networks.

In this section, we give formal definition of the three performance metrics: the cache-hit probability, the cache-aided throughput, and the average content delivery delay. Note that in D2D caching networks, to increase the probability of self-request, user devices tend to cache the most popular files with higher probability. However, this will decrease the content diversity in local caches. Therefore, the optimal caching probabilities will strongly depend on the objective of the optimization problem.

5.3.1 Cache-Hit Probability

The cache-hit probability represents the probability that a random active user finds its requested file in local caches, either by *self-request* or by *D2D cache hit*. It is the most commonly used performance metric in the early studies of wireless caching schemes.

Denoting by p_{self} the self-request probability, it is the probability that the requested file of a random user can be found within its own cache. It is given by

$$p_{\mathrm{self}} = \sum_{i=1}^{N} \mathbb{P}[f_i \text{ requested}] \cdot \mathbb{P}[f_i \text{ cached}] = \sum_{i=1}^{N} p_i q_i. \tag{5.2}$$

If the user cannot find the requested file in its own device, it will search among the potential D2D helpers within distance $R_{\rm d}$. When file f_i is requested, as a result of independent thinning, the distribution of devices that have cached file f_i follows a homogeneous PPP Φ_i with intensity $q_i(1-\rho)\lambda_{\rm u}$. The number of points N from Φ_i within a circular area of radius $R_{\rm d}$ follows a Poisson distribution, with $\mathbb{P}[N=n] = \frac{[\pi(1-\rho)\lambda_{\rm u} q_i R_{\rm d}^2]^n}{n!} e^{-\pi(1-\rho)\lambda_{\rm u} q_i R_{\rm d}^2}$ [24, 25]. The void probability, which is the probability of finding the zero point from Φ_i within the searching distance $R_{\rm d}$, is $\mathbb{P}[N=0] = e^{-\pi(1-\rho)\lambda_{\rm u} q_i R_{\rm d}^2}$. Therefore, the probability for a random user device to find f_i cached in the potential D2D helpers within distance $R_{\rm d}$ is given by [2]:

$$p_{\text{hit},i}^{\rm d} = 1 - e^{-\pi(1-\rho)\lambda_{\rm u} q_i R_{\rm d}^2}. \tag{5.3}$$

Averaging over all the files in the content library $\mathcal{F}$, the D2D cache-hit probability is given by

$$\begin{aligned} p_{\text{hit}}^{\rm d} &= \sum_{i=1}^{N} \mathbb{P}[f_i \text{ requested}] \cdot \mathbb{P}[f_i \text{ not cached}] \cdot \mathbb{P}[f_i \text{ found in nearby devices}] \\ &= \sum_{i=1}^{N} p_i \,(1-q_i)\, p_{\text{hit},i}^{\rm d}. \end{aligned} \tag{5.4}$$

Plugging (5.3) into (5.4), we obtain

$$p_{\text{hit}}^{\rm d} = \sum_{i=1}^{N} p_i \,(1-q_i) \left(1 - e^{-\pi(1-\rho)\lambda_{\rm u} q_i R_{\rm d}^2}\right). \tag{5.5}$$

The cache-hit probability is then the sum of the two probabilities, i.e., $p_{\text{hit}} = p_{\text{self}} + p_{\text{hit}}^{\rm d}$, which gives

$$p_{\text{hit}} = 1 - \sum_{i=1}^{N} p_i \,(1-q_i)\, e^{-\pi(1-\rho)\lambda_{\rm u} q_i R_{\rm d}^2}. \tag{5.6}$$

Note that the cache-hit probability characterizes only the possibility of finding the requested file within a certain distance. Due to the signal attenuation and the perturbation of interference and noise, the transmission of the file might fail, and the success probability is affected by the channel statistics, the transmission distance, and some other network parameters.

5.3.2 Cache-Aided Throughput

The second performance metric is named cache-aided throughput or the density of cache-served requests, which measures the average density of successfully served user requests by locally cached content without connecting to the macro base station. This metric takes into account both the cache-hit probability and the successful content transmission probability.

Assume that the transmission of each file with equal size takes the same amount of time, one slot for instance. In the self-request case, the request is automatically served with probability one, while in the D2D cache hit case, the success probability of content delivery depends on the received signal-to-interference-plus-noise ratio (SINR). Let $p^{\mathrm{d}}_{\mathrm{suc},i}$ denote the success probability of D2D transmission for file f_i. We have the cache-aided throughput given by

$$\mathcal{T} = \rho\lambda_{\mathrm{u}} \left[\sum_{i=1}^{N} p_i q_i \cdot 1 + \sum_{i=1}^{N} p_i (1-q_i) p^{\mathrm{d}}_{\mathrm{hit},i} \cdot p^{\mathrm{d}}_{\mathrm{suc},i} \right], \tag{5.7}$$

where $\rho\lambda_{\mathrm{u}}$ is the density of active user requests in a given time slot. Assume that the SINR needs to exceed a certain threshold θ in order to have successful D2D transmission, the successful transmission probability of file f_i is given by

$$p^{\mathrm{d}}_{\mathrm{suc},i} = \mathbb{P}[\mathrm{SINR}_i > \theta]. \tag{5.8}$$

Without loss of generality, conditioning on a typical receiver k at the origin, its received SINR is given by

$$\mathrm{SINR}_i = \frac{P_{\mathrm{d}} |h_{k,k}|^2 d_i^{-\alpha}}{\sigma^2 + \sum_{j \in \Phi^{\mathrm{d}}_{\mathrm{t}} \setminus \{k\}} P_{\mathrm{d}} |h_{j,k}|^2 d_{j,k}^{-\alpha}}, \tag{5.9}$$

where $\Phi^{\mathrm{d}}_{\mathrm{t}}$ denotes the set of active D2D transmitters; $h_{j,k}$ denotes the small-scale channel fading from transmitter j to receiver k, which follows $\mathcal{CN}(0,1)$ (Rayleigh fading); $d_{j,k}^{-\alpha}$ denotes the power-law pathloss, where $d_{j,k}$ is the distance between transmitter j and receiver k, and α is the pathloss exponent; P_{d} and σ^2 denote the device transmission power and the background noise power, respectively.

Note that depending on which file f_i is requested, the typical D2D link distance d_i follows different distributions. Conditioning on $d_i \leq R_{\mathrm{d}}$, due to the maximum D2D searching distance, the probability density function (PDF) of d_i is given by

$$f_{d_i}(r) = \begin{cases} \frac{2\pi(1-\rho)\lambda_{\mathrm{u}} q_i r}{1-\exp\left[-\pi(1-\rho)\lambda_{\mathrm{u}} q_i R_{\mathrm{d}}^2\right]} e^{-\pi(1-\rho)\lambda_{\mathrm{u}} q_i r^2} & 0 \leq r \leq R_{\mathrm{d}} \\ 0 & r > R_{\mathrm{d}}. \end{cases} \tag{5.10}$$

The interference distribution in (5.9) depends on the density of concurrently active D2D transmissions links. A randomly generated file request from a user in $\Phi^{\mathrm{u}}_{\mathrm{r}}$ will be served through D2D transmission with probability $p^{\mathrm{d}}_{\mathrm{hit}}$, as given in (5.5). Thus the density of cache-assisted D2D transmissions is $\rho\lambda_{\mathrm{u}} p^{\mathrm{d}}_{\mathrm{hit}}$. We approximate the distribution of the active D2D transmitters $\Phi^{\mathrm{d}}_{\mathrm{t}}$ by a homogeneous PPP with intensity $\rho\lambda_{\mathrm{u}} p^{\mathrm{d}}_{\mathrm{hit}}$.

The D2D success probability can be obtained as

$$\begin{aligned} p^{\mathrm{d}}_{\mathrm{suc,i}} &= \mathbb{P}\left[\frac{P_{\mathrm{d}} |h_{k,k}|^2 d_i^{-\alpha}}{\sigma^2 + \sum_{k \in \Phi^{\mathrm{d}}_{\mathrm{t}} \setminus \{k\}} P_{\mathrm{d}} |h_{j,k}|^2 d_{j,k}^{-\alpha}} > \theta \right] \\ &= \mathbb{P}\left[|h_{k,k}|^2 > \frac{\theta d_i^{\alpha}}{P_{\mathrm{d}}} \left(\sigma^2 + \sum_{k \in \Phi^{\mathrm{d}}_{\mathrm{t}} \setminus \{k\}} P_{\mathrm{d}} |h_{j,k}|^2 d_{j,k}^{-\alpha} \right) \right] \end{aligned}$$

$$\overset{(a)}{=}\mathbb{E}_{d_i}\left[\mathcal{L}_{I_\mathrm{d}}\left(\theta d_i^{\alpha}\right)\cdot\exp\left(-\theta\sigma^2 d_i^{\alpha}/P_\mathrm{d}\right)\right]$$
$$\overset{(b)}{=}\mathbb{E}_{d_i}\left[\exp\left(-\frac{\pi\rho\lambda_\mathrm{u}p_\mathrm{hit}^\mathrm{d}d_i^2\theta^{\frac{2}{\alpha}}}{\mathrm{sinc}(2/\alpha)}\right)\exp\left(-\theta\sigma^2 d_i^{\alpha}/P_\mathrm{d}\right)\right]$$
$$=\int_0^{\infty} f_{d_i}(r)\exp\left(-\frac{\pi\rho\lambda_\mathrm{u}p_\mathrm{hit}^\mathrm{d}r^2\theta^{\frac{2}{\alpha}}}{\mathrm{sinc}(2/\alpha)}\right)e^{-\frac{\theta\sigma^2 r^{\alpha}}{P_\mathrm{d}}}\,\mathrm{d}r, \quad (5.11)$$

where $\mathcal{L}_{I_d}(s) = \mathbb{E}\left[\exp(-sI_d)\right]$ is the Laplace transform of D2D interference $I_d = \sum_{k\in\Phi_\mathrm{t}^\mathrm{d}\setminus\{k\}} |h_{j,k}|^2 d_{j,k}^{-\alpha}$. Here (a) follows from the complementary cumulative distribution function (CCDF) of $|h_{k,k}|^2$, which is exponentially distributed with unit mean value; (b) follows from the probability generating functional (PGFL) of PPP [26].

Substituting (5.3), (5.11), and (5.10) into (5.7), we obtain the cache-aided throughput averaged over all the files in the content library.

5.3.3 Average Content Delivery Delay

To characterize the delay that a user experiences between the time of requesting a file and the time of successfully receiving it, we first need to specify the retransmission policy. When a user finds its requested file cached inside its own device, the delay is zero. Otherwise, depending on whether a user can locate its requested file within its neighborhood, the following cases might happen:

- If a user cannot locate the requested content within its searching range:
  - If in the next time slot, it is still an active receiver waiting to be served, then the request goes to the nearest BS;
  - If in the next time slot, it acts as a potential transmitter, then wait.
- If a user finds the requested file within nearby D2D transmitters, but the transmission fails:
  - If in the next time slot, it is still an active receiver, it searches again within its neighborhood for potential D2D transmitters that have the requested content;
  - If in the next time slot, it acts as a potential transmitters, then wait.

Note that in the literature of wireless caching systems, the inverse of the transmission success probability has been commonly considered as the delay to receive a file. This way of calculating delay assumes that the transmitter keeps transmitting the file until it is successfully received. In the random network we consider in this work, the network topology is changing in each time slot. It is not guaranteed that the same potential transmitter is still within the searching range of the receiver in the next time slot. In this case, the re-searching and retransmission policy is critical for the delay performance and the delay cannot be simply considered the inverse of the transmission success probability.

Denote by $D_{\text{miss},i}$ and $D_{\text{fail},i}$ the delay when the file i is requested and it is not stored locally and the delay when the transmission of file i fails, respectively. We have the average delay for receiving file i as follows:

$$D_i = p^{\text{d}}_{\text{hit},i} \cdot p^{\text{d}}_{\text{suc},i} + (1 - p^{\text{d}}_{\text{hit},i})(1 + D_{\text{miss},i}) + p^{\text{d}}_{\text{hit},i}(1 - p^{\text{d}}_{\text{suc},i})(1 + D_{\text{fail},i}). \quad (5.12)$$

We define D_{BS} as the delay for a BS to retrieve the content from the core network and deliver it to the requesting user. In a dense network with overloaded user request, usually we have $D_{\text{BS}} \gg 1$. The delay when a cache miss happens is

$$D_{\text{miss},i} = \rho \cdot D_{\text{BS}} + (1 - \rho)(1 + D_{\text{miss},i}). \quad (5.13)$$

Then we obtain

$$D_{\text{miss},i} = \frac{1 + \rho(D_{\text{BS}} - 1)}{\rho}, \quad (5.14)$$

which is the same for all the files. The delay when the first transmission fails is given by

$$\begin{aligned} D_{\text{fail},i} &= \rho \cdot p^{\text{d}}_{\text{hit},i} p^{\text{d}}_{\text{suc},i} + (1 - \rho)(1 + D_{\text{fail},i}) + \rho \cdot p^{\text{d}}_{\text{hit},i}(1 - p^{\text{d}}_{\text{suc},i})(1 + D_{\text{fail},i}) \\ &\quad + \rho \cdot (1 - p^{\text{d}}_{\text{hit},i})(1 + D_{\text{miss},i}). \end{aligned} \quad (5.15)$$

After several steps of derivations, we obtain

$$D_{\text{fail},i} = \frac{1 + \rho(1 - p^{\text{d}}_{\text{hit},i})D_{\text{miss},i}}{\rho - \rho \cdot p^{\text{d}}_{\text{hit},i}(1 - p^{\text{d}}_{\text{suc},i})}. \quad (5.16)$$

Plugging (5.14) and (5.16) into (5.12), we obtain the delay to successfully receive file i. Averaging over all the files in the content library, the average delay to receive a requested file is

$$D = \sum_{i=1}^{N} p_i D_i. \quad (5.17)$$

5.4 Optimization of Probabilistic Caching Placement

In this section, we simplify the expressions of the three performance metrics defined in Section 5.3 and study three optimization problems to find the optimal caching probabilities $\mathbf{q} = [q_1, \ldots, q_N]$ for the considered network.

5.4.1 Cache-Hit Maximization

In the literature of probabilistic caching, the most commonly considered optimization problem is to maximize the cache-hit probability (or offloading probability) under the cache capability constraints. Based on (5.6), the optimization problem for maximizing the cache-hit probability can be defined as

$$\max_{\mathbf{q}} 1 - \sum_{i=1}^{N} p_i (1 - q_i) e^{-\pi(1-\rho)\lambda_u q_i R_d^2} \tag{5.18}$$
$$\text{s.t. } 0 \leq q_i \leq 1 \text{ for } i = 1, \ldots, N$$
$$\sum_{i=1}^{N} q_i \leq M_d.$$

The second order derivative of the objective function is strictly negative, thus p_{hit} is a concave function of q_i for all $i = 1, \ldots, N$, and (5.18) is a convex optimization problem.

Define μ as the nonnegative Lagrangian multiplier, we consider the following Lagrangian function

$$\mathcal{L}(\mathbf{q}, \mu) = -1 + \sum_{i=1}^{N} p_i (1 - q_i) e^{-\pi(1-\rho)\lambda_u q_i R_d^2} + \mu \left(\sum_{i=1}^{N} q_i - M_d \right). \tag{5.19}$$

We apply the Karush–Kuhn–Tucker (KKT) conditions to solve this optimization problem. From $\frac{\partial \mathcal{L}}{\partial q_i} = 0$, we have

$$- p_i e^{-\pi(1-\rho)\lambda_u q_i R_d^2} \left[1 + (1 - q_i)\pi(1 - \rho)\lambda_u R_d^2 \right] + \mu = 0, \tag{5.20}$$
$$\Longrightarrow \frac{\mu}{p_i} \exp\left[1 + \pi(1 - \rho)\lambda_u q_i R_d^2 \right] = (1 - q_i)\pi(1 - \rho)\lambda_u R_d^2 + 1. \tag{5.21}$$

Since this equation involves both polynomial and exponential function, its solution can be written in closed form with the help of the Lambert function. For a general type of equation that has the form $p^{ax+b} = cx + d$, where x is the variable and a, b, c, d, p are constant, when $p > 0$ and $a, c \neq 0$, the solution is

$$x = -\frac{\mathcal{W}\left(-\frac{a \ln p}{c} p^{b - \frac{ad}{c}} \right)}{a \ln p} - \frac{d}{c}, \tag{5.22}$$

where $\mathcal{W}$ denotes the Lambert W function [27]. Based on this, we obtain the solution to (5.21) as

$$q_i(\mu) = -\frac{\mathcal{W}\left\{ \frac{\mu}{p_i} \exp\left[1 + \pi(1 - \rho)\lambda_u R_d^2 \right] \right\}}{\pi(1 - \rho)\lambda_u R_d^2} + \frac{1}{\pi(1 - \rho)\lambda_u R_d^2} + 1. \tag{5.23}$$

Followed by the additional condition $0 \leq q_i \leq 1$, and defining $[x]^+ = \max\{x, 0\}$, we have

$$q_i^\star = \min\left\{ [q_i(\mu^\star)]^+, 1 \right\}. \tag{5.24}$$

Here, the optimal value of $\mu^\star$ can be obtained by the bisection search method subject to the condition $\sum_{i=1}^{N} q_i^\star = M_d$.

5.4.2 Cache-Aided Throughput Maximization

As we can see from (5.7) and (5.11), the expression of $\mathcal{T}$ involves integrals that are difficult to solve in optimization problems, even with numerical evaluation. To overcome this problem, we consider the following approximation

$$\mathbb{E}_{d_i}[\exp\left(-\eta d_i^{\delta}\right)] \approx \exp\left(-\eta \mathbb{E}[d_i^2]^{\delta/2}\right), \tag{5.25}$$

then the success probability $p_{\text{suc},i}^{\text{d}}$ in (5.11) can be approximated by

$$\hat{p}_{\text{suc},i}^{\text{d}} \approx \exp\left[-\frac{\pi\rho\lambda_{\text{u}}p_{\text{hit}}^{\text{d}}\mathbb{E}[d_i^2]\theta^{2/\alpha}}{\text{sinc}(2/\alpha)}\right]\exp\left[-\frac{\theta\sigma^2\mathbb{E}[d_i^2]^{\alpha/2}}{P_{\text{d}}}\right]. \tag{5.26}$$

Note that (5.26) does not guarantee to provide a tight approximation on the success probability, but it simplifies the expression such that it is easier to be solved by standard optimization solvers. From the PDF of d_i in (5.10), we can obtain $\mathbb{E}[d_i^2]$ as follows.

$$\begin{aligned}\mathbb{E}[d_i^2] &= \int_0^{R_{\text{d}}} r^2 \frac{2\pi(1-\rho)\lambda_{\text{u}}q_i r}{1-e^{-\pi(1-\rho)\lambda_{\text{u}}q_i R_{\text{d}}^2}} e^{-\pi(1-\rho)\lambda_{\text{u}}q_i r^2}\text{d}r \\ &= \frac{1}{\pi(1-\rho)\lambda_{\text{u}}q_i} - \frac{R_{\text{d}}^2}{e^{\pi(1-\rho)\lambda_{\text{u}}q_i R_{\text{d}}^2}-1}.\end{aligned} \tag{5.27}$$

When $q_i \to 0$, we obtain $\lim_{q_i\to 0}\mathbb{E}[d_i^2] = R_{\text{d}}^2/2$ by applying L'Hôpital's rule.

Then we have an approximated expression for the cache-aided throughput, given as

$$\hat{\mathcal{T}} = \rho\lambda_{\text{u}}\left[\sum_{i=1}^{N} p_i q_i + \sum_{i=1}^{N} p_i(1-q_i)p_{\text{hit},i}^{\text{d}} \cdot \hat{p}_{\text{suc},i}^{\text{d}}\right], \tag{5.28}$$

where $\hat{p}_{\text{suc},i}^{\text{d}}$ is given in (5.26). Our objective is to find $\mathbf{q}^{\star} = \max_{\mathbf{q}} \hat{\mathcal{T}}$, subject to $0 \le q_i \le 1$ and $\sum_{i=1}^{N} q_i \le M_{\text{d}}$.

Since this optimization problem is non-convex, it is difficult to provide any analytical solution to this problem. In the numerical evaluation section, we solve this problem numerically with simulated annealing.

5.4.3 Delay Minimization

Due to the complicated form of the delay as a function of the caching probability vector $\mathbf{q}$, we resort to numerical methods of finding the optimal solution by using simulated annealing. Note that in some cases, this method might not be able to find the global optimal point.

5.5 Numerical and Simulation Results

In this section, we evaluate the optimal caching probabilities obtained from the previously defined three optimization problems. The user density we consider falls in the range $\lambda_u \in [10^{-4}, 10^{-3}]/\text{m}^2$. At every time slot, $\rho = 50\%$ of the users will generate a random request for a file in $\mathcal{F}$, according to the request probability vector $\mathbf{p} = [p_1, \ldots, p_N]$, which follows the Zipf distribution with skewness parameter $\gamma \in \{0.5, 1.2\}$. The remaining $1 - \rho = 50\%$ of users will be potential D2D helpers to serve the user requests locally. The device cache capacity is $M_d = 2$ files. The content catalog size is $N = 20$ files.[1] The maximum D2D searching distance is $R_d = 75$ m. The D2D transmission power and the background noise power are $P_d = 0.1$ mW and $\sigma^2 = -110$ dB, respectively. The target SINR for successful D2D transmissions is $\theta = 0$ dB. In case of a cache miss event, the delay to receive a file from the BS is $D_{BS} = 5$ timeslots.

In Figures 5.1, 5.2, 5.3, and 5.4, we compare the caching probability vector $\mathbf{q}^{\star}$ obtained in the three cases: cache-hit-optimal (Section 5.4.1), throughput-optimal (Section 5.4.2) and delay-optimal (Section 5.3.3). The results are presented for both

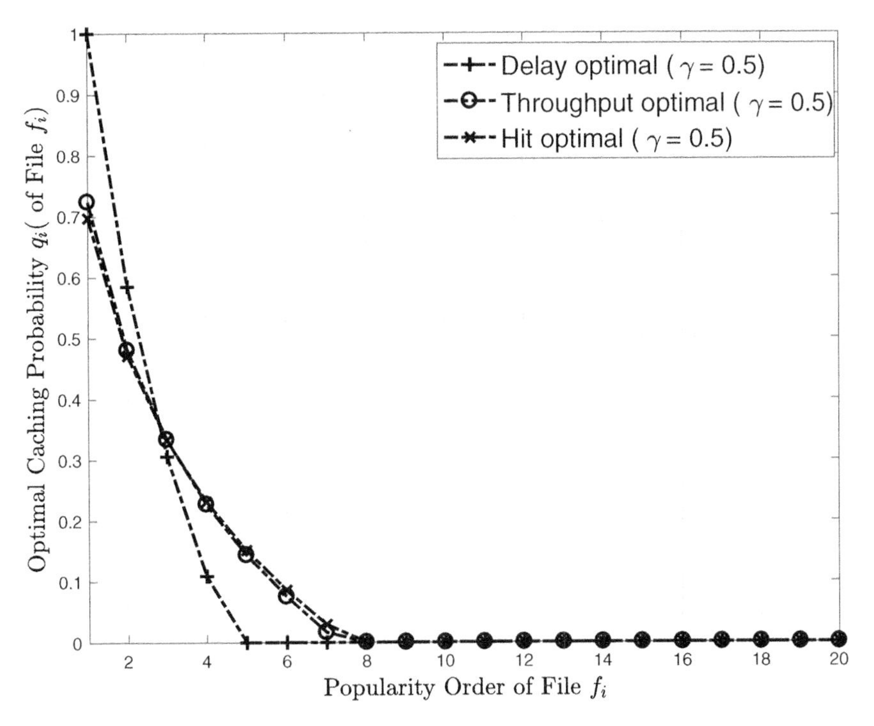

Figure 5.1 Optimal caching probabilities in sparse network, $\lambda_u = 10^{-4}/\text{m}^2$. $\gamma = 0.5$.

[1] In reality, the size of the content catalog is usually very large. Here we consider $N = 20$ files, mainly to avoid high complexity to solve the optimization problems. Similar choices of small content catalog size can also be found in [2, 16].

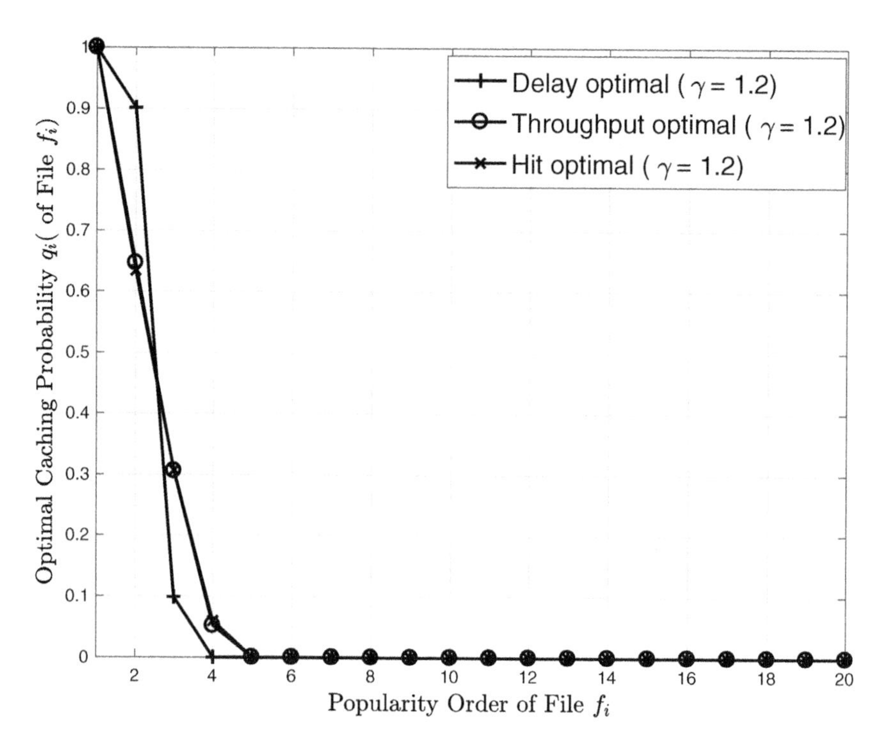

Figure 5.2 Optimal caching probabilities in sparse network, $\lambda_u = 10^{-4}/m^2$. $\gamma = 1.2$.

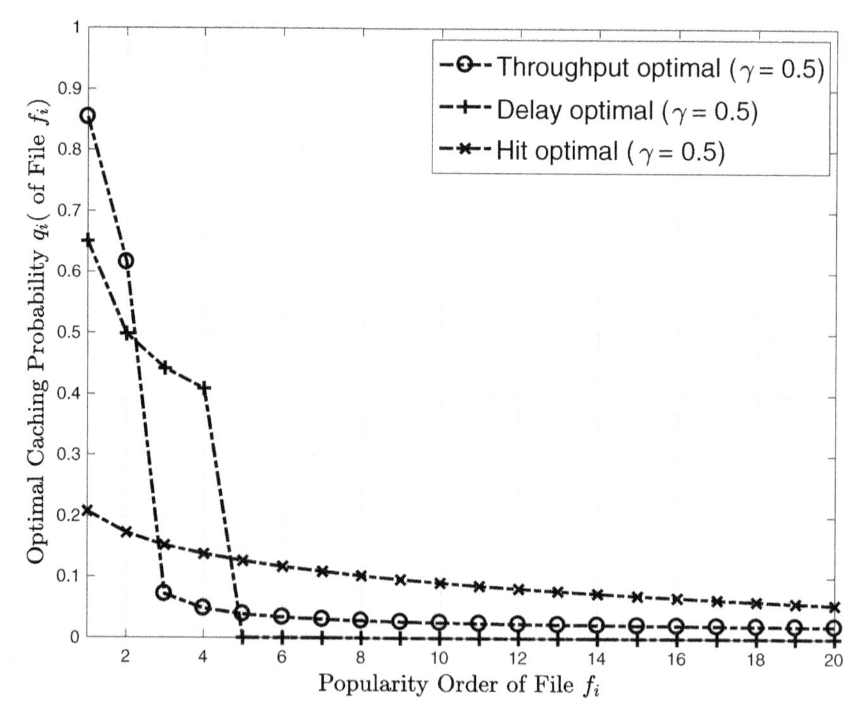

Figure 5.3 Optimal caching probabilities in dense network, $\lambda_u = 10^{-3}/m^2$. $\gamma = 0.5$.

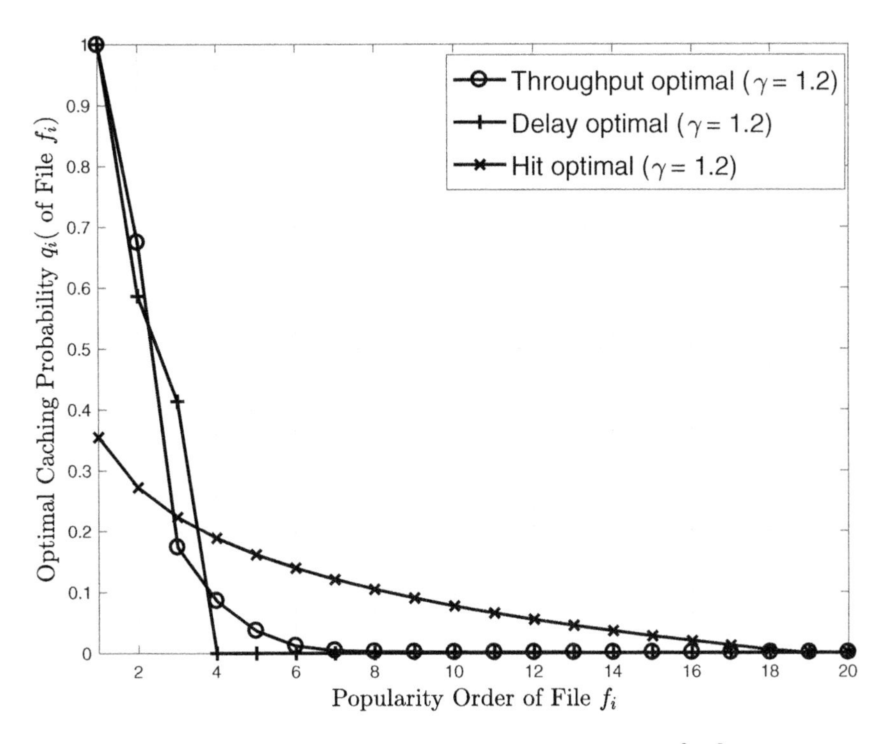

Figure 5.4 Optimal caching probabilities in dense network, $\lambda_u = 10^{-3}/m^2$. $\gamma = 1.2$.

sparse and dense networks. The optimal caching probabilities of file f_i for $i = 1, \ldots, N$ are presented as a function of the file popularity order i.

Our first remark is that the throughput-optimal and cache-hit-optimal caching probabilities are very close in sparse D2D network, while with dense users devices, the optimal caching probabilities in these two cases are quite different. To be more specific, with the throughput-optimal strategy, each device tends to cache the most popular files with higher probability. For instance, in Figure 5.3, the caching probabilities of the two most popular files $q_1^\star$ and $q_2^\star$ in the throughput-optimal case are much higher than in the cache-hit-optimal case. The delay-optimal caching probabilities are even more concentrated on the most popular files than the two other cases. The explanation behind this observation is that *in dense D2D networks, due to the high interference level, a D2D connection has low success probability. Therefore, users' caching strategy tends to be more "selfish" such that self-request is more preferable than cache-assisted D2D transmission.*

In Figures 5.5 and 5.6, we present the theoretical and simulated cache-aided throughputs obtained with the throughput-optimal caching probabilities. First, we notice that the theoretical and simulated throughput results are very close, which shows that the approximation we used in (5.26) gives negligible error on the cache-aided throughput optimization problem. To compare the throughput performance of different caching strategies, we also present the simulated cache-aided throughput when applying the cache-hit-optimal caching probabilities. A baseline scheme, "cache the most popular content" (MPC), is also used for comparison.

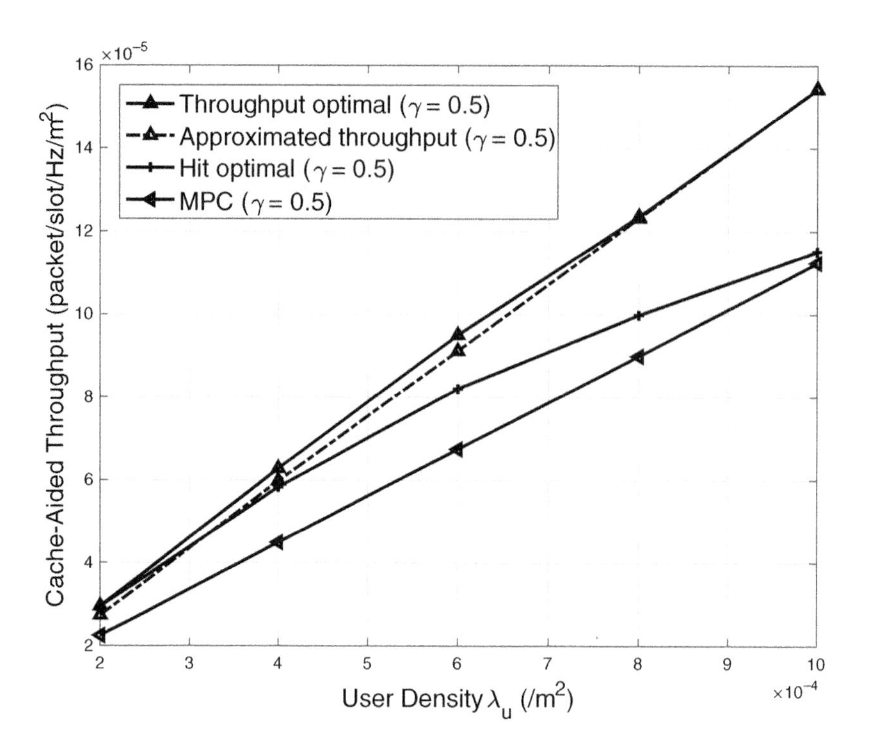

Figure 5.5 Simulated cache-aided throughput vs. user device density λ_{u}. $\gamma = 0.5$.

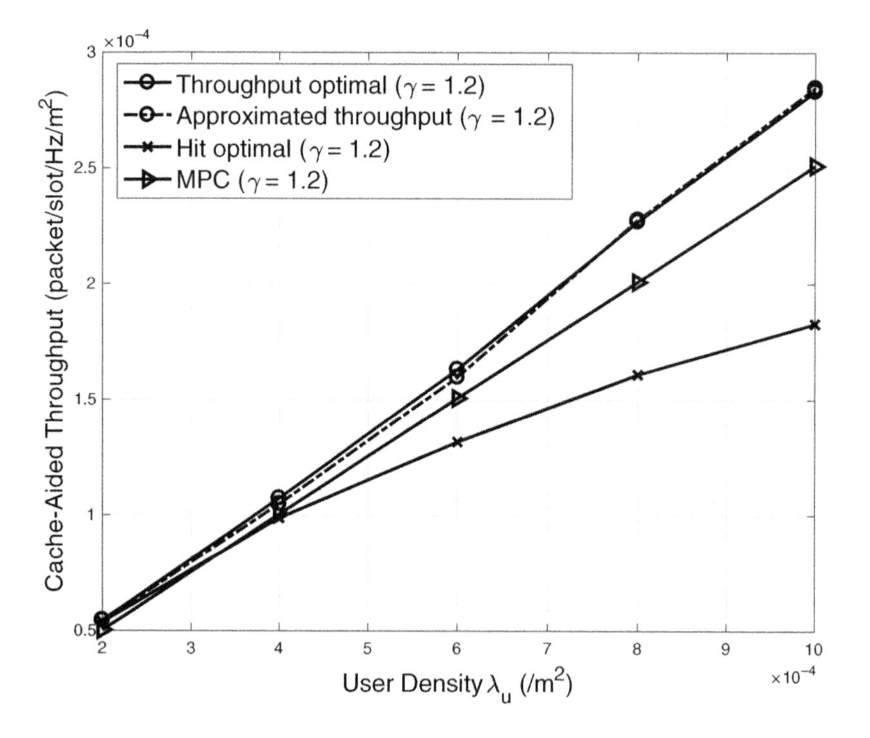

Figure 5.6 Simulated cache-aided throughput vs. user device density λ_{u}. $\gamma = 1.2$.

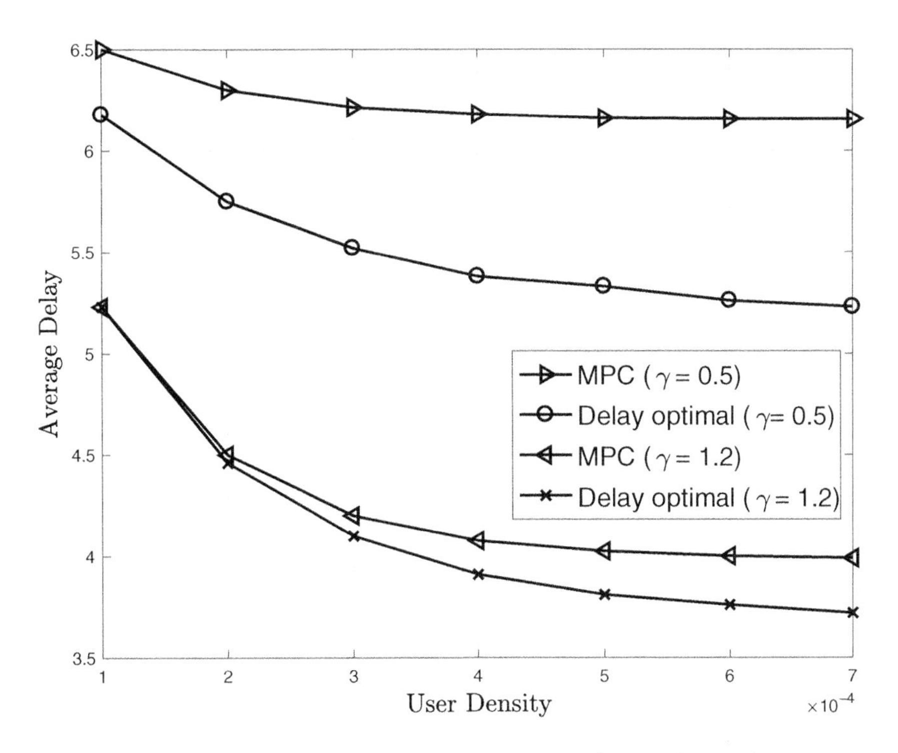

Figure 5.7 Minimized average delay vs. user device density λ_u. $\gamma = \{0.5, 1.2\}$.

Since the throughput-optimal strategy takes into account the D2D success probability, the achieved cache-aided throughput is significantly improved compared to the cache-hit-optimal and the MPC strategies. The improvement is even more notable in dense D2D network. In the case with dense users and highly concentrated content popularity (i.e, $\gamma = 1.2$), the cache-aided throughput obtained with MPC gives better performance than the cache-hit-optimal case, meaning that it is more beneficial to increase the chance of "self-request" than increasing the total cache-hit ratio. These results validate the necessity of considering the D2D transmission reliability while searching for the optimal content placement.

In Figure 5.7 we plot the numerical values of the optimized average delay to receive a requested file as a function of the user device density. As expected, with larger γ, the probability to have cache hit is higher, thus the average delay to receive a packet is smaller. We also notice that with smaller γ, the gain of using delay-optimal caching probabilities compared to the baseline MPC case is more notable.

5.6 Conclusions

In this chapter, we investigated probabilistic caching placement in stochastic wireless D2D networks where user device distribution follows a homogeneous PPP. The optimal caching probabilities were presented for three different optimization objectives: maximizing cache-hit probability, maximizing the density of successfully served user requests, and minimizing the average delay to receive a requested file. We showed that considering only cache-hit probability as the main performance metric leads to overconfident evaluation of the percentage of content requests that can be served locally. The success probability of content delivery is another critical factor that affects much the performance of wireless D2D caching. An optimal caching strategy should take into account both factors to improve the throughput and delay performance in cache-enabled wireless networks.

References

[1] "Cisco visual networking index: Global mobile data traffic forecast update, 2015–2020 white paper," Feb. 2015, www.cisco.com/c/dam/m/en_in/innovation/enterprise/assets/mobile-white-paper-c11-520862.pdf.

[2] J. Rao, H. Feng, C. Yang, Z. Chen, and B. Xia, "Optimal caching placement for D2D assisted wireless caching networks," in *IEEE International Conference on Communications (ICC)*, Kuala Lumpur, Malaysia, May 2016, pp. 1–6.

[3] E. Bastug, M. Bennis, and M. Debbah, "Living on the edge: the role of proactive caching in 5G wireless networks," *IEEE Communications Magazine*, vol. 52, no. 8, pp. 82–89, Aug. 2014.

[4] K. Shanmugam, N. Golrezaei, A. G. Dimakis, A. F. Molisch, and G. Caire, "Femtocaching: Wireless content delivery through distributed caching helpers," *IEEE Transactions on Information Theory*, vol. 59, no. 12, pp. 8402–8413, Dec. 2013.

[5] M. Ji, G. Caire, and A. F. Molisch, "Fundamental limits of caching in wireless D2D networks," *IEEE Transactions on Information Theory*, vol. 62, no. 2, pp. 849–869, Feb. 2016.

[6] N. Golrezaei, A. Dimakis, and A. Molisch, "Scaling behavior for device-to-device communication with distributed caching," vol. 60, no. 7, pp. 4286–4298, July 2014.

[7] B. Blaszczyszyn and A. Giovanidis, "Optimal geographic caching in cellular networks," in *IEEE International Conference on Communications (ICC)*, IEEE, 2015, pp. 3358–3363.

[8] D. Moltchanov, "Distance distributions in random networks," *Ad Hoc Networks*, vol. 10, pp. 1146–1166, 2012.

[9] S. H. Chae, J. Y. Ryu, T. Q. S. Quek, and W. Choi, "Cooperative transmission via caching helpers," in *IEEE Global Conference on Communications (Globecom)*, San Diego, CA, Dec. 2015, pp. 1–6.

[10] J. Song, H. Song, and W. Choi, "Optimal caching placement of caching system with helpers," in *IEEE International Conference on Communications (ICC)*, London, June 2015, pp. 1825–1830.

[11] Y. Chen, M. Ding, J. Li, Z. Lin, G. Mao, and L. Hanzo, "Probabilistic small-cell caching: Performance analysis and optimization," *IEEE Transactions on Vehicular Technology*, vol. 66, no. 5, pp. 4341–4354, May 2017.

[12] Z. Chen and M. Kountouris, "D2D caching vs. small cell caching: where to cache content in a wireless network?" in *IEEE 17th International Workshop on Signal Processing Advances in Wireless Communications (SPAWC)*, July 2016, pp. 1–6.

[13] B. Chen, C. Yang, and Z. Xiong, "Optimal caching and scheduling for cache-enabled D2D communications," *IEEE Communications Letters*, vol. 21, no. 5, pp. 1155–1158, May 2017.
[14] D. Malak, M. Al-Shalash, and J. G. Andrews, "Spatially correlated content caching for device-to-device communications," *IEEE Transactions on Wireless Communications*, vol. 17, no. 1, pp. 56–70, Jan. 2018.
[15] W. Bao, D. Yuan, K. Shi, W. Ju, and A. Y. Zomaya, "Ins and outs: optimal caching and re-caching policies in mobile networks," in *18th ACM International Symposium on Mobile Ad Hoc Networking and Computing (MobiHoc)*, ACM, 2018, pp. 41–50.
[16] S. H. Chae and W. Choi, "Caching placement in stochastic wireless caching helper networks: channel selection diversity via caching," *IEEE Transactions on Wireless Communications*, vol. 15, no. 10, pp. 6626–6637, Oct. 2016.
[17] D. Liu and C. Yang, "Caching policy toward maximal success probability and area spectral efficiency of cache-enabled hetnets," *IEEE Transactions on Communications*, vol. 65, no. 6, pp. 2699–2714, June 2017.
[18] D. Malak, M. Al-Shalash, and J. G. Andrews, "Optimizing content caching to maximize the density of successful receptions in device-to-device networking," *IEEE Transactions on Communications*, vol. 64, no. 10, pp. 4365–4380, Oct. 2016.
[19] Z. Chen, N. Pappas, and M. Kountouris, "Probabilistic caching in wireless D2D networks: Cache hit optimal versus throughput optimal," *IEEE Communications Letters*, vol. 21, no. 3, pp. 584–587, Mar. 2017.
[20] K. Li, C. Yang, Z. Chen, and M. Tao, "Optimization and analysis of probabilistic caching in n -tier heterogeneous networks," *IEEE Transactions on Wireless Communications*, vol. 17, no. 2, pp. 1283–1297, Feb. 2018.
[21] W. Wen, Y. Cui, F. Zheng, S. Jin, and Y. Jiang, "Random caching based cooperative transmission in heterogeneous wireless networks," *IEEE Transactions on Communications*, vol. 66, no. 7, pp. 2809–2825, July 2018.
[22] H. J. Kang and C. G. Kang, "Mobile device-to-device (D2D) content delivery networking: a design and optimization framework," *Journal of Communications and Networks*, vol. 16, no. 5, pp. 568–577, Oct. 2014.
[23] L. Breslau, P. Cao, L. Fan, G. Phillips, and S. Shenker, "Web caching and Zipf-like distributions: evidence and implications," in *Proceedings of the IEEE INFOCOM '99*, vol. 1, Mar. 1999, pp. 126–134, vol. 1.
[24] A. Baddeley, "Spatial point processes and their applications," *in Stochastic Geometry, Lecture Notes in Mathematics*, vol. 1892, Berlin, Heidelberg: Springer, 2007.
[25] F. Baccelli and B. Blaszczyszyn, "Stochastic geometry and wireless networks: Volume I theory," *Foundations and Trends in Networking*, vol. 3, no. 3–4, pp. 249–449, Mar. 2009.
[26] M. Haenggi and R. K. Ganti, "Interference in large wireless networks," *Foundations and Trends in Networking*, vol. 3, no. 2, pp. 127–248, Feb. 2009.
[27] R. M. Corless, G. H. Gonnet, D. E. G. Hare, D. J. Jeffrey, and D. E. Knuth, "On the Lambert W function," *Advances in Computational Mathematics*, vol. 5, no. 1, pp. 329–359, 1996.

6 Joint Policies for Caching, Routing, and Channel Selection in Next-Generation Wireless Edge Systems

Jacob Chakareski

Selecting joint policies for caching, routing, and channel selection in coordinated multi-cellular systems for edge-based video delivery is challenging, due to the involved combinatorial complexity. Therefore, related work to date has studied caching and routing in such cellular networks independently from the problem of channel selection and interference induced when two nearby wireless links transmit on the same frequency/channel. We overcome this challenge for the first time by leveraging two novel concepts in this context: column generation, a mathematical framework for large-scale optimization that enables adapting the number of variables considered in an optimization problem, and conflict graph, a mathematical framework for effective modeling of interference in wireless networks. Integrating these two, we then pursue a novel problem formulation that makes up a master–slave problem structure for selecting optimal joint policies for caching, routing, and channel selection in this setting. The master problem makes up a subset of the original problem variables and adaptively decides whether to introduce an additional variable via a slave subproblem that characterizes the reduction in the value of the objective function with the addition of a new variable. The search for the optimal joint policies concludes when no new variable can be introduced to further lower the value of the objective function. Since leveraging a conflict graph still induces an exponential complexity in the number of wireless links that can be prospectively activated simultaneously, we also investigate an effective method to quickly approximate the optimal solution within an ϵ deviation, at lower computational complexity. We comprehensively evaluate the performance characteristics of our analytical advances via realistic simulation experiments, demonstrating a close to 50% improvement in expected video streaming data rate relative to a competitive state-of-the-art method. Such performance benefits will enable up to a 5 dB improvement in the video streaming quality of experience delivered to the mobile user.

Supported in part by NSF Awards CCF-1528030, ECCS-1711592, CNS-1836909, and CNS-1821875, and research gifts from Adobe Systems and Tencent Research.

6.1 Background

We ever more often use our mobile devices and cellular networks to access the internet and related services, such as social networking and video streaming. These outcomes have led to online video becoming the overwhelmingly dominant source of mobile and internet network traffic [1], thus requiring new methods for effective allocation of system resources. Storing the most popular video content in edge-based servers collocated with small base stations, in emerging cellular systems made up of a collection of small cells embedded within a bigger macro-cell, has been investigated recently as one such viable strategy. This will enable local delivery of this content to mobile clients in such systems, thereby avoiding the costlier alternative of streaming content from a back-end internet server. A fundamental problem in this context then is the assignment of the most popular video files to the storage units of each edge server, known also as caches. The problem features constraints and unique aspects that need to be taken into consideration, such as the limited storage of an edge server and the content popularity that may vary across the small cells serving the mobile users.

6.2 Related Work and Our Advances

The highlighted problem has been investigated using a variety of approaches. Related work has studied this setting using different approaches. Concretely, the study in [2] facilitates information theory to explore the minimization of the induced cost when streaming video in caching-enabled small-cell systems. The study in [3] instead facilitates game theory and online learning to characterize the statistical risk associated with learning the popularity of different video files requested by mobile clients served by a small base station. As elaborated earlier, several shortcomings of existing work necessitate the present investigation. In particular, cooperative operation of small-cell base stations in video streaming with local caching has not been studied. We have recently shown that such cooperation can help avoid the cost of video delivery from a distant back-end server, by streaming the requested content from an adjacent small cell via backhaul links that interconnect the small cells [4]. Moreover, to date, interference has not been integrated into the analysis, due to the challenges it introduces, as explained earlier. That is, the state-of-the-art considers that transmitting links in the system do not interfere with each other and that the related channel selection has been solved externally. Finally, the impact of a small-cell base station not having the requested video content stored locally is handled uniformly across the macro cell, by existing work. However, integrating the location of the specific small cell relative to the macro base station, when addressing such instances, can enhance performance.

We overcome these shortcomings and challenges by exploring joint policies for caching, routing, and channel selection in next-generation cellular systems. Our approach advances existing methods notably, as it helps avoid unlikely and performance-penalizing assumptions about the system, it introduces cooperation among small cells and enhances the system's performance. We demonstrate through experimental evaluation that these advances enable much higher expected video streaming throughput and thus much higher quality of experience for the mobile user. We proceed by closely reviewing existing work.

Caching at the cell level has been studied considerably. The study in [5] formulates a basic cost model that facilitates determining the benefits of caching at a base station, for network operators. The study in [6] examines base station video caching for improving the quality of experience (QoE) of users, among other aspects. The study in [4] investigates collaborative caching among base stations of different macro cells to minimize the overall file delivery cost. The cellular system under investigation comprises backhaul connections that interlink the base stations. However, installing and maintaining such connections are costly, and if the connections are wireless, facilitating them reduces the serving capacity of a macro cell. Thus an addition transmission efficiency is induced.

The work in [3] investigates the caching policy selection at a single base station, aiming to minimize the file retrieval latency by learning the online popularity of the cell's contents. The work in [7] investigates the same problem by using different system performance measures, such as the data delivery rate to a user or the outage probability, while integrating aspects such as the base station's location and the request rate for the content. The study in [8] integrates network coding to augment the caching performance of small cells. Finally, the study in [9] utilizes information theory to investigate the minimization of the cost of streaming video in small cells enabled with caching capabilities. We note that these investigations explore only the caching policy selection.

In the broader area of caching, novel integration of caching with packet scheduling at the application layer has been explored in regard to rate-distortion optimized network edge proxy-driven video streaming, demonstrating advances over conventional sender-driven or receiver-driven methods [10–12]. Similarly, novel integration of caching with edge computing and viewport-driven 360° streaming has been explored recently for next-generation mobile virtual reality applications delivered over 5G cooperative cellular systems [13].

The present investigation is most closely related to [2, 14, 15]. In particular, the study in [15] leverages a conflict graph to explore the problem of routing and channel selection in wireless multi-hop networks. The objective under consideration is to maximize the achieved throughput. No caching capabilities have been integrated into the analysis. Small-cell caching and transmission link scheduling were investigated independently in [2]. The objective of interest is maximizing the user serving rate. The study in [14] integrates the network bandwidth assigned to a small cell in the analysis of the caching policy selection. However, it disregards the aspect of interference that arises during simultaneous transmission of adjacent links on the same channel.

The rest of this chapter is organized as follows. We proceed by describing the formulation of the architecture of our system and the related models that will be needed. In Section 6.4, we then characterize the problem of interest. We leverage column generation and a conflict graph to formulate in Section 6.5 an algorithmic strategy that effectively computes the optimal solution within a small factor of ϵ approximation. We explore the performance benefits induced by our analytical advances in Section 6.6. We present our concluding remarks in Section 6.8.

6.3 System Modeling

6.3.1 Network Setting Characterization

We formally characterize the setting we investigate as illustrated in Figure 6.1. As considered in earlier studies as well, there is a macro cell featuring a base station (MBS) and a set of N small base stations (SBSs), spatially distributed around it. Each SBS can serve a small subset of mobile users in its vicinity, as indicated by the dashed circles in Figure 6.1. Let $\mathcal{N} = \{1, 2, \ldots, n, \ldots, N, MBS\}$ denote the set of base stations embedded in the macro cell. Similarly, we define $\mathcal{J} = \{1, 2, \ldots, j, \ldots, J\}$ to be the set of J video files that the mobile users can request. We denote $\mathcal{K} = \{1, 2, \ldots, k, \ldots, K\}$ to be the set of K mobile users in the macro cell. We denote with a_n and a_k the number of transmit/receive antennas for SBS n and user k, respectively.

We assume that each SBS n comprises a storage capacity C_n^{SBS}. We also assume that the MBS can store all J files. Hence a user requesting file j can always be served by the MBS, as the last option. We consider that user k generates α_{kj} requests for video file j per unit of time. Let S_j denote the data volume of file j. There are $C + 1$ transmission channels $\mathcal{C} = \{0, 1, 2, \ldots, c, \ldots C\}$, with $c > 0$ indicating a secondary channel, and $c = 0$ indicating a primary channel. Let W_c denote the transmission bandwidth of channel c. We assume that the mobile users and the base stations can use the secondary channels, and that only the MBS can use the primary channel for communication. In our setting, we assume a one-hop transmission, i.e., a base station can deliver a file

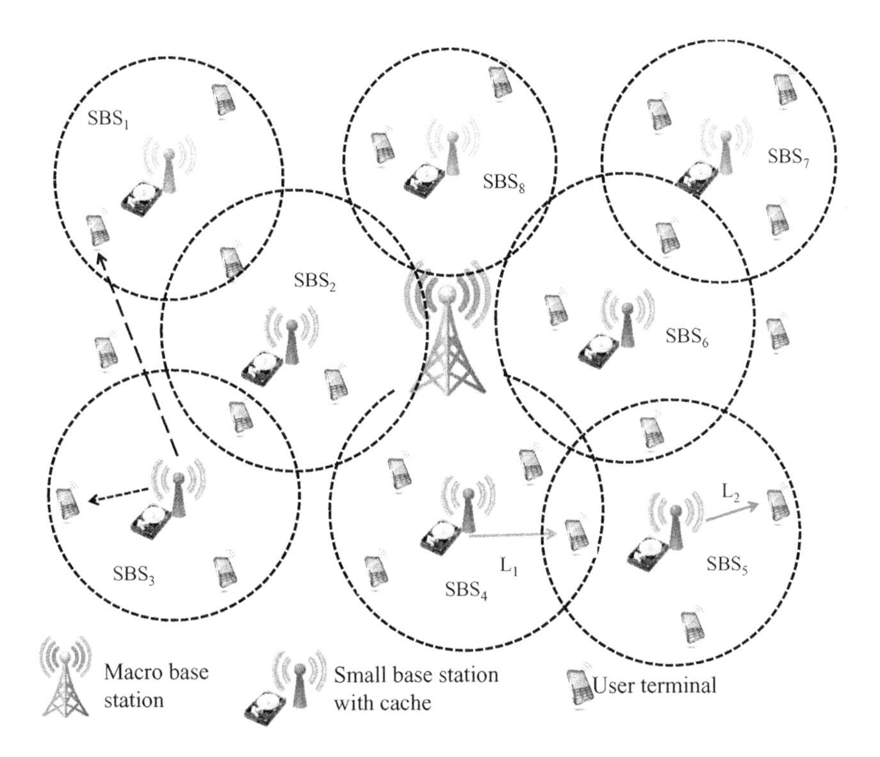

Figure 6.1 The system model under investigation.

requested by a user, only if the latter is within its transmission range. In the following, we use *transmitter* to refer to a base station and *receiver* to refer to a user, given that we investigate only downlink transmission in the considered setting.

6.3.2 Network Coding

It has been demonstrated that using network coding leads to higher network throughput [16] by enabling efficient utilization of resources and exact optimization methods that can be implemented at lower complexity (typically in polynomial time). To leverage these benefits in our setting, we assume that a video file is streamed via its network coded packets, each constructed as $\sum_{m=1}^{M} \kappa_m q_m$. Here, q_i denotes one of the original data packets making up this file, and k_m denotes the random coefficients necessary in the construction. We note that the summation is carried out over a finite Galois field [17]. Selecting the latter to be sufficiently large will enable the constructed network coded packets to be linearly independent. Then a user can easily recover the original video file data by having delivered any M different network coded packets.

6.3.3 Transmission and Interference Ranges and Capacity of a Link

For a given communication channel c, we set the transmission range for a base station n to be constant, by setting its transmit power P_n^c to be constants. This will simultaneously set the interference range for the base station to a constant value. In our analysis, we adopt a model introduced in [18] that considers a transmission over a channel c to be successful when the received power at the destination user associated with this transmission exceeds a threshold P_T^c. Similarly, the model considers that the thereby induced interference to another nearby user receiving on the same channel is nonnegligible when the received power associated with this transmission exceeds an interference threshold P_I^c. Following this model, we can formulate the transmission range TR_n^c and interference ranges IR_n^c of the nth SBS as:

$$TR_n^c = (gP_n^c/P_T^c)^{1/\gamma},$$
$$IR_n^c = (gP_n^c/P_I^c)^{1/\gamma}.$$

Here, we denote with γ the pathloss. Moreover, the constant g relates to the electromagnetic wave wavelength, the transmitter and receiver antenna profiles, and other factors used in the communication.

Now, we denote with $\hat{c}_{nk}^c$ the transmission capacity of a link between user k and SBS n, established over channel c. Leveraging Shannon–Hartley's theorem, we can formulate this quantity as:

$$\hat{c}_{nk}^c = W_c \log_2(1 + \frac{G_{nk}P_n^c}{\eta}).$$

We denote with η the receiver's white noise. Moreover, we denote with $G_{nk} = g \cdot (d_{nk})^{-\gamma}$ the propagation gain of transmission power over Euclidean distance of d_{nk} between SBS n and user k. Due to the model we leveraged [18], we formulate $\hat{c}_{nk}^c$ via the related signal-to-noise ratio (SNR). In particular, a small base station is precluded from transmitting over a channel to a given user, if the latter is within the interference range of a different base station communicating over the same channel. Thus the signal-to-interference-plus-noise ratio (SINR) is irrelevant in this case.

6.3.4 Capturing Interference via a Conflict Graph and Its Independent Sets

As introduced earlier, a conflict graph will enable us to effectively formulate link interference in our system. In particular, let $G(V, E)$ denote a graph, whose set of vertices V makes up the triples $((n,k),c)$, for $n \in \mathcal{N}$, $k \in \mathcal{K}$, and $c \in \mathcal{C}$. Such triplets refer to the prospective links that can be established in our setting between SBS n and user k, over a channel c. Two vertices in G are connected, if the two respective triplets share a common channel c and the receiver k in one triplet is within interference range of the transmitter n in other triplet. In other words, the two links represented by the two respective triplets cannot transmit in parallel.

Now, we define as I a set of communication triplets that do not feature interconnecting edges in G, i.e., they represent transmitting links that can be activated in parallel, as they do not cause interference among them. Such a set is also denoted as a set of independent communication links or, briefly, an independent set. A desirable objective is to find the largest independent set, as it can maximize the transmission efficiency of the system. Such a set is also denoted as maximal. In our analytical modeling, we can formulate any prospective I as an ordered (indexed) vector of uniform length, featuring binary entries for every communication triplet $((n,k),c)$. An entry of 1 would indicate that the respective triplet/link is a member of I, and an entry of zero would indicate the opposite.

For concreteness, a small example of conflict graph and independent set construction is illustrated in Figure 6.2. The respective communication setting comprises three users, two small base stations, and two transmission channels, as shown in the figure. Only user 2 can receive communication from SBS 1 over both channels. The conflict graph captures for example that the communication triplets $((1,1),1)$ and $((2,2),1)$ cannot be activated in parallel, as they interfere with each other, indicated by the edge

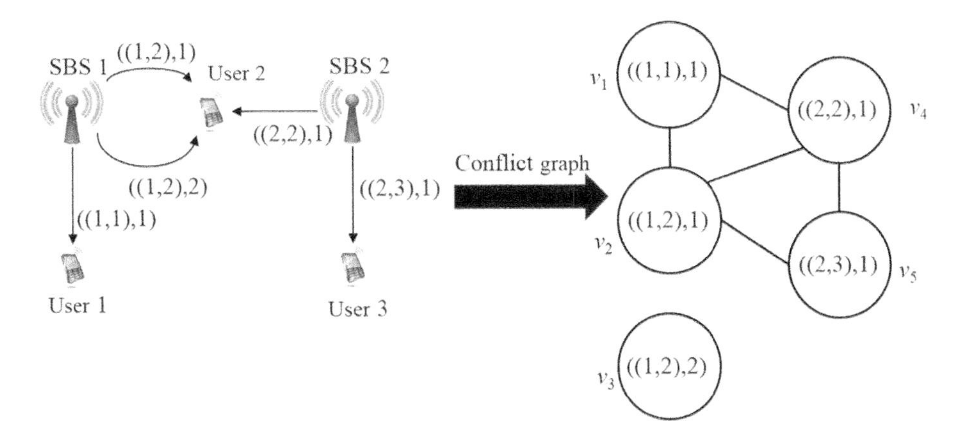

Figure 6.2 An example of a conflict graph and independent set construction.

between the respective nodes in the graph. Finally, leveraging the conflict graph in Figure 6.2, one can pursue the construction of the associated independent sets. The set $I_1 = (1,0,0,0,1)$ represents one such example. Moreover, we point out that the set $I_2 = (1,0,1,0,1)$ is maximal.

6.4 Formulation of Joint Caching, Routing, and Channel Selection Policy Problem

We begin by introducing the following required variables:

- X_{nj}: proportion of network coded packets of video file j cached by SBS n.
- Y_{nj}^k: proportion of network coded packets of video file j streamed to user k from SBS n.
- Z_{nk}^c: aggregate volume of data streamed to user k by small base station n over channel c.
- f_i: fraction of time during which the communication triplets comprising independent set I_i are selected for simultaneous transmission.
- $\mathcal{I}$: collection of all prospective sets I_i.

The problem of identifying the optimal joint policy for caching, routing, and channel selection in this context can be framed as linear program (LP):

$$\delta^* = \min \sum_{1 \le i \le |\mathcal{I}|} f_i, \tag{6.1}$$

subject to:

$$\sum_j S_j X_{nj} \le C_n^{SBS}, \qquad \forall n, \tag{6.2}$$

$$\sum_n Y_{nj}^k \ge 1_{\{\alpha_{kj}>0\}}, \qquad \forall k, j, \tag{6.3}$$

$$Y_{nj}^k \le X_{nj}, \qquad \forall n, k, j, \tag{6.4}$$

$$\sum_j Y_{nj}^k \alpha_{kj} S_j \le \sum_c Z_{nk}^c, \qquad \forall n, k, \tag{6.5}$$

$$Z_{nk}^c \le \sum_{1 \le i \le |\mathcal{I}|} f_i \hat{c}_{nk}^c(I_i), \qquad \forall n, k, c, \tag{6.6}$$

$$\sum_{1 \le i \le |\mathcal{I}|} f_i \le 1. \tag{6.7}$$

In words, our objective is to minimize the duration of time necessary to stream the requested video files to the mobile users. We explain the necessary problem constraints in the following. To not exceed the storage limit of SBS n, we introduce (6.2). To ensure recovery of the original video data, we require that the aggregate proportion of the respective network coded packets received by the user is at least one. Hence we introduce (6.3). Next, a base station will be able to stream a video file to a user only if that file is stored by the base station. This condition is captured by constraint (6.4).

Put together, (6.5) and (6.6) maintain that the data volume streamed to user k by small base station n over channel c cannot exceed the activation time-weighted cumulative transmission capacity of the communication triplet $((n,k),c)$ they comprise, over all independent sets in which $((n,k),c)$ appears. Last, we require that a feasible choice for the aggregate normalized time of activation across all sets I_i should not be greater than unity. This is ensured by (6.7).

We can reformulate the cost function to pursue other objectives of interest. For example, if we are interested instead in maximizing the aggregate network throughput, we can recast the objective in (6.1) as max $\sum_{n,k,c} Z_{nk}^c$.

6.5 Column Generation for Efficient Approximation Solution

An inherent challenge to solving problems (6.1)–(6.7) is its high computational complexity, due to the huge number of variables it will feature in a realistic setting, induced by the exponentially combinatorial nature of $|\mathcal{I}|$, as explained earlier. Interestingly, the majority of the problem variables will not have an impact on its objective function, as they will necessarily be set to zero [19]. We leverage this property of our problem, to investigate its solution via column generation, which will enable us to adaptively explore a small subset of variables that can make impact on the objective. In particular, we partition (6.1)–(6.7) into a hierarchical master–slave problem, made up fo two components. We denote the former as RMP, or regulated master subproblem, and the latter as PP, or pricing subproblem. We initiate RMP with a very small subset of variables. We then leverage the optimal dual solution of RMP in the formulation of PP to seek a new variable to be introduced to RMP next, such that this variable is associated with the highest reduction of the present objective function of RMP. Once we identify such a variable, we reintroduce it into RMP and repeat the steps, until we meet an ϵ bound on the thereby produced approximation solution is met. A high-level illustration of the described optimization procedure is included in Figure 6.3.

6.5.1 Formulation of Regulated Master Subproblem

We initiate RMP with a subset $\mathcal{I}' \subset \mathcal{I}$. We construct $\mathcal{I}'$ by selecting one communication triplet $((n,k),c)$ for every member element of $\mathcal{I}'$, in which case $\mathcal{I}'$ obtains the form of an identity matrix with its rows reshuffled. We then cast RMP accordingly as

$$\delta = \min \sum_{1 \le i \le |\mathcal{I}'|} f_i, \tag{6.8}$$

given (6.2)–(6.5), and

$$Z_{nk}^c \le \sum_{1 \le i \le |\mathcal{I}'|} f_i \hat{c}_{nk}^c(I_i) \quad \forall n,k,c, \tag{6.9}$$

$$\sum_{1 \le i \le |\mathcal{I}'|} f_i \le 1. \tag{6.10}$$

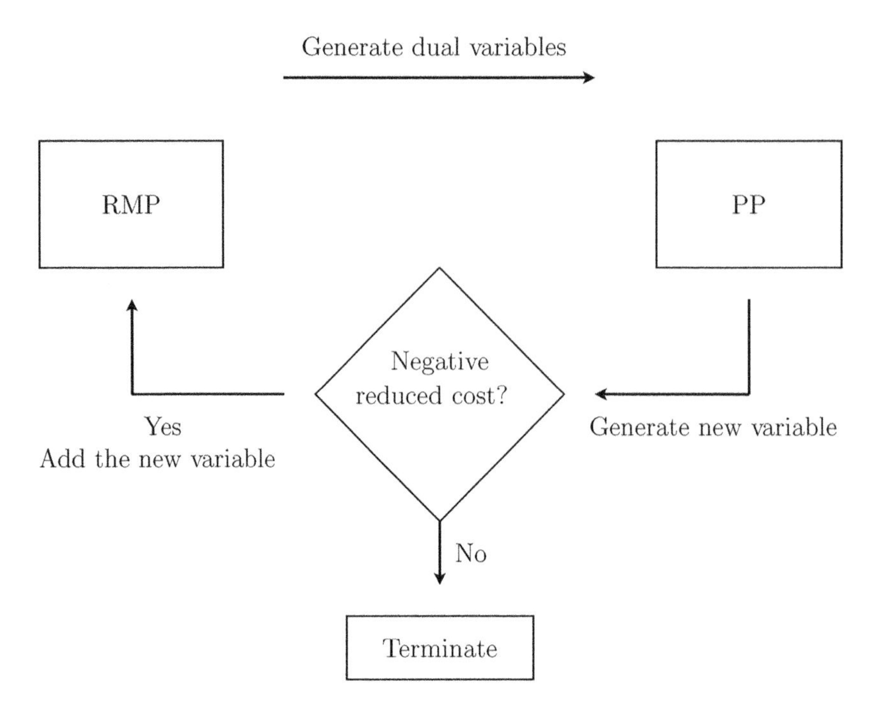

Figure 6.3 Hierarchical master–slave optimization procedure enabled by column generation.

We note that the problem also represents a linear program; however, relative to (6.1)–(6.7), it features a much smaller size. This will allow us to solve it exactly within a reasonable (polynomial) amount of time [19]. In turn, that will will enable us to produce its optimal dual solution. Unfortunately, the solution to RMP can serve only as an upper bound to (6.1)–(6.7), since it examines only a subset $\mathcal{I}'$. We can gradually produce a better (smaller) upper bound, by introducing new member element into $\mathcal{I}'$ one element at a time.

This is where we facilitate the slave subproblem PP. We are interested in identifying the new variable f_i that can be added to $\mathcal{I}'$ such it leads to the highest reduction of the objective in (6.8). We present the formulation of PP next.

6.5.2 Formulation of Slave Pricing Subproblem

We pursue the independent set I_i not yet included in $\mathcal{I}'$ that enables the highest reduction of the objective in (6.8). Concretely, we formulate the objective function value reduction induced by such new I_i as [19]:

$$\omega_i = 1 - \sum_{n,k,c} \lambda_{nk}^c \hat{c}_{nk}^c t_{nk}^c.$$

Here, we denote with λ_{nk}^c the dual problem variables associated with constraint (6.9). Moreover, we introduce a new variable, of binary nature, denoted as t_{nk}^c, which captures the action of including the triplet $((n,k),c)$ in I_i. To identify the desired I_i, we formulate and pursue the following problem (PP):

$$\min_{I_i \in \mathcal{I}/\mathcal{I}'} \omega_i, \tag{6.11}$$

which can be equivalently written as

$$\max_{I_i \in \mathcal{I}/\mathcal{I}'} \beta_i = \sum_{n,k,c} \lambda_{nk}^c \hat{c}_{nk}^c t_{nk}^c. \tag{6.12}$$

We denote the optimal solutions of (6.11) and (6.12), as u_i^* and β_i^*, respectively. Given that our declared objective is to identify the new I_i with the highest objective function reduction, the entire iterative procedure of master–slave optimization enabled by column generation will conclude when no further set I_i can be identified that leads to any objective reduction, i.e., when it holds that $u_i^* \geq 0$ or $\beta_i^* \leq 1$, for any new I_i.

To ensure that the desired I_i identified by PP is feasible, i.e., it does not comprise two interfering triplets $((n,k),c))$, we introduce these conditions into PP:

$$\sum_k t_{nk}^c \leq 1 \qquad \forall n, c \tag{6.13}$$

$$\sum_n t_{nk}^c \leq 1 \qquad \forall k, c \tag{6.14}$$

$$\sum_{k,c} t_{nk}^{c'} \leq a_n \qquad \forall n \tag{6.15}$$

$$\sum_{n,c} t_{nk}^c \leq a_k \qquad \forall k \tag{6.16}$$

$$t_{nk}^c + \sum_{\substack{n' \neq n | k \in \mathcal{F}_{n'} \\ k' \neq k}} t_{n'k'}^c \leq 1 \qquad \forall n, k, c \tag{6.17}$$

$$t_{nk}^c \in \{0,1\} \qquad \forall n, k, c. \tag{6.18}$$

Concretely, constraint (6.13) ensures that one channel cannot be used by a single transmitter to communicate over multiple links. Constraint (6.14) ensures the equivalent for a receiver. Next, (6.15) indicates that the number of antennas a transmitter has limits the number of communication links it can leverage. Again, (6.16) ensures the equivalent for a receiver. Moreover, constraint (6.17) ensures that a receiver k of an active communication triplet $((n,k),c)$ is not interfered by another prospectively interfering transmitter n', where $\mathcal{F}_{n'}$ denotes the collection of prospectively interfered receivers for n'. Finally, constraint (6.18) captures the binary nature of the variable t_{nk}^c.

6.5.3 An Algorithm for an Approximation Solution with ϵ Guarantees

The analytical advances described heretofore can facilitate arriving at the optimal solution with reasonable complexity. However, the challenge of having to examine a huge number of independent sets in the pricing slave subproblem still poses a concern, due to the prospectively exponentially large size of $|\mathcal{I}|$, as we explained earlier. Still, the study in [20] has established that one can quickly reach the optimal solution, with a very fine

Algorithm 4 ϵ-Bounded Approximation Algorithm

Input: Approximation factor ϵ, initial subset of independent sets $\mathcal{I}'$, $\delta^u = \infty$, $\delta^l = 0$
Output: $\delta^u, \delta^l, f_i^*, X_{nj}, Y_{nj}^k$
while $\frac{\delta^l}{\delta^u} < \frac{1}{1+\epsilon}$ and $\beta_i^* > 1$ **do**
  Solve RMP to obtain its optimal solution δ^u and the dual optimal solution λ_{nk}^c
  Using λ_{nk}^c, solve PP to generate an independent set I_i and obtain β_i^*
  Update $\mathcal{I}' = \mathcal{I}' \cup I_i$
  $\omega_i^* = 1 - \beta_i^*$

  $\delta^l = \max\{\delta^u + \Phi\omega_i^*, 0\}$
end while
if $\delta^u \leq 1$ **then**
  Demand can be supported
  Cache the files according to X_{nj}
  Route the files according to Y_{nj}^k
elseif$\delta^u > 1 + \epsilon$ or $\delta^l > 1$
  Demand cannot be supported
  set $\epsilon = 0$ to see if demand can be supported
end if

approximation. We leverage this result to characterize an algorithm that can achieve that for us in this setting. We first introduce some necessary terminology.

We call a solution δ to (6.1)–(6.7) to be within an ϵ approximation of the optimal solution, if the following condition holds: $(1-\epsilon)\delta^* \leq \delta \leq (1+\epsilon)\delta^*$. In Algorithm 4 (see Section 6.6.2), we provide a formal characterization of our algorithm. Now, following this definition, we can prove this desired property of Algorithm 4, via the following theorem.

THEOREM 6.1 *We denote with δ^u and δ^l the approximation solution bounds (high and low) produced by Algorithm 4. They are within an ϵ approximation of the optimal solution to* (6.1) *to* (6.7).

Proof When Algorithm 4 exits, one of the following two cases took place: (1) $\beta_i \leq 1$, i.e., there is no further independent set I_i to be added to the RMP such that it leads to a reduction of the objective function. Hence we have reached δ^*, or (2) $\frac{\delta^l}{\delta^u} \geq \frac{1}{1+\epsilon}$, which means the following inequalities hold:

$$\delta \leq \delta^u \leq (1+\epsilon)\delta^l \leq (1+\epsilon)\delta^*, \tag{6.19}$$

$$\delta \geq \delta^l \geq (1-\epsilon)\delta^u \geq (1-\epsilon)\delta^*. \tag{6.20}$$

Given the earlier condition for an ϵ approximation of the optimal solution, and the analysis carried out earlier, we can establish that indeed δ represents a solution that meets that condition. □

The last item that remains is how to produce the low and high approximation solution bounds δ^u and δ^l. To this end, we first recall that solving RMP leads to an upper bound to (6.1)–(6.7), as explained earlier. We can leverage the solution of the former to establish the value of δ^u. Moreover, we can formulate δ^l as [19]:

$$\delta^l = \max\{\delta^u + \Phi\omega_i^*, 0\}. \tag{6.21}$$

In particular, we recall that for a feasible solution of RMP, it must hold that $\sum_{1 \leq i \leq |\mathcal{I}'|} f_i \leq 1$. Therefore, we can establish the value of Φ to one, since following a result from [19], the latter must represent an upper bound to the former. Moreover, in (6.21), we set ω_i^* to represent the optimal solution of the pricing slave subproblem. The maximum operator therein is taken to ensure that the selected lower bound leads to feasible (nonnegative) solutions.

6.6 Experimental Evaluation

6.6.1 Outline

We carry out a select few experiments to illustrate the performance benefits of our framework. A comprehensive assessment of this nature can be found in [21]. Similarly, the study in [21] includes a detailed discussion of all relevant implementation aspects that would arise in a practical deployment. Finally, a rigorous complexity analysis of the optimization framework we formulate here is also included in [21].

6.6.2 Experimental Setup

The macro cell setting we examine is illustrated in Figure 6.1. We set the macro cell diameter to 800 m. We distribute the mobile users uniformly across the spatial area of the cell. Each SBS and user can leverage 5 channels from a pool of 10 available secondary channels, selected at random. We set the transmission bandwidth of each such channel to 400 KHz. The primary channel's bandwidth has been set to 1 MHz.

We set the popularity factors of the video files in the system induced by the mobile users' requests for streaming to exhibit a Zipf distribution featuring the parameter $\zeta = 0.8$ [22]. Concretely, in a sorted collection of files, according to their popularity, the popularity factor of a video file ranked as the -th most popular is identified as $\frac{1/m^{\zeta}}{\sum_{j=1}^{|\mathcal{J}|} 1/j^{\zeta}}$. The average data volume of a video file streamed in the system is set to 400 MB. Finally, to terminate the execution of Algorithm 4, we set $\epsilon = 0.03$.

Table 6.1 summarizes our simulation parameters. We benchmark the performance of our framework to that of the femtocaching system proposed in [2].

6.6.3 Experimental Results and Discussion

6.6.3.1 Streaming Data Rate against Storage Space

Here we investigate the enabled streaming data rate for a user, while we vary the storage space at a small base station. In Figure 6.4, we show these results. All base stations feature the same storage space. We can see that the enabled data rate increases with

Table 6.1 Main Parameters and Their Values used in Our Simulations

Parameter	Value
Diameter of macro cell	800 m
Transmission bandwidth of primary channel	1 MHz
Transmission bandwidth of secondary channel	400 KHz
Number of secondary channels	10
Number of secondary channels available at SBS and mobile user	5
Average data volume of a video file	400 MB
Number of different video files in the system	200 (unless stated otherwise)
Distribution of video file popularity	Zipf (parameter $\zeta = 0.8$)
Average cache size	4 GB (unless stated otherwise)
Number of small base stations	14 (unless stated otherwise)
Number of mobile users	200 (unless stated otherwise)
The transmission range of a base station	100 m (unless stated otherwise)
The interference range of a base station	$2\times$ of transmission range
Approximation fidelity ϵ	0.03

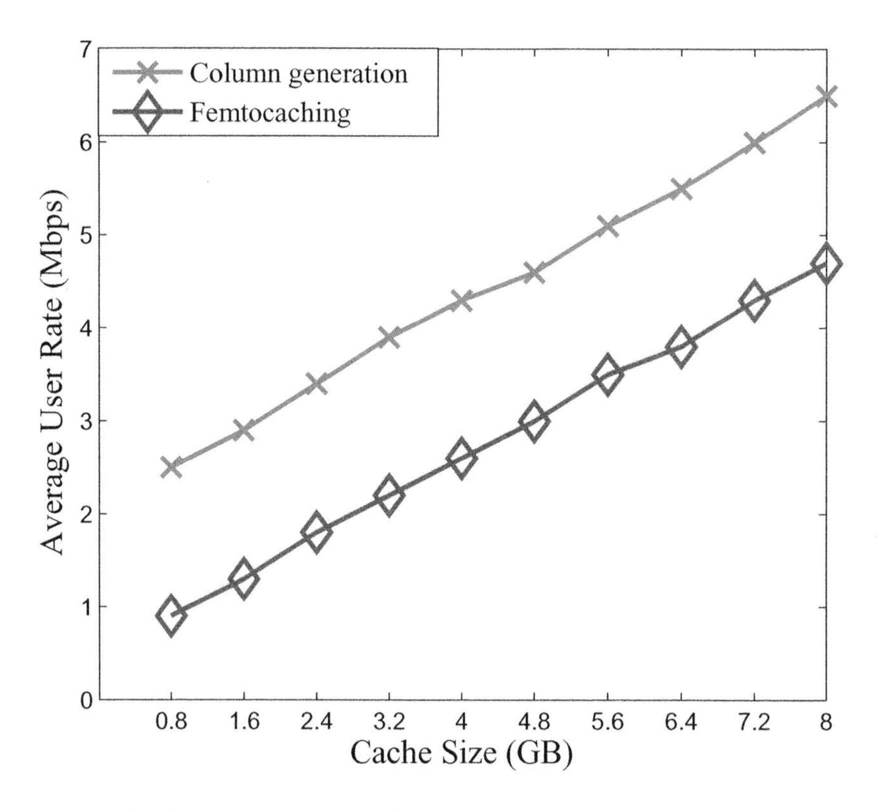

Figure 6.4 Enabled streaming data rate for a user against storage space at a small base station.

the storage capacity of an SBS. Larger storage space means more video files can be cached and served locally, thereby avoiding the need to stream from the macro base station. In turn, the former can then lead to more parallel communication triplets being established in parallel from different small base stations. Hence more users can be streamed video files and at higher data rates. We can also see from Figure 6.4 that our system outperforms femtocaching by 40%, which is quite considerable. This is enabled by allowing parallel transmissions from different small base stations over the same channel, as long as that does not lead to interference.

6.6.3.2 Streaming Data Rate against Number of Video Files

In these experiments, we vary the number of video files in the system and measure the impact on the enabled streaming data rate to a mobile user. These results are shown in Figure 6.5. We can see that video file diversity in the system has an adverse impact on the enabled streaming data rate. Since the users can select from a broader set of files, it will lead to a higher volume of data being sent over one channel. This in turn will increase the amount of time required to stream the video files to all requesting users. Still, we emphasize that relative to femtocaching, we enable performance gains of 39%, for lower video file diversity, and 42%, for higher video file diversity, as observed from Figure 6.5. Again, these performance advances stem from the same reason explained earlier.

6.6.3.3 Streaming Data Rate against Transmission Range of SBS

Here we explore the impact of the small base stations' transmission rage on the enabled streaming data rate for a user. These results are included in Figure 6.6. As expected,

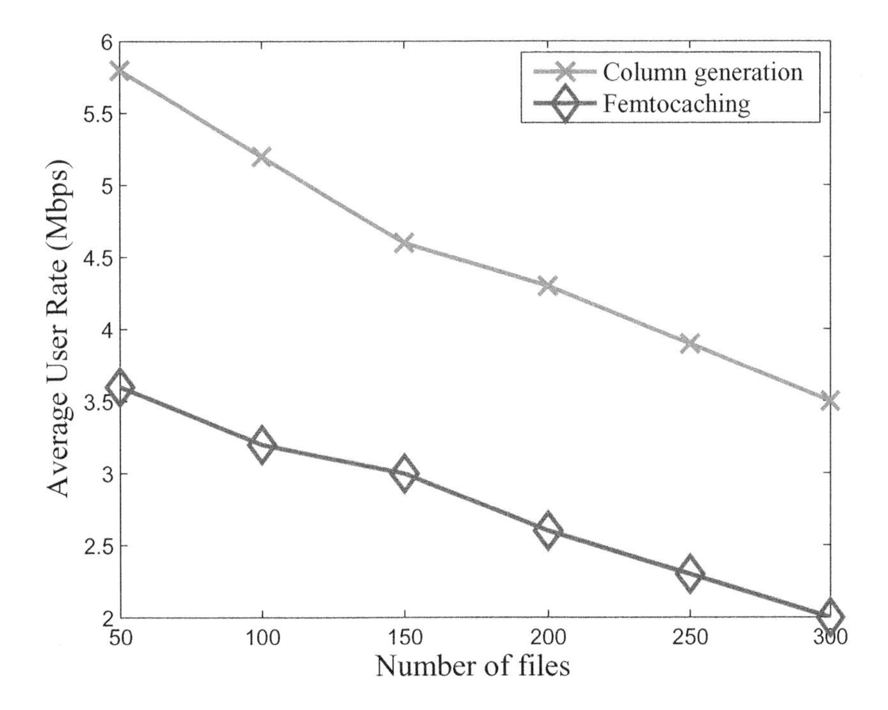

Figure 6.5 Enabled streaming data rate for a user against video file diversity in the system.

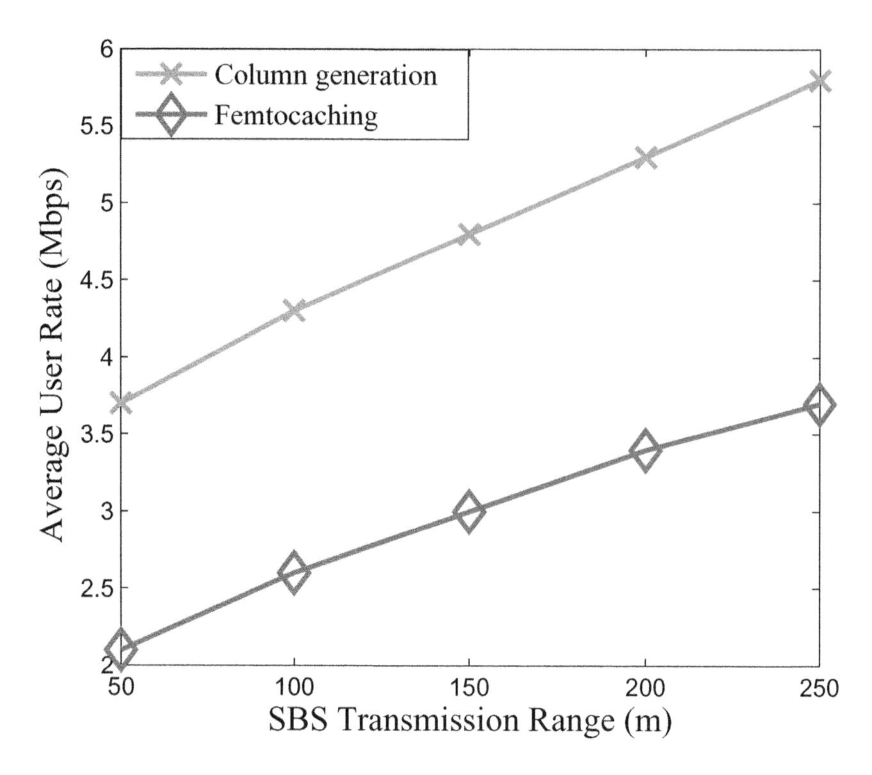

Figure 6.6 Enabled streaming data rate for a user against transmission range of an SBS.

extending the transmission range helps boost the streaming data rate at which the video files can be delivered to the mobile users. Concretely, a longer transmission range for a small base station means a higher number of mobile users can be streamed video data directly from it. Otherwise, some may be required to be served by the macro base station in which case the used communication channels could not be reused by any SBS across the cell. Hence the aggregate streaming capacity and enabled data rate for the mobile users would be lower. We observe from Figure 6.6 that considerable performance advances of 34–46% are enabled relative to femtocaching, induced by the unique capability introduced by our framework, as explained earlier.

6.7 Benefits for Video Quality of Streaming Application

Our framework leads to considerable benefits in enabled streaming data rate for a mobile user. These in turn will map to respective benefits in the delivered video quality of the video streaming application running on the mobile devices of the users. Concretely, the streaming data rate gains of up to 46% demonstrated here can lead to up to 4–5 dB of respective video quality gains [23, 24], in the most popular video streaming settings considered presently [25, 26], inclusive of emerging related applications such as 360° video streaming to mobile virtual reality (VR) devices in 5G systems [13, 27–30]. In addition, the performance advances introduced by our framework can also enable higher temporal–spatial resolution of the streaming video content and lower initial playback/buffering delays for the related applications. All of these benefits can similarly augment the quality of experience for the mobile user [31].

It is expected that the further performance benefits can be harnessed within the setting of our framework, if the video streaming content is represented using scalable video coding [32–34]. Then, higher utilization of resources and more effective performance trade-offs can be pursued [12, 35–38]. This motivates future work based on the findings presented in this chapter.

6.8 Concluding Remarks

We investigated selecting joint policies for caching, routing, and channel selection in coordinated multi-cellular systems for edge-based video delivery. To overcome the challenges of this problem, arising from its large-scale nature, and thereby considerably advance the state-of-the-art, we leverage two novel concepts in this context: column generation, a mathematical framework for large-scale optimization that enables adapting the number of variables considered in an optimization problem, and conflict graph, a mathematical framework for effective modeling of interference in wireless networks. Integrating these two, we then pursue a novel problem formulation that comprises a master–slave problem structure for selecting optimal joint policies for caching, routing, and channel selection in the this setting. The master problem comprises a subset of the original problem variables and adaptively decides whether to introduce an additional variable via a slave subproblem, which characterizes the reduction in the value of the objective function with the addition of a new variable. The search for the optimal joint policies concludes when no new variable can be introduced to further lower the value of the objective function. Since leveraging a conflict graph still induces an exponential complexity in the number of wireless links that can be prospectively activated simultaneously, we also investigated an effective method to quickly approximate the optimal solution within an ϵ deviation, at lower computational complexity. We comprehensively evaluated the performance characteristics of our analytical advances via realistic simulation experiments, demonstrating a close to 50% improvement in expected streaming data rate enabled for a mobile user relative to a competitive state-of-the-art method, known popularly as femtocaching. Such performance benefits will enable up to 5 dB improvement in video streaming QoE delivered to the mobile user.

References

[1] "Global mobile data traffic forecast update, 2013–2018," in *Cisco Visual Networking Index*, Cisco Inc., Feb. 2014.

[2] N. Golrezaei, K. Shanmugam, A. G. Dimakis, A. F. Molisch, and G. Caire, "Femtocaching: wireless video content delivery through distributed caching helpers," in *INFOCOM, 2012 Proceedings*, IEEE, 2012, pp. 1107–1115.

[3] P. Blasco and D. Gunduz, "Learning-based optimization of cache content in a small cell base station," in *Proc. Int'l Conf. Communication*, IEEE, Jun. 2014, pp. 1897–1903.

[4] A. Khreishah and J. Chakareski, "Collaborative caching for multicell-coordinated systems," in *IEEE INFOCOM CNTCV Workshop*, 2015.

[5] J. Erman, A. Gerber, M. Hajiaghayi, D. Pei, S. Sen, and O. Spatscheck, "To cache or not to cache: the 3G case," *Internet Computing, IEEE*, vol. 15, no. 2, pp. 27–34, 2011.

[6] H. Ahlehagh and S. Dey, "Video caching in radio access network: impact on delay and capacity," in *Wireless Communications and Networking Conference (WCNC), 2012*, IEEE, 2012, pp. 2276–2281.

[7] E. Baştuğ, M. Bennis, and M. Debbah, "Cache-enabled small cell networks: modeling and tradeoffs," in *Proceeding of the International Symposium of Wireless Communications Systems*, Barcelona, Spain, IEEE, Aug. 2014, pp. 649–653.

[8] P. Ostovari, A. Khreishah, and J. Wu, "Cache content placement using triangular network coding," in *Wireless Communications and Networking Conference (WCNC), 2013*, IEEE, 2013, pp. 1375–1380.

[9] N. Karamchandani, U. Niesen, M. A. Maddah-Ali, and S. Diggavi, "Hierarchical coded caching," in *2014 IEEE International Symposium on Information Theory (ISIT)*, 2014, pp. 2142–2146.

[10] J. Chakareski and P. Chou, "RaDiO edge: rate-distortion optimized proxy-driven streaming from the network edge," *IEEE/ACM Transactions on Networking*, vol. 14, no. 6, pp. 1302–1312, Dec. 2006.

[11] J. Chakareski, P. Chou, and B. Girod, "Computing rate-distortion optimized policies for hybrid receiver/sender driven streaming of multimedia," in *Proceedings of the Asilomar Conference on Signals, Systems, and Computers*, vol. 2, IEEE, Nov. 2002, pp. 1310–1314.

[12] J. Chakareski, "In-network packet scheduling and rate allocation: a content delivery perspective," *IEEE Transactions on Multimedia*, vol. 13, no. 5, pp. 1092–1102, Oct. 2011.

[13] J. Chakareski, "VR/AR immersive communication: caching, edge computing, and transmission trade-offs," in *Proceedings of the SIGCOMM Workshop on Virtual Reality and Augmented Reality Network (VR/AR Network)*, ACM, Aug. 2017.

[14] K. Poularakis, G. Iosifidis, and L. Tassiulas, "Approximation algorithms for mobile data caching in small cell networks," *IEEE Transactions on Communications*, vol. 62, no. 10, pp. 3665–3677, 2014.

[15] K. Jain, J. Padhye, V. N. Padmanabhan, and L. Qiu, "Impact of interference on multi-hop wireless network performance," *Wireless networks*, vol. 11, no. 4, pp. 471–487, 2005.

[16] T. Ho and D. Lun, *Network Coding: An Introduction*. Cambridge University Press, 2008.

[17] T. Ho, M. Médard, R. Koetter, D. R. Karger, M. Effros, J. Shi, and B. Leong, "A random linear network coding approach to multicast," *IEEE Transactions on Information Theory*, vol. 52, no. 10, pp. 4413–4430, 2006.

[18] P. Gupta and P. R. Kumar, "The capacity of wireless networks," *IEEE Transactions on Information Theory*, vol. 46, no. 2, pp. 388–404, 2000.

[19] D. Bertsimas and J. N. Tsitsiklis, *Introduction to Linear Optimization*, vol. 6, Belmont, MA: Athena Scientific, 1997.

[20] L. S. Lasdon, *Optimization Theory for Large Systems*, North Chelmsford, MA: Courier Corporation, 2013.

[21] A. Khreishah, J. Chakareski, and A. Gharaibeh, "Joint caching, routing, and channel assignment for collaborative small-cell cellular networks," *IEEE Journal of Selected Areas in Communications*, vol. 34, no. 8, pp. 2275–2284, Aug. 2016.

[22] L. Breslau, P. Cao, L. Fan, G. Phillips, and S. Shenker, "Web caching and Zipf-like distributions: evidence and implications," in *INFOCOM '99. Eighteenth Annual Joint*

Conference of the IEEE Computer and Communications Societies, vol. 1, IEEE, 1999, pp. 126–134.
[23] J. Chakareski, "Informative state-based video communication," *IEEE Transactions on Image Processing*, vol. 22, no. 6, pp. 2115–2127, June 2013.
[24] J. Chakareski and P. Frossard, "Adaptive systems for improved media streaming experience," *IEEE Communications Magazine*, vol. 45, no. 1, pp. 77–83, Jan. 2007.
[25] J. Chakareski, "Uplink scheduling of visual sensors: when view popularity matters," *IEEE Transactions on Communications*, vol. 2, no. 63, pp. 510–519, Feb. 2015.
[26] J. Chakareski, "Joint source-channel rate allocation and client clustering for scalable multistream IPTV," *IEEE Transactions on Image Processing*, vol. 24, no. 8, pp. 2429–2439, Aug. 2015.
[27] J. Chakareski, R. Aksu, X. Corbillon, G. Simon, and V. Swaminathan, "Viewport-driven rate-distortion optimized 360° video streaming," in *Proceedings of the International Conference on Communications*, IEEE, May 2018.
[28] X. Corbillon, A. Devlic, G. Simon, and J. Chakareski, "Viewport-adaptive navigable 360-degree video delivery," in *Proceedings of the International Conference on Communications*, IEEE, May 2017.
[29] X. Corbillon, A. Devlic, G. Simon, and J. Chakareski, "Optimal set of 360-degree videos for viewport-adaptive streaming," in *Proceedings of the International Conference on Multimedia*, ACM, Oct. 2017, pp. 934–951.
[30] J. Chakareski, "UAV-IoT for next generation virtual reality," *IEEE Transactions on Image Processing*, vol. 28, no. 12, pp. 5977–5990, Dec. 2019.
[31] R. Matos, N. Coutinho, C. Marques, S. Sargento, J. Chakareski, and A. Kassler, "Quality of experience based routing in multi-service wireless mesh networks," in *Workshop on Realizing Advanced Video Optimized Wireless Networks at the International Conference on Communications*, IEEE, June 2012.
[32] ITU-T and ISO/IEC JTC 1, "Advanced video coding for generic audiovisual services, amendment 3: scalable video coding," *Draft ITU-T Recommendation H.264 - ISO/IEC 14496-10(AVC)*, Apr. 2005.
[33] H. Schwarz, D. Marpe, and T. Wiegand, "Overview of the scalable video coding extension of the H.264/AVC standard," *IEEE Transactions on Circuits and Systems for Video Technology*, vol. 17, no. 9, pp. 1103–1120, Sept. 2007.
[34] J. Chakareski, S. Han, and B. Girod, "Layered coding vs. multiple descriptions for video streaming over multiple paths," *Multimedia Systems Journal*, vol. 10, no. 1 pp. 275–285, Jan. 2005.
[35] Q. Gong, J. W. Woods, K. Kar, and J. Chakareski, "Fine-grained scalable video caching," in *Proceedings of the International Symposium of Multimedia*, IEEE, Dec. 2015, pp. 101–106.
[36] R. Aksu, J. Chakareski, and V. Swaminathan, "Viewport-driven rate-distortion optimized scalable live 360° video network multicast," in *Proceedings of the ICME International Workshop on Hot Topics in 3D (Hot3D)*, IEEE, July 2018.
[37] J. Chakareski, "Wireless streaming of interactive multi-view video via network compression and path diversity," *IEEE Transactions on Communications*, vol. 62, no. 4, pp. 1350–1357, Apr. 2014.
[38] Q. Gong, J. W. Woods, K. Kar, and J. Chakareski, "Multiple-cache pairing for fine-grained scalable video caching and networking," in *Proceedings of the International Conference on Future Internet Technologies*, ACM, Aug. 2019.

Part II

Proactive Caching

7 Learning Popularity for Proactive Caching in Cellular Networks

Khai Nguyen Doan, Thang Van Nguyen, and Tony Q.S. Quek

Video data have been shown to dominate a significant portion of mobile data traffic, which has a strong influence on a backhaul congestion issue in cellular networks. To tackle the problem, proactive caching is considered as a prominent candidate in terms of cost efficiency. In this chapter, we study an end-to-end cache placement strategy based on a popularity prediction of all videos, including published and unpublished ones. For dealing with unpublished videos with unknown statistical information, features from the video content are extracted and condensed into a high-dimensional vector. This type of vector is then mapped to a lower-dimensional space. This process not only alleviates the computational burden but also creates a new vector that is more meaningful and understandable. At this stage, different types of prediction models can be trained to anticipate the popularity using information from published videos as training data. In addition, this method also includes the process of updating the popularity of the published video set based on predictions with an expert advice scheme. Regarding this, the upper bound of the expected cumulative loss is analyzed to gain more insight into the theoretical performance of the presented method.

7.1 Introduction

The current network infrastructure has faced a dramatic growth in data demand due to the development of electronic devices. This is one of the biggest reasons for congestion in wireless networks, especially for mobile data traffic in which video data make up a significant portion and have been recorded to have a significant impact [1, 2]. Fortunately, caching has been proposed as an effective method for tackling this problem. Specifically, storing multiple copies of the most popular content items across access points, e.g., base stations (BSs), provides a fast and cost-effective manner to serve users. Although there are several points from which caching can be deployed, BSs are an attractive choice because they are capable of serving multiple users on wide areas. In the case of a cache hit,[1] the long-distance communications from users to content server can be altered by short-range links to a nearby BS.

[1] The event that user's requested items can be found in local cache memories.

7.1.1 Background and Motivation

Caching is an attractive field that appeals to many research works. These works target different aspects of caching advantages, such as improving the cache-hit ratio, reducing the delivery power cost, and latency. For example, the authors of [3–7] sought to design optimal cache placement policies either to enhance the user's quality of service (QoS) or increase the energy efficiency. However, there are many studies that are based on an assumption of having precise content popularity, where using a Zipf distribution is the most common assumption. Meanwhile, this information is not available in practice and requires appropriate estimation or prediction schemes. For this reason, another research direction is formed that anticipates the popularity of online content items (specifically, videos). Regarding this, the log-transformed popularity of YouTube videos at different times was discovered to have a strong correlation in the long run by authors of [8]. This work also suggested a method that gave out predictions by observing the number of views. In another work [9], the prediction is made by exploiting K-spectral clustering technique. Besides that, in [10], the authors made use of both social influence among users and video correlation to foresee future demand. In addition, [11] enhanced the cache-hit rate via optimizing the cache replacement where learning short-term popularity of content items was their approach.

There are very few works addressing the complete process from analyzing raw videos to caching. In addition, published videos are the only focus. Meanwhile, considering both published and unpublished video sets is a promising way to get even closer to the network traffic offloading target. This is supported by previous research [12, 13], which found out that there is a considerably large data volume uploaded to the YouTube server every day. Further more, in a short time after publishing, many videos have their views increased remarkably fast. Therefore, this chapter examines a caching strategy that takes into account all videos that have not reached users' awareness as a way to push caching effectiveness to a new high. Unlike published videos, unpublished ones require different approaches, because a lot of importantly related information is not available.

7.1.2 Approach and Main Outcomes

The main goal behind constructing the mentioned caching procedure is to minimize the average backhaul load, which is always a highly important issue in current cellular networks [14]. In particular, our consideration is a cellular network having many users who are served by a BS given connection to the content server via backhaul links. At the considered time, this server stores many videos that have been published to the community. Concurrently, a set of new videos are available that are not acknowledged by any user. In off-peak time, when caching is executed, a machine-learning approach is taken to forecast if any of these new videos are worth fetching to cache memory before to users' requests. Regarding this, the published video set will serve as training data. In addition, since the statistical information of published videos may vary over time, a *prediction with expert advice* method is presented to keep the knowledge about this video set up to date.

7.1.3 Optimal Caching Policy

Our study is based on a system having a BS communicating with K users. The BS has connection to a content server through backhaul links. In our model, videos in the server are encoded with the fountain coding method [15], and users are required to gather enough encoded data pieces to obtain the desired video files. We denote s_v as the total size of data segments required to reconstruct video file v. In our context, caching will be deployed at the BS.

The server originally contains a set of videos that has already been published to the community. Besides that, there is another set that is newly uploaded and has not reached user awareness yet. We denote V the total number of available videos, including those that have reached and have not reached user's awareness; α_v is a caching variable indicating the proportion of video v stored in cache memory. We also use p_v as the popularity of video v. In order to minimize the average backhaul load, we come up with the following problem formulation:

$$\min_{\alpha_i} \sum_{v=1}^{V} (1-\alpha_v)\, s_v p_v \tag{7.1}$$

$$\text{s.t.} \quad \sum_{v=1}^{V} \alpha_v s_v = M, \tag{7.2}$$

$$0 \le \alpha_v \le 1, \tag{7.3}$$

where the function in (7.1) is the average load put on the backhaul. The cache memory at BS can keep up to M data units as implied by (7.2). The optimal solution can be presented by

$$\alpha_v = \begin{cases} \mathcal{C}_1(s_1), & \text{for } v = \mathcal{M}(1) \\ \mathcal{C}_w\left(\left\{s_{\mathcal{M}(k)}\right\}_{k=1}^{w}\right), & \text{for } v = \mathcal{M}(w), w = 2, \ldots, V, \end{cases} \tag{7.4}$$

where

$$\begin{aligned} \mathcal{C}_1(x_1) &= \min\left\{1, \frac{M}{x_1}\right\}, \\ \mathcal{C}_w\left(\{x_k\}_{k=1}^{w}\right) &= \min\left\{1, \frac{1}{x_w}\left(M - \sum\nolimits_{k=1}^{w-1} x_k \mathcal{C}_k(x_k)\right)\right\}, \quad w = 2, \ldots, V, \end{aligned} \tag{7.5}$$

and

$$\mathcal{M}(w) = \underset{k=1,\ldots,V}{\operatorname{argmax}^{w}}\, p_k. \tag{7.6}$$

Note that (7.6) is actually extracting the wth greatest member in the sequence $\{p_k\}_{k=1}^{V}$.

The solution (7.4)–(7.6) requires knowledge of popularity from all videos that need to be predicted. In terms of this, Figure 7.1 shows every step for estimating how popular a video can be. At the "analyzing" stage, there are many substeps and all will be presented clearly in the subsequent section.

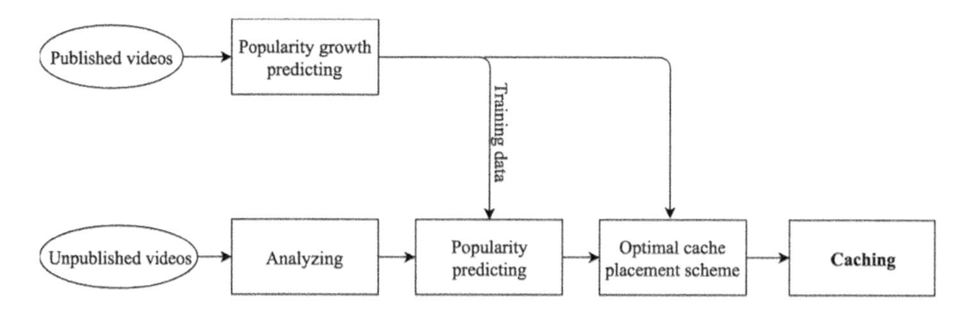

Figure 7.1 A general diagram of the step-by-step process for obtaining popularity values from raw videos.

7.2 Learning and Predicting Popularity of Unpublished Videos

In terms of unpublished videos, analyzing video frames to anticipate the content popularity should be an appropriate choice when all statistical data are missing. However, since there is a diversity in video types and user preferences nowadays, the features collected from analyzing video frames should be sufficiently sophisticated. In addition, this needs to be done without human assistance due to a vast amount of movies uploaded daily. Moreover, processing a video should not take too much time in order to ensure that those videos can be published on time. Regarding this, conventional methods like tagging fail to adapt. Because of these strict constraints, deep neural networks appear to be a potential solution in the video feature extraction step, which are also due to recent breakthroughs in video and image domains. In Figure 7.1, the "analyzing" block consists of three main steps, which are discussed in the next three sections.

7.2.1 Feature Extraction with Deep Neural Networks

The first step is called *feature extraction*. There are many designs for neural networks that can be applied to this step to extract feature vectors from videos, including those designed for image processing such as GoogleNet and AlexNet. Input images for those networks can be produced by sampling from a given video. However, in this section we would like to introduce a deep neural network model that sets itself apart from other models by the capability of learning spatial and temporal features all at once. This model is named the three-dimensional convolutional neural network. Readers may refer to [16] for the details of the convolutional neural network (CNN). Regarding the model structure, in order to perpetuate the original temporal information, regular two-dimensional (2D) neuron network filters are altered by 3D ones. This turns out to be the main difference between this propose and conventional CNNs. More implementation and structural details can be found in the original paper [17].

7.2.2 Feature Clustering

After extracting features, the analyzing stage moves to the *feature clustering* step. This step is done by merging a number of features into a group where overlapping between groups can possibly exist. The term *overlapping*, here, means that some features can concurrently present in more than one group. A group of features may represent for a type of video. For instance, a group consisting of features like fight scenes, guns, and daggers may represent the action movie category. Therefore, for the ease of presentation, we call a cluster of features a *video category* (VC). From the design, as a VC is formed by a large number of features, it is easier to distinguish a video from the others. Nevertheless, this results in a high computational load. Criteria and rules for designing the number of features in each VC may vary across applications and depend on designer experiences. Let us consider the case where G VCs are formed, and G is expected to be smaller than the dimension of extracted feature vectors from the previous step.

First, the motivation for having this step is that feature vectors extracted in the previous step are high-dimensional vectors containing a large number of features that are considerably diverse. Hence clustering helps reduce the amount of video properties to a manageable degree. Second, the high-dimensional vectors provided from neural networks are not fully understood. Thus having this step provides a basis for the next one in which the original high-dimensional vector is mapped to another space having G dimension. This next step will be detailed in the subsequent subsection. This process not only helps reduce the computational burden but also creates a more understandable feature vector type for videos.

7.2.3 Probability Estimation in Multi-class Classification

The final step in the analyzing stage is called *feature mapping*, which maps vectors of extracted features to a lower-dimensional space (G-dimension to be specific). Let us call the resulted vector the *representation vector*, where each element implies how likely the video has features from a VC. Therefore, this new vector type is able to reveal to us the characteristic of a video, and as the video becomes popular, we are able to know which sets of features make it attractive.

This task can be converted into a multi-class classification problem where there are G given classes, each corresponding to a VC. Then, a video is classified to these VCs, however, we stop at the second-last step of the classification process to obtain the probability vectors rather than a label. Each element in these vectors tells the probability that a video belongs to a VC. Let us denote $\beta_{ij}(\boldsymbol{x})$ the probability that a video having feature vector $\boldsymbol{x}$ will belong to the VC i given only VC i and j. To this end, by providing $\beta_{ij}(\boldsymbol{x})$, the desired vectors can be obtained following the method proposed in [18], which is based on a support vector machine (SVM).

Denote r_i the probability that a certain video belongs to VC i, it can be obtained by solving the following equation system.

$$r_i = \frac{1}{G-1} \sum_{j=1,\, j\neq i}^{G} \beta_{ij}\left(r_i + r_j\right), \tag{7.7}$$

which is constrained by

$$\sum_{i=1}^{G} r_i = 1. \quad (7.8)$$

$$r_i \geq 0, \forall i. \quad (7.9)$$

As has been proved in [17], the unique solution of $r_i, \forall i$ is available given that $\beta_{ij} > 0, \forall i \neq j$. To this end, there can be several approximation methods for finding β_{ij}. Hereafter, we describe a simple method for estimating this quantity. For the ease of understanding, we recall that the *separating hyperplane* in the SVM method is the one that has the greatest distance to the closest data points of all classes. Let us consider a classification problem involving class i and j only. We denote δ_{iw} the distance between training data point w of class i to the separating hyperplane and

$$\delta_i^{\max} = \max_w \delta_{iw}, \text{ for } w \text{ in class } i. \quad (7.10)$$

Assume that the considered video has its feature vector separated from the hyperplane a distance of δ. Then, the estimation of β_{ij} is as follows.

$$\beta_{ij} = 0.5 + \min\left(\frac{\delta}{\delta_k^{\max}}, 0.5\right), \quad (7.11)$$

where $k = i$ if the feature vector of the considered video is on class i side and $k = j$ if it is on class j side.

The next step is the "popularity estimating" stage (see Figure 7.1), where a regression model will be trained to predict video popularity.

7.2.4 Performance Evaluation

Recall in Section 7.2.1 we introduced C3D as a prominent candidate in the video feature extraction task; however, this is not the only option for this stage. Besides that, there are a variety of regression models that can be applied to predict the video popularity based on representation vectors computed in the previous section. Therefore, this section is devoted to evaluating our learning procedure performance under different choices of feature extractors and regression models.

The figures showed in this section illustrate the performance of the presented method. They are results from experiments conducted on a combination of YUPENN [19] and UCF101 [20] datasets. YUPENN is a collection of 14 video clip types from a total of 420 videos, and UCF101 consists of 13,320 videos and 101 types. The two sets are mixed into a single one, and we consider each video type as a VC; thus $G = 115$. Next, the video set is separated to form unpublished and published sets in the proportion of 30% and 70%, respectively. The separation task guarantees that both subsets have all 115 VCs.

The offload ratio is presented as a function of network size, adjusted by the number of users in Figure 7.2. In this experiment, we investigate two different sizes of cache memory. The small scale corresponds to 20% of the server content (here we assume that

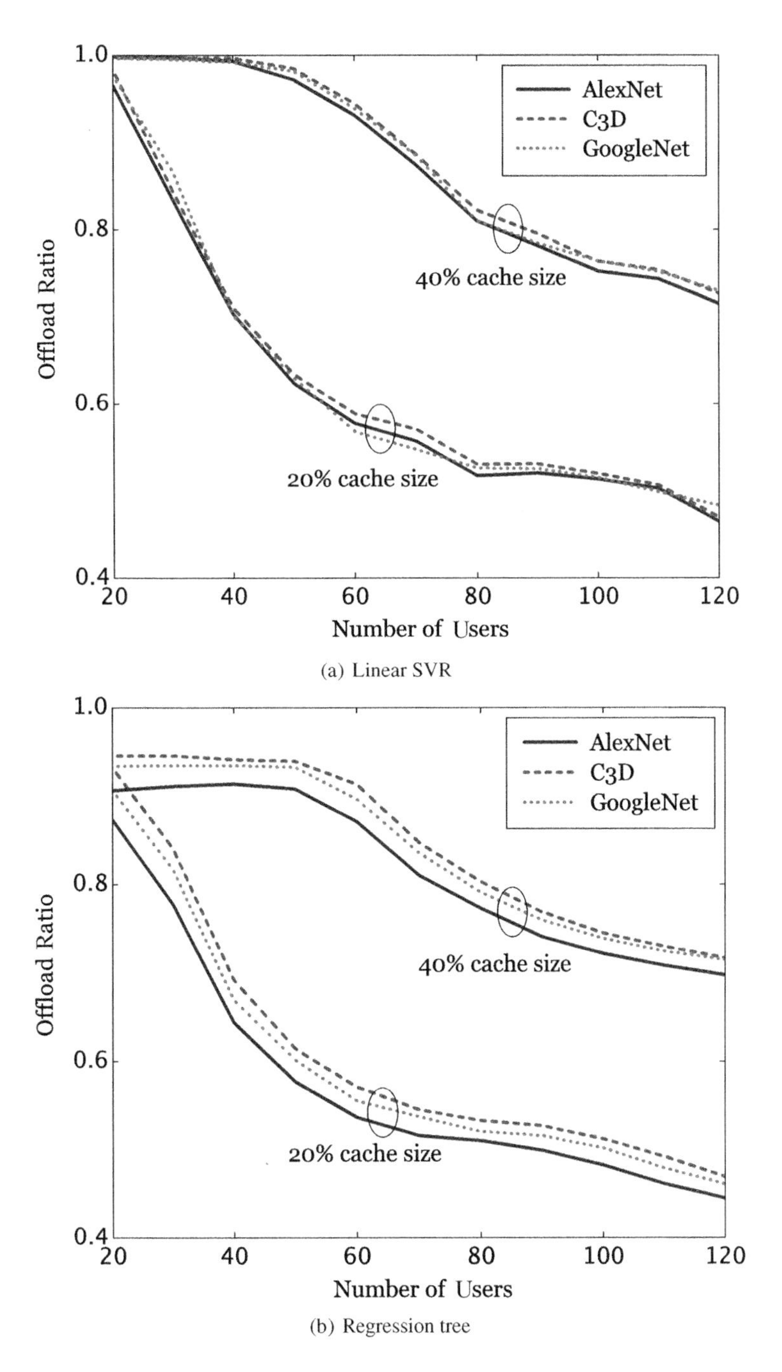

Figure 7.2 The effect from the population growth in the cell on the offloading task under small and large cache capacity scales. A (a) linear SVR and (b) regression tree are used as the popularity predictor, respectively.

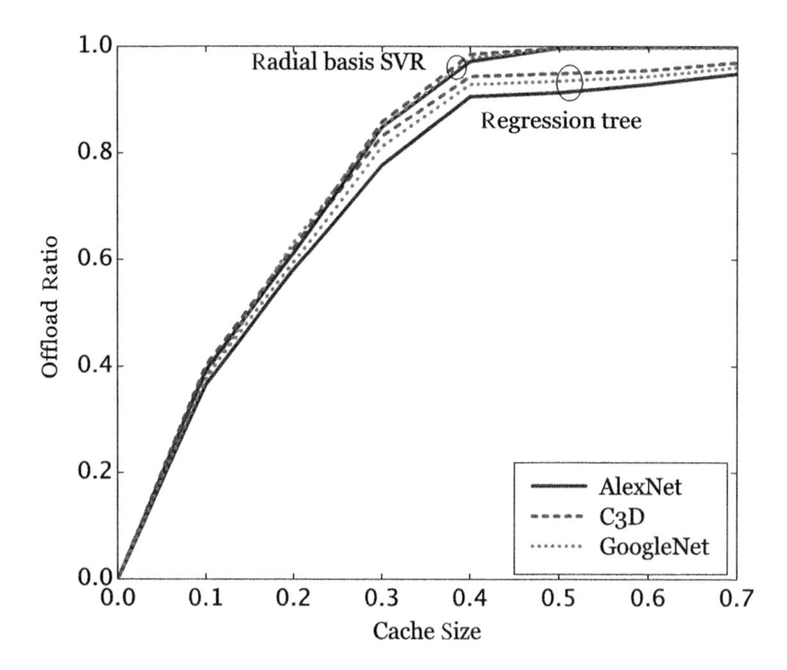

Figure 7.3 The positive impact from the cache size enhancement on the offloading effectiveness.

the server stores only the mentioned videos set and no other data), and 40% for the large scale. Then, all combinations between three neural network models and two regression models are examined. Those neural networks are C3D, AlexNet, and GoogleNet, while a linear support vector regression (SVR) and regression tree are chosen regression methods. As shown, the linear SVR predictor performs well with our representation vectors formed in Section 7.2.3. The performance of both the linear SVR and regression tree can be further improved by tuning and finding a better set of parameters. The figure also suggests that C3D should be a better option for feature extractor than AlexNet and GoogleNet. This example implies that combining linear SVR and C3D is a reasonable approach in predicting unpublished video popularity.

In Figure 7.3, another influence on the offload ratio that is from cache capacity is investigated. Although similar to the previous experiment, in this one, the linear function of SVR is replaced by a radial basis—a nonlinear kernel. It can be seen that the presented method is efficient, especially in the case of limited cache capacity. As an example, with only 10% and 20% of cache capacity, the offloading performance jumps up to 35% and 55%, respectively.

Assume that each user has the most favorite VC that receives the highest request probability from that user compared to other VCs. For each user we define the gap between the request probability of the most favorite VC and the average request probability of other VCs as the *preference intensity* (PI) of that user. This quantity tells us how easy videos can be cached to serve a certain user. Specifically, if this value is high, the user has a clear preference; then caching videos in the user's most favorite VC can serve most of its requests. Meanwhile, if this value is low, the user seems to requests

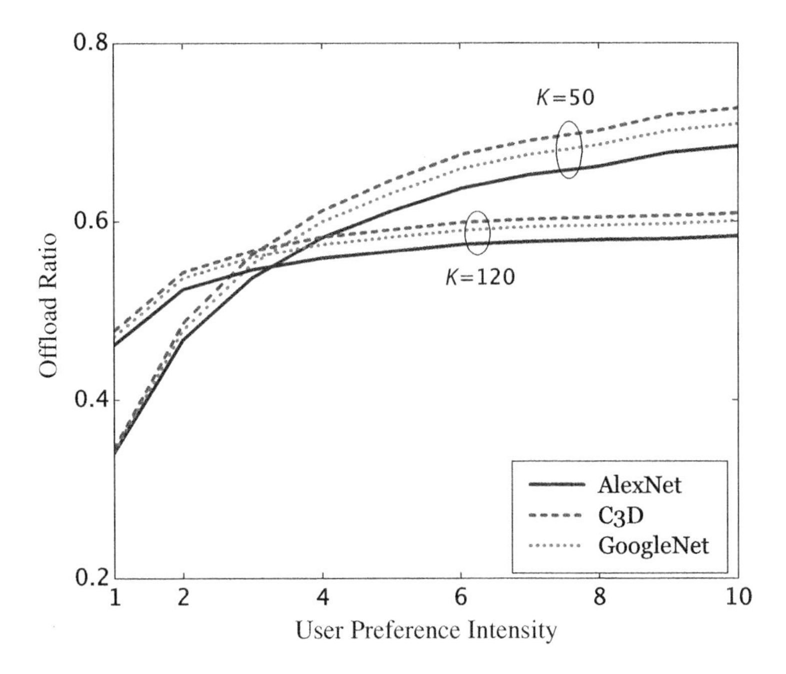

Figure 7.4 Graph illustrating how the difficulty of caching users' favorite videos is driven by their PI.

videos from all VCs evenly, which makes the caching job much harder. In Figure 7.4, to evaluate such an effect, users are randomly given favorite VCs, and we represent PI by the number of times that their request probability to favorite VCs is higher than that to other VCs (horizontal axis). Two different user populations, 50 and 120, are involved in the experiment with a cache that can store up to 30% of all videos. On the one hand, we can see that it is hard to cache videos from mostly requested VCs when the PI is low, which limits the offloaded data amount. On the other hand, when the PI is large enough for our predictor to define the users' favorite VCs, the offloading performance seems to reach its peak and the impact from increasing PI turns negligible.

In addition, there is a swap in offloading performance of two cases $K = 50$ and $K = 120$ (K denotes the number of users as mentioned in Section 7.1.3) when the PI passes a certain value, which is 4 with our setup. This can be explained as a double-sided effect from adding more users. When the PI takes small values, the user request probability to its favorite VCs is not significantly higher than of other VCs. Then, by adding more users to the system, there is a chance for some VCs to have their popularity added, hence becoming far higher than other VCs. Therefore, in this case, caching those VCs in a crowded system is more efficient, which explains why the 120-user system is better than the 50-user one when the user PI is less than 4.[2] However, there is a second effect that is more obvious when the user PI is high. This effect is that the added set of new users has different preferences from the first set, which originally entered in the system.

[2] The value 4 may applicable only in the presented setup. When the setup is changed, this value may change as well.

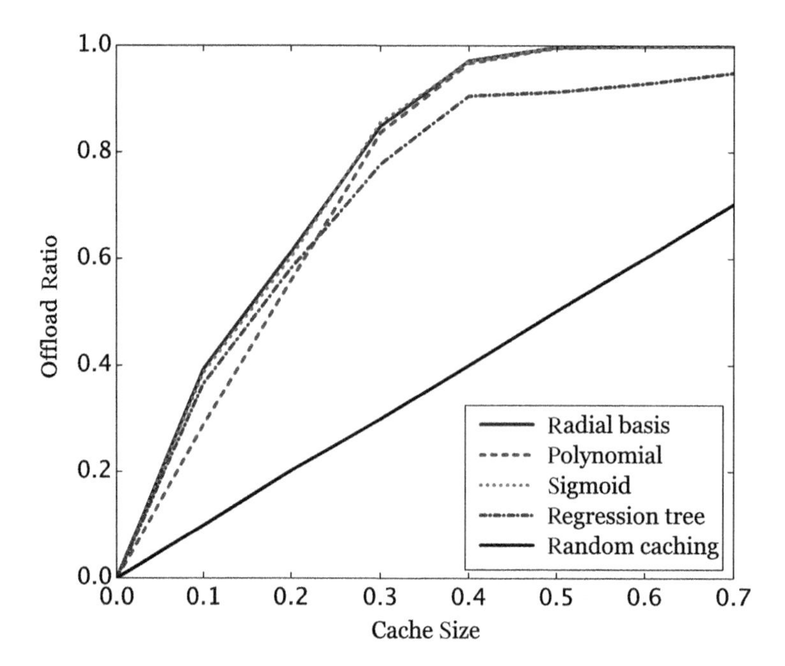

Figure 7.5 Performance of the proposed caching strategy with respect to the cache size condition with the involvement of AlexNet and various regression models. Random caching is added as a lower bound.

This effect creates more popular VCs; meanwhile, the cache capacity stays unchanged. Thus in this high-user-PI case, the offloading task in a crowded system seems to be more difficult, which can be seen as a turnaround when the user PI exceeds 4. To this point, there should be a pair of cell population and user PI such that the offloading process works at its optimum. Therefore, as the PI can be obtained (e.g., learning from history data), the cell coverage can be adjusted to have a suitable number of users within. This may be an interesting idea for future research.

In our previous experiments, AlexNet's performance is a little bit below that of other feature extractors. Hence it is chosen for the experiment shown in Figure 7.5, implying that other choices of neural network models should result in better offload ratios. This experiment focuses on examining the effect of predictor selections with 5 different options, as shown in the figure. "Random caching," where videos are cached in a completely random manner, is added and serves as a lower bound for performance evaluation. The presented method is showed to outperform the benchmark random caching, with a significant gap. Meanwhile, all the predictors seem to have similar performance in this context.

7.3 Published Set Popularity Updating

The analysis in the "popularity growth predicting" block in Figure 7.1 is presented in this section for keeping the published set information up to date. This stage is necessary due to the uncertainty of user preference.

Specifically, in each time interval, e.g., a day, the popularity of published videos in the next interval will be predicted and used to make predictions for the unpublished group. Our updating, however, will follow the *prediction with expert advice* method [21]. In practice, each expert represents a predictor associated with the same or different sources of information. For instance, one predictor is based on people's download history, the other predictor is based on unusual events, such as a scandal involving some singers or actors. Such unusual events can attract immediate community attention, which may reduce the attention paid on other content items. Hence each predictor can output a different result and each of them can be considered as advice from an expert. Then the main predictor will combine all of this advice to produce a final prediction.

In particular, let us consider a slotted time frame where a slot is equivalent to the time interval mentioned earlier. Using our notations, $p_v(t)$ will be referred to as the popularity of the video v in time slot t, and $\hat{p}_v(t)$ is a prediction of it. However, for the ease of presentation, in this section, we consider the prediction for one specific video. The process will be exactly similar when considering other videos in the published set. Therefore, the video index will be dropped, i.e., $p(t)$ and $\hat{p}(t)$ are used instead. Let assume we have E experts and each expert i gives a prediction $e_i(t)$ about $p(t)$ that is treated as a piece of advice. Then, our final prediction will be a weighted average of all such advice:

$$\hat{p}(t) = \frac{\sum_{n=1}^{E} \gamma_i(t-1)\, e_i(t)}{\sum_{n=1}^{E} \gamma_i(t-1)}. \tag{7.12}$$

For the rest of this chapter, this will be called the weighted average prediction (WAP) method. In (7.12), the nonnegative parameter $\gamma_i(t)$ is the weight given to each expert i. We will show later that this parameter is decided according to the cumulative loss of each expert i and is expressed as

$$L_i(t) = \sum_{k=1}^{t} L_{inst}(k, e_i(k)), \tag{7.13}$$

with $L_{inst}(t,x)$ indicating the instantaneous loss at time t for a prediction x. Here x may come from either experts or our predictor. This function is given by

$$L_{inst}(t,x) = (p(t) - x)^2. \tag{7.14}$$

Similarly, the cumulative loss of our predictor at t is formally presented as

$$L(t) = \sum_{k=1}^{t} L_{inst}\left(k, \hat{p}(k)\right). \tag{7.15}$$

As mentioned about the weights, this parameter set will be updated at every time slot. The general updating rule is to maintain the *cumulative regret* of the final prediction to all experts at the lowest possible level. The cumulative regret to expert i counting up to the tth slot is

$$R_i(t) = L(t) - L_i(t). \tag{7.16}$$

Based on the finding in [8] that the number of views of videos in the logarithm form at different time instants are linearly correlated. Thus our analysis follows the popularity variation hereafter:

$$p(t) = n\left(t_{ref}, t\right) p\left(t_{ref}\right) + \epsilon(t), \tag{7.17}$$

where $\epsilon(t)$ is an instantaneous variation following a 0-mean distribution. This quantity is driven by many factors such as the popularity of other videos, news, and government restriction. Some may have a positive effect, making a particular video more popular, while some others may not. Therefore, it is reasonable to have an approximation that $\epsilon(t) \sim N\left(0, \sigma^2\right)$.[3] The reference time of the considered video is t_{ref} which is usually the time when it is published. The parameter $n\left(t_{ref}, t\right)$ can be estimated from history data. Specifically, this estimation process can be described as follows.

Assume we want to estimate $n\left(t_{ref} + \tau, t_{ref}\right)$. Let $t_{v,ref}$ indicate the reference time of video v. Denote $\mathcal{S} = \{v | t_{v,ref} + \tau < t_{present}\}$, where $t_{present}$ is the current time slot. $\mathcal{S}$ is the index set of videos that have their reference slot earlier than the current time slot by at least τ slots. $\mathcal{S}$ will be used as data to estimate $n\left(t_{ref} + \tau, t_{ref}\right)$. According to [22], this estimation is expressed as

$$\hat{n}\left(t_{ref} + \tau, t_{ref}\right) = \frac{\sum\limits_{v \in \mathcal{S}} \left(p_v\left(t_{v,ref} + \tau\right) - \bar{p}_v\left(t_{v,ref} + \tau\right)\right)\left(p_v\left(t_{v,ref}\right) - \bar{p}\left(t_{v,ref}\right)\right)}{\sum\limits_{v \in \mathcal{S}} \left(p_v\left(t_{v,ref}\right) - \bar{p}\left(t_{v,ref}\right)\right)^2}, \tag{7.18}$$

where

$$\bar{p}(t) = \frac{1}{|\mathcal{S}|} \sum_{v \in \mathcal{S}} p_v(t). \tag{7.19}$$

The notation $|\bullet|$ implies the cardinality of a set. To this end, the prediction of expert i about the given video takes the following form:

$$e_i(t) = \hat{n}\left(t, t_{ref}\right) p\left(t_{ref}\right) + \phi_i(t). \tag{7.20}$$

The first right-hand-side term in (7.20) corresponds to trend prediction, while the second term is the instantaneous change prediction. In practice, different experts associated with different information sources may provide different advice regarding the instantaneous variation.

Now equation (7.13) can be expressed as

$$L_n(t) = \sum_{k=1}^{t} \left(e_i(k) - p(k)\right)^2 = \sum_{k=1}^{t} \left(\mathcal{D}_i(k) + \phi_i(k) - \epsilon(k)\right)^2, \tag{7.21}$$

with $\mathcal{D}_i(t)$ denoting the trending popularity estimation error from expert i at time t. For the presentation of subsequent subsections, we use $\star$ to denote the index of the best expert that has the smallest cumulative loss at the considered time, i.e.,

$$L_\star(t) = \min_i L_i(t). \tag{7.22}$$

[3] Normal distribution with mean 0 and variance σ^2.

7.3.1 Cumulative Loss Expectation

Different WAP methods are presented in [21] characterized by different ways to define the weight set $\gamma_i, \forall i \in \{1, \ldots, E\}$. Moreover, the difference between cumulative loss of our predictor and that of expert $\star$ is showed to be upper-bounded by a function of t and E. Let us denote that function as $h(t, E)$. Then, we have

$$L(t) \leq h(t, E) + L_\star(t), \tag{7.23}$$

where f takes different forms for different WAP methods. To this end, we can further analyze the upper bound as follows.

$$\begin{aligned} L(t) &\leq h(t, E) + L_\star(t) \\ &\leq h(t, E) + \min\left(L_i(t), L_j(t)\right), \forall i \neq j, \end{aligned} \tag{7.24}$$

which leads to

$$\mathbb{E}[L(t)] \leq h(t, E) + \min\left(\mathbb{E}\left[\min_{i \neq j}\left(L_i(t), L_j(t)\right)\right]\right), \tag{7.25}$$

where $\mathbb{E}[\bullet]$ denotes the expectation. The expectation term on the right-hand side will be derived in the next section as a two-expert scenario.

7.3.2 Two-Expert Scenario

This section is devoted to analyzing the two-expert case, so we will use indexes 1 and 2 instead of i and j. In addition, the distribution $\epsilon(t)$ is assumed to have variance 1 for simplicity. The result for the more general case of variance σ^2 can be obtained straightforwardly and will be mentioned later.

From (7.21), we have

$$\begin{aligned} L_1(t) - L_2(t) = 2\sum_{k=1}^{t}\left(\mathcal{D}_1(k) + \phi_1(k) - \mathcal{D}_2(k) - \phi_2(k)\right)\epsilon(k) \\ + \sum_{k=1}^{t}\left(\mathcal{D}_1^2(k) + \phi_1^2(k) - \mathcal{D}_2^2(k) - \phi_2^2(k)\right). \end{aligned} \tag{7.26}$$

Hence, under the condition of a given $L_2(t)$, the cumulative loss of the first expert will follow a Gaussian distribution with mean

$$\mu = \sum_{k=1}^{t}\left(\mathcal{D}_1^2(k) + \phi_1^2(k) - \mathcal{D}_2^2(k) - \phi_2^2(k)\right) \tag{7.27}$$

and standard deviation

$$\kappa = 2\sum_{k=1}^{t}\left(\mathcal{D}_1(k) + \phi_1(k) - \mathcal{D}_2(k) - \phi_2(k)\right)\epsilon(k). \tag{7.28}$$

Therefore, the probability density function (PDF) of $L_1(t)$ conditioned on $L_2(t)$ is written as

$$f_{L_1(t)|L_2(t)}(x_1|x_2) = \kappa\sqrt{2\pi}\exp\left(-\frac{1}{2\kappa^2}(x_1 - x_2 - \mu)^2\right). \tag{7.29}$$

In addition, we can figure out from (7.21) that $L_2(t)$ follows a noncentrality chi-squared distribution, thus

$$f_{L_2(t)}(x_2) = \frac{1}{2}\exp\left(-\frac{x_2 + d}{2}\right)\left(\frac{x_2}{d}\right)^{\frac{t}{4}-\frac{1}{2}} I_{\frac{t}{2}-1}\left(\sqrt{x_2 d}\right), \tag{7.30}$$

where

$$I_{\frac{t}{2}-1}\left(\sqrt{x_2 d}\right) = \sum_{k=0}^{+\infty} \frac{\left(\sqrt{d}/2\right)^{2k+t/2-1}}{k!\,\Gamma(k + t/2)} x_2^{k+t/4-1/2} \tag{7.31}$$

is the modified Bessel function of the first kind, and

$$d = \sum_{k=1}^{T} \phi_2^2(t). \tag{7.32}$$

To this end, we have

$$\begin{aligned} f_{L_1(t),L_2(t)}(x_1,x_2) &= f_{L_1(t)|L_2(t)}(x_1|x_2) \times f_{L_2(t)}(x_2) \\ &= \exp\left(-\frac{(x_1 - x_2 - \mu)^2}{2\kappa^2} - \frac{x_2 + d}{2}\right)\frac{I_{\frac{t}{2}-1}\left(\sqrt{x_2 d}\right)}{2\kappa\sqrt{2\pi}}\left(\frac{x_2}{d}\right)^{\frac{t}{4}-\frac{1}{2}}. \end{aligned} \tag{7.33}$$

Then the closed-form of $\mathbb{E}[\min(L_1(t), L_2(t))]$ can be derived following Theorem 7.1.

THEOREM 7.1 *Two random variables $L_1(t)$ and $L_2(t)$ have the joint PDF as in (7.33), then, the expectation of their minimum can be expressed as*

$$\mathbb{E}[\min(L_1(t), L_2(t))] = \frac{\exp\left(-\frac{d}{2}\right)}{4}\sum_{k=0}^{+\infty}\frac{2^{1-2k-t/2}d^k}{k!\,\Gamma(k + t/2)}\Delta_k, \tag{7.34}$$

with

$$\begin{aligned} \Delta_k = \sum_{n=0}^{k+\frac{t}{2}-1} &\left\{\binom{k + \frac{t}{2} - 1}{n}(-1)^{k+\frac{t}{2}-n}\left(\frac{1}{2\kappa^2}\right)^{\frac{n+1}{2}}\Gamma\left(\frac{n+1}{2}, \frac{\left(2\mu + \kappa^2\right)^2}{8\kappa^2}\right)\right. \\ &\left.\times \frac{\left(k + \frac{t}{2} - n - 1\right)!}{\kappa^{2k+2-2n}}\left(\mu + \frac{\kappa^2}{2} + 2\left(k + \frac{t}{2} - n\right)\right)\right\}\frac{(-2)^{k+\frac{t}{2}}\kappa^{2k+t+2}}{\exp\left(\frac{\kappa^2}{8}\right)}. \end{aligned} \tag{7.35}$$

Proof: See Section 7.5 (Appendix).

Now, for the case of σ^2-variance noise, the expression (7.21) can be modified correspondingly as follows

$$L_i(t)/\sigma^2 = \sum_{k=1}^{t}\left(\left(\mathcal{D}_i(k) + \phi_i(k)\right)/\sigma - \epsilon(k)\right). \tag{7.36}$$

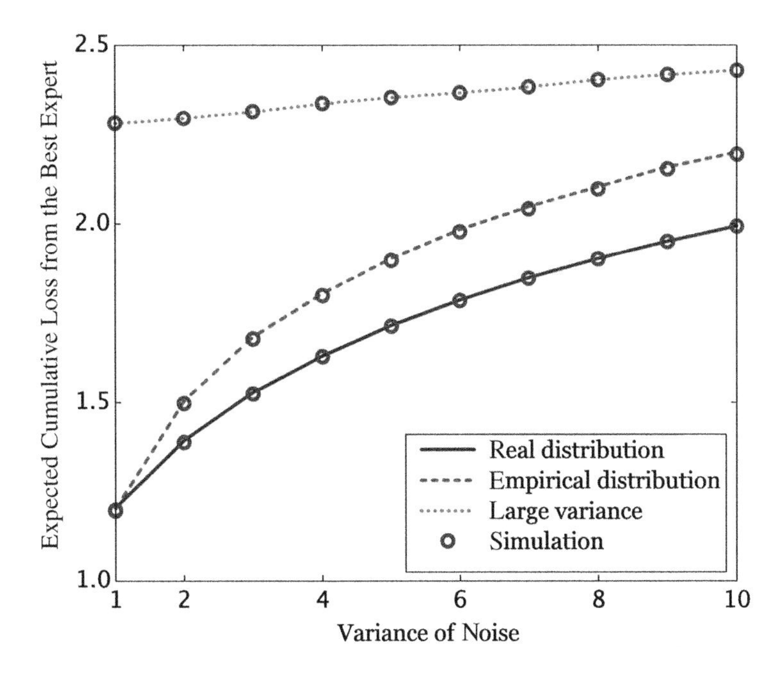

Figure 7.6 Influence of σ^2 on the best expert in two-expert scenario, with the vertical axis in a logarithm scale. Three different cases about the knowledge of distribution of the instantaneous change $\epsilon(t)$ are considered.

Note that $\epsilon(k)$ still follows the standard normal distribution. Therefore, by treating $L_i(t)/\sigma$ as $L_i'(t)$, $\mathcal{D}_i(t)/\sigma$ and $\phi_i(t)/\sigma$ as $\mathcal{D}_i'(t)$ and $\phi_i'(t)$, respectively, we obtain $\mathbb{E}\left[L_1'(t), L_2'(t)\right]$. This, by multiplying with σ^2, gives us the desired result.

Figure 7.6 presents, in a two-expert scenario, the influence of σ^2 on the expected cumulative loss of expert $\star$. This figure also verifies that our result in Theorem 7.1 matches that from the simulation. The three considered cases are when the true distribution of $\epsilon(t)$ is known, when the empirical distribution is constructed from observations and used instead, and when a very large variance Gaussian distribution is taken. The third one represents the case when experts make random guesses. Finally, the situation of multiple experts is showed in Figure 7.7. The upper bound curves are obtained by plugging the final result of this section to the second term on the right-hand side of (7.25).

7.4 Summary

In this chapter, we discuss a backhaul load alleviation method in cellular networks. The load minimization task is formulated in the form of linear programming with a closed-form solution provided. However, the solution still requires the knowledge of video popularity. Therefore, a raw-video-to-cache-content framework is presented, which estimates the probability that videos from both published and unpublished types are requested by users. This multi-step process extracts user preference from statistical

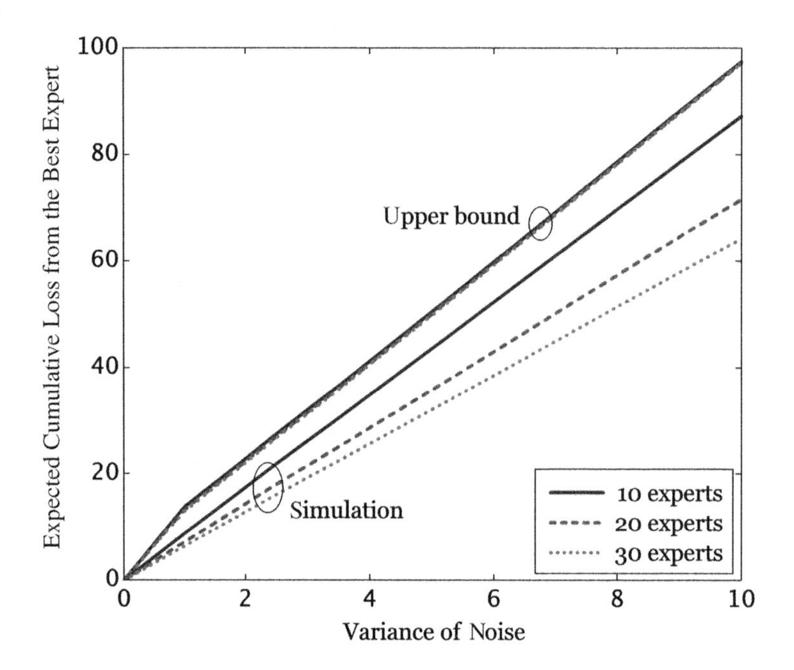

Figure 7.7 The derived upper bound with actual loss obtained via simulation in a multi-expert scenario.

information of published videos to predict the popularity of an unpublished set. In addition, due to the uncertainty of user preference, the popularity of published videos needs to be kept updating, which is done based on predictions with the expert advice method. Finally, the effectiveness of the whole procedure is illustrated via various experimental results.

7.5 Appendix: Proof of Theorem 7.1

Using (7.33), we have

$$\mathbb{P}\left(L_1(t) \geq y, L_2(t) \geq y\right) = \int_y^{+\infty}\int_y^{+\infty} f_{L_1(t),L_2(t)}(x_1,x_2)\,dx_1dx_2 \tag{7.37}$$

$$= \frac{1}{4}\int_y^{+\infty}\left(\frac{x_2}{d}\right)^{\frac{t}{4}-\frac{1}{2}} I_{\frac{t}{2}-1}\left(\sqrt{x_2 d}\right)\exp\left(-\frac{x_2+d}{2}\right)\operatorname{erfc}\left(\frac{y-x_2-\mu}{\kappa\sqrt{2}}\right)dx_2 \tag{7.38}$$

$$= \frac{1}{4}\exp\left(-d/2\right)\sum_{k=0}^{+\infty}\frac{d^k}{k!\,\Gamma\left(k+t/2\right)2^{2k+t/2-1}}\mathcal{J}\left(k,y\right), \tag{7.39}$$

where

$$\mathcal{J}(k,y) = \int_{y}^{+\infty} x_2^{k+\frac{t}{2}-1} \exp\left(-\frac{x_2}{2}\right) \operatorname{erfc}\left(\frac{y-x_2-\mu}{\kappa\sqrt{2}}\right) dx_2. \tag{7.40}$$

To this point, based on Leibniz's integral rule, taking the derivative of the CDF expression in (7.39) gives us the PDF as follows

$$\begin{aligned}\frac{\partial \mathcal{J}(k,y)}{\partial y} &= -\frac{2}{\kappa\sqrt{2\pi}} \int_{y}^{+\infty} x_2^{k+\frac{t}{2}-1} \exp\left\{-\left(\frac{y-x_2-\mu}{\kappa\sqrt{2}}\right)^2 - \frac{x_2}{2}\right\} dx_2 \\ &\quad - y^{k+\frac{t}{2}-1} \exp\left(-\frac{y}{2}\right) \operatorname{erfc}\left(-\frac{\mu}{\kappa\sqrt{2}}\right) \qquad (7.41)\\ &= -\frac{2}{\kappa\sqrt{2\pi}} \exp\left\{-\frac{(\mu-y)^2}{2\kappa^2}\right\} \mathcal{F}\left(y, \frac{1}{2\kappa^2}, \frac{\mu-y}{\kappa^2}+\frac{1}{2}, k+\frac{t}{2}-1\right) \\ &\quad - y^{k+\frac{t}{2}-1} \exp\left(-\frac{y}{2}\right) \operatorname{erfc}\left(-\frac{\mu}{\kappa\sqrt{2}}\right), \qquad (7.42)\end{aligned}$$

in which

$$\frac{\partial \operatorname{erfc}\left((y-x_2-\mu)/\left(\kappa\sqrt{2}\right)\right)}{\partial y} = -\frac{2}{\kappa\sqrt{2\pi}} \exp\left\{-\left(\frac{y-x_2-\mu}{\kappa\sqrt{2}}\right)^2\right\} \tag{7.43}$$

and

$$\begin{aligned}\mathcal{F}(n,m,q,l) &\triangleq \int_{n}^{+\infty} x^l \exp\left(-mx^2 - qx\right) dx \\ &= \frac{\exp\left(\frac{q^2}{4m}\right)}{2(-m)^{l+1}} \sum_{k=0}^{l} \binom{l}{k} m^{\frac{k+1}{2}} \left(\frac{q}{2}\right)^{l-k} \Gamma\left(\frac{k+1}{2}, \frac{(q+2mn)^2}{4m}\right).\end{aligned} \tag{7.44}$$

From (7.42), we can obtain

$$\mathbb{E}\left[\min\left(L_1(t), L_2(t)\right)\right] = \frac{1}{4} \exp\left(-\frac{d}{2}\right) \sum\nolimits_{k=0}^{+\infty} \frac{2^{1-2k-t/2} d^k}{k!\,\Gamma(k+t/2)} \Delta_k, \tag{7.45}$$

with

$$\begin{aligned}\Delta_k &= \frac{2}{\pi\kappa^2} \int_{0}^{+\infty} y \exp\left(\frac{-(\mu-y)^2}{2\kappa^2}\right) \mathcal{F}\left(y, \frac{1}{2\kappa^2}, \frac{\mu-y}{\kappa^2}+\frac{1}{2}, k+\frac{t}{2}-1\right) dy \\ &\quad + \Gamma\left(k+\frac{t}{2}\right) \left(2\kappa^2\right)^{\frac{k}{2}+\frac{t}{4}}.\end{aligned} \tag{7.46}$$

Note that with four given arguments as in (7.46), the expression (7.44) can be computed as follows:

$$\mathcal{F}\left(y, \frac{1}{2\kappa^2}, \frac{\mu - y}{\kappa^2} + \frac{1}{2}, k + \frac{t}{2} - 1\right) = \frac{\left(-2\kappa^2\right)^{k+\frac{t}{2}}}{2} \exp\left(\left(\frac{\mu - y}{\kappa\sqrt{2}} + \frac{\kappa}{2\sqrt{2}}\right)^2\right)$$
$$\times \sum_{n=0}^{k+\frac{t}{2}-1} \binom{k + \frac{t}{2} - 1}{n} \frac{\Gamma\left(\frac{n+1}{2}, \left(\frac{\mu}{\kappa\sqrt{2}} + \frac{\kappa}{2\sqrt{2}}\right)^2\right)}{\left(\frac{\mu - y}{2\kappa^2} + \frac{1}{4}\right)^{n+1-k-\frac{t}{2}}} \left(\frac{1}{2\kappa^2}\right)^{\frac{n+1}{2}}. \tag{7.47}$$

It can be seen that combining (7.47) with (7.46) yields an expression containing a sequence of sum of integrals with respect to y. Terms in that sequence have the same form, and each of them can be computed as follows:

$$\int_0^{+\infty} y\exp\left(-\frac{(\mu - y)^2}{2\kappa^2}\right) \exp\left(\left(\frac{\mu - y}{\kappa\sqrt{2}} + \frac{\kappa}{2\sqrt{2}}\right)^2\right) \left(\frac{\mu - y}{2\kappa^2} + \frac{1}{4}\right)^{k+\frac{t}{2}-n-1} dy$$
$$= (-1)^{k+\frac{t}{2}-n} 2\kappa^2 \exp\left(-\frac{\kappa^2}{8}\right) \frac{\left(k + \frac{t}{2} - n - 1\right)!}{\kappa^{2k+t-2n}} \left(\mu + \frac{\kappa^2}{2} + 2k + 2t - 2n\right), \tag{7.48}$$

where the left-hand side of (7.48) is the form of each integral term in the sequence obtained by plugging (7.47) into (7.46). Obtaining (7.45) to (7.48) completes our proof.

References

[1] "Cisco visual networking index: global mobile data traffic forecast update, 2016-2021," Cisco, Feb. 2017.

[2] "YouTube statistics," www.youtube.com/yt/press/statistics.html.

[3] Z. Chen, J. Lee, T. Q. S. Quek, and M. Kountouris, "Cooperative caching and transmission design in cluster-centric small cell networks," *IEEE Trans. Wireless Commun.*, vol. 16, no. 5, pp. 3401–3415, May 2017.

[4] J. Song, M. Sheng, T. Q. S. Quek, C. Xu, and X. Wang, "Learning based content caching and sharing for wireless networks," *IEEE Trans. Commun.*, vol 65, no. 10, pp. 4309–4324, 2017.

[5] H. T. Cheng and W. Zhuang, "Simple channel sensing order in cognitive radio networks," *IEEE J. Sel. Areas Commun.*, vol. 29, no. 4, pp. 676–688, Apr. 2011.

[6] R. Fan and H. Jiang, "Optimal multi-channel cooperative sensing in cognitive radio networks," *IEEE Trans. Wireless Commun.*, vol. 9, no. 3, pp. 1128–1138, Mar. 2010.

[7] X. Kang, Y.-C. Liang, H. K. Garg, and L. Zhang, "Sensing-based spectrum sharing in cognitive radio networks," *IEEE Trans. Veh. Technol.*, vol. 58, no. 8, pp. 4649–4654, Oct. 2009.

[8] G. Szabo and B. A. Huberman, "Predicting the popularity of online contents," *Commun. ACM*, vol. 53, no. 8, pp. 80–88, Aug. 2010.

[9] F. Figueiredo, "On the prediction of popularity of trends and hits for user generated videos," in *Proc. ACM Web Search and Data Mining Conf.*, Feb. 2013, pp. 741–746.

[10] Y. Wu, C. Wu, B. Li, L. Zhang, Z. Li, and F. Lau, "Scaling social media applications into geo-distributed clouds," *IEEE/ACM Trans. Netw.*, vol. 23, no. 3, pp. 689–702, June 2015.

[11] S. Li, J. Xu, M. V. D. Schaar, and W. Li, "Popularity-driven content caching," in *Proc. IEEE INFOCOM Conf.*, July 2016, pp. 1–9.

[12] Google, "36 mind blowing YouTube facts, figures and statistics—2017," https://fortunelords.com/youtube-statistics, Jan. 2017.

[13] F. Figueiredo, F. Benevenuto, and J. M. Almeida, "The tube over time: characterizing popularity growth of YouTube videos," in *Proc. ACM Web Search and Data Mining Conf.*, Feb. 2011, pp. 745–754.

[14] K. N. Doan, T. V. Nguyen, T. Q. S. Quek, and H. Shin, "Content-aware proactive caching for backhaul offloading in cellular network," *IEEE Trans. Commun.*, vol. 17, no. 5, pp. 3128–3140, May 2018.

[15] D. J. C. MacKay, "Fountain codes," *IEEE Proc. Commun.*, vol. 152, no. 6, pp. 1062–1068, Dec. 2005.

[16] C. M. Bishop, *Pattern Recognition and Machine Learning (Information Science and Statistics)*, Secaucus, NJ: Springer-Verlag, 2006.

[17] D. Tran, L. Bourdev, R. Fergus, L. Torresani, and M. Paluri, "Learning spatiotemporal features with 3D convolutional networks," in *Proc. IEEE International Conf. on Computer Vision*, Dec. 2015, pp. 4489–4497.

[18] T. Wu, C. Lin, and R. Weng, "Probability estimates for multi-class classification by pairwise coupling," *J. Mach. Learn. Res.*, vol. 5, pp. 975–1005, Dec 2004.

[19] K. Derpanis, M. Lecce, K. Daniilidis, and R. Wildes, "Dynamic scene understanding: the role of orientation features in space and time in scene classification," in *Proc. IEEE Conf. on Computer Vision and Pattern Recognition*, June 2012, pp. 1306–1313.

[20] K. Soomro, A. R. Zamir, and M. Shah, "UCF101: a dataset of 101 human actions classes from videos in the wild," *CoRR*, vol. abs/1212.0402, 2012.

[21] N. Cesa-Bianchi and G. Lugosi, *Prediction, Learning, and Games*, New York: Cambridge University Press, 2006.

[22] L. Wassereman, *All of Statistics: A Concise Course in Statistical Inference*, New York: Springer, 2010.

8 Wireless Edge Caching for Mobile Social Networks

Yuris Mulya Saputra, Dinh Thai Hoang, Diep Nguyen, Eryk Dutkiewicz, and Dusit Niyato

Wireless edge caching for mobile social networks (MSNs) has emerged as one of the prospective solutions for providing reliable and low-latency communication services for mobile users on social networking. In this chapter, we first give an overview of MSNs, including their development and challenges. We then discuss mobile edge caching (MEC) paradigms to address emerging issues for MSNs, e.g., service delay, users' experience, and economic efficiency. In addition to the advantages, the development of MEC networks also has some key challenges such as hierarchical architecture of MEC networks, proactive caching, privacy, and security issues. We thus review recent advanced approaches to cope with challenges when MEC networks are deployed in MSNs. Afterward, we discuss a novel MEC framework that can address some emerging issues for caching at the edges in MSNs. Specifically, the framework can authenticate MSN users based on public-key cryptography and predict their content demands utilizing a matrix factorization method. Based on the prediction, an optimal content caching policy for an MEC node is presented to minimize the average latency of all MSN users under the MEC nodes' storage capacity constraints. Furthermore, this framework provides an optimal business model to maximize the revenue for MSN service providers based on the demands of the MSN users and the obtained optimal caching policy. Through performance evaluation, we show that the considered framework outperforms other conventional policies in terms of the average delay and revenue and is expected to be the potential solution for future mobile social networking.

8.1 Introduction

The vast evolution of internet technology has promoted the development of online social networks (e.g., Twitter, Facebook, Instagram, and LinkedIn) to connect people with common interests. Following the significant improvement of mobile communication systems (e.g., cellular networks, WLAN, and device-to-device [D2D] communications) as well as the proliferation of smart devices (e.g., smartphones and tablets), the social networking services have also improved their capability in mobile social networks. A mobile social network (MSN) is a new model for mobile users to share contents and communicate through their mobile devices. The development of MSNs brings many advantages to mobile users. Specifically, MSNs can create the virtual and physical associations between MSN users and their social behaviors/relationships [1].

In general, MSNs are deployed in a centralized architecture in which all contents are stored at the content servers. This architecture enables effective resource management, thereby providing high-quality services to mobile users based on the existing network infrastructure. Nonetheless, the recent tremendous popularity of mobile applications for social networks (e.g., Instagram, Facebook, Twitter, WhatsApp, and YouTube) induces a huge demand on bulky and diverse contents (e.g., photo sharing and video streaming). Consequently, the development of MSNs leads to an explosion of mobile data traffic on the backhaul link and leads to a significant challenge for MSN service providers. In particular, a serious congestion may occur in the network, and thus MSN users can suffer long transmission latency. One possible solution is to deploy more macro base stations (MBSs) to mitigate the network traffic on the backhaul link. However, the expensive cost for upgrading the backbone network becomes a main challenge for network infrastructure providers and mobile network operators. Furthermore, the additional deployment of MBSs cannot deal with content-duplicate problems. For example, multiple MSN users may require popular contents very often concurrently. Consequently, the available bandwidth of the backhaul link is not effectively utilized due to duplicate requests.

Recently, mobile edge caching (MEC) networks have been introduced as an alternative architecture to address the aforementioned problem. An MEC network is a distributed caching network in which MEC nodes are deployed near mobile users to help them access their favorite content easier. In other words, instead of downloading content from content servers, mobile users can download requested content from MEC nodes that are located nearby to reduce the service delay. In mobile communication networks, the service delay is one of the critical parameters that determines QoS and users' satisfaction. In practice, two types of the delay are considered when a user requires a content. The first delay is called *content preprocessing time* [2], which occurs when some MSN contents have diverse data types (e.g., image processing and video transcoding). Additionally, this delay can be used when MSN users are using different kinds of smart devices. For example, when a video content is requested by two mobile users using a smartphone and a tablet, the MEC node may need to transcode for the video in order to adapt with the requested devices' screens. After the preprocessing time process, the MEC node will transmit the content to the requested user during transmission time. The transmission time is calculated based on the allocated bandwidth between the requesting user and its source node (e.g., its associated MEC node) and the content size. This type of the delay is known as *content downloading time*.

The general model of an MEC network is illustrated in Figure 8.1. This new paradigm aims to achieve low latency, high reliability, and high network effectiveness due to nearby access connections. Furthermore, the deployment of MEC networks can reduce network congestion on the backhaul links significantly. The reason is that duplicate contents can be downloaded from nearby MEC nodes. As a result, the users' experience can be greatly improved due to nearby reliable wireless access connections and low latency services. In addition, the costs of mobile network operators and the network infrastructure providers can be minimized because the network traffic can be reduced

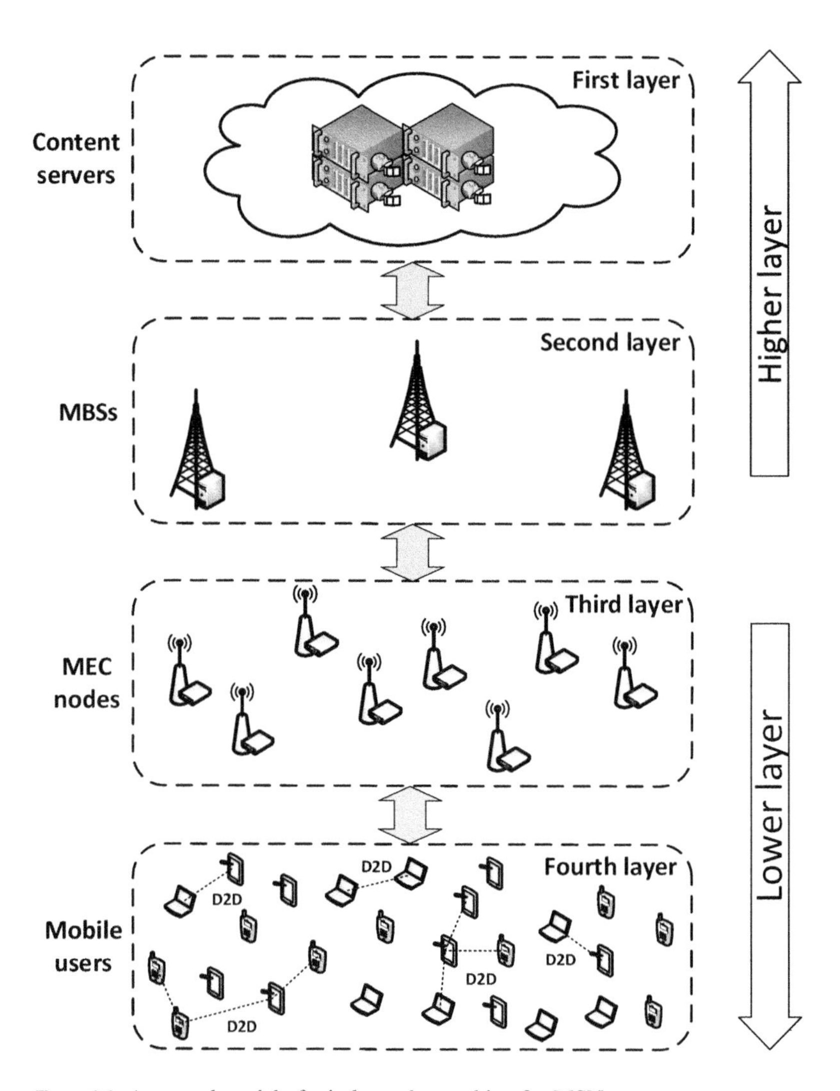

Figure 8.1 A general model of wireless edge caching for MSNs.

significantly. Overall, the potential benefits of utilizing mobile edge caching for MSNs can be summarized as follows:

- *Benefits to MSN users*: MEC nodes usually provide fast and reliable wireless network connections, e.g., WiFi connection. Thus energy consumption for communications on the mobile devices is greatly reduced. Moreover, by distributing popular content closer to the mobile users, the service delay can be significantly reduced. Therefore, this leads to a great user satisfaction on social network applications.
- *Benefits to network infrastructure providers and mobile network operators*: When MEC networks are deployed, the network traffic can be decreased significantly. This can help network infrastructure providers and mobile network operators reduce expensive operational costs on the backhaul links [3].

The reason is that they do not need to upgrade their infrastructure to meet users' demands.

- *Benefits to MSN service providers*: Since the users' experience are improved, more users may join the MSN services. Therefore, MSN service providers' profits can be increased. Furthermore, MSN service providers can achieve the optimum revenues by optimizing the MEC deployment under their users' demands.
- *Benefits to nearby local devices*: The MEC networks architecture can open potential applications through D2D communications in an ad hoc manner [4]. In particular, several MSN users' devices can collaborate with other devices to create social ties/relationship, social neighbors, and communities in an opportunistic way. For example, MSN users with common interests can share favorite entertainment program content to each other. As such, a reliable local connection such as Bluetooth, WiFi, or WiFi Direct can be directly utilized to ensure the transmission efficient of MSN contents and provide low-cost deployment [1].

To that end, the major contributions of the chapter can be summarized as follows:

- Provide an overview about MSNs and the development of MEC networks for MSNs.
- Discuss emerging MEC models, challenges, and potential solutions for the development of MEC networks for future MSNs.
- Introduce a dynamic edge caching approach that can simultaneously address users' privacy, delay tolerance, and economic efficiency.
- Highlight open issues and some potential future research directions of MEC for MSNs.

The rest of this chapter is organized as follows. In Section 8.2, we discuss challenges together with potential solutions for the development of MEC for MSNs. Then, in Section 8.3, we introduce a novel MEC framework that can simultaneously address some emerging issues in MSNs such as privacy, quality of services, and economic efficiency. Finally, Section 8.4 summarizes this chapter and highlights some open issues.

8.2 Edge Caching for Mobile Social Networks: Challenges and Solutions

8.2.1 Hierarchical Social-Network Content Caching

A general model of an MEC network for MSNs is illustrated in Figure 8.1. In this model, MEC servers are deployed in a hierarchical architecture to connect the content servers, the MBSs, the MEC nodes, and the MSN users. In particular, the hierarchical caching architecture includes four main layers. The first layer contains the content servers that store all MSN content. In a centralized architecture, this layer is the most crucial one because all content can be downloaded from the content servers only through the backhaul links. However, as mentioned in Section 8.1, the centralized architecture induces the heavy network traffic on the backhaul links. Hence, we can use an alternative distributed architecture to mitigate the workloads at the first layer. For example, some favorite MSN content can be stored at the second layer together with MBSs. Moreover, the MSN content can be stored at the MEC nodes that connect directly to the MBSs at the third layer. These MEC nodes are designed to accommodate particular purposes for MSN users with similar interests. To maximize the number of cached contents in MSNs, we can also exploit the local connections (e.g., Bluetooth, WiFi, and WiFi Direct) of smart devices via D2D communication links at the fourth layer. Based on the aforementioned distributed architecture, the network traffic on the backhaul link can be reduced significantly. Furthermore, this architecture can improve users' experience, reduce operational cost for network infrastructure providers and mobile network operators, and raise higher profits for MSN service providers.

To optimize the network traffic distribution in the hierarchical MEC networks, two techniques can be applied. First, storage capacity of a node at the higher layer is always set to be larger than that of the one at the lower layer [5]. For example, the capacity of an MBS is larger than that of an MEC node. The reason is that MBSs need to act as a central controller to maintain communications and distribute the required content between the first and third layer. Second, the peak-hour data traffic can be aggregated opportunistically according to different layers [6]. As such, if workloads cannot be served by nodes at the lower layer due to their limited capacities, nodes at the higher layers can handle those unhandled workloads. Hence a more proportional burden among nodes can be achieved. For example, suppose that storage capacity S is equally allocated into four MEC nodes and thus each MEC node has $\frac{S}{4}$ to cache MSN contents. Then, if we decrease the storage capacity of each MEC node to $\frac{S}{8}$ and apply the remaining $\frac{S}{2}$ to an MBS in the higher layer, each MEC node can serve up to $\frac{S}{8} + \frac{S}{2} = \frac{5S}{8}$ mobile users. However, this condition applies only when workloads of each MEC node has not occurred simultaneously.

The efficiency of hierarchical communications on distributing MSN workloads using a *coded caching* was further investigated in [7]. The objective is to minimize transmission rates among different layers. In particular, the content servers at the first layer are represented as the root nodes, while nodes at the second and third layers can operate as parent and child nodes, respectively. Generally, an *uncoded caching* mechanism can be used to store the same parts of contents at some nodes. For example, the first half part of two contents c_1 and c_2 are stored at two parent nodes n_1 and n_2 equally. When mobile user u_1 and u_2 request full-size content c_1 through n_1 and c_2 through n_2, respectively, the root node can transmit the second half part of those contents to the parent nodes simultaneously. In particular, the root node needs to send the second half part of c_1 to n_1 and the second half part of c_2 to n_2 via unicast transmission. As a result, the transmission rate between a root node and a parent node can be denoted by $\frac{R}{2}$, thus the total transmission rate is $\frac{R}{2} + \frac{R}{2} = R$. To reduce the total transmission rate to $\frac{R}{2}$, a coded caching mechanism using a multicasting opportunity approach is applied. In this mechanism, parent nodes n_1 and n_2 can first cache the first and second half part of the contents, respectively. Then, the root node can send common coded message to n_1 and n_2 to reconstruct full-size content via multicast transmissions. Based on the aforementioned mechanism, the coded caching between a parent node and its corresponding child nodes can also be applied to minimize the total transmission rate. It is shown that this mechanism can achieve the optimal communication rates under the limited caching storage capacity and heavy network congestion.

Although the workload distribution of MSN contents can be greatly improved, how and where to effectively cache MSN contents on different layers remain challenges. To address these issues, we need to consider the storage capacity of nodes in all layers. This aims to balance workloads between the computation and service latency in the hierarchical architecture. For example, if popular contents are mostly stored at the higher layer, it will decrease the contention at the lower layer and maximize the workload aggregation. However, this leads to a longer transmission latency. In practice, the consequence can be worse when wireless channels and mobility of MSN users dynamically change over time due to demands and social relationships. Therefore, strategies to select caching nodes and to deliver MSN nodes to mobile users need to be further examined. (i.e., *content placement and delivery* mechanisms).

8.2.2 Social-Aware Content Caching Placement and Delivery

Content placement and delivery for MSNs are always critical issues in caching strategies. The primary reason is that the distributed architecture of wireless edge caching for MSNs has the following key challenges in managing the caching policy:

- *Small coverage*: D2D communication ranges between MSN users and MEC nodes are often very limited. Specifically, Bluetooth and WiFi Direct for D2D connections have coverage ranges up to 10 and 60 m, respectively [8]. Accordingly, D2D connections are often deployed among MSN users in the proximity, e.g., a viral video is shared between two mobile users at the bus stop. Meanwhile, WiFi connections using IEEE 802.11b/g/n for MEC nodes have operation ranges up to 70 m (indoor) and 250 m (outdoor). Thus the MEC nodes are mostly deployed in specific areas to support users with similar interests. For example, MEC nodes can be deployed at commercial areas (e.g., city market and mall) to provide popular advertisement content for mobile customers. Therefore, one needs to determine the optimal caching strategy based on the social-aware and location-based demand of MSN users.
- *Diverse contents and various connections*: The optimal caching strategies are often very complicated due to the hierarchical architecture of MEC networks and the diversity of MSN content. In particular, an MEC node should identify the demand of mobile users within its coverage. Furthermore, the MEC node needs to observe the content size, preprocessing time, bandwidth capacity, and network connections to find the optimal caching strategy.
- *Limited storage capacity*: In practice, MEC nodes and mobile devices have limited storage capacity. Hence the MEC nodes and the users' devices need to select the optimal content placement and delivery mechanisms. Moreover, since the number of mobile users and data traffic have been growing [9], solutions to optimize storage capacity as well as collaborative caching among MEC nodes and mobile devices need to be further investigated.

In practice, content popularity and data traffic fluctuate due to the behavior of MSN users [10]. This behavior is influenced by several features of the social network relationship [1]:

- *Social tie*: This feature represents connections among MSN users and/or MEC nodes to share knowledge, information, feelings, and experiences. In this way, according to the range of interactions and exchanges between two users/nodes, social ties can be latent, strong, or weak.
- *Community*: This term indicates a set of MSN users with common interests based on the degree of their social ties.
- *Centrality*: When a user or its associated MEC node has a high influence on other users' decisions, that user becomes a central node. Commonly, content requested by the central node is likely to be requested by other users as well. To determine the central node, we need to consider its *degree* (i.e., the number of social ties), *closeness* (i.e., the distance among nodes), *betweenness* (i.e., the routing path selection), and *eigenvector* (i.e., the influence of nodes estimation) [11].
- *Social selfishness*: This condition takes place when a user treats other users differently based on the degree of social tie.

Based on the aforementioned challenges, the behavior of MSN users, and their social relationships, we can determine the optimal content placement and delivery solutions. The authors in [12] proposed a content placement and delivery policy based on the social relationship priority. For example, if one of the nearest MSN users caches a requested content, the requesting user can download the content via D2D links. Otherwise, the requested content can be downloaded from the serving MEC node. The lowest priority is applied when none of the available nodes caches the content. It is demonstrated that this policy can efficiently improve the ratio of delivery rate and bandwidth consumption up to 25% on the backhaul links.

Following the work in [12], the authors in [13] developed a new mechanism to allow mobile users to share MSN contents in a more efficient way. This mechanism aims to provide a content sharing scheme when the mobile users have limited D2D communication duration due to the restricted energy, i.e, battery life of their mobile devices. Consider that some mobile users are deployed as *content helpers* to cache the contents in their devices. Meanwhile, the other users act as *content requesters*, which request the contents from the content helpers via D2D links. This set of content helpers is very important for recovering the cached contents when some of the content helpers become invalid due to the aforementioned limitation. For example, if a content helper fails to find a requested content for a content requester, this invalid content helper can act as a relay to deliver content. Thus the content requester can download the content from other valid content helpers through the invalid content helper. This multi-hop mechanism leads to a larger caching capacity in the network and helps improve the cache-hit ratio when a content requester cannot obtain contents from one-hop D2D communications.

Another approach for choosing the important nodes for caching contents using D2D and MBS communications was introduced in [14]. Particularly, two social-network layers are introduced. An MBS social network is deployed at the first layer to provide sharing capability among MBSs and the *Indian buffet model* is adopted at the second layer to express the social relationship among mobile users. To select very important users and MBSs, their social-tie values from real data collection are considered. Specifically, the very important users and MBSs are chosen when their social-tie values are higher than a specified social-tie value threshold. These nodes can cache the popular content and provide traffic diversity to other requesting users and MBSs. It is shown that this approach can achieve high throughput by optimizing the D2D and MBS communication range and reducing the data traffic by offloading the traffic to the second layer.

The time duration of the social-aware content placement and delivery is also a significant factor that needs to be investigated. In particular, the authors in [7] showed that the content placement and delivery should be implemented in different time. Typically, the content placement should be performed when the network traffic is low. In the content placement process, all cache-enabled nodes can store MSN content based on their social relationship prediction. Conversely, the content delivery process should take place when the network traffic is high, and thus MSN users can request content accordingly. Following the aforementioned implementation, a content placement using time schedule along with an adaptive location caching was proposed in [15]. In particular, content cached in an MBS is adaptively changed based on the regularity estimation of the mobile users in the serving area. In this approach, the MBS will cache the specific contents according to the access history of the mobile users over specific time duration. Since MSN users are dynamically moving to different locations over time, mobility pattern of the social group can be adopted. For example, during different periods, a city park may be occupied by dissimilar groups of MSN users. A group of sport-related users may come to the park in the morning. Then, a group of students may use the park in the afternoon after completing their schoolwork. In the end of the day, a group of family-related users can walk through the park in the evening or at night. As a result, the MSN content should be updated dynamically over different time periods to satisfy the demand of dissimilar social groups. Based on their performance results and considering the time needed for content placement and delivery, it will be possible to increase the data offloading ratio and improve the caching efficiency.

The content sharing schemes also play an important role in social-aware content placement and delivery. The authors in [16] introduced a cost-sharing service using a sequential game formulation. In particular, a *Stackelberg* game is adopted to represent the interaction between representative users and a caching node. The caching node acts as the leader of the game, while the representative users are the followers. The idea is that the representative users in each community compete to obtain contents from MEC nodes or MBSs. Specifically, the users first complete the payment process based on the requested content and specific space of the caching node's storage capacity. Then, the users in the same social group will share the payment and the cached contents. Using this technique, a lower delay and higher hit ratio of distributing video content can be achieved. Similarly, [17] considered the placement cost for content caching at MEC nodes and MSN users along with an additional accessing cost for downloading required contents. These mechanisms aim to address an incentive problem due to additional costs in sharing contents among MSN users. It is shown that the proposed system can achieve optimal social group utility and outperform the baseline approaches (i.e., selfish, physical reciprocity, and social trust caching schemes). Alternatively, a sharing mechanism using seeding strategy and an opportunistic approach for MSN users was proposed in [18]. Specifically, several MSN users are selected to seed/cache the same content based on two parameters: (1) their sharing impacts in online social networking and (2) their dynamic positions in offline MSNs. These online and offline social interactions will efficiently choose MSN users who work as *followees* to share the MSN contents with their associated users who act as *followers*. Based on the performance evaluation, the mechanism can reduce the network traffic significantly while compensating the service delay of all users.

Another point of view for content placement and delivery was presented in [19]. Instead of downloading MSN content from MEC nodes or content servers, a user can generate a specific content independently. In particular, the user will first upload the generated content to the associated MEC node or to the content servers. Then, the user will suggest the content to communities and/or friends through social media networks. Subsequently, the probability of other MSN users requiring that content will be increased. It is shown that the proposed scheme can improve the service delay and cache-hit ratio compared with the conventional methods (i.e., random and auction methods).

8.2.3 Proactive and Cooperative Social-Network Caching

Proactive caching and cooperative caching are two typical schemes to increase the cache-hit rate and/or to reduce service delay of mobile users requesting MSN contents. In the following, issues and approaches for these schemes will be discussed.

8.2.3.1 Proactive Caching

The first scheme is *proactive caching*, which predicts demands of the MSN content and its popularity. Using proactive caching schemes, we can obtain the optimal caching decision and maximize the cache-hit rate based on the current status of users' requests. As a result, quality of service (QoS) for the MSN users can be enhanced.

Typically, proactive caching mechanisms are developed based on machine learning–based prediction. The authors in [20] introduced a learning-based approach to obtaining initial information of popular MSN contents. In particular, the demand estimation is determined based on the number of requests. Then, the prediction result from the estimation is applied to find the optimal social-aware caching strategy. However, to reduce workloads on the backhaul link, we need to determine the optimal time for their prediction. Hence [11] showed that MSN content can be proactively predicted during off-peak periods to enhance the number of contents cached at the MEC nodes. This prediction is generated based on frequency of accesses, association among users, and content patterns. In practice, a user is likely to appreciate MSN content that is highly recommended by other important users with similar interests. Therefore, the future correlation prediction in the network can be determined based on the interdependence of MSN users through their social links and behavior. It is demonstrated that using machine learning–based prediction in proactive caching can increase the cache-hit rate because the requested users can download more preferred contents efficiently.

In general, popularity of MSN contents is subject to change over a specified time duration due to the diversity of content requests. Therefore, users' preferences play a significant role for future prediction. This parameter may rely on the context of users [21]. Specifically, the popularity of contents is observed based on the context information of associated MSN users. This information includes users' location, individual characteristics (e.g., age, gender, personality, and mood), and users' devices' specification at a particular time. Based on the context information, the MSN service providers can provide service priority to their active users. For example, an MSN service provider will prioritize a group of MSN users requesting similar content instead of MSN users requiring the diverse content. By adding priority factors on similar content, the content delivery can be optimized. Alternatively, the lower-priority content is transmitted right after the higher-priority content is successfully downloaded by the requesting users. As a result, the cache-hit rate can be increased. Another prediction scheme using an alternative content-aware caching was introduced in [22]. In particular, the popularity of unpublished MSN content is determined according to the social influence among active users. In this scheme, the published contents are used as the training dataset. This mechanism aims to address the problem when the popularity of content increases quickly in a short period due to some reason (e.g., viral videos and copyright-violated videos). The point is that the curiosity of MSN users may trigger a tendency to try searching desired content immediately. The simulation results then demonstrate that the proposed scheme can minimize the video loads on the backhaul links effectively.

8.2.3.2 Cooperative Caching

To further leverage the optimization of edge caching for MSNs, the second mechanism, called *cooperative caching*, is investigated. This mechanism is defined as a collaboration among active nodes to cache and share MSN contents based on users' social interactions and nodes' storage capacity. In particular, the cooperative caching aims to decrease more service latency for MSN users in the network. For example, requested contents can be downloaded from the nearest adjacent nodes instead of directly downloading from the content servers. Cooperative caching mechanisms mainly focus on the collaboration with nodes at the same layer. For example, MSN users can request content from other nearby MSN users via D2D links or from adjacent MEC nodes in the network.

An example of the cooperative caching using D2D links was presented by the authors in [23]. This work proposes a fairness-aware cooperative caching to treat each user fairly based on access probabilities. Specifically, each user's device has the same probability of being a caching node to satisfy individual fairness. Another example of D2D cooperative caching using social wireless networks (SWNETs) was introduced in [24]. In particular, MSN users build a cooperative caching network to share common interests in electronic books based on Amazon Kindle system. This approach aims to minimize the cost of downloading the content from a service provider. Consequently, it leverages the sharing process locally via SWNET. To guarantee that the method is legally implemented, a peer-to-peer incentive technique is applied. Particularly, commercial benefits are shared among the cache-enabled devices, the service provider, and the MSN content providers. To improve D2D cooperative caching performance, a cooperative coded caching was proposed in [13]. As storage capacity of MSN users' devices is limited, MSN contents are cached in several devices using a coded caching mechanism. In this mechanism, fragments of the same contents are stored at different devices and transmitted using multi-hop cooperative caching efficiently. It is observed that by increasing the utility of D2D links for cooperative caching, we can obtain lower service latency and higher cache hit ratio for the requested contents.

Cooperative caching also can be performed through MEC nodes. The authors in [25] proposed a cooperative beam-forming scheme using multiple MEC nodes with limited storage capacity. The purpose is to transfer preferred MSN contents to MSN users when their signal strength is sufficient. In particular, the beam-forming cooperation among the MEC nodes occurs when multiple nodes cache the same content requested by a user. In another work [26], the authors introduced a scalable video coding caching mechanism for cooperative operators. As such, the operators cooperate using their adjacent MEC nodes to avoid long service delay from the content servers. To provide diverse qualities for each layer, this approach stores dissimilar layers of a video in different MEC nodes. When a user requests a full video, MEC nodes that contain a *base layer* (i.e., the most important information) will transfer that information first. Afterward, the *enhancement layer* (i.e., less important information) from other MEC nodes can be transmitted to provide the higher-quality video. Furthermore, another distributed caching and cooperation employing femto base stations was presented in [27]. Particularly, a group of femto base stations dynamically cooperates to send MSN content to a requesting user based on a per-user request. Finally, a combined cooperative caching mechanism to maximize the diversity of MSN content and enhance cache-hit probability was studied in [28]. In this work, several MEC nodes are reserved for caching the most popular content, while other MEC nodes can cache the less popular content. Similar to the effectiveness of cooperative caching via D2D links, the cooperative caching through MEC nodes will provide faster content delivery and better cache-hit ratio compared with the non-cooperative caching scheme.

8.2.4 Delay Tolerance Social-Network Caching Policies

Based on proactive and cooperative caching policies, we can obtain the optimal caching policy that meets the delay requirement or minimizes the service delay. The delay tolerance policy for multi-layer video contents was studied in [16]. In practice, a video content typically brings a substantial delay due to its vast data size. Thus to maximize the satisfaction and minimize the payment cost for mobile users in the social group, several layers of the video can be cached at different MEC nodes or users' devices. Similarly, the authors in [18] introduced an optimization technique to minimize the network traffic while meeting the delay condition of all MSN users. This technique will find the best content placement strategy for the selected users through online and offline social MSNs. Since each user may access mobile applications in MSNs with various frequencies of accesses, the service delay of each user may be different. Different from [16, 18], a preset delay of a requested content was introduced in [15]. Specifically, the delay is applied to choose different types of communications to send the requested content. For example, when a user requests content that can be likely transmitted within the preset delay, D2D communications can be used for this user directly. Otherwise, the user can download the content from an MEC node or an MBS. Similarly, the authors in [23] pointed out that the lifetime of each cached content may be dissimilar. The content is considered to be valid if the buffer delay does not exceed its lifetime threshold. Otherwise, the content can be considered to be useless. It is shown that the delay requirements to share and download content are important to improve the users' experience and reduce the service costs.

Another way to minimize the delay for MSN services is to further exploit optimal caching policies. For example, the workload caching problem to minimize the total delay for MSNs was introduced in [6]. In particular, MSN content is adaptively stored at different layers of MEC nodes. This aims to find computational loads for the delay minimization. Taking into account only one node for each layer, the minimum service delay is obtained. This is achieved by considering the content size, network bandwidth, amount of computations, and other nodes at the different layers. Another approach for minimizing the total delay cost of all MSN users was proposed in [5]. In this work, the delay is considered to be equivalent with the bandwidth consumption and users' experience. To address optimization problem for the delay minimization, a binary decision variable is applied to represent which server will deliver the contents. Similar to [5, 6], the authors in [29] also proposed an optimization approach to reduce the average downloading delay of all MSN users using D2D communications and femto base station (FBS) connections. Particularly, there are four main parameters used to decide content placement: network topology, content request probability, storage capacity, and bandwidth allocation. To minimize the delay, a joint strategy using two variables for caching decision and transmission policy was developed. The first decision variable represents whether the desired content should be cached at a specific node or not. Meanwhile, the second one indicates whether a particular node is chosen to send the requested content to its corresponding user or not. It is observed that the aforementioned approaches can reduce the average delay and increase the local cache hit rate efficiently.

In [25], a comparison between the centralized and distributed approaches of cached MSN contents was presented. Both approaches consider a delay minimization problem under limited storage capacity. In the centralized approach, the central controller is responsible for performing the content placement. Then to solve the optimization problem, the centralized approach adopts a greedy algorithm. On the other hand, there exists no central controller in the distributed approach, and thus the content placement is locally carried out by base stations. In the distributed approach, a distributed algorithm with low-complexity is applied to locally store the contents at the base stations. Although the centralized approach can achieve service delay performance comparable to the distributed approach, the computation time and communication overhead of distributed approach can be significantly reduced compared with those of the centralized approach.

8.2.5 Privacy and Security for Edge Caching in Mobile Social Networks

Privacy and security are also main concerns for the development of edge caching in MSNs due to many factors, such as low-security protection at MEC nodes and sensitive data of MSN users. For example, attackers can breach users' information by accessing to the MEC nodes illegally. As such, personal information of users (e.g., location, phone number, email address, and interests) can be exploited illegally when they register, login, and access MEC nodes. Furthermore, viruses can be spread among MSN users when they download content from untrustworthy users or MEC nodes. This is even worse when many social networking applications are lack of sufficient privacy protection [4]. Therefore, solutions to protect users when they access MSN contents need to be taken into consideration.

8.2.5.1 Privacy

One approach to protect the privacy of MSN users was introduced in [30]. In particular, an interest-matching mechanism to find a suitable community for the requesting user is considered. Under this mechanism, a requesting user can share only required information with other users in the same group to minimize the privacy disclosure. Another approach to avoid disclosing users' information when they share MSN contents was proposed in [31]. Instead of generating a real-time location for social relationships, a dynamic fake location for anonymous traces is applied. The purpose is to increase the location privacy without relying on any third-party application. To improve the system privacy, two privacy protection methods were deployed in [32]. In particular, location and friendship privacy protections are proposed to protect the current locations and friendship information of mobile users. In these methods, a novel cryptography technique, *functional pseudonym* is introduced to provide anonymous user's identity and location. Additionally, the technique can be used for the mobile users' public key. Based on this technique, a pair of mobile users can derive a secure shared key to represent their social ties. As a result, the current location of the mobile users and their social friendship cannot be accessed by unintended nodes including the MSN service provider. The performance results show that these methods can provide efficient system privacy for location sharing and sustainability improvement for the mobile users.

Although the secure privacy mechanisms are necessary for the MSN users, the MSN users' access history is still important for MEC nodes to perform the proactive caching process. In particular, MEC nodes need to predict the expected popular content and other users' preferences through using users' access history. Subsequently, the privacy for mobile users in MSNs should be safely considered. A mechanism for addressing the privacy issue for proactive caching was proposed in [33]. The main idea is when a user u_s requests a content from another user u_d, user u_s cannot identify if the requested content is downloaded from the storage of user u_d or not. Likewise, user u_d cannot know if the requesting user u_s requires the content through the service nodes or not. However, user u_d still knows that the requesting user is its neighbor. Alternatively, both users can provide a *plausible deniability* factor to improve the privacy among the participating users. Likewise, another method to provide a secure proactive caching using *blind cache* was introduced in [34]. In particular, blind cache is defined as an encryption method that exchanges the key of encrypted contents from content providers to MSN users. During the encryption process, the requested contents are stored at the MEC nodes through the content providers. It is proposed that the secure proactive caching mechanisms are essential for obtaining the frequency of access prediction without revealing MSN users' private information.

8.2.5.2 Security

To provide different security levels for MEC nodes, the authors in [19] presented a trust operation scheme to review the security of MEC nodes. Since MEC nodes are typically deployed by third-party providers, they can be openly accessed. Thus dissimilar levels of security may be applied for different MEC nodes. In particular, the MEC nodes are separated into four types based on their security levels. The first MEC node is called *honest node*, which serves MSN users in a secure way when contents are requested. This node has the highest level of security during the caching service process. The second MEC node is called *selfish node*. Although this node has the same security level as the honest node, the selfish node may refuse to give caching service for MSN users. The third MEC node, which provides a lower-level secure caching, is called *speculative node*. In a particular time, this node can provide a secure caching service for MSN users; however, this node can also conduct a malicious attack at another time. The fourth MEC node, which only carries out the malicious attacks to the MEC networks, is called *malicious node*. This node has the lowest level of security because its lack of sufficient secure protection. To provide a sufficient trust for the MSN users in the proposed scheme, the probabilities of honest and selfish nodes are always higher than those of speculative and malicious nodes. Based on the performance evaluation, the proposed scheme can provide a lower service delay and a higher secured caching ratio.

8.3 Dynamic Edge Caching Approach for Mobile Social Networks

In this section, we discuss a novel dynamic edge caching approach for MSNs to address some important issues such as users' privacy, delay tolerance, and economic efficiency. In particular, we describe a novel security scheme that can be applied at an MEC node to protect the privacy of MSN users. In this scheme, the authorized users are authenticated without using a centralized node and/or a security third party. In other words, the users' content access history can be accessed by the MEC node without disclosing any personal information. Then, to help the MEC node update and predict content demands of the users dynamically, a proactive caching approach using matrix factorization method is adopted. Based on this prediction, a mixed integer linear programming optimization problem is applied to obtain an optimal caching policy that minimizes the average delay for the MSN users. In addition, we discuss the optimal business model for the MSN service provider to obtain its revenue maximization.

Figure 8.2 [35] shows the considered dynamic caching framework with four main steps, i.e., authentication, dynamic demand prediction, optimal caching strategy, and business model for MSN service provider.

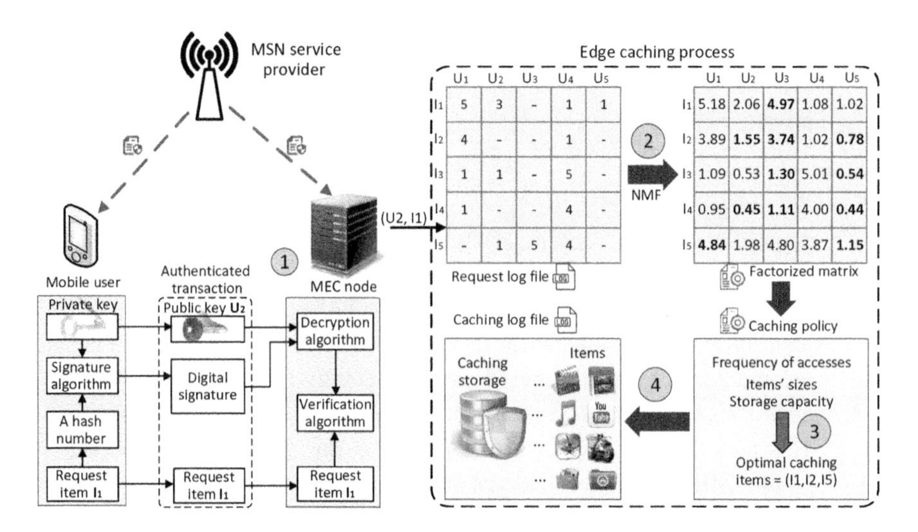

Figure 8.2 General scheme of the proposed wireless edge caching for MSNs.

8.3.1 Authentication

In this procedure, the public-key cryptography (also known as *asymmetric cryptography*) is applied. In fact, this cryptography method has been successfully implemented in many blockchain networks [36], e.g., Bitcoin, Coloredcoins, and Bitcoin cash. In particular, asymmetric cryptography consists of pairs of keys, including public keys, which may be distributed widely in MEC networks, and private keys, which are kept only by the owner. This accomplishes two functions: (1) authentication, where the public key is used to verify the holder, who sends the information, and (2) encryption, where only the paired private key holder can decrypt the information encrypted with the public key. Thus public key algorithms, unlike private key algorithms (i.e., *symmetric cryptography*), do not require a secure channel for exchanging the secret key between a sender and a receiver.

In this framework, when a mobile user associates to MSN services, the MSN service provider will assign a digital wallet to the user. This wallet aims to create a pair of public and private keys for the user. The public key can be sent to MEC nodes for verification, while the private key of the user is kept secretly by that particular user. The following procedures explain how the pair of keys is used to help the mobile user authenticate at an MEC node.

- **Step 1**: The mobile user sends an access request with the public key to the MEC node.
- **Step 2**: The MEC node verifies the public key of the user and generates a smart contract.
- **Step 3**: The MEC node sends the smart contract and its public key to the mobile user.
- **Step 4**: The mobile user verifies the public key of the MEC node, adds the content request to the contract, and signs the contract with a digital signature generated from the private key.

- **Step 5**: The mobile user encrypts the signed contract with the MEC node's public key and sends back to the MEC node.
- **Step 6**: The MEC node verifies the signed contract using its private key and the user's public key.
- **Step 7**: The MEC node sends the content to the mobile user if the contract authentication is completed.

In this process, we note that the authentication process at Step 2 is to verify that the user is allowed to use MSN services. Meanwhile, the authentication process at Step 6 is to ensure that the requested content is from the verified mobile user. Furthermore, the signature of the mobile user is dissimilar at disparate times with different content requests, and thus the private information of the user is secured. This method has the following advantages:

- Only the MEC node can decode the contract of the user. Thus only the MEC node knows the user's requested content.
- The MEC node needs to know only the public key of the user. Therefore, the private information of the user is guaranteed.
- The user can be authenticated without sharing private information. As a result, we can decrease the disclosure and spoofing risks by MEC nodes.
- The user and the MEC node can perform authentication processes even when they are not connected to the internet.
- This technique can be implemented without using a centralized node and/or a security third-party (i.e., in a decentralized way).

As a result, the proposed method can be implemented effectively for MSNs.

8.3.2 Dynamic Demand Prediction

If a content request from an authorized MSN user is verified, the user's access history is generated in the *request log file*. As shown in Figure 8.2, the file contains a two-dimensional table with two associated parameters. These parameters include the user's public keys in the columns with the corresponding requested contents in the rows. Each cell in this table indicates the frequency of access of content accessed by a user. If a particular content has never been requested by the user before, the corresponding cell is empty. However, this table may encounter some problems related to dynamic demand and big data of MSN users. Specifically, when the number of records increases over time, there is a huge amount of data stored in this table. Furthermore, the demand of users may change over time [37], e.g., some contents may be accessed during a short period of time. Thus we can implement the following schemes to address these problems. When a new user connects to the network or a new content is requested, a new column or row is added to the table, respectively. However, if a user does not request any available content in a particular period, e.g., four months, the user's information will be deleted from the existing table. Likewise, if any content has not been requested by any user in a certain period, information of the content will be removed from the table.

In this way, this policy can handle the big data problem, and the dynamic demand of users as well as the popularity of contents.

Based on the information provided in the request log file, the MEC nodes can predict the content demands of MSN users in the near future. To make the prediction, the nonnegative matrix factorization (NMF) [38] can be adopted. The NMF has been widely known as an analysis tool to extract meaningful features from a set of high-dimensional and nonnegative data vectors automatically. The reason for the popularity of using the NMF technique is because of its ability to automatically extract sparse and easily interpretable factors, as shown in [39, 40].

To clarify the benefits of the NMF, we compare the prediction matrices of the NMF with another well-known matrix factorization: singular value decomposition (SVD), which was used for proactive caching in [41]. Given an original table on the left-hand side, we can predict the demand of mobile users through the NMF and the SVD techniques as shown in Figure 8.3. For example, Figure 8.3b shows the result of using the SVD technique to find the prediction matrix, and we observe that there are some elements in the matrix with negative numbers (numbers in circles) that provide no information about demands of the users or the popularity of content. The reason is that the SVD technique shows approximated ranks of elements, and thus the elements can be negative numbers. In contrast, the NMF technique incorporates the nonnegativity elements, thus increasing the interpretability of the elements correspondingly by generating the parts-based representation [39]. As a result, elements of the prediction matrix obtained by the NMF technique can be used to predict the demand of the mobile users. For nonnegative matrices with sparse information, it has been demonstrated in some studies, e.g., [39, 40], that the NMF technique outperforms the SVD technique.

Suppose that the request log file is presented as an original matrix $\mathbf{M}$ with C rows and U columns. This matrix represents nonnegative values of current frequency of accesses of user u on content c, where u and c indicate the indices with $u \in [1, U]$ and $c \in [1, C]$. Basically, this NMF finds two new matrices $\mathbf{A}$ with C rows and N columns and $\mathbf{B}$ with N rows and U columns such that:

$$\mathbf{A} \cdot \mathbf{B} = \hat{\mathbf{M}} \approx \mathbf{M}, \tag{8.1}$$

where the elements of $\mathbf{A}$ and $\mathbf{B}$ contain nonnegative values. In particular, each row of $\mathbf{A}$ and $\mathbf{B}$ indicates a correlation strength for a user with its corresponding content and a content with its corresponding user, respectively. To find $\mathbf{A}$ and $\mathbf{B}$, the NMF utilizes an iterative process to obtain a dot product such that the difference between final factorized matrix $\hat{\mathbf{M}}$ and original matrix $\mathbf{M}$ nearly converges to a difference tolerance $\delta = |\mathbf{M} - \hat{\mathbf{M}}|$. Particularly, we set $m_{c,u}$, $a_{c,n}$, and $b_{n,u}$ as the matrix elements of $\mathbf{M}$, $\mathbf{A}$, and $\mathbf{B}$, respectively. Hence a prediction error ξ for each user and corresponding content can be calculated as follows:

$$\xi_{c,u} = m_{c,u} - \sum_{n=1}^{N} a_{c,n} b_{n,u}. \tag{8.2}$$

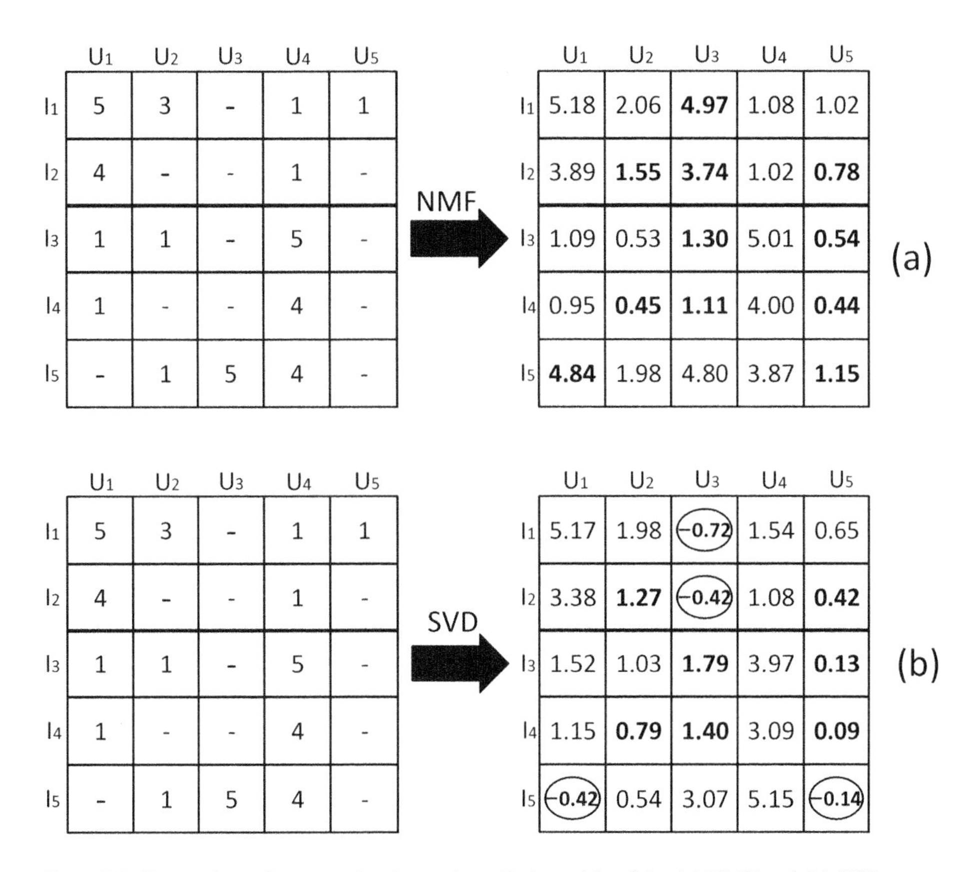

(a)

	U1	U2	U3	U4	U5
I1	5	3	-	1	1
I2	4	-	-	1	-
I3	1	1	-	5	-
I4	1	-	-	4	-
I5	-	1	5	4	-

NMF →

	U1	U2	U3	U4	U5
I1	5.18	2.06	**4.97**	1.08	1.02
I2	3.89	**1.55**	**3.74**	1.02	**0.78**
I3	1.09	0.53	**1.30**	5.01	**0.54**
I4	0.95	**0.45**	**1.11**	4.00	**0.44**
I5	**4.84**	1.98	4.80	3.87	**1.15**

(b)

	U1	U2	U3	U4	U5
I1	5	3	-	1	1
I2	4	-	-	1	-
I3	1	1	-	5	-
I4	1	-	-	4	-
I5	-	1	5	4	-

SVD →

	U1	U2	U3	U4	U5
I1	5.17	1.98	**-0.72**	1.54	0.65
I2	3.38	**1.27**	**-0.42**	1.08	**0.42**
I3	1.52	1.03	**1.79**	3.97	**0.13**
I4	1.15	**0.79**	**1.40**	3.09	**0.09**
I5	**-0.42**	0.54	3.07	5.15	**-0.14**

Figure 8.3 Comparisons between the demand prediction table of the (a) NMF and (b) SVD techniques.

To update the rules iteratively, a gradient descent technique is adopted to modify the element $a_{c,n}$ and $b_{n,u}$ such that:

$$\hat{a}_{c,n} = a_{c,n} + \mu \left(\xi_{c,u} a_{c,n} - \eta b_{n,u} \right), \tag{8.3}$$

and

$$\hat{b}_{n,u} = b_{n,u} + \mu \left(\xi_{c,u} b_{n,u} - \eta a_{c,n} \right), \tag{8.4}$$

where μ and η are applied to handle the rate of updated elements and the magnitudes of original elements, respectively.

To minimize the final prediction error, we can adopt a Frobenius form, which is expressed as $||\mathbf{M} - \mathbf{AB}||_{\mathbf{F}}^2$. Then, the regularized squared error is obtained as:

$$\min \sum_{c \in \mathcal{C}, u \in \mathcal{U}} \left(m_{c,u} - \sum_{n=1}^{N} \hat{a}_{c,n} \hat{b}_{n,u} \right)^2 + \eta \sum_{n=1}^{N} \left(||\hat{a}_{c,n}||^2 + ||\hat{b}_{n,u}||^2 \right), \tag{8.5}$$

$$\text{s.t. } \hat{a}_{c,n} \geq 0, \hat{b}_{n,u} \geq 0. \tag{8.6}$$

This procedure terminates when the prediction error converges or the specified number of iterations is attained [42]. Based on the aforementioned process, we can obtain the factorized matrix $\hat{\mathbf{M}}$ from $\mathbf{A}$ and $\mathbf{B}$.

8.3.3 Optimal Caching Strategy

After the frequency of access prediction matrix is obtained by the NMF technique, the optimal caching strategy for the MEC node can be derived. Specifically, we denote $\hat{\mathbf{M}}$ with C rows and U columns corresponding to C contents and U users. Different content may have disparate data sizes $\mathbf{z} = \{z_1, \ldots, z_C\}$. Furthermore, different content may require dissimilar preprocessing time before the content is downloaded by the users. For example, mobile devices typically have small screen with different resolutions. Hence, before the users download the requested MSN content (e.g., videos or images), the content may need to be adjusted into the appropriate size to fit the device. Furthermore, when MSN users request contents from an MEC node at different locations, the users may have different signal strengths to receive the contents. In this case, the MEC node needs to adjust the content size using particular compression methods, e.g., JPEG and MPEG standards, to maintain users' experience. Since the MEC node has limitation on the hardware and computing resources, its content preprocessing time may be different from the remote content server. Therefore, we denote $\mathbf{t}^{mec} = \{t_1^{mec}, \ldots, t_C^{mec}\}$ and $\mathbf{t}^{cs} = \{t_1^{cs}, \ldots, t_C^{cs}\}$ as the content preprocessing time at the MEC nodes and the content server, respectively.

Suppose that the allocated bandwidth between a user u to the MEC node and the MEC node to the content server are denoted by b_u and B, respectively. Then, the delay d_α when a user u wants to download a content c from the MEC node is

$$d_\alpha = \frac{z_c}{b_u} + t_c^{mec}, \tag{8.7}$$

and the delay d_β when a user u needs to download the content c from the content server is

$$d_\beta = \frac{z_c}{b_u} + \frac{z_c}{B} + t_c^{cs}. \tag{8.8}$$

If we define Z as storage capacity of the MEC node, the problem formulation for minimizing the total delay of all MSN users can be expressed as follows:

$$\min_{\mathbf{X}} f(\mathbf{X}) = \sum_{c=1}^{C} \sum_{u=1}^{U} \hat{m}_{c,u} \left(x_c d_\alpha + (1 - x_c)\, d_\beta \right), \tag{8.9}$$

$$\text{s.t.} \quad \sum_{c=1}^{C} x_c z_c \leq Z, \tag{8.10}$$

$$\text{and } x_c \in \{0,1\}, \ \forall c \in \{1,2,\ldots,C\}, \tag{8.11}$$

where $\mathbf{X} = [x_1, \ldots, x_C]^\top$ is a vector of binary decision variables and $\hat{m}_{c,u}$ is the element of matrix $\hat{\mathbf{M}}$ at row c and column u. Based on (8.9), we can divide the objective function into two terms. In particular, the first term indicates the total delay of requested contents downloaded from the MEC node. Meanwhile, the second term represents the transmission delay when the requested contents are downloaded from the content server. Furthermore, the constraint in (8.10) implies that the number of cached MSN contents at the MEC node cannot exceed its storage capacity. Then, the second constraint is to ensure that variables are binary. Specifically, the values 1 and 0 correspond to decisions that a content is *cached* and *not cached* at the MEC node, respectively. According to (8.11), since $\mathbf{X}$ are binary variables, a mixed integer nonlinear programming (MILP) can be adopted to find the caching policy for the MEC node.

8.3.4 Business Model of MSN Service Provider

Practically, when the MSN service provider deploys MEC nodes to deliver contents, it often collaborates with third-party partners (e.g., Microsoft Azure [3] and Akamai). This is to reduce the implementation and deployment costs of MEC nodes. Nevertheless, the deployment of MEC nodes is not free of charge. In particular, MEC nodes with various storage capacities can be offered at different prices. Thus we denote $\mathcal{S} = \{S_1, \ldots, S_i, \ldots, S_I\}$ as the set of bundles available for the MSN service provider to choose from. Each bundle has a renting cost C_i with a corresponding caching capacity Z_i for a certain period. Based on the caching capacity Z_i, the optimal caching policy to minimize the total delay of all MSN users can be obtained by solving (8.9). Generally, when the caching capacity is bigger, the number of cached contents can be increased, and thus the average delay can be reduced. However, in this case, we need to trade off between the delay minimization and renting cost.

To obtain the average revenue $\overline{\mathcal{R}}$ at the MEC node, we denote s_i as the cost for a user to download a data unit (e.g., \$0.1 per 10 MB) over an average threshold delay d_i (e.g., 1 second per 10 MB). Then a representation of caching storage capacity function can be expressed as follows:

$$s_i = s^0 + \gamma\left(d^0 - \hat{d}_i^*\right), \tag{8.12}$$

where s^0 and d^0 are the base cost and average delay when there is no caching at the MEC node, respectively. Meanwhile, $\gamma > 0$ and $\hat{d}_i^*$ represent a conversion parameter and optimal average delay obtained from (8.9), respectively. The average demand $\overline{\mathcal{P}}$ at the MEC node over the period is determined from frequency of access prediction $\hat{\mathbf{M}}$ and size of each content z_c as follows:

$$\overline{\mathcal{P}} = \sum_{c=1}^{C}\sum_{u=1}^{U} \hat{m}_{c,u} z_c. \tag{8.13}$$

Then, based on (8.12) and (8.13), the average revenue at the MEC node is computed as:

$$\overline{\mathcal{R}} = s_i \times \overline{\mathcal{P}}. \tag{8.14}$$

When the storage capacity is bigger, the average delay becomes lower, and thus the offered price becomes higher. Then the average revenue of the MSN service provider for bundle S_i is computed as follows:

$$\overline{\mathcal{R}}(S_i) = s_i \times \overline{\mathcal{P}} - C_i. \tag{8.15}$$

If the service bundle S_i is known, the corresponding offered price s_i and renting cost C_i can be obtained. Therefore, the optimal service bundle $\hat{S}$ to maximize the average revenue of the MSN service provider can be determined accordingly.

8.3.5 Performance Evaluation

We present numerical results under the following parameters to evaluate the performance of the proposed approaches. The available bandwidth between a user and the MEC node is 5 Mbps. The bandwidth 500 Mbps is allocated for the connection between the MEC node and the content server (i.e., backhaul link). The number of MSN contents is 10. Each content has a different data size and frequency of access, which are shown in Table 8.1. Particularly, a set of frequency of accesses is normalized, and it can be obtained from the matrix $\hat{\mathbf{M}}$ using the aforementioned NMF technique.

To evaluate progressive trend of the average delay of all MSN users, the storage capacity is varied from 0 to 600 MB. We also exploit the performance with and without content preprocessing time to show the impact of the preprocessing time. The content preprocessing time for the MEC node and the content server is shown in Table 8.1. Furthermore, to evaluate performance of the proposed solution, the *most frequently accessed* (MFA) and *noncaching* policies are used. In particular, for the MFA policy, content with a high frequency of access has a higher priority to be cached at the MEC node. Conversely, noncaching policy is to represent that there is no content cached at the MEC node.

We first discuss the average delay in two cases, i.e., without and with preprocessing time, as shown in Figure 8.4. Particularly, in Figure 8.4a, we show the average delay of mobile users when the preprocessing time of all content is zero. In this case, the average delay obtained by the MFA policy is relatively close to that of the optimal

Table 8.1 The Size, Frequency of Access, and Preprocessing Time of the MSN Content at the MEC Node and Content Server

Content	Size (MB)	Frequency of Access	t_c^{mec} (s)	t_c^{cs} (s)
1	70	38	1.5	0.5
2	50	32	0	0
3	5	41	0	0
4	10	42	0	0
5	40	39	0.5	0.1
6	75	28	0	0
7	20	37	0	0
8	85	40	2	0.5
9	47	22	0	0
10	30	12	0	0

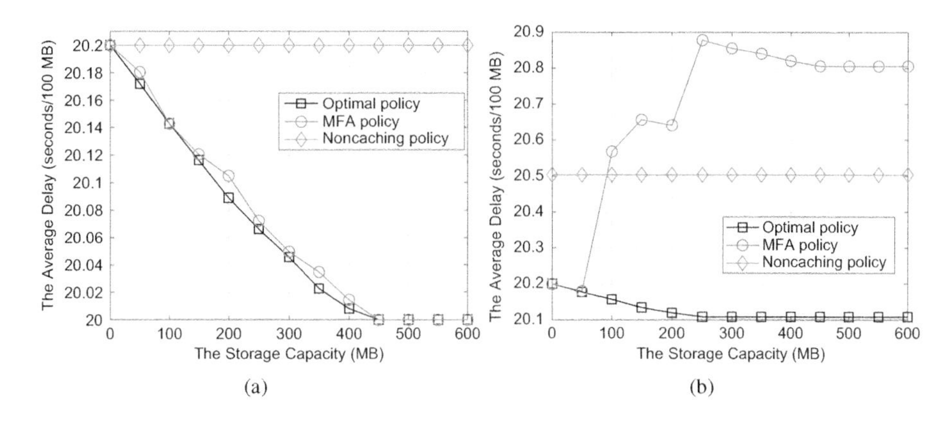

Figure 8.4 The average delay of all MSN users as the storage capacity increases (a) without considering preprocessing time and (b) with preprocessing time.

policy. The reason is that when we solve Eq. (8.9) to find the optimal caching policy without considering the content preprocessing time, the frequency of access is the most important factor. However, when we solve (8.9) with preprocessing time taken into consideration, the average delay of the mobile users obtained by the optimal policy is much lower than that of the MFA policy, as shown in Figure 8.4b. The reason is that the MFA policy considers only the frequency-of-access factor. Consequently, for the solutions without accounting for the content preprocessing time as the MFA policy, the average delay of mobile users is even higher than that of noncaching policy. In this case, some content may require a long time to be preprocessed at the MEC node due to its hardware constraints, and hence they should be preprocessed at the content server instead of the MEC node. This implies that the preprocessing time is also an important factor that significantly impacts the average delay of mobile users.

Given the users' demand at the MEC node, the offered MSN service price, and the caching cost, we then analyze the business model for the MSN service provider. The base price is set at 0.01 monetary unit (MU) per 1 MB. Furthermore, γ is set at 0.05. We set the caching costs at 30 MUs and 7.5 MUs for the first 50 MB and the next 50 MB, respectively. As observed in Figure 8.5a, users with a full demand (i.e., 100% demand) occurs at the MEC node. In particular, considering the caching costs and the users' demand at this location, the MSN service provider will rent 400 MB to obtain the maximum revenue. However, if the users' demand reduces into 70%, the MSN service provider will rent only 350 MB as seen in Figure 8.5b. Based on the results in Figure 8.5, the proposed optimal policy outperforms the MFA and noncaching policies in obtaining the average revenue.

Next, we explain the convergence of the MFA and the optimal policy. In Figure 8.4, the average delay obtained by both policies will converge. In particular, as the storage capacity increases, the average delays obtained by the MFA and the optimal policy in Figure 8.4a converge to 20 s at 450 MB. Meanwhile, the convergence of the average delays to 20.8 s at 450 MB and 20.11 s at 250 MB are obtained by the MFA and

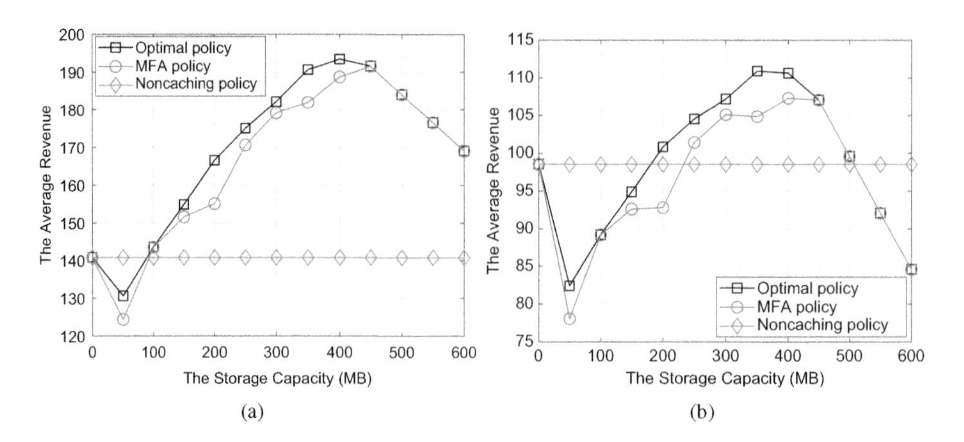

Figure 8.5 The average revenue as the storage capacity increases (a) with full demand and (b) with 70% demand of MSN users at the MEC node.

the optimal policy in Figure 8.4b, respectively. The reason is that when the storage capacity of the MEC node is very large, it can contain all MSN contents. Hence when we keep increasing the storage capacity, the average delay will not change because now all content will be downloaded directly from the MEC node. However, the average revenue obtained by the MFA and optimal policy will not converge when we increase the storage capacity as shown in Figure 8.5. The reason can be explained through the average revenue function: $\overline{\mathcal{R}}(S_i) = s_i \times \overline{\mathcal{P}} - C_i$. As explained in Section 8.3.4, given a service bundle S_i, we can obtain the offered price s_i and corresponding renting cost C_i. Thus, given the average demand $\overline{\mathcal{P}}$ for the MEC node, we can find the optimal service bundle that maximizes the average revenue for the MSN service provider. As a result, as shown in Figure 8.5, if we keep increasing the storage capacity, the average revenue will be reducing.

8.4 Conclusions and Open Issues

In this chapter, we first provided an overview of MSNs, including their development and challenges. We then presented wireless edge caching, a new caching model, to address many problems for MSNs such as delay, quality of service, and economic efficiency. Although mobile edge caching networks bring many benefits to MSN users, MSN service providers, mobile network operators, and network infrastructure providers, they also place some new challenges. These include hierarchical architecture of MEC networks, content placement, proactive caching, privacy, and security problems at MEC nodes. We then reviewed recent approaches in the literature to address the problems when deploying MEC for MSNs. After that, we discussed the MEC framework, which can address some urgent issues for edge caching in MSNs. This framework includes the authentication method using public key cryptography, the proactive caching scheme using NMF algorithm, the optimal caching policy, and a business strategy using advanced optimization techniques. Some numerical results were presented to show the potential of the considered framework.

However, there are still some open issues that need to be further investigated:

- *Implementation of MEC networks*: To enhance users' experience and minimize the service delay, the MSN service provider can implement more MEC nodes in its serving area. However, this will increase deployment costs. Furthermore, due to dynamic mobility of MSN users, the MSN service provider needs to observe the users' demands in different locations. Therefore, the efficiency of MEC networks deployment should be further investigated.
- *Collaborative caching in MEC networks*: For a cooperative caching network, an MEC node can collaborate to share content with other MEC nodes. In this way, the MEC node needs to determine whether to cache the requested content or not, and if the MEC node decides to not cache the requested content, where to download them from. Nonetheless, the popularity of MSN content and associations among MEC nodes change dynamically over time. Hence how to create a dynamic protocol to (1) cooperatively cache and update the content information from other MEC nodes and then (2) find an optimal caching strategy in MEC networks remains as an open issue.
- *Payment management in MEC networks*: When a user requests content from an MEC node, that user needs to pay for the request. Typically, this payment information is sent to the MSN service provider in a centralized way. However, this may not be efficient when a huge number of MSN users request at the same time. Thus we can adopt blockchain technology [43] to manage users' payment and information in a decentralized architecture. In particular, we can secure the payment transactions through mining processes without relying on a central controller. However, how to verify the miners and how to incentive MEC nodes to mine information are still open issues for future researches.

References

[1] N. Vastardis and K. Yang, "Mobile social networks: architectures, social properties, and key research challenges," *IEEE Communications Surveys & Tutorials*, vol. 15, no. 3, 2013, pp. 1355–1371.

[2] T. X. Tran, A. Hajisami, P. Pandey, and D. Pompili, "Collaborative mobile edge computing in 5G networks: new paradigms, scenarios, and challenges," *IEEE Communications Magazine*, vol. 55, no. 4, Apr. 2017, pp. 54–61.

[3] "Azure pricing," Microsoft Azure, https://azure.microsoft.com/en-us/pricing/details/managed-cache/, accessed June 2018.

[4] X. Hu, T. H. S. Chu, V. C. M. Leung, E. C. H. Ngai, P. Kruchten, and H. C. B. Chan, "A survey on mobile social networks: applications, platforms, system architectures, and future research directions," *IEEE Communications Surveys & Tutorials*, vol. 17, no. 3, Aug. 2015, pp. 1557–1581.

[5] T. X. Tran, A. Hajisami, and D. Pompili, "Cooperative hierarchical caching in 5G cloud radio access networks," *IEEE Network*, vol. 31, no. 4, July 2017, pp. 35–41.

[6] L. Tong, Y. Li, and W. Gao, "A hierarchical edge cloud architecture for mobile computing," *IEEE INFOCOM*, Apr. 2016, pp. 1–9.
[7] N. Karamchandani, U. Niesen, M. A. Maddah-Ali, and S. N. Diggavi, "Hierarchical coded caching," *IEEE Transactions on Information Theory*, vol. 62, no. 6, June 2016, pp. 3212–3229.
[8] E. Geier, "Discovering the upcoming Wi-Fi direct standard," Cisco, Aug. 2010, www.ciscopress.com/articles/article.asp?p=1620205&seqNum=2.
[9] "Cisco mobile visual networking index (VNI)," Cisco, Feb. 2017, https://newsroom.cisco.com/press-release-content?articleId=1819296.
[10] Y. Wu, S. Yao, Y. Yang, T. Zhou, H. Qian, H. Hu, and M. Hamalainen, "Challenges of mobile social device caching," *IEEE Access*, vol. 4, Jan. 2017, pp. 8938–8947.
[11] E. Bastug, M. Bennis, and M. Debbah, "Living on the edge: the role of proactive caching in 5G wireless networks," *IEEE Communications Magazine*, vol. 52, no. 8, Aug. 2014, pp. 82–89.
[12] X. Wang, S. Leng, and K. Yang, "Social-aware edge caching in fog radio access networks," *IEEE Access*, vol. 5, June 2017, pp. 8492–8501.
[13] L. Wang, H. Wu, Z. Han, P. Zhang, and H. V. Poor, "Multi-hop cooperative caching in social IoT using matching theory," *IEEE Transactions on Wireless Communications*, vol. 17, no. 4, Apr. 2018, pp. 2127–2145.
[14] X. Zhang, Y. Li, Y. Zhang, J. Zhang, H. Li, S. Wang, and D. Wang, "Information caching strategy for cyber social computing based wireless networks," *IEEE Transactions on Emerging Topics in Computing*, vol. 5, no. 5, July 2017, pp. 391–402.
[15] R. Wang, X. Peng, J. Zhang, and K. B. Letaief, "Mobility-aware caching for content-centric wireless networks: modeling and methodology," *IEEE Communications Magazine*, vol. 54, no. 8, Aug. 2016, pp. 77–83.
[16] Z. Su, Q. Xu, F. Hou, Q. Yang, and Q. Qi, "Edge caching for layered video contents in mobile social networks," *IEEE Transactions on Multimedia*, vol. 19, no. 10, Oct. 2017, pp. 2210–2221.
[17] K. Zhu, W. Zhi, X. Chen, and L. Zhang, "Socially motivated data caching in ultra-dense small cell networks," *IEEE Network*, vol. 31, no. 4, July 2017, pp. 42–48.
[18] X. Wang, M. Chen, Z. Han, D. O. Wu, and T. T. Kwon, "TOSS: traffic offloading by social network service-based opportunistic sharing in mobile social networks," *IEEE INFOCOM*, Apr. 2014, pp. 2346–2354.
[19] Q. Xu, Z. Su, Q. Zheng, M. Luo, and B. Dong, "Secure content delivery with edge nodes to save caching resources for mobile users in green cities," *IEEE Transactions on Industrial Informatics*, vol. 14, no. 6, June 2018, pp. 2550–2559.
[20] B. N. Bharath, K. G. Nagananda, and H. V. Poor, "A learning-based approach to caching in heterogenous small cell networks," *IEEE Transactions on Communications*, vol. 64, no. 4, Apr. 2016, pp. 1674–1686.
[21] S. Muller, O. Atan, M. van der Schaar, and A. Klein, "Context-aware proactive content caching with service differentiation in wireless networks," *IEEE Transactions on Wireless Communications*, vol. 16, no. 2, Feb. 2017, pp. 1024–1036.
[22] K. N. Doan, T. Van Nguyen, T. Q. S. Quek, and H. Shin, "Content-aware proactive caching for backhaul offloading in cellular network," *IEEE Transactions on Wireless Communications*, vol. 17, no. 5, May 2018, pp. 3128–3140.
[23] D. Wei, K. Zhu, and X. Wang, "Fairness-aware cooperative caching scheme for mobile social networks," *IEEE ICC*, June 2014, pp. 2484–2489.

[24] M. Taghizadeh, K. Micinski, S. Biswas, C. Ofria, and E. Torng, "Distributed cooperative caching in social wireless networks," *IEEE Transactions on Mobile Computing*, vol. 12, no. 6, June 2013, pp. 1037–1053.

[25] J. Liu, B. Bai, J. Zhang, and K. B. Letaief, "Cache placement in fog-RANs: from centralized to distributed algorithms," *IEEE Transactions on Wireless Communications*, vol. 16, no. 11, Nov. 2017, pp. 7039–7051.

[26] K. Poularakis, G. Iosifidis, A. Argyriou, I. Koutsopoulos, and L. Tassiulas, "Caching and operator cooperation policies for layered video content delivery," *IEEE INFOCOM*, Apr. 2016, pp. 1–9.

[27] W. C. Ao and K. Psounis, "Fast content delivery via distributed caching and small cell cooperation," in *IEEE Transactions on Mobile Computing*, vol. 17, no. 5, pp. 1048–1061, 1 May 2018.

[28] Z. Chen, J. Lee, T. Q. S. Quek, and M. Kountouris, "Cooperative caching and transmission design in cluster-centric small cell networks," *IEEE Transactions on Wireless Communications*, vol. 16, no. 5, May 2017, pp. 3401–3415.

[29] W. Jiang, G. Feng, and S. Qin, "Optimal cooperative content caching and delivery policy for heterogeneous cellular networks," *IEEE Transactions on Mobile Computing*, vol. 16, no. 5, May 2017, pp. 1382–1393.

[30] M. Li, N. Cao, S. Yu, and W. Lou, "FindU: privacy-preserving personal profile matching in mobile social networks," *IEEE INFOCOM*, Apr. 2011, pp. 2435–2443.

[31] W. Chang, J. Wu, and C. C. Tan, "Friendship-based location privacy in mobile social networks," *International Journal of Security and Networks*, vol. 6, no. 4, Jan. 2011, pp. 226–236.

[32] J. Son, D. Kim, M. Z. A. Bhuiyan, R. Tashakkori, J. Seo, and D. H. Lee, "Privacy enhanced location sharing for mobile online social networks," *IEEE Transactions on Sustainable Computing*, 2018, pp. 1–1.

[33] S. Nikolaou, R. Van Renesse, and N. Schiper, "Proactive cache placement on cooperative client caches for online social networks," *IEEE Transactions on Parallel and Distributed Systems*, vol. 27, no. 4, Apr. 2016, pp. 1174–1186.

[34] G. Eriksson, J. Mattsson, N. Mitra, and Z. Sarker, "Blind cache: a solution to content delivery challenges in an all-encrypted web," *Ericsson White Paper*, 2016.

[35] D. T. Hoang, D. Niyato, D. Nguyen, E. Dutkiewicz, P. Wang, and Z. Han, "A dynamic edge caching framework for mobile 5G networks," *IEEE Wireless Communications*, vol. 25, no. 5, pp. 95–103, 2018.

[36] K. Christidis and M. Devetsikiotis, "Blockchains and smart contracts for the internet of things," *IEEE Access*, vol. 4, June 2016, pp. 2292–2303.

[37] M. Cha, H. Kwak, R. Rodriguez, Y. Y. Ahn and S. Moon, "Analyzing the video popularity characteristics of large-scale user generated content systems," *IEEE/ACM Transactions on Networking*, vol. 17, no. 5, Oct. 2009, pp. 1357–1370.

[38] Y. Koren, R. Bell, and C. Volinsky, "Matrix factorization techniques for recommender systems," *Computer*, vol. 42, no. 8, Aug. 2009, pp. 30–37.

[39] Y. X. Wang and Y. J. Zhang, "Nonnegative matrix factorization: a comprehensive review," *IEEE Transactions on Knowledge and Data Engineering*, vol. 25, no. 6, June 2013, pp. 1336–1353.

[40] N. Gillis. "The why and how of nonnegative matrix factorization." *Regularization, Optimization, Kernels, and Support Vector Machines*, Machine Learning and Pattern Recognition Series, New York: Chapman & Hall/CRC, Oct. 2014.

[41] E. Zeydan, E. Bastug, M. Bennis, M. A. Kader, I. A. Karatepe, A. S. Er, and M. Debbah, "Big data caching for networking: moving from cloud to edge," *IEEE Communications Magazine*, vol. 54, no. 9, Sept. 2016, pp. 36–42.

[42] Y. Mao, L. K. Saul, and J. M. Smith, "IDES: an internet distance estimation service for large networks," *IEEE Journal on Selected Areas in Communications*, vol. 24, no. 12, Dec. 2006, pp. 2273–2284.

[43] F. Tschorsch and B. Scheuermann, "Bitcoin and beyond: a technical survey on decentralized digital currencies," *IEEE Communications Surveys & Tutorials*, vol. 18, no. 3, Jan. 2016, pp. 2084–2123.

9 A Proactive and Big Data–Enabled Caching Analysis Perspective

Engin Zeydan, Ejder Baştuğ, Mehdi Bennis, and Mérouane Debbah

Large-scale data analysis is becoming an important source of information for mobile network operators (MNOs). MNOs can now investigate the feasibility of possible new technological advances such as storage/memory utilization, context-awareness and edge/cloud computing using analytic platforms designed for big data processing. Within this context, studying caching from a mobile data traffic analytical perspective can offer rich insights on evaluating the potential benefits and gains of proactive caching at base stations. In this chapter, we study how data collected from MNOs can be leveraged using machine learning tools in order to infer insights into the benefits of caching. Through our practical architecture, a vast amount of data can be harnessed for content popularity estimations and placing strategic contents at base stations (BSs). Our results demonstrate several gains in terms of both content demand satisfaction and backhaul offloading rates while utilizing real-world data sets collected from a major MNO.

9.1 Introduction

The continuous increase in mobile data traffic demand due to video, social networks, and over-the-top (OTT) applications are pushing mobile network operators to search for new ways to handle their growing complex networks. This increase of traffic from diverse domains (e.g., Internet of things [IoT], healthcare, autonomous cars, user-created content, smart metering) have different data structures (e.g., structured/nonstructured, semi-structured) and is usually called big data [1]. While big data brings "big blessings," there are compelling challenges in handling large-scale data sets due to the tremendous volume and dimensionality of the data. A primary challenge of big data analytics for decision making is to move through huge chunks of data to uncover hidden patterns. In fact, the time of gathering and storing data in remote standalone servers for offline decision making is outdated. Instead, mobile network operators are looking to decentralized and scalable network architectures in which anticipatory resource management plays an important role exploiting recent advances in storage/memory, context-awareness, and local/edge/cloud computing [3, 4, 6]. In the world of wireless, big data provides

The research carried out in this chapter has been supported by the projects SHARING (Finland grant no. 128010), TUBITAK TEYDEB 1509 (Turkey grant no. 9120067), BESTCOM, as well as the ERC Starting Grant 305123 MORE (Advanced Mathematical Tools for Complex Network Engineering). Some of results here has appeared in parts in [36–38].

plenty of new information sets (e.g., social geodata, user velocity, location) to network planning that can be inter-connected to better understand user and network behaviors. In addition, public data sets available on social networks like Twitter and Facebook give additional information about network lifecycle, which could be further leveraged. The associated advantages are finer granularity of user location information and the ability to identify and estimate clustering of users efficiently—for example, in particular events. Definitely, enormous potential associated with big data has stimulated a great research interest from industry, academics, and government (for more details see [5]), and will keep going in the upcoming years.

Concurrently, mobile cellular networks are shifting toward the next generation of 5G (and beyond) wireless communications, in which ultra-dense networks, device-to-device and millimeter wave communications, edge caching, and massive multiple-input multiple-output (massive MIMO) technologies bring a crucial role for their evolution (see [6] and references therein). Instead of relying on a classical base station–centric network paradigm that assumes *dumb* terminals and *reactive* network optimization approaches, 5G (and beyond) networks will be surely disruptive in the sense of being context aware, user-centric, and proactive/anticipatory in essence. While continued improvement in spectral efficiency is anticipated, the sophistication of air interfaces of existing systems (LTE-Advanced) has shown no major improvements in terms of gains in spectral efficiency. Additionally, extra measures such as the brute force deployment of mobile network infrastructure (e.g., increasing the number of cells) and more spectrum licensing are excessively expensive. Therefore, more creative solutions are indeed necessary.

In this chapter, given these motivations and challenges, we focus on optimization of 5G (and beyond) wireless networks through proposing a proactive caching architecture. We leverage large amount of available data together with the tools utilized for big data analytics and machine learning. Particularly, we examine the proactive caching gains in both backhaul offloading and content demand satisfaction metrics. The approaches and tools used in machine learning are utilized to model and estimate the users behavior in both spatial and temporal dimensions for the proactive cache placement problem. Together with caching critical contents at the edge of a network, specifically at the base stations, resources of mobile networks are used more efficiently and further enhancements on the users' experience is achieved. Notwithstanding, given the high dimensionality of data, their sparsity, and lack of large-scale measurements, the prediction of content popularity coupled with users' spatiotemporal behavior is a nontrivial problem. In this respect, we introduce a platform that can be used to parallelize the computation effort so that the execution of the content estimation algorithms for cache placement can be done at the base stations. As a real-world practical study a huge amount of data, gathered from a telecom operator in Turkey with more than 16.2 million active subscribers, are investigated for several caching scenarios. Specifically, the mobile users' activity traces are gathered in hourly time intervals from large-scale areas covered with many base stations. The analysis results are obtained using the big data platform on the telecom operator's premises under privacy and regulation constrains. The cache placements at the base stations have been numerically studied to observe both the users' experience and backhaul offloadings improvements.

The rest of this chapter is structured as follows. The requirements, challenges, and benefits of big data analytics in 5G (and beyond) networks are described in Section 9.2. The proactive edge caching concept and its network model are given in Section 9.3. A content popularity prediction on a big data platform that showcases a practical case study and users' content access behavior characterization is given in Section 9.4. Afterward, numerical studies for base stations that are cache enabled and technical discussions are given in Section 9.5. We provide a conclusion and future directions in Section 9.6.

9.2 Big Data Analytics for Telcos: Requirements, Challenges, and Benefits

Today the requirements on networking over telco infrastructure are transforming toward a software-defined paradigm with the aim of being more scalable and flexible for big data. Following this trend, tomorrow's networks are going to be big and are expected to be even more complex and interconnected. Because of this, the data centers of mobile operators (MOs) as well as their network infrastructures will need to track traffic patterns of tens of millions of users/devices (e.g., location information, generated traffic demand and pattern of usage, device capabilities). These data can be then used for a more detailed analysis and better network optimization outcomes.

9.2.1 Big Data Networking Challenges and Trends

Recently, generated data traffic and their corresponding patterns inside the data centers of MOs have indeed changed dramatically. Big data has allowed large traffic exchange between gateway devices at backhaul. Even though wireless technologies have evolved significantly toward 5G, this rapid jump has not been observed in backhaul connections of mobile cellular networks. Therefore, the intra-traffic in mobile backhaul has been getting larger than the inter-traffic between mobile backhaul devices and end-users over the years. In fact in existing networks of telecommunication providers, the data produced by gateway and backhaul elements also contribute to traffic load within the operator's infrastructure in addition to traditional operational data generated by management of mobile users' traffic via mobile backhaul or data fetched from a number of different backend, database and cache servers. Indeed, user terminal (UT) level interactions require various data exchanges with hundreds of routers, switches and servers that resides between the backhaul and core network domains. For instance, the intranet data traffic might go up to 930× for an original 1 KB HTTP request of users [7]. This is in contrast to the traditional mobile architecture of telecommunication providers where wireless devices, such as UT and wireless access nodes, are assumed to be bottlenecks that lack the necessary computational overhead instead of the backhaul infrastructure. In addition, as big data still remains a grand challenge for today's mobile infrastructures, moving toward such big data–driven networks in *cloud environments* is challenging.

In this regard, multi-access edge computing (formerly mobile edge computing, sometimes called "fog" computing) is yet another rising technology where edge devices have *cloud-computing like features* within the radio access network to perform functionalities such as communication, storage and control [3]. Notwithstanding for 5G (and beyond) networks, it could be remarked that installing distributed cloud computing capabilities on top of or near each unique BS site (particularly at locations where relative traffic volume is low) may also yield a high deployment cost comparable to solutions with centralized computing capabilities due to the existence of many sites in a traditional MO. Furthermore, for proper analysis to model and predict the users' spatiotemporal content access patterns in 5G (and beyond) mobile networks, traffic handled at a centralized location requires horizontal scale-out across servers and racks. This is possible only inside the core network locations of MOs rather than distributed locations with possibly less traffic.

9.2.2 When Big Data Analytics Meets Caching

Recent advances in standardization networking as well as communications have enabled big data to receive great popularity, notably for possible usage in data centers and mobile network operators' infrastructures. Together with the big data challenges in the networking area, it is becoming clear that the only option to deal with the surge in data traffic is via developing methods to handle data appropriately and enabling data to move from cloud to the edge. Lately, a big data management solution named Hadoop has been quite successful in providing dramatic cost reduction over conventional tier-one database architectures, processing various data formats and executing parallel computing over multiple servers. In addition, more sophisticated machine learning analytic techniques together with nonrelational databases (i.e., NoSQL databases) that exploit big data have created a better understanding of big data for MOs.

It is evident that placing content closer to the edge is critical whenever connectivity failures happen during streaming and/or fetching activities. To alleviate this, placement of data closer to demanding users so that the distance of content to users can be reduced as well as placing the right content and applications toward the edge give better user experience. For example, analyzing users' content access patterns (through the core networks of MOs) and caching proactively at the edge (i.e., at base stations (BSs)) using the distributed data processing engine provided by the Hadoop platform can relieve the backhaul network usage and improve users' quality of experience (QoE) in terms of reduction in latency. In the following, we detail the related system model, then describe our big data processing platform based on the Hadoop framework and its interconnection with caching at the edge, as one way of handling and exploiting the existing big data over the infrastructure of MOs.

9.3 System Model

In this chapter, we assume that there are M small base stations (SBSs) from the set $\mathcal{M} = \{1, \ldots, M\}$ and N UTs from the set $\mathcal{N} = \{1, \ldots, N\}$ in a network deployment scenario. A broadband internet connection is provided to each SBS m by a wired backhaul connection that has C_m Mbps of capacity and the broadband internet service is provided to wireless users over a wireless link that has C'_m Mbps of capacity. Assuming a densely deployed SBSs scenario where backhaul capacity has its limitations [6], we further assume that $C_m < C'_m$. Moreover, consider that each user $n \in \mathcal{N}$ is associated with single SBS and unicast sessions[1] are used to serve each user. Specifically, suppose that the contents (i.e., text, videos, images) are demanded by UTs from a library $\mathcal{F} = \{1, \ldots, F\}$, where each content $f \in \mathcal{F}$ has $L(f)$ MB of length and $B(f)$ Mbps of bitrate requirement, with

$$L_{\min} = \min_{f \in \mathcal{F}} \{L(f)\} > 0 \tag{9.1}$$

$$L_{\max} = \max_{f \in \mathcal{F}} \{L(f)\} < \infty \tag{9.2}$$

and

$$B_{\min} = \min_{f \in \mathcal{F}} \{B(f)\} > 0 \tag{9.3}$$

$$B_{\max} = \max_{f \in \mathcal{F}} \{B(f)\} < \infty. \tag{9.4}$$

The content demands of users indeed can be characterized by a Zipf-like distribution $P_{\mathcal{F}}(f), \forall f \in \mathcal{F}$ such as [10]:

$$P_{\mathcal{F}}(f) = \frac{\Omega}{f^{\alpha}}, \tag{9.5}$$

where

$$\Omega = \Big(\sum_{i=1}^{F} \frac{1}{i^{\alpha}}\Big)^{-1}.$$

The shape of the distribution is modeled by the α parameter in (9.5). Such power laws (distributions) characterize several real-world conditions, for example the file distribution in the web proxies [10] and the traffic dynamics of mobile user devices [11]. Bigger α values signify a steeper distribution, which means that the popularity of a slight portion of content is higher than the remaining content catalog. However, the smaller values mean higher uniformity with almost identical contents popularity. The practical α value varies based on users' content access patterns and the deployment scenarios of SBSs (i.e., enterprise, rural, urban, suburban, and home environments), and next sections will provide its value using the proposed practical setup.

Provided that our global content popularity is decreasingly ordered, the mth SBS's content popularity matrix at time t is specifically modeled by $\mathbf{P}^m(t) \in \mathbb{R}^{N \times F}$, where each component $P^m_{n,f}(t)$ quantizes to the probability of the nth user demands for content f. In other words, the matrix $\mathbf{P}^m(t)$ signifies the local content popularity distribution experienced at mth the base station and time t, while the global content popularity distribution of all contents in decreasing order is modeled via the Zipf distribution $P_{\mathcal{F}}(f), \forall f \in \mathcal{F}$.

[1] One can also extend the unicast service model to the multicast case. See [8, 9] for more details.

In our considered deployment scenario, each SBS is assumed to have a limited storage size of S_m and content that is selected strategically from $\mathcal{F}$ is proactively cached by each SBS in the course of off-peak hours. By following this approach, the bottlenecks due to the finite-backhaul capacity are mitigated when users' content demands are delivered during peak hours. The number of satisfied requests as well as the backhaul load are of high significance and are given in the following. Assume that in the course of T seconds, the number of demanded content is D and is selected from the set $\mathcal{D} = \{1, \ldots, D\}$. Consider that content delivery is commenced instantly just after SBS receives the content demand $d \in \mathcal{D}$. Later on, the demand d is assumed *satisfied* if the content delivery rate is equal or higher than the content bitrate at final delivery, that is:

$$\frac{L(f_d)}{\tau'(f_d) - \tau(f_d)} \geq B(f_d), \tag{9.6}$$

where f_d models the demanded content; $L(f_d)$ and $B(f_d)$ are content length and bitrate, respectively; $\tau(f_d)$ is the content demand appearance time; and $\tau'(f_d)$ is the final delivery time.[2] Defining the condition in (9.6) is due to the fact that, if the rate of delivery rate is not equal or lower than the requested content bitrate, disruption at the time of playback (or download) happens. This results in lower QoE experienced by users.[3] For this reason, the scenarios where this condition holds are more preferable for higher QoE. In (9.6), observe also that the final delivery time for demand d, given by $\tau'(d)$, is dependent on network load, backhaul, and wireless links' capacities as well as content availability at SBSs. Under these motivations and given the definition of satisfied demands used here, the average demand *satisfaction ratio* of users is therefore defined for all demands sets, particularly:

$$\eta(\mathcal{D}) = \frac{1}{D} \sum_{d \in \mathcal{D}} \mathbb{1} \left\{ \frac{L(f_d)}{\tau'(f_d) - \tau(f_d)} \geq B(f_d) \right\}, \tag{9.7}$$

where $\mathbb{1}\{\ldots\}$ denotes the indicator function, which is 1 if the condition is satisfied and 0 if not. Moreover, let $R_d(t)$ Mbps denote the instantaneous backhaul rate for the demand d at time t, with $R_d(t) \leq C_m$ and $\forall m \in \mathcal{M}$, then the average *backhaul load* can be written as:

$$\rho(\mathcal{D}) = \frac{1}{D} \sum_{d \in \mathcal{D}} \frac{1}{L(f_d)} \sum_{t=\tau(f_d)}^{\tau'(f_d)} R_d(t). \tag{9.8}$$

In Eq. (9.8) the outer sum is used to sum over all demands set whereas the inner sum yields the total data over backhaul for demand d, which is no more than the size of demanded content $L(f_d)$. The instantaneous backhaul rate for demand d, modeled by $R_d(t)$, excessively depends on system load, backhaul link capacity, and contents that are strategically cached at the base stations.

Indeed, by performing a content caching operation at the small base stations (SBSs), the experienced delays when accessing the content diminish, particularly during peak

[2] Future information (i.e., content's start and final delivery times) can also be leveraged during proactive resource allocation (e.g., see [12]).

[3] In general, the bitrate requirement of video content is normally between 1.5 and 68 Mbps [13].

hours, hence producing better satisfaction ratios with a smaller backhaul load. To detail this, assume that SBSs' cache decision matrix is denoted as $\mathbf{X}(t) \in \{0,1\}^{M\times F}$, where each component of the matrix $x_{m,f}(t)$ is 1 if the mth SBS caches content f at time t, and 0 if it does not. Thereupon, the formal problem of offloading backhaul under a specific constraint of demand satisfaction is modeled as follows:

$$\begin{aligned}
&\underset{\mathbf{X}(t),\mathbf{P}^m(t)}{\text{minimize}} \quad && \rho(\mathcal{D}) && (9.9)\\
&\text{subject to} && L_{\min} \le L(f_d) \le L_{\max}, \quad \forall d \in \mathcal{D}, && (9.9a)\\
& && B_{\min} \le B(f_d) \le B_{\max}, \quad \forall d \in \mathcal{D}, && (9.9b)\\
& && R_d(t) \le C_m, \quad \forall t, \forall d \in \mathcal{D}, \forall m \in \mathcal{M}, && (9.9c)\\
& && R'_d(t) \le C'_m, \quad \forall t, \forall d \in \mathcal{D}, \forall m \in \mathcal{M}, && (9.9d)\\
& && \sum_{f\in\mathcal{F}} L(f) x_{m,f}(t) \le S_m, \quad \forall t, \forall m \in \mathcal{M}, && (9.9e)\\
& && \sum_{n\in\mathcal{N}} \sum_{f\in\mathcal{F}} P^m_{n,f}(t) = 1, \quad \forall t, \forall m \in \mathcal{M}, && (9.9f)\\
& && x_{m,f}(t) \in \{0,1\}, \quad \forall t, \forall f \in \mathcal{F}, \forall m \in \mathcal{M}, && (9.9g)\\
& && \eta_{\min} \le \eta(\mathcal{D}), && (9.9h)
\end{aligned}$$

where $R'_d(t)$ Mbps is the wireless link's instantaneous rate for demand d and $\eta_{\min}$ denotes the minimum targeted ratio of satisfaction. Moreover, the constraints in (9.9a) and (9.9b) are used to restrict the content length and bitrate in the catalog to obtain an attainable solution, the constraints in (9.9c) and (9.9d) represent the capacity constraints over backhaul and wireless links, (9.9e) represents the caching capacity for storage, (9.9f) confirms the content popularity matrix as a probability measure, (9.9g) represents caching binary decision variables, and finally the constraint in (9.9h) models the QoE satisfaction ratio.

To solve with the problem just defined, joint optimization of $\mathbf{X}(t)$ (decision matrix for caching) and $\mathbf{P}^m(t)$ (the estimate of content popularity matrix) is necessary. Unfortunately, solving the problem (9.9) is not straightforward as:

1. Wireless/backhaul links' and SBSs' storage capacities are limited.
2. Very large contents in the catalog as well as a high number of users with unexplored rating values[4] exist in real-world scenarios.
3. Non-tractability for optimal uncoded[5] cache placement exists for a given demand [15–17],
4. The sparse content popularity/rating matrix of SBSs $\mathbf{P}^m(t)$ needs to be tracked, learned, and predicted by SBSs while making the cache placement.

[4] The term *rating* is used to define the empirical content popularity/probability value and is used interchangeably in this chapter.

[5] From an information-theoretical perspective, two groups can be formed with the caching placement such as "coding" and "uncoded" groups (see [14]) for details).

To deal with these problems, our scenario is restricted to the case when cache placement is done at the time of off-peak hours. Therefore $\mathbf{X}(t)$ is fixed at the time of the delivery of content during peak hours and modeled by $\mathbf{X}$. In addition, in the course of T time slots, the content popularity matrix becomes stationary and equivalent between the base stations, therefore $\mathbf{P}^m(t)$ is denoted by $\mathbf{P}$.

After considering these assumptions, another assumption is to decompose the problem into two separate subproblems in which the content popularity matrix $\mathbf{P}$ is first predicted, and later exploited for the caching placement $\mathbf{X}$ appropriately for solutions. Indeed, if a sufficient number of users' ratings can be collected at the SBSs, a k-rank approximate popularity matrix $\mathbf{P} \approx \mathbf{N}^T\mathbf{F}$ can be built, by jointly learning the factor matrices $\mathbf{N} \in \mathbb{R}^{k\times N}$ and $\mathbf{F} \in \mathbb{R}^{k\times F}$, which can minimize the following cost function:

$$\underset{\mathbf{P}}{\text{minimize}} \sum_{P_{ij}\in\mathcal{P}} \left(\mathbf{n}_i^T\mathbf{f}_j - P_{ij}\right)^2 + \mu\left(||\mathbf{N}||_F^2 + ||\mathbf{F}||_F^2\right), \tag{9.10}$$

where the aggregation is applied over the analogous pairs of user/content rating P_{ij} inside the training set $\mathcal{P}$. The $\mathbf{n}_i$ and $\mathbf{f}_j$ vectors represent the ith and jth columns of $\mathbf{N}$ and $\mathbf{F}$ matrices, respectively, and $||.||_F^2$ means the Frobenius norm. The μ value is utilized to balance the regularization and the training data fitting terms. Hence a better estimate of $\mathbf{P}$ due to the high correspondence between the content factor matrix $\mathbf{F}$ and user factor matrix $\mathbf{N}$ can be achieved. Indeed, the problem in (9.10) can be considered to be a problem of regularized least squares in which the formulation has embedded the matrix factorization. Among many methods to solve these problems, the matrix factorization methods are usually applied and have various utilization areas inside the recommendation systems (e.g., inside the video recommendation systems of Netflix). In the considered setup explained in the rest of the chapter, a regularized sparse singular value decomposition (SVD) method is applied to find a solution to the problem in an algorithmic manner by exploiting the least-squares type of the considered problem. The surveys of these considered methods, which are also called collaborative filtering (CF) tools, are available in [18] and [19]. Just after content popularity matrix $\mathbf{P}$ estimation is completed, the caching placement $\mathbf{X}$ may be done accordingly.

During practical deployment scenarios, the estimation of $\mathbf{P}$ in (9.10) can be done by collecting and analyzing the huge amount of existing data using the on-premise *big-data platforms* of the MO. Hence *cache-enabled base stations*, with cache decisions that are modeled by $\mathbf{X}$, can cache the strategic and most popular contents from an estimation of $\mathbf{P}$. With this way of operation, the minimization of the backhaul offloading problem in (9.9) can be accomplished and users' content demands are better satisfied. Figure 9.1 illustrates the considered system model that includes the overall envisioned practical setup. In the next section, as a real-world practical case study, we first describe our big data–enabled platform, which analyzes the users' traffic characteristics by gathering large chunks of data onto the platform. Then, the collected data are utilized to predict the content popularity matrix $\mathbf{P}$, which will later be used for the cache placement $\mathbf{X}$. The details are described in the next sections.

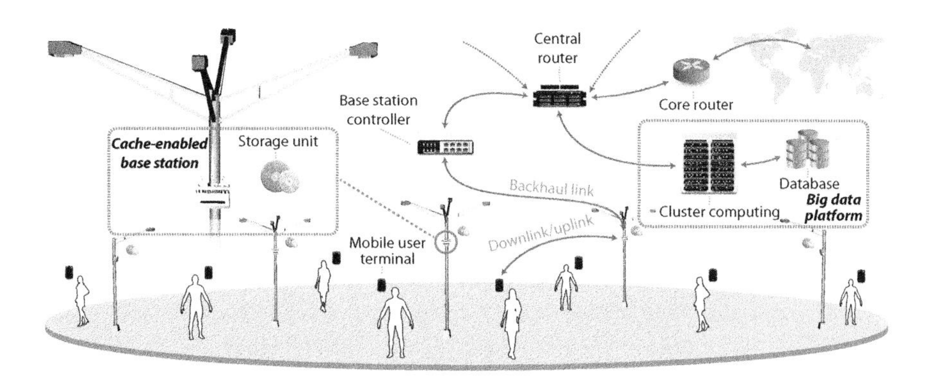

Figure 9.1 The considered network model. Users' demands are tracked/predicted by a *big data platform*, whereas *cache-enabled base stations* cache the predicted strategic/popular content by exploiting the big data platform.

9.4 Big Data Platform

This section details our big data analysis platform for processing users' huge amount of data traffic. One of main targets of this platform is to collect users' data traffic and excerpt meaningful analysis results to accomplish proactive cache placement at base stations. Assuming that Apache Hadoop is deployed and connected to the core network gateways of the MO, the requirements of the considered platform can be debated as follows:

1. *Massive processing power for a huge amount of data in a short time:* In order to cache the contents proactively, a big data analysis platform connected to the core network infrastructure needs to gather and combine data from various sources and ensure quick and reliable ways to extract smart insights. Therefore, for a comprehensive analysis of streaming data after mirroring the interface using tools for network analysis, the collected/gathered data should be ingested into a storage platform that is designed for big data, such as Hadoop distributed file system (HDFS), using techniques from enterprise data integration, such as spring integration.
2. *Data cleansing, parsing, and formatting:* During the data analysis process, data cleansing plays a key role. Indeed, before application of any statistical analysis and machine learning methods on data, data have to be cleansed. In general, this mechanism can consume more time than the analysis used within machine learning approaches for the data scientists. In fact, several steps are required for data cleansing process. In the first step, raw data should be cleaned. There can be, for example inappropriate, malfunctioning, and inconsistent packets with incorrect character encoding inside the raw data itself, which requires elimination. Then the next (second) step is to extract the relevant fields/properties from the raw data. During this step, both the data and the control packet headers that are required for our analysis are determined based on modeling and data analysis requirements.

Finally, the serialization and deserialization of the parsed/extracted data need to be done accordingly (e.g., in Parquet or Avro format) for HDFS storage.

3. *Data analysis:* Given the processed information in HDFS, high-level query languages (e.g., [HiveQL] and Pig Latin) can be used for various data analytics purposes to exploit control/data planes' headers as well as payload information. The goal of such a procedure is to relate data and control packets with each other, e.g., the position or mobile subscriber integrated services for digital network number (MSISDN) of mobile users (which is available only in control packets rather than data packets) to the demanded content (which is available only in data packets) via MapReduce operations. Note that the concept of MapReduce is naturally embedded under HiveQL and Pig Latin.
4. *Statistical analysis and visualizations:* After prediction of the spatio temporal user content access patterns are done using machine learning analysis, the results can be recorded and recycled. In addition, the reformatting of analysis results can be done using appropriate extract, transform and load (ETL) tools or via different kinds of analytics engines such as Apache Spark's MLlib for more advanced data analytics purposes. Moreover, visualizations such as tables and graphs can be utilized to present the analysis results in a visual way to obtain better insights.

Together with a big data processing platform described earlier, the strategic contents of users can be deduced from a massive data set through application of machine learning methods, which lies at the core of recommendation systems. Subsequently, a simulation exercise for characterizing the potential benefits of caching at BSs are performed.

9.4.1 Platform Description

The big data platform that we use is installed in the core network of the operator. The aim of this platform, as explained before, is to record users' traffic and extract necessary information that will be exploited for estimation of content popularity. Regarding the infrastructure, the operator's network is made of multiple districts (about 10 regional areas) scattered around Turkey. The total average throughput spanning over many regional areas is made of more than 15 billion packets in uplink as well as 20 billion packets in the downlink per day. These figures add up to a daily use of almost 80 TB of data flow in uplink and downlink in the core network of operator. The data consumption behavior leads to an exponential usage in data traffic of a mobile network operator. For instance, in 2012, approximately 7 TB of daily traffic (together in uplink and downlink) was observed.

The data traces that are discussed hereafter were gathered from one of the major network regions of the operator (which has mobile traffic from several base stations) and are stored on a server equipped with a high-speed link up to 200 Mbps at peak times/hours. As part of capturing the internet traffic by this server available on the platform, a mechanism is established to mirror/copy real-world Gn interface data.[6]

[6] Gn is an interface representing the serving GPRS support node (SGSN) and gateway GPRS support node (GGSN). Packets in the network, transmitted from a user device to the packet data network (PDN), i.e.

Once this mechanism of Gn interface is established, network traffic is then forwarded to the server on the platform. In order to establish a technical analysis, we have gathered mobile traffic of roughly 7 hours (between 12 PM and 7 PM, Saturday March 21, 2015). The big data platform, based on Hadoop, is then used to analyze this mobile traffic.

Given the existing platforms, Hadoop appears to be one of commonly used solutions as open source software [20]. A storage module (named HDFS) as well as a computation module (named MapReduce) are integral parts of Hadoop. While HDFS can operate in both centralized and distributed fashions, MapReduce naturally considers a distributed mechanism where the the jobs can be executed in parallel on many computers.

As mentioned, the precision/accuracy of our data collection was examined within the operator's network. In particular, the big data platform was implemented using a modified version of Hadoop distributed by Cloudera [21], consisting of four computers (one being a cluster node), with each computer having the following computational power and equipment: 32 core CPU (namely, Intel Xeon CPU E5-2670 performing at 2.6 GHz), 132 GB of RAM, and 20 TB of storage. This big data system is utilized to infer the meaningful information from processing raw data and is detailed in Section 9.4.2.

9.4.2 Data Extraction Procedures

First of all, the raw data is fed to a *tshark* tool (namely, the Wireshark command line utility [22]) for extraction of relevant fields, such as:

- The *service area* within *location area*, represented by service area code (SAC) and location area code (LAC) fields, respectively. They identify an area of single or multiple base stations, where in practice, tens or even hundreds of base stations can operate in a specific location area.
- The requested content related field such as hypertext transfer protocol (HTTP) request uniform resource identifier (URI).
- Tunnel end point identifiers such as the tunnel end point identifier (TEID) and TEID-DATA, for control and data plane respectively.[7]
- Arrival time of requests, which could be approximately captured by a field like FRAME TIME.

Note that *control* packets with fields like ID (CELL-ID), LAC, and TEID-DATA, contain information about future data packets, whereas *data* packets have fields such as HTTP-URI and TEID.

Next, after both control and data packets are obtained, the data are then transferred to HDFS for more comprehensive analysis. With HDFS in hand, one can experiment with various data analytics over the collected data, using tools like Hive Query language (HiveQL) [23]. In order to find more information about the requested contents

internet, go through SGSN and GGSN in which the GPRS tunneling protocol (GTP) is the main protocol in network packets going through the Gn interface.

[7] A TEID uniquely identifies a tunnel end point on the receiving end of the GTP tunnel. A local TEID value is given at the receiving end of a GTP tunnel for sending messages through the tunnel.

(namely, HTTP-URIs) at a specific location and time, one can match CELL-ID-LAC fields over the same TEID (data panel) and TEID-DATA (control panel) fields. Due to limited number of rows we could have obtained for exact matching (with each row storing HTTP-URI, CELL-ID-LAC fields, and other fields), we have continued with HTTP-URIs and TEID matchings that approximately represent request–location pairs, therefore providing us more rows for further analysis.

After the matching operation, the data are then stored in a temporary table called the *traces-table-temp*, using HiveQL. Note that this temporary table contains HTTP Request-URI, FRAME TIME, and TEID fields. The final stage in this process is to find the size/length of the content (namely, lengths of HTTP-URI), which was done by implementing a *size calculator* application via HTTPClient API [24]. The final table is called the *traces-table*, where each row has fields of FRAME TIME, HTTP Request-URIs, TEID, and SIZE. The total size of this final table is roughly 420,000 rows, after iterating over 4 million rows (which were available in the temporary table) and discarding all SIZE queries returning zero or null. We would like to remark that, in a particular session with a specific TEID, one could expect several HTTP request-URIs. While each TEID belongs to a particular user, on the other hand, a user can have several TEIDs with multiple HTTP request-URIs. All the procedures of data extraction mentioned here, with the help of the big data platform, are summarized in Figure 9.2. We remark that this data extraction procedure is developed for our setup of proactive caching. Nevertheless, similar approaches in terms of platform and exploitation of big data analytics for mobile network operators could be observed in [25–30].

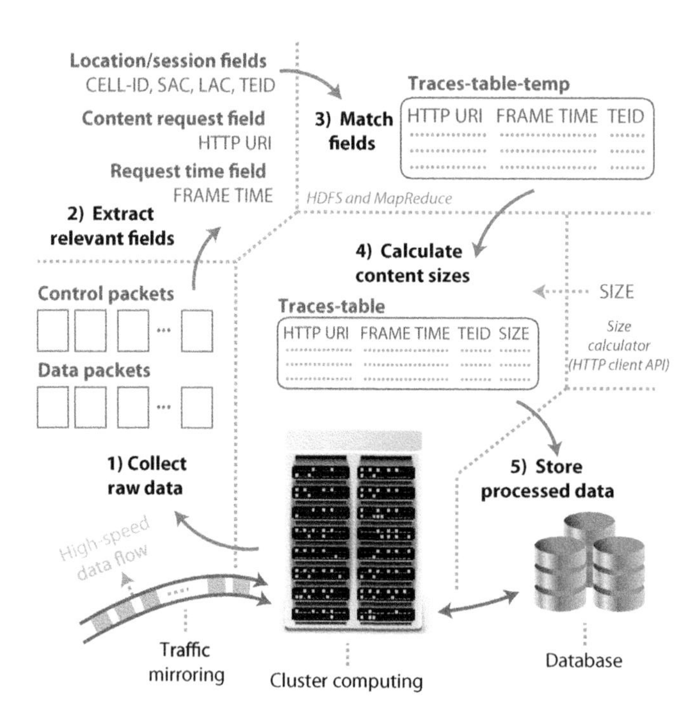

Figure 9.2 A summary of the data extraction procedures on the big data platform.

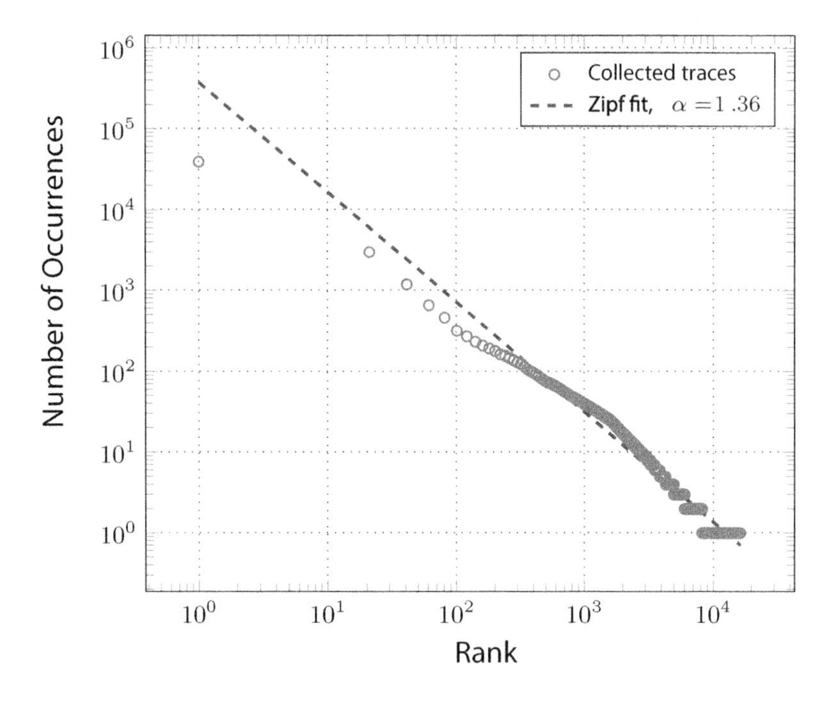

Figure 9.3 Global content popularity distribution.

9.4.3 Traffic Characteristics

The information stored in the *traces-table*, representing the global content popularity distribution (namely, global popularity distribution of HTTP-URI), is shown in Figure 9.3, with popularities in a decreasing ranked order. Based on our gathered/processed data, we note that the popularity behavior of files can be characterized by a Zipf law, where the steepness parameter α equals 1.36.[8] In the figure, the Zipf line is computed in a least-squares fashion, and the parameter α is then obtained from the slope of the Zipf line. Additionally, the cumulative size distribution of content is depicted in Figure 9.4, using the same support of global content popularity (with a decreasing order) and taking into account the size of content cumulatively. The content up to the 41st most-popular content has a total/cumulative size of 0.1 GB, whereas a dramatical jump in cumulative size appears afterward. One could remark that this is an indication that most content in our catalog has a small size, and bigger content/files are relatively less demanded.

While our results here shed light on the characterization of content popularity in base stations, we would like state that a more detailed study can be conducted in the future. Additionally, while our work differs from existing works in the sense that we take into account mobile traffic from a large geographical area and leverage the data for traffic

[8] The practical values of the steepness/shape parameter α can different in other scenarios. For example, the shape of the parameter of content popularities in the YouTube catalog is observed to be between 1.5 and 2.5 [31, 32].

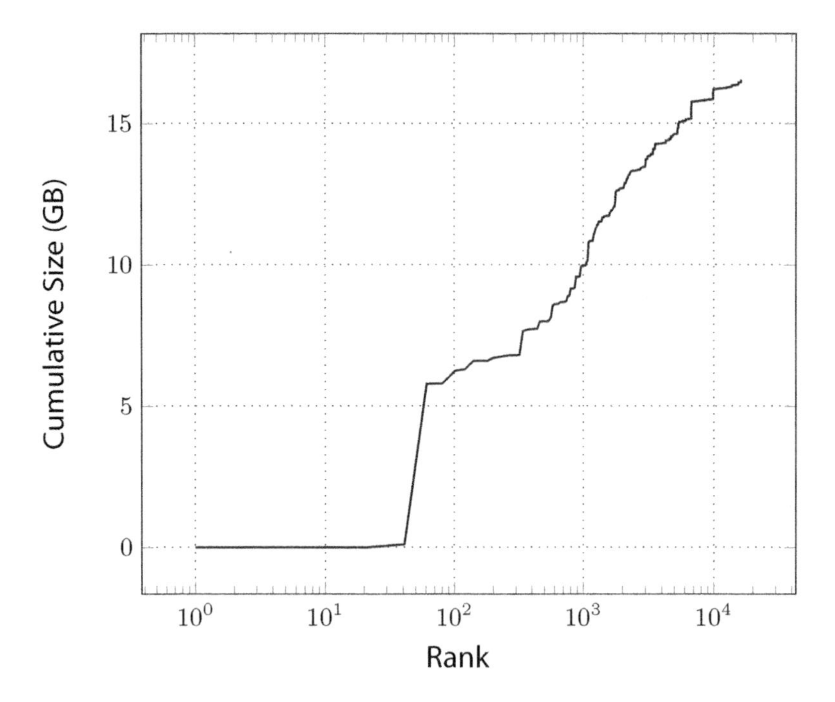

Figure 9.4 Cumulative size distribution.

characterization and caching, we note that some interesting studies could be found in the context of web caching [10], campus network [33], mobile users in Mexico [34] and others. In Section 9.5, we utilize information in the *traces-table* and perform a simulation study to asses the impact of cache-enabled mobile networks.

9.5 Numerical Results and Discussions

The parameter setting considered in our simulation setup is summarized in Table 9.1. In addition, we assume for sake of simplicity that wireless, backhaul, and storage capacities are identical with each other.

In the simulation study, there are in total D number of content requests found in the processed/collected data (called the *traces-table*), spanning over 6 hours 47 minutes. The arrival time of each demand (FRAME TIME), demanded content (HTTP-URI), and content size (SIZE) are taken from the *traces-table*. Then, these content demands are mapped to M base stations pseudo-randomly. For the backhaul offloading problem we stated in (9.9), the content popularity matrix $\mathbf{P}$ and cache placement strategy $\mathbf{X}$ are calculated separately. More precisely, we have considered two different approaches for constructing the content popularity matrix $\mathbf{P}$:

- *Ground truth*: We built the matrix $\mathbf{P}$ (namely, the content popularity matrix) from all available data in the *traces-table* without relying on the problem stated in (9.10). This matrix in fact quantifies how popular files are across base stations,

Table 9.1 List of Simulation Parameters

Parameter	Description	Value
T	Total number of time units/slots	6 h 47 min
D	Number of content requests	422, 529
F	Number of files	16, 419
M	Number of small cells	16
$L_{\min}$	Minimum size of a content	1 byte
$L_{\max}$	Maximum size of a content	6.024 GB
$B(f)$	Bitrate of content f	4 Mbps
$\sum_m C_m$	Cumulative backhaul link capacity	3.8 Mbps
$\sum_m \sum_n C'_m$	Cumulative wireless link capacity	120 Mbps

with columns representing the content and rows indicating the base stations. It is observed that the matrix **P** has rating density of 6.42%.

- *Collaborative filtering*: We have estimated the content popularity matrix **P**, via attempting to solve the problem in (9.10). For this, we have uniformly and randomly picked 10% of ratings from the *traces-table* for the test set, and the remaining 90% is used for training via regularized SVD (see CF approaches [19, 35]).

Given both methods for constructing the content popularity matrix **P**, the content cache replacement decision (**X**) is then performed by picking the most popular files greedily at all SBSs subject to storage constraints (see [15] for more details). Assuming that these selected files/contents are proactively available at SBSs at $t = 0$, the content demand of users is then satisfied over the time. The measurements of performance metrics, such as backhaul load and content demand satisfaction, are recorded for analysis.

Figure 9.5 represents the changes of users' demand for satisfaction as the storage size increases, where (1) 100% of storage size is representing the whole file/content catalog (17.7 GB), and (2) 0% means no storage capacity, for both collaborative filtering and ground truth approaches. The users' content demand satisfaction substantially improves with the increase in storage size, whereas the performance gap between the methods is arguably small until 87% of caching/storage capacity (potentially due to estimation errors). As an example, with 40% of storage capacity, one can observe that the CF method can achieve 69% of users' demand satisfaction whereas the ground truth could go up to 92%.

Figure 9.6 presents the changes of backhaul usage/load as the storage size at each base station increases. The higher backhaul load reduction is observed at the base stations while increasing the storage size. For example, both ground truth and CF methods can have offloading up to 98%. On one hand, one can also observe that the performance of the ground truth approach is higher since all available information from the *traces-table* is used for the construction of content popularity matrix. On the other hand, after a specific storage size capacity, the figure shows a sharp decrease of backhaul load in both approaches, mostly because of the fact that relatively less popular

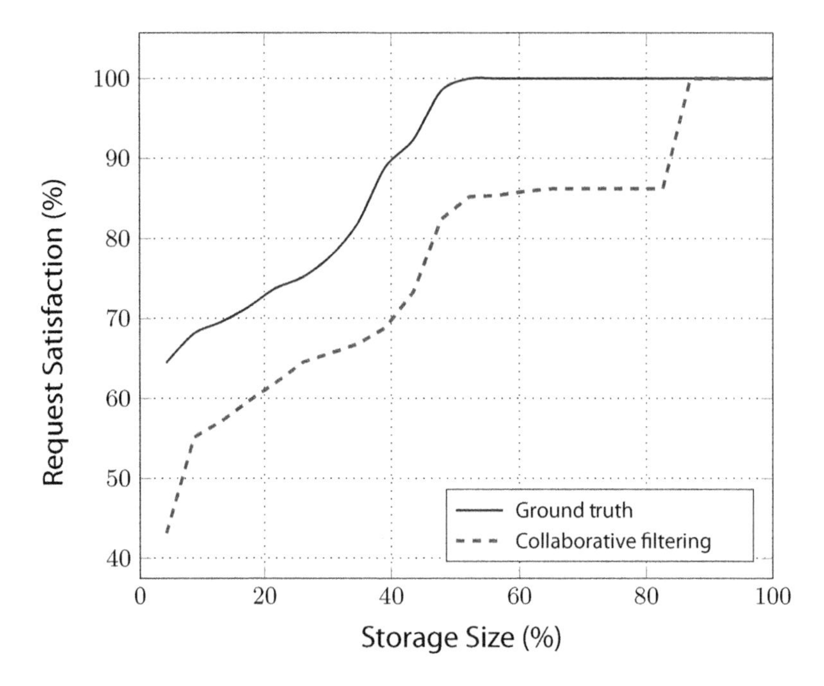

Figure 9.5 Change of satisfaction with respect to the storage size.

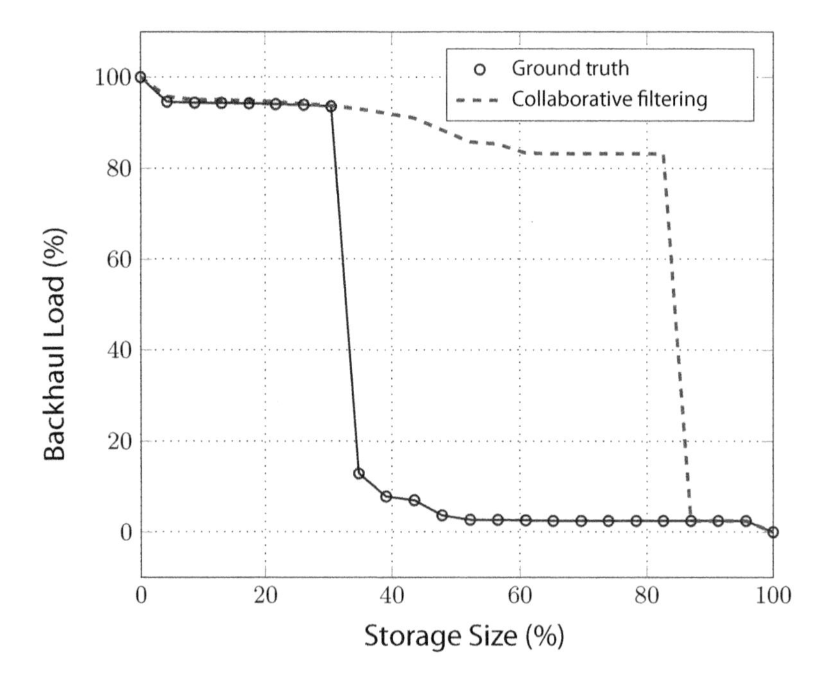

Figure 9.6 Change of backhaul usage with respect to the storage size.

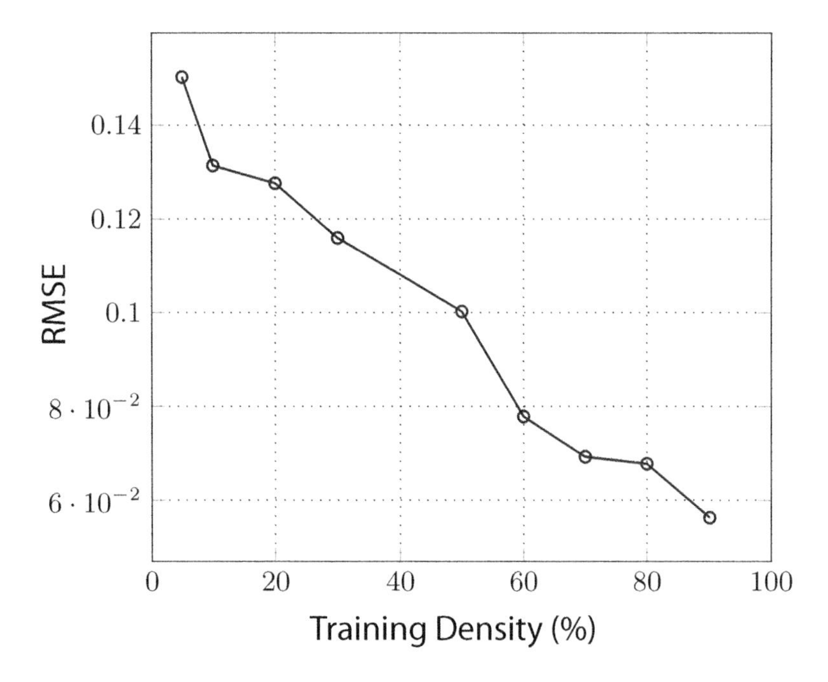

Figure 9.7 Change of RMSE with respect to the training density.

(but large-size) files have less impact compared to highly popular files in our setting. While not considered in the cache placement methods of this numerical study, this behavior shows the importance of considering content sizes (see Figure 9.4) when designing placement algorithms, in order to more efficiently use the storage size at base stations.

Finally, Figure 9.7 shows the evolution of root-mean-square error (RMSE) with respect to the training density for CF approach. Note that the training density in the evaluations described is 90%, whereas the aim in this figure is to assess its impact on performance and show the validity of those evaluations. The performance of the CF method is evidently better when the training density is increasing, allowing the CF method to better estimate, thus leading to smaller estimation errors.

9.6 Conclusions

Our study in this chapter focused on a proactive caching method that leverages tools from machine learning and big data platforms, with the aim of improving performance of cache-enabled base stations in 5G (and beyond) wireless networks. The experimental test bed as well as data extraction procedures enabled us to have an estimation of content popularity matrix in a real-world setup, and the cache performance benefits in terms of network backhaul offloadings and users' content request satisfaction were provided.

Regarding future directions of the work, one could think of a more detailed characterization of the mobile data traffic that takes into account joint spatiotemporal variation of content access patterns. As far as learning for cache placement is concerned, online machine learning mechanisms, such as deep reinforcement learning, could be considered. For cache placement, the development of new randomized/deterministic algorithms are needed and should take into account not only content popularity but also content lengths and other factors; therefore, higher backhaul offloading and demand satisfaction could be achieved. Finally, another line of work would be to extend the big data platform such that real-time processing (instead of offline) could be performed. In this regard, the recent tools of the Hadoop ecosystem, like Apache Spark and related libraries such as Spark Streaming, could be considered as well as MLLib for machine learning analysis.

References

[1] C. Lynch, "Big data: how do your data grow?" *Nature*, vol. 455, no. 7209, pp. 28–29, 2008.

[2] J. Andrews, S. Buzzi, W. Choi, S. Hanly, A. Lozano, A. Soong, and J. Zhang, "What will 5G be?" *IEEE Journal on Selected Areas in Communications*, vol. 32, no. 6, pp. 1065–1082, June 2014.

[3] F. Bonomi, R. Milito, J. Zhu, and S. Addepalli, "Fog computing and its role in the internet of things," in *Proceedings of the First Edition of the MCC Workshop on Mobile Cloud Computing*, ACM, 2012, pp. 13–16.

[4] T. H. Luan, L. Gao, Z. Li, Y. Xiang, and L. Sun, "Fog computing: focusing on mobile users at the edge," arXiv:1502.01815, 2015.

[5] H. Hu, Y. Wen, T.-S. Chua, and X. Li, "Toward scalable systems for big data analytics: a technology tutorial," *IEEE Access*, vol. 2, pp. 652–687, 2014.

[6] E. Baştuğ, M. Bennis, and M. Debbah, "Living on the edge: the role of proactive caching in 5G wireless networks," *IEEE Communications Magazine*, vol. 52, no. 8, pp. 82–89, Aug. 2014.

[7] N. Farrington and A. Andreyev, "Facebook's data center network architecture," in *Proceedings of IEEE Optical Interconnects Conference*, May 2013, pp. 49–50.

[8] K. Poularakis, G. Iosifidis, V. Sourlas, and L. Tassiulas, "Multicast-aware caching for small cell networks," in *IEEE Wireless Communications and Networking Conference (WCNC'2014)*, IEEE, 2014, pp. 2300–2305.

[9] B. Zhou, Y. Cui, and M. Tao, "Optimal dynamic multicast scheduling for cache-enabled content-centric wireless networks," arXiv:1504.04428, 2015.

[10] L. Breslau, P. Cao, L. Fan, G. Phillips, and S. Shenker, "Web caching and Zipf-like distributions: evidence and implications," in *IEEE Eighteenth Annual Joint Conference of the IEEE Computer and Communications Societies (INFOCOM'99)*, IEEE, 1999, vol. 1, pp. 126–134.

[11] M. Z. Shafiq, L. Ji, A. X. Liu, and J. Wang, "Characterizing and modeling internet traffic dynamics of cellular devices," in *Proceedings of the ACM SIGMETRICS Joint International Conference on Measurement and Modeling of Computer Systems*, ACM, 2011, pp. 305–316.

[12] J. Tadrous, A. Eryilmaz, and H. E. Gamal, "Proactive data download and user demand shaping for data networks," arXiv: 1304.5745, 2014.

[13] "Recommended upload encoding settings (advanced)," Google, 2015, https://goo.gl/KJXfhh.

[14] M. A. Maddah-Ali and U. Niesen, "Fundamental limits of caching," *IEEE Transactions on Information Theory*, vol. 60, no. 5, pp. 2856–2867, May 2014.

[15] E. Baştuğ, J.-L. Guénégo, and M. Debbah, "Proactive small cell networks," in *20th International Conference on Telecommunications (ICT'13)*, May 2013, pp. 1–5.

[16] K. Poularakis, G. Iosifidis, and L. Tassiulas, "Approximation algorithms for mobile data caching in small cell networks," *IEEE Transactions on Communications*, vol. 62, no. 10, pp. 3665–3677, Oct. 2014.

[17] N. Golrezaei, A. F. Molisch, A. G. Dimakis, and G. Caire, "Femtocaching and device-to-device collaboration: a new architecture for wireless video distribution," *IEEE Communications Magazine*, vol. 51, no. 4, pp. 142–149, 2013.

[18] Y. Koren, R. Bell, and C. Volinsky, "Matrix factorization techniques for recommended systems," *Computer*, no. 8, pp. 30–37, Aug. 2009.

[19] J. Lee, M. Sun, and G. Lebanon, "A comparative study of collaborative filtering algorithms," arXiv: 1205.3193, 2012.

[20] "Apache Hadoop," Apache, 2015, http://hadoop.apache.org/.

[21] "Cloudera," www.cloudera.com/content/cloudera/en/documentation.html, accessed Apr.

[22] "The Wireshark network analyzer 1.12.2," www.wireshark.org/docs/man-pages/tshark.html, accessed Apr. 2015.

[23] "Apache Hive TM," Apache, https://hive.apache.org/, accessed Apr. 2015.

[24] O. Kalnichevski, J. Moore, and J. van Gurp, "HttpClient tutorial," Apache, https://hc.apache.org/httpcomponents-client-ga/tutorial/pdf/httpclient-tutorial.pdf, accessed Apr. 2015.

[25] Y. Dong, Q. Ke, Y. Cai, B. Wu, and B. Wang, "Teledata: data mining, social network analysis and statistics analysis system based on cloud computing in telecommunication industry," in *Proceedings of the Third International Workshop on Cloud Data Management*, ACM, 2011, pp. 41–48.

[26] H.-D. J. Jeong, W. Hyun, J. Lim, and I. You, "Anomaly teletraffic intrusion detection systems on hadoop-based platforms: a survey of some problems and solutions," in *15th International Conference on Network-Based Information Systems (NBiS)*. IEEE, 2012, pp. 766–770.

[27] J. Magnusson and T. Kvernvik, "Subscriber classification within telecom networks utilizing big data technologies and machine learning," in *Proceedings of the 1st International Workshop on Big Data, Streams and Heterogeneous Source Mining: Algorithms, Systems, Programming Models and Applications*, ACM, 2012, pp. 77–84.

[28] W. Indyk, T. Kajdanowicz, P. Kazienko, and S. Plamowski, "MapReduce approach to collective classification for networks," in L. Rutkowski, M. Korytkowski, R. Scherer, R. Tadeusiewicz, L. A. Zadeh, and J. M. Zurada, eds., *Artificial Intelligence and Soft Computing*. Berlin and Heidelberg: Springer, 2012, pp. 656–663.

[29] O. F. Celebi, E. Zeydan, O. F. Kurt, O. Dedeoglu, O. Ileri, B. A. Sungur, A. Akan, and S. Ergut, "On use of big data for enhancing network coverage analysis," in *20th International Conference on Telecommunications (ICT'13)*, May 2013.

[30] I. A. Karatepe and E. Zeydan, "Anomaly detection in cellular network data using big data analytics," in *Proceedings of European Wireless 2014*, VDE, 2014, pp. 1–5.

[31] M. Cha, H. Kwak, P. Rodriguez, Y.-Y. Ahn, and S. Moon, "I tube, you tube, everybody tubes: analyzing the world's largest user generated content video system," in *Proceedings of the 7th ACM SIGCOMM Conference on Internet Measurement*, ACM, 2007, pp. 1–14.

[32] D. Rossi and G. Rossini, "On sizing CCN content stores by exploiting topological information," in *IEEE Conference on Computer Communications Workshops (INFOCOM WKSHPS)*, 2012, pp. 280–285.

[33] M. Zink, K. Suh, Y. Gu, and J. Kurose, "Characteristics of YouTube network traffic at a campus network–measurements, models, and implications," *Computer Networks*, vol. 53, no. 4, pp. 501–514, 2009.

[34] E. Mucelli Rezende Oliveira, A. Carneiro Viana, K. P. Naveen, and C. Sarraute, "Measurement-driven mobile data traffic modeling in a large metropolitan area," INRIA, Research Report RR-8613, Oct. 2014.

[35] A. Paterek, "Improving regularized singular value decomposition for collaborative filtering," in *Proceedings of KDD CUP and Workshop*, vol. 2007, 2007, pp. 5–8.

[36] E. Zeydan, E. Bastug, M. Bennis, M. A. Kader, I. A. Karatepe, A. S. Er, and M. Debbah, "Big data caching for networking: moving from cloud to edge," *IEEE Communications Magazine*, vol. 54, no. 9, pp. 36–42, 2016.

[37] E. Baştuğ, M. Bennis, E. Zeydan, M. A. Kader, I. A. Karatepe, A. S. Er, and M. Debbah, "Big data meets telcos: a proactive caching perspective," *Journal of Communications and Networks*, vol. 17, no. 6, pp. 549–557, 2015.

[38] M. A. Kader, E. Bastug, M. Bennis, E. Zeydan, A. Karatepe, A. S. Er, and M. Debbah, "Leveraging big data analytics for cache-enabled wireless networks," in *Globecom Workshops (GC Wkshps)*, IEEE, 2015, pp. 1–6.

10 Mobility-Aware Caching in Cellular Networks

Shankar Krishnan, Mehrnaz Afshang, and Harpreet S. Dhillon

Driven by the inherent spatiotemporal correlation in wireless data demand, cellular network design is becoming increasingly *content centric*. An integral component of this new paradigm is the network's ability to cache popular content at its *edge*, which includes base stations, access points [1], and handheld devices [2]. This additionally reduces latency, which is one of the key challenges facing the next generation of cellular networks. As discussed in the earlier chapters, the huge size of a typical library of popular files and relatively smaller storage capacities of edge devices, especially small cell base stations (SCBSs) and handheld devices, makes it necessary to carefully determine the set of files (cache) that should be placed on each device. Compared to a wireless network for which caching mechanisms are fairly well understood, a distinctive feature of content-centric wireless networks is the mobility of the end users, which needs to be included in the system design. Inspired by this, in this chapter we investigate the impact of mobility on edge caching. In particular, we determine the optimal caching policies in both static and mobile user scenarios and show that the optimal solutions in the two scenarios are significantly different. In particular, while it is preferable for the SCBSs to cache the most popular files in the static scenario, the files can be cache much more evenly in the mobile case.

10.1 Optimal Caching in Static Networks

With the increasing consumption of internet content, the number of *popular* files accessed by the users is growing, which makes a strong case for wireless edge caching. Said differently, whenever some content becomes popular, say a YouTube video goes viral, it is beneficial to cache it in edge devices, such as SCBSs, rather than retrieving it each time from the internet. Determining the content that should be cached at each SCBS is, however, the main challenge while designing such a network. Due to the amount of content available on the internet, this problem seems hopeless at first. However, as already discussed in earlier chapters, popular content forms a small fraction of the total content, while a large amount of content remains unpopular and sparsely requested [4]. This allows one to focus on the library of popular files, where each file is associated with a popularity distribution, which can be empirically determined and is often modeled as Zipf distribution [5].

This chapter is an expanded version of [3].

In this chapter, we specifically focus on small cell caching, which is perhaps the most realistic use case of edge caching. The problem of small cell caching has two key dimensions: (1) learning the library of popular files and (2) determining the subset of library contents to be placed on each SCBS in order to optimize system performance. In this chapter, we focus on the latter, assuming that the library of popular files has already been determined. Note that while the library of popular files is a small fraction of the total content available on the internet, it is still, in general, not possible to cache the whole library at each SCBS due to the limited capacities of cache storage units. Therefore, one can cache only a subset of the library on each SCBS. This means that even if the file requested by a particular user is a part of the library of popular files, the user may still not be able to download it because it may not be cached at any SCBS located in the user's vicinity.

There are two basic approaches for determining the optimal subset of library content to be placed at each SCBS/device, namely the deterministic and probabilistic. In [6] and [7], deterministic content placement was analyzed, where optimal placement of the popular content was determined by exploiting information about the node locations along with the statistical or instantaneous channel states. In deterministic content placement, the optimal strategy is to cache the same copy of popular contents in all SCBSs. However, in reality, the geographic locations of the users, wireless channel states, and content popularity are all the time varying, and thus the optimal cache needs to be frequently updated in the deterministic content placement strategy. In order to overcome this, probabilistic content placement was proposed and analyzed in the context of device-to-device (D2D) network caching in [8], where each mobile terminal caches a specific subset of the library with a given caching probability. In this chapter, we also focus on probabilistic content placement.

The performance of wireless caching systems depends heavily on the adopted cache placement strategy. Most prior works focus on optimizing cache placement, deterministic or probabilistic, by maximizing some performance metric of interest. This metric, say successful reception probability or download time, is optimized by constructing a suitable utility function. For instance, [9] and [10] provide an optimal probabilistic placement policy, which guarantees maximum total hit probability in cache-enabled cellular networks. To achieve optimality, this policy exploits multi-coverage regions and delivers considerable performance improvement. Similarly, considering both coded and uncoded cases, [6] studies optimal cache placement that minimizes download time. A common assumption in all these works is that the user locations are stationary. In the resulting static setup, optimal cache is predominantly determined by the request probabilities of the files in the library. However, as we discuss in detail in this chapter, optimal cache placement is significantly different when the mobility of the users is taken into account.

10.2 Mobility in Cellular Networks

A large fraction of users in a wireless network are in general mobile [11]. User mobility impacts (1) resource management aspects, such as channel allocation schemes, call blocking rate, and traffic volume per cell, and (2) radio propagation aspects, such as signal strength variation and time dispersion of signals. For efficient design and dimensioning of the network, it is necessary to analyze mobility-aware metrics such as handoff rate [12], handoff probability [13–15], and sojourn time (or dwell time) [16]. Along with these classical metrics, the transformation toward a *content-centric* design of cellular networks renders it important to investigate the impact of mobility on optimal caching strategies in cellular networks [3, 17–20]. For instance, consider the content request pattern of a mobile user who moves over large distances, e.g., from office to home, in a certain time period. This user at home may be interested in a completely different content compared to when the user is in office. Considering different interest levels over requested content in different locations, a large-scale mobility model, such as the clustering algorithm proposed in [21], needs to be considered, where users are grouped into clusters based on their request profile. In such large-scale mobility scenarios, determining the optimal content to be cached depends on determining the library of files that are popular in that specific region (say home or office) [22]. Since our focus is not on determining the library of popular files, this large-scale mobility case does not come under the purview of our analysis. Instead, we focus on small-scale mobility scenarios, which correspond to the cases where the users move over smaller distances, such as within a mall or a university. In such scenarios, it is reasonable to assume that the library of popular files is not impacted by the user mobility. For instance, users likely to access a similar set of files irrespective of where they are in the university. The main question here is to determine the distribution of the files that should be placed in the cache and how it is impacted by user mobility.

The existing methodologies for mobility-aware network performance analysis can be classified in two major categories: trajectory-based and association-based approaches. Trajectory-based approaches consider explicit mobility models that emulate movement patterns of mobile users. These models are either obtained using logs of connectivity information of mobile users, e.g., see [23], or are based on mathematical models, such as random walk mobility [24], random way point mobility [25], and random direction mobility [16]. On the other hand, association-based mobility approach, as the name suggests, is conducive for answering mobility-related questions related to the association of users to the BSs. For instance, if one wants to know whether there was a handover when a mobile moved from a given location to another location, one simply needs to determine whether that user is associated with the same or different base station at the two locations. Since association-based mobility approaches are known to be much more tractable than the trajectory-based approaches, we also focus on them in this chapter. More finer details about the mobility model are provided in the next section.

10.3 Overview of System Model

We consider a cache-enabled cellular network in which the locations of SBSs are modeled as a homogeneous Poisson point processes (PPP) Φ with density λ [26]. It is assumed that a maximum of L files can be cached by each SCBS, while the total number of files in the library is denoted by K. We denote by $\mathrm{P}_{\mathrm{R}_i}$ the probability that the ith file, $\mathcal{F}_i$, will be requested. The files are ordered based on their popularity, with the indexing $i = 1$ and $i = K$ corresponding to the most popular and least popular files, respectively. It is assumed that the users' content requests follow Zipf's law, i.e., $\mathrm{P}_{\mathrm{R}_i} = i^{-\gamma}/\sum_{j=1}^{K} j^{-\gamma}$, where $\gamma > 0$ is the Zipf parameter [5]. It is also assumed that each SCBS caches file $\mathcal{F}_i$ with probability b_i, which is independent of the other SCBSs. Thus $\sum_{i=1}^{K} b_i = L$.

10.3.1 Mobility Model

As noted, we consider the association-based mobility approach in this chapter. This circumvents the need for going into the trajectory details for each mobile user and hence results in more tractable analytical analysis compared to the trajectory-based approaches. This approach is also more conducive for the *large-system* analysis of the considered setup. In order to visualize this system, let us assume that a user follows an arbitrary trajectory as illustrated in Figure 10.1. The mobile user attempts to download a file of interest at multiple discrete points along the trajectory, as shown in Figure 10.1. Since each SCBS caches only a subset of files in the library, it is quite likely that none of the SCBSs in the user's vicinity has cached the file that is currently being requested by the user. However, if the user's equipment is allowed to access the caches at multiple locations on the trajectory, it becomes more likely that it will run into an SCBS, which will have the requested file in its cache. This is fundamentally what makes the static and mobile scenarios different from the caching perspective.

In a dense cellular network, even a displacement by a small distance (small-scale mobility) may take the user in a completely new local neighborhood of small cells.

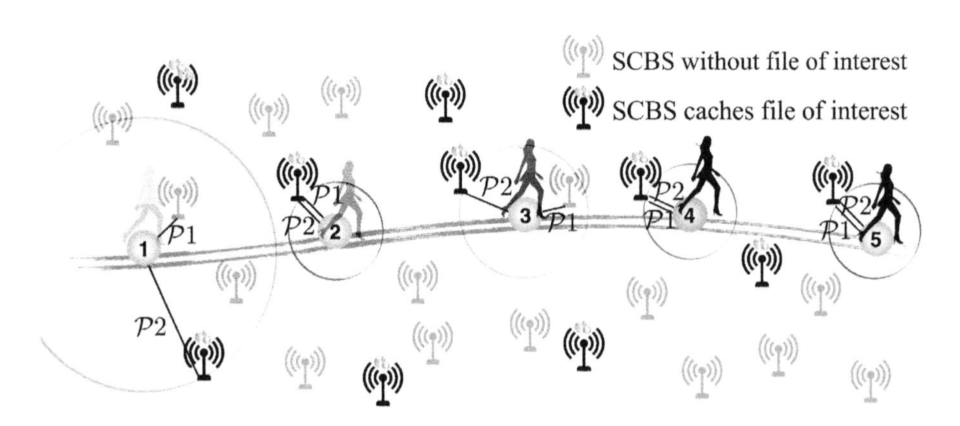

Figure 10.1 Illustration of the system model in which a user moves from location 1 to 5 while trying to obtain the file of interest cached in certain SCBS using Policy 1 ($\mathcal{P}1$) and Policy 2 ($\mathcal{P}2$). In Policy 1, the user connects to the closest SCBS, while in Policy 2, the user connects to the closest SCBS that caches its requested file.

As a result, a user may perceive a completely different set of serving and interfering SCBSs when it tries to access files at the two locations (original and new). This is evident from Figure 10.1 where the user sees a completely new neighborhood of cached SCBSs while moving from location 1 to 2. The file of interest is cached at an SCBS far from the user location 1 while it is much closer at location 2. Again, the user is served by an entirely different SCBS at location 3 compared to locations 1 or 2. The user displacement, even though small with respect to its previous location, can be perceived as an *infinite-mobility scenario* due to its perceived new neighborhood at the new location [27, 28]. We first provide formal insights under the infinite mobility assumption and then show in Section 10.5 that the insights hold for the general case (finite-mobility scenario) as well. For completeness, note that because of the infinite mobility assumption, we do not need to make any specific assumptions about the user trajectories in our analysis.

For the setup studied in this chapter, it is assumed that the typical user can attempt to receive its desired file $\mathcal{F}_i$ from SCBSs within at most n transmission attempts. This is equivalent to $n - 1$ SCBS retransmissions. In the kth transmission, the signal-to-interference ratio (SIR) at the typical user is

$$\mathtt{SIR}_{i,k} = \frac{h_{xk}\|x\|^{-\alpha}}{\sum\limits_{y\in\Phi\setminus\{x\}} h_{yk}\|y\|^{-\alpha}}, \tag{10.1}$$

where fading gains are assumed to be independent across transmission attempts. Here, $\{h_{xk}, h_{yk}\} \sim \exp(1)$ represent Rayleigh fading channel gains from the serving SCBS $x \in \Phi$ and interferer $\{y\}$ in the kth transmission attempt, and $\|\cdot\|^{-\alpha}$ (with $\alpha > 2$) is standard power-law pathloss.

10.3.2 Cell Selection Policy

For the typical user, a straightforward choice of selecting the serving SCBS is to connect to the SCBS with maximum average received power, irrespective of the cached file in that SCBS. This would correspond to the closest SCBS to the typical user. We refer this policy as Policy 1 (*cache-agnostic policy*) and denote it by $\mathcal{P}1$. However, in cache-enabled networks, the closest SCBS may not necessarily cache the user's file of interest. To address this drawback in cache-enabled networks, we define Policy 2 (*cache-aware policy*) and denote it by $\mathcal{P}2$ in which the user connects to the closest SCBS that has cached its file of interest, instead of blindly connecting to the closest SCBS by distance. This, however, requires the knowledge of the cache contents of SCBSs in the user's vicinity, which can perhaps be obtained with the assistance of the macro cells. That said, in Policy 2, the file transfer may not necessarily succeed, as the SCBS with the cached file of interest may not always be close enough to the typical user.

We determine optimal caching probabilities that maximize the *hit probability* (HP) for this system model in the next section. The notation used in this chapter is summarized in Table 10.1.

Table 10.1 Summary of Notation

Notation	Description
Φ, λ	A homogeneous PPP modeling the locations of SCBSs, density of SCBSs
$\mathtt{p}_{\mathtt{R}_i}$	The probability that the ith file, $\mathcal{F}_i$, will be requested
K	The total number of files in the library
L	The maximum number of files that can be cached by each SCBS
$\mathcal{P}1, \mathcal{P}2$	Cache-agnostic policy, cache-aware policy
$p_{\mathtt{S}_{i,n}}, p_{\mathtt{C}_{i,n}}, p_{\mathtt{O}_{i,n}}$	Success probability, coverage probability, outage probability
$p_{\mathtt{C}}^{(1)}, p_{\mathtt{C}}^{(2)}$	Coverage probability under $\mathcal{P}1$, Coverage probability under $\mathcal{P}2$
α	Pathloss exponent, $\alpha > 2$
h_{xk}, h_{yk}	Exponential fading coefficients with unity mean
T	$\mathtt{SIR}$ threshold for successful demodulation and decoding
$\mathtt{p}_{\mathtt{hit}}$	Hit probability

10.4 Optimal Caching in Cellular Networks

In a given transmission attempt, a file is successfully received only if the user is in the *coverage* of an SCBS that has its requested file in its cache. The coverage probability of file $\mathcal{F}_i$ in the kth transmission is defined as $\mathbb{P}(\mathtt{SIR}_{i,k} > T)$, where T is $\mathtt{SIR}$ threshold for successful decoding. Let S_i be the event that file $\mathcal{F}_i$ is successfully received within n transmission attempts. The *success probability* is $p_{\mathtt{S}_{i,n}} = \mathbb{P}(S_i)$. It is assumed that the fading gains are independent and identically distributed over the n transmission attempts. Therefore, the success probability of file $\mathcal{F}_i$ in each transmission attempt is the same, which we denote by $p_{\mathtt{S}_i}$. If all the n transmission attempts are unsuccessful (i.e., the requested file is not received successfully in any of the transmissions), the user is said to be in outage of file $\mathcal{F}_i$. The outage probability of file $\mathcal{F}_i$ after n transmission attempts is $p_{\mathtt{O}_{i,n}} = 1 - p_{\mathtt{S}_{i,n}}$. In this chapter, we study network performance in terms of HP, which is mathematically defined as the sum of the probabilities of successfully receiving each file in the library within n transmission attempts, weighted by their request probabilities, which can be expressed as:

$$\mathtt{p}_{\mathtt{hit}} = \sum_{i=1}^{K} \mathtt{p}_{\mathtt{R}_i} p_{\mathtt{S}_{i,n}} = \sum_{i=1}^{K} \mathtt{p}_{\mathtt{R}_i} (1 - p_{\mathtt{O}_{i,n}}). \tag{10.2}$$

In the next two sections, we maximize the hit probability and obtain optimal caching probabilities $\{b_i\}$ for two scenarios: a *mobile user* and a *static user*.

10.4.1 Mobile User

The success (or outage) probability in each transmission attempt does not depend on the prior transmission attempts under infinite mobility assumption. Hence the outage probability of file $\mathcal{F}_i$ after n transmission attempts is the product of the outage probabilities in each transmission attempt, i.e., $p_{\mathrm{O}_{i,n}} = (1 - p_{\mathrm{S}_i})^n$. In order to maximize the HP, we formulate an optimization problem as follows:

$$\max_{\{b_i\}} \sum_{i=1}^{K} \mathrm{P}_{\mathrm{R}_i}(1 - (1 - p_{\mathrm{S}_i})^n), \tag{10.3}$$
$$\text{s.t.} \sum_{i=1}^{K} b_i = L \text{ and } 0 \le b_i \le 1,\ i = 1, \dots K.$$

10.4.1.1 Policy 1

As described in Section 10.3, the user connects to the closest SCBS that maximizes its average received power under Policy 1. A successful reception of the file depends on the probability that the closest SCBS caches the file requested by the user and that the SIR is larger than a given threshold. In a given transmission under Policy 1, the probability of successfully receiving file $\mathcal{F}_i$ is the product of its caching probability b_i and coverage probability, which can be written as $p_{\mathrm{C}_i}^{(1)}$, i.e., $p_{\mathrm{S}_i} = b_i p_{\mathrm{C}_i}^{(1)}$. It is important to note that coverage probability (when the user is served by the closest SCBS) does not depend on the density of SCBSs under an interference-limited regime as shown in [29]. For a given transmission under Policy 1, the probability of successfully receiving file $\mathcal{F}_i$ is therefore,

$$\begin{aligned} p_{\mathrm{S}_i} &= b_i p_{\mathrm{C}_i}^{(1)} = b_i \mathbb{P}(\mathtt{SIR}_{i,k} \ge T) = b_i \mathbb{P}\Bigg(\frac{h_{xk}\|x\|^{-\alpha}}{\sum_{y \in \Phi \setminus \{x\}} h_{yk}\|y\|^{-\alpha}} > T \Bigg) \\ &\overset{(a)}{=} b_i \mathbb{E}\Big[\exp\Big(-T\|x\|^{\alpha} \sum_{y \in \Phi \setminus \{x\}} h_{yk}\|y\|^{-\alpha} \Big) \Big] \\ &\overset{(b)}{=} b_i \mathbb{E}\Big[\prod_{y \in \Phi \setminus \{x\}} \frac{1}{1 + T\|x\|^{\alpha}\|y\|^{-\alpha}} \Big] \\ &\overset{(c)}{=} b_i \int_0^{\infty} \exp\Big(-2\pi\lambda \int_x^{\infty} \Big(1 - \frac{1}{1 + Tr^{\alpha}u^{-\alpha}}\Big) u\mathrm{d}u \Big) 2\pi\lambda r \exp(-\pi\lambda r^2)\mathrm{d}r \\ &= \frac{b_i}{1 + \rho_1(T, \alpha)}, \end{aligned} \tag{10.4}$$

where

$$\rho_1(T, \alpha) = T^{2/\alpha} \int_{T^{-2/\alpha}}^{\infty} \frac{\mathrm{d}u}{1 + u^{\alpha/2}}. \tag{10.5}$$

Steps (a) and (b) follow from $\{h_{xk}, h_{yk}\} \sim \exp(1)$. Step (c) follows from the probability generating functional (PGFL) of PPP, where $\|x\| = r$ and $\|y\| = u$. Differentiating (10.3) with respect to b_i, we obtain

$$\frac{\mathrm{d}}{\mathrm{d}b_i} \sum_{i=1}^{K} \mathrm{P}_{\mathrm{R}_i}(1 - (1 - b_i p_{\mathrm{C}_i})^n) = \sum_{i=1}^{K} \mathrm{P}_{\mathrm{R}_i} p_{\mathrm{C}_i} n(1 - b_i p_{\mathrm{C}_i})^{n-1} \ge 0, \tag{10.6}$$

and hence the objective function is concave, which implies that the Karush–Kuhn–Tucker (KKT) conditions provide necessary and sufficient conditions for optimality. The Lagrangian function corresponding to problem (10.3) is

$$\mathcal{L}(\mathbf{b}, \nu, \boldsymbol{\mu}, \mathbf{w}) = \sum_{i=1}^{K} \mathrm{P}_{\mathrm{R}_i}(1 - (1 - b_i p_{\mathrm{c}_i}^{(1)})^n) + \nu\left(\sum_{i=1}^{K} b_i - L\right) - \sum_{i=1}^{K} \mu_i b_i + \sum_{i=1}^{K} w_i(b_i - 1),$$

where $\boldsymbol{\mu}, \mathbf{w} \in \mathbb{R}_+^K$, and $\nu \in \mathbb{R}$. Let $\mathbf{b}^*, \nu^*, \boldsymbol{\mu}^*$, and $\mathbf{w}^*$ be primal and dual optimal. The KKT conditions for problem (10.3) state that

$$\sum_{i=1}^{K} b_i^* = L, \tag{10.7}$$

$$0 \le b_i^* \le 1, \mu_i^* \ge 0,\ w_i^* \ge 0, \mu_i^* b_i^* = 0, \qquad \forall i = 1, \dots K \tag{10.8}$$

$$w_i^*(b_i^* - 1) = 0, \qquad \forall i = 1, \dots K, \tag{10.9}$$

$$\mathrm{P}_{\mathrm{R}_i} n(1 - b_i^* p_{\mathrm{c}_i}^{(1)})^{n-1} p_{\mathrm{c}_i}^{(1)} + \nu^* - \mu_i^* + w_i^* = 0, \qquad \forall i = 1, \dots K. \tag{10.10}$$

Based on this, the optimal cache placement under Policy 1 is given next.

THEOREM 10.1 *Under Policy 1 with a maximum of n transmission attempts, the optimal caching probability of file $\mathcal{F}_i$ denoted by b_i^*, which maximizes the HP for a mobile user, is*

$$b_i^* = \begin{cases} 0, & \nu^* < -\mathrm{P}_{\mathrm{R}_i} n p_{\mathrm{c}_i}^{(1)} \\ 1, & \nu^* > -\mathrm{P}_{\mathrm{R}_i} n p_{\mathrm{c}_i}^{(1)} (1 - p_{\mathrm{c}_i}^{(1)})^{n-1}, \\ \frac{1}{p_{\mathrm{c}_i}^{(1)}}\Big[1 - \big(\frac{-\nu^*}{\mathrm{P}_{\mathrm{R}_i} n p_{\mathrm{c}_i}^{(1)}}\big)^{\frac{1}{n-1}}\Big], & \text{otherwise} \end{cases} \tag{10.11}$$

where $\nu^* = -\mathrm{P}_{\mathrm{R}_i} n(1 - b_i^* p_{\mathrm{c}_i}^{(1)})^{n-1} p_{\mathrm{c}_i}^{(1)}$ can be obtained as the solution of the constraint $\sum_{i=1}^{K} b_i^* = L$.

Proof From (10.8) and (10.10), we have

$$w_i^* = b_i^*[-\mathrm{P}_{\mathrm{R}_i} n(1 - b_i^* p_{\mathrm{c}_i}^{(1)})^{n-1} p_{\mathrm{c}_i}^{(1)} - \nu^*], \tag{10.12}$$

which when inserted into (10.9) gives

$$b_i^*(b_i^* - 1)[-\mathrm{P}_{\mathrm{R}_i} n(1 - b_i^* p_{\mathrm{c}_i}^{(1)})^{n-1} p_{\mathrm{c}_i}^{(1)} - \nu^*] = 0. \tag{10.13}$$

From (10.13), it can be seen that $0 < b_i^* < 1$ only if,

$$\nu^* = -\mathrm{P}_{\mathrm{R}_i} n(1 - b_i^* p_{\mathrm{c}_i}^{(1)})^{n-1} p_{\mathrm{c}_i}^{(1)}. \tag{10.14}$$

Given that $0 \le b_i^* \le 1$, we have

$$\nu^* \in [-\mathrm{P}_{\mathrm{R}_i} n p_{\mathrm{c}_i}^{(1)}, -\mathrm{P}_{\mathrm{R}_i} n p_{\mathrm{c}_i}^{(1)} (1 - p_{\mathrm{c}_i}^{(1)})^{n-1}]. \tag{10.15}$$

For this interval, solving for ν^* using the constraint $\sum_{i=1}^{K} b_i^* = L$, we get,

$$\sum_{i=1}^{K} \frac{1}{p_{\mathrm{c}_i}^{(1)}} \left[1 - \left(\frac{-\nu^*}{\mathrm{P}_{\mathrm{R}_i} n p_{\mathrm{c}_i}^{(1)}}\right)^{\frac{1}{n-1}}\right] = L$$

$$\left(\frac{-\nu^*}{n p_{\mathrm{c}}^{(1)}}\right)^{\frac{1}{n-1}} \overset{(a)}{=} \frac{K - L p_{\mathrm{c}}^{(1)}}{\sum_{j=1}^{K} \left(\frac{1}{\mathrm{P}_{\mathrm{R}_j}}\right)^{\frac{1}{n-1}}}, \tag{10.16}$$

where (a) results by using $p_{c_i}^{(1)} = p_c^{(1)}, \forall i = 1, \ldots K$ and rearranging a few terms. Also, it can be seen that for $v^* < -\text{P}_{\text{R}_i} n p_{c_i}^{(1)}, b_i^* = 0$ and if $v^* > -\text{P}_{\text{R}_i} n p_{c_i}^{(1)} (1 - p_{c_i}^{(1)})^{n-1}$, $b_i^* = 1$. □

We now specialize Theorem 10.1 to the simple case of unitary storage space ($L = 1$) in the SCBS and two files in the library ($K = 2$).

COROLLARY *For $K = 2$, the optimal value (b_1^*, b_2^*) obtained as solution of the optimization problem (10.3) is*

$$b_1^* = \begin{cases} 1, & n < 1 + \frac{\gamma}{\log_2\left(\frac{1}{1-p_{c_i}^{(1)}}\right)} \\ \frac{a-1+p_{c_i}^{(1)}}{(a+1)p_{c_i}^{(1)}}, & otherwise \end{cases}, \tag{10.17}$$

where $a = 2^{\frac{\gamma}{n-1}}$, γ is the Zipf parameter and $b_2^* = 1 - b_1^*$.

Proof Using Eq. (10.16) for the case of $K = 2$ and $L = 1$, we obtain,

$$\left(\frac{-v^*}{n p_c^{(1)}}\right)^{\frac{1}{n-1}} = \frac{2 - p_c^{(1)}}{\left(\frac{1}{\text{P}_{\text{R}_1}}\right)^{\frac{1}{n-1}} + \left(\frac{1}{\text{P}_{\text{R}_2}}\right)^{\frac{1}{n-1}}}. \tag{10.18}$$

Rearranging a few terms in the intervals of Theorem 10.1, we obtain

$$b_1^* = \begin{cases} 0, & \left(\frac{-v^*}{np_c^{(1)}}\right)^{\frac{1}{n-1}} > \text{P}_{\text{R}_1}^{\frac{1}{n-1}} \\ 1, & \left(\frac{-v^*}{np_c^{(1)}}\right)^{\frac{1}{n-1}} < \text{P}_{\text{R}_1}^{\frac{1}{n-1}} (1 - p_c^{(1)}) \\ \frac{1}{p_c^{(1)}}\left[1 - \left(\frac{1}{\text{P}_{\text{R}_1}}\right)^{\frac{1}{n-1}} \left(\frac{-v^*}{np_c^{(1)}}\right)^{\frac{1}{n-1}}\right], & \text{otherwise} \end{cases}$$

$$\stackrel{(a)}{=} \begin{cases} 0, & 1 - p_c^{(1)} > \left(\frac{\text{P}_{\text{R}_1}}{\text{P}_{\text{R}_2}}\right)^{\frac{1}{n-1}} \\ 1, & \frac{1}{1-p_c^{(1)}} < \left(\frac{\text{P}_{\text{R}_1}}{\text{P}_{\text{R}_2}}\right)^{\frac{1}{n-1}} \\ \frac{1}{p_c^{(1)}}\left[1 - \frac{2-p_c^{(1)}}{1+\left(\frac{\text{P}_{\text{R}_1}}{\text{P}_{\text{R}_2}}\right)^{\frac{1}{n-1}}}\right], & \text{otherwise} \end{cases}$$

$$\stackrel{(b)}{=} \begin{cases} 0, & n < 1 + \frac{\gamma}{\log_2(1-p_c^{(1)})} \\ 1, & n < 1 + \frac{\gamma}{\log_2\left(\frac{1}{1-p_c^{(1)}}\right)}, \\ \frac{a-1+p_{c_1}^{(1)}}{(a+1)p_{c_1}^{(1)}}, & \text{otherwise} \end{cases}$$

where (a) follows by using Eq. (10.18) and rearranging a few terms. Step (b) is obtained by using the Zipf's law $\text{P}_{\text{R}_i} = i^{-\gamma} / \sum_{j=1}^{K} j^{-\gamma}$, where $\gamma > 0$ is the Zipf parameter and using $a = 2^{\frac{\gamma}{n-1}}$. The final result follows by ignoring the interval corresponding to

$b_1^* = 0$ as it happens only when the number of transmission attempts $n < 1$, which is not possible. □

From Corollary 10.4.1.1, it can be seen that it is optimal to cache only the most popular file $\mathcal{F}_1$ if $n < 1 + \frac{\gamma}{\log_2(\frac{1}{1-p_c^{(1)}})}$, and hence for a single transmission attempt (i.e., $n = 1$ scenario), it is optimal to cache the most popular file.

10.4.1.2 Policy 2

In this Policy 2, the user connects to the closest SCBS that has cached the file of interest instead of just connecting to the closest SCBS based on received signal strength. In Policy 2, the success probability is not weighted by the file's caching probability as the user always connects to the SCBS that contains its requested file. Hence, under Policy 2, the success probability of getting file $\mathcal{F}_i$ is the same as its coverage probability, which is denoted by $p_{c_i}^{(2)}$. The $p_{c_i}^{(2)}$ for a similar scenario is obtained in [30, theorem 1]. Denote by Φ_1 the set of SCBSs that have $\mathcal{F}_i$ in their caches and its complement by $\Phi_2 \equiv \Phi \setminus \Phi_1$. We have

$$p_{s_i} = p_{c_i}^{(2)} = \mathbb{P}\left(\frac{h_{xk}\|x\|^{-\alpha}}{\sum\limits_{y\in\Phi_1\setminus\{x\}} h_{yk}\|y\|^{-\alpha} + \sum\limits_{y\in\Phi_2} h_{yk}\|y\|^{-\alpha}} \geq T\right)$$

$$= \int_0^{\infty} \exp\left(- 2\pi(1-b_i)\lambda \int_0^{\infty}\left(1 - \frac{u^{-\alpha}}{Tr^{-\alpha}+u^{-\alpha}}\right)u\mathrm{d}u\right) \tag{10.19}$$

$$\times \exp\left(- 2\pi b_i\lambda \int_r^{\infty}\left(1 - \frac{u^{-\alpha}}{Tr^{-\alpha}+u^{-\alpha}}\right)u\mathrm{d}u\right)2\pi\lambda b_i r\exp(-\pi\lambda b_i r^2)\mathrm{d}r$$

$$= \frac{b_i}{b_i + \rho_1(T,\alpha) + (1-b_i)\rho_2(T,\alpha)}, \tag{10.20}$$

where

$$\rho_2(T,\alpha) = T^{2/\alpha}\int_0^{T^{-2/\alpha}} \frac{\mathrm{d}u}{1+u^{\alpha/2}}, \tag{10.21}$$

and $\rho_1(T,\alpha)$ is defined in (10.5). In Policy 2, the solution of the optimization problem (10.3) is obtained on the same lines as Theorem 10.1 (Policy 1) by using the success probability p_{s_i}. The optimal solution is provided next.

THEOREM 10.2 *Under Policy 2 with a maximum of n transmission attempts, the optimal caching probability of file $\mathcal{F}_i$ denoted by b_i^*, which maximizes the hit probability for a mobile user, is*

$$b_i^* = \begin{cases} 0, & v^* < \frac{-p_{R_i} n}{C} \\ 1, & v^* > \frac{-p_{R_i} nC(B+C-1)^{n-1}}{(B+C)^{n+1}} \\ \phi(v^*), & \textit{otherwise} \end{cases}, \tag{10.22}$$

where $\phi(v^*)$ is the solution over b_i of

$$\frac{p_{R_i} nC((B-1)b_i + C)^{n-1}}{(Bb_i + C)^{n+1}} + v^* = 0, \tag{10.23}$$

$B = 1 - \rho_2(T,\alpha)$, $C = \rho_1(T,\alpha) + \rho_2(T,\alpha)$, and v^* can be obtained as the solution of the constraint $\sum_{i=1}^{K} b_i^* = L$.

The success probability p_{S_i} for Policy 2 (given by (10.20)) is more complicated than Policy 1 (given by (10.4)). Therefore, it is more difficult to obtain the optimal solution since it requires solving the polynomial equalities of the form (10.23), which may not have closed-form solutions. Thereby, we limit our further discussion on the optimal solution under Policy 2 only to the extreme cases ($n = 1$ and $n \to \infty$)

COROLLARY *(Single transmission, Policy 2) Under Policy 2 with $n = 1$ and $L = 1$, the optimal caching probability of file $\mathcal{F}_i$ denoted by b_i^*, which maximizes the HP for a mobile user, is*

$$b_i^* = \left[\frac{\sqrt{\frac{p_{R_i}}{\epsilon}} - (\rho_1(T,\alpha) + \rho_2(T,\alpha))}{1 - \rho_2(T,\alpha)}\right]^+, i = 1, \ldots K, \tag{10.24}$$

where $[x]^+ = \max(0,x)$, $\sqrt{\epsilon} = \frac{\sum_{i=1}^{K^*} \sqrt{p_{R_i}}}{(K^*-1)\rho_1(T,\alpha) + K^*\rho_2(T,\alpha) + 1}$ and K^*, $1 \leq K^* \leq K$, satisfies the constraint that $0 \leq b_i^* \leq 1$. Here $\rho_1(T,\alpha)$ and $\rho_2(T,\alpha)$ are defined in (10.5) and (10.21), respectively.

Proof The result is obtained by substituting $n = 1$ in Theorem 10.2, solving for v^* using the constraint $\sum_{i=1}^{K} b_i^* = 1$, along with simple algebraic manipulation. □

COROLLARY *(Large transmission attempt scenario) In a scenario with large number of transmission attempts (approaching infinite attempts asymptotically) for the mobile user case (Policies 1 and 2), it is optimal to cache the files uniformly, i.e., $\lim_{n\to\infty} b_i{}^* = \frac{L}{K}$, where K is the total number of popular files in the library.*

Proof It can be seen from (10.4) and (10.20) that p_{S_i} is a monotonically increasing function of b_i for caching policies 1 and 2. The generalized optimization function for (10.3) can hence be written as

$$\max_{\{b_i\}} \sum_{i=1}^{K} p_{R_i}(1 - (1 - f(b_i))^n), \text{ s.t. } \sum_{i=1}^{K} b_i = L. \tag{10.25}$$

After taking the differential of (10.25) with respect to $\{b_i\}_{i=1,\ldots K}$, we get

$$p_{R_i} n(1 - (1 - f(b_i))^{n-1}) = 0 \qquad \forall i = 1, \ldots K, \text{ i.e.,}$$

$$p_{R_i} n(1 - (1 - f(b_i))^{n-1}) = p_{R_j} n(1 - (1 - f(b_j))^{n-1}), \qquad \forall i \neq j,$$

$$\frac{1 - f(b_i^*)}{1 - f(b_j^*)} = \left(-\frac{f'(b_j^*)P_{R_j}}{f'(b_i^*)P_{R_i}}\right)^{\frac{1}{n-1}}, \qquad \forall i \neq j.$$

For $n \to \infty$, we therefore get $f(b_i^*) = f(b_K^*)$, or equivalently $b_i^* = b_j^*$. With $\sum_{i=1}^K b_i^* = L$, the optimal caching strategy is therefore to cache the files uniformly for a mobile user scenario encountering large number of retransmissions. □

10.4.2 Static User

For comparison purpose, we also consider a static scenario, where the user attempts to receive its file of interest within n transmission attempts while being stationary at a certain location in the network. Since the user does not move, it sees exactly the same set of transmitters across n transmission attempts, resulting in the temporal coupling in the success probabilities. Hence, the probability that a file will be successfully received in a given transmission depends on its success probability in the prior transmission attempts. Let $S_{i,k}$ denote the event that file $\mathcal{F}_i$ is in coverage during the kth transmission attempt. The probability that file $\mathcal{F}_i$ is in coverage (at least) once in n transmission attempts, denoted by $p_{\mathrm{c}_{i,n}}$ is given as

$$p_{\mathrm{c}_{i,n}} = \mathbb{P}(\cup_{k=1}^n S_{i,k}) = \mathbb{P}(\cup_{k=1}^n (\mathtt{SIR}_{i,k} > T))$$
$$\stackrel{(a)}{=} \sum_{k=1}^{n} \binom{n}{k} (-1)^{k+1} \mathrm{P}_{i,k}, \tag{10.26}$$

where (a) follows from the inclusion–exclusion principle and $\mathrm{P}_{i,k} = \mathbb{P}(\cap_{j=1}^k (\mathtt{SIR}_{i,j} > T))$ is defined as the joint coverage probability of file $\mathcal{F}_i$ in k transmission attempts.

Similar to the mobile user scenario, we now formulate an optimization problem to maximize the HP for a static user under $\mathcal{P}1$ and $\mathcal{P}2$.

10.4.2.1 Policy 1

As we know that the file of interest $\mathcal{F}_i$ is cached with probability b_i in the closest SCBS, the success probability in n transmission attempts can be derived by multiplying coverage probability $p_{\mathrm{c}_{i,n}}$ by b_i, i.e., $p_{\mathrm{s}_{i,n}} = b_i p_{\mathrm{c}_{i,n}}$. We therefore obtain the following optimization problem after using (10.26) in the earlier result and substituting in (10.2).

$$\max_{\{b_i\}} \sum_{i=1}^{K} \mathrm{P}_{\mathrm{R}_i} \sum_{k=1}^{n} \binom{n}{k} (-1)^{k+1} b_i \mathrm{P}_{i,k}^{(1)}, \tag{10.27}$$

$$\text{s.t.} \quad \sum_{i=1}^{K} b_i = L, \tag{10.28}$$

where $\mathrm{P}_{i,k}^{(1)}$, the joint coverage probability of file $\mathcal{F}_i$ in k transmission attempts under $\mathcal{P}1$, is derived next.

$$\mathrm{P}_{i,k}^{(1)} = \mathbb{E}_{R_1}\left[\mathbb{P}\left(\bigcap_{j\in\{1\ldots k\}} \frac{h_{xj} r_1^{-\alpha}}{\sum_{y\in\Phi\setminus\{x\}} h_{yj}\|y\|^{-\alpha}} > T \middle| r_1\right)\right]$$
$$\stackrel{(a)}{=} \mathbb{E}_{R_1}\left[\prod_{j=1}^{k} \exp\left(-T r_1^{\alpha} \sum_{y\in\Phi\setminus\{x\}} h_{yj}\|y\|^{-\alpha}\right)\right]$$

$$\stackrel{(b)}{=} \mathbb{E}_{R_1}\Big[\prod_{y\in\Phi\backslash\{x\}}\Big(\frac{1}{1+Tr_1^{\alpha}\|y\|^{-\alpha}}\Big)^k\Big]$$

$$\mathrm{P}_{i,k}^{(1)} = \int_0^{\infty} \exp\Big(-2\pi\lambda\int_{r_1}^{\infty}\Big(1-(\frac{u^{\alpha}}{Tr^{\alpha}+u^{\alpha}})^k\Big)u\mathrm{d}u\Big)f_{R_1}(r_1)\mathrm{d}r_1,$$

where (a) follows from $h_{xj} \sim \exp(1)$ and the assumption of independent and identical distribution fading across the k transmission attempts and (b) follows from $h_{yj} \sim \exp(1)$. The final result follows from the PGFL of PPP Φ followed by converting the coordinates from Cartesian to polar and deconditioning with respect to R_1, where R_1 represents the distance of the closest SCBS from the typical user. Here, the probability density of R_1 is given from the null probability of PPP as $f_{R_1}(r_1) = 2\lambda\pi r_1 e^{-\lambda\pi r_1^2}$ [31].

10.4.2.2 Policy 2

In Policy 2, as the user always connects to the closest SCBS that has the file of interest in its cache, the success probability of getting $\mathcal{F}_i$ is the same as the coverage probability, i.e., $p_{\mathrm{s}_{i,n}} = p_{\mathrm{c}_{i,n}}$. Thereby, simply using (10.26) in (10.2), we get the following optimization problem.

$$\max_{\{b_i\}} \sum_{i=1}^{K} \mathrm{P}_{\mathrm{R}_i} \sum_{k=1}^{n} \binom{n}{k}(-1)^{k+1}\mathrm{P}_{i,k}^{(2)}, \qquad \text{s.t.} \quad \sum_{i=1}^{K} b_i = L, \tag{10.29}$$

where $\mathrm{P}_{i,k}^{(2)}$, the joint coverage probability of file $\mathcal{F}_i$ in k transmission attempts under Policy 2, is obtained using a similar derivation approach as Policy 1, which is presented next.

$$\mathrm{P}_{i,k}^{(2)} = \int_0^{\infty} \exp\Big(-2\pi(1-b_i)\lambda\int_0^{\infty}\Big(1-(\frac{u^{\alpha}}{Tr_2{}^{\alpha}+u^{\alpha}})^k\Big)u\mathrm{d}u\Big)$$
$$\times \exp\Big(-2\pi b_i\lambda\int_{r_2}^{\infty}\Big(1-(\frac{u^{\alpha}}{Tr_2{}^{\alpha}+u^{\alpha}})^k\Big)u\mathrm{d}u\Big)f_{R_2}(r_2)\mathrm{d}r_2,$$

where R_2 represents the distance of the typical user to the closest SCBS that has cached the file of interest $\mathcal{F}_i$. As the caching probability of file $\mathcal{F}_i$ in the network is b_i, the distribution of R_2 is therefore given by the closest point of the PPP of intensity $b_i\lambda$, and its probability distribution is given by $2\lambda\pi r_2 e^{-b_i\lambda\pi r_2^2}$. The key difference to be noted while analyzing Policy 2 is that the interference field is now divided into two regions: (1) interference from those SCBSs that have cached the file of interest $\mathcal{F}_i$, which constitutes a PPP of intensity $b_i\lambda$ outside a radius r_2 (closest distance of $\mathcal{F}_i$) and (2) interference from the remaining SCBSs (not having cached $\mathcal{F}_i$), which constitutes a PPP of intensity $(1-b_i)\lambda$ in $\mathbb{R}^2$.

Similar to the mobile user scenarios, we can obtain the optimal caching probabilities of a static user under $\mathcal{P}1$ and $\mathcal{P}2$, i.e., the solutions of the optimization problem (10.27) and (10.29). In the next section, we provide more insights on the optimal caching probabilities for a static user.

10.5 Results and Discussion

For the setup described thus far in this chapter, we now numerically analyze the effect of mobility on the HP-optimal caching for the two policies. For brevity, we first limit our focus to $\mathcal{P}1$ in the results presented in Figures 10.2 and 10.3. Similar observations can be drawn for $\mathcal{P}2$ as well. To complete the picture, the primary difference between the two policies (in terms of the resulting optimal caching probabilities) is highlighted later in the section. For the purpose of numerical results, we assume that SCBSs are distributed according to a homogeneous PPP with density 150 SCBSs/km^2. For this setup, we consider a library of K files, with an $\texttt{SIR}$ threshold of $\beta = 0$ dB and Zipf parameter of $\gamma = 1.2$. Please note that the asterisk denotes the optimal caching probability in all figures.

10.5.1 Mobility in Ultra-dense Networks

Figure 10.2 plots the HP for different caching probabilities of file $\mathcal{F}_1$ for a 2-file library scenario ($K = 2$) and various levels of mobility. For the setup described (SCBS density of 150 BSs/km^2), the average cell "radius" of the SCBSs comes out to be approximately 40 m. Following $\mathcal{P}1$, the user tries to access the file of interest at two locations on its trajectory (say x meters apart). We consider $x = 0$ m to model the static case, a large value of $x = 400$ m to model the *infinite mobility case*, and the intermediate values

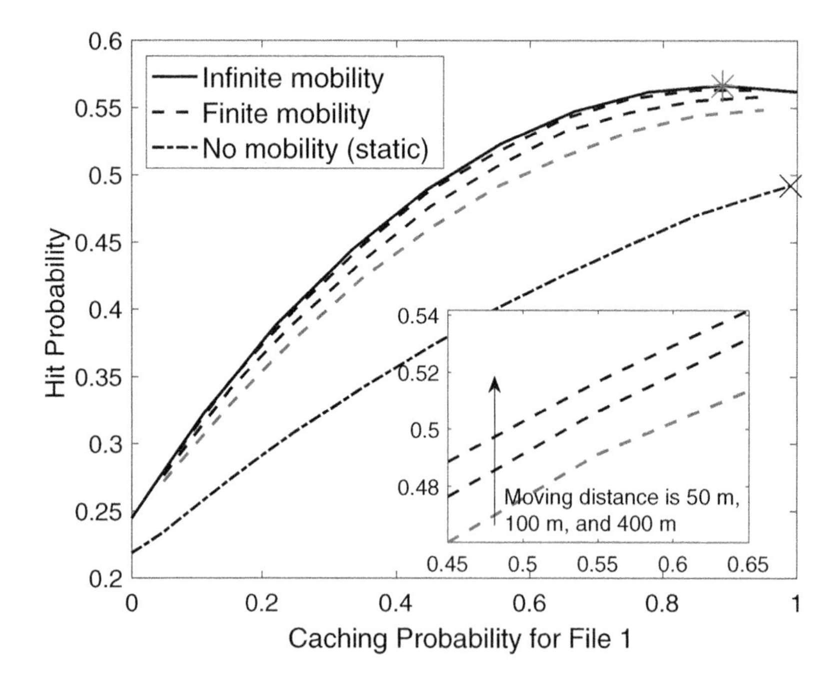

Figure 10.2 Effect of mobility in ultra-dense networks when number of attempts to access the file of interest is $n = 2$.

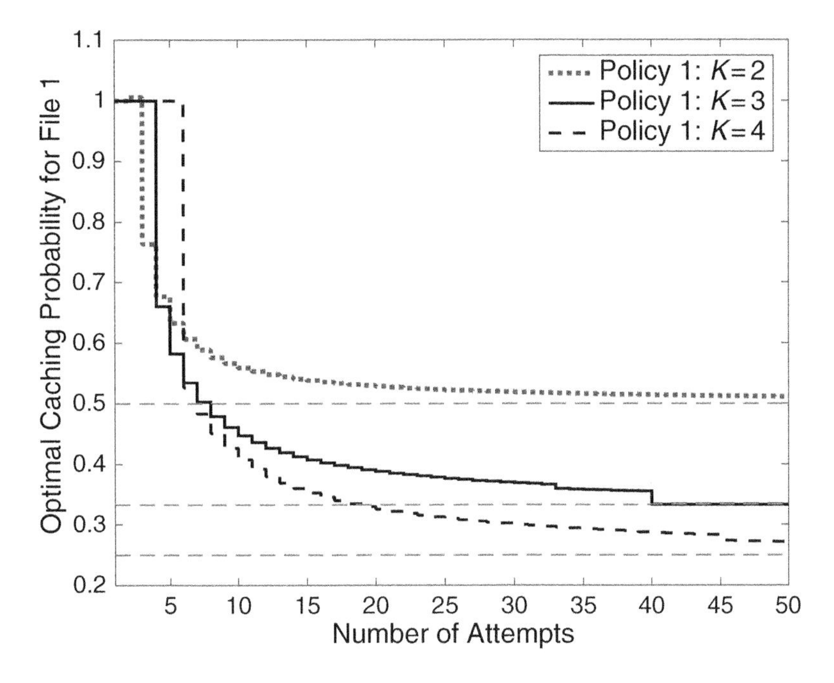

Figure 10.3 Effect of number of attempts n on the optimal caching probability.

of $x = 50$ m and 100 m to model the *finite mobility cases*. It can be observed from Figure 10.2 that the HP for all levels of mobility is significantly higher compared to the static case for all caching probabilities of file $\mathcal{F}_1$. As discussed in the previous sections, this is due to the fact that mobility allows a user to come across more *unique* SCBSs, thus providing access to more unique *caches*, which increases the probability of finding one that has its file of interest. Even a displacement of 50 m (approximately the cell radius) gives substantial gains in HP and can be approximated by the infinite-mobile scenario for a dense network. Therefore, for the simplicity of exposition, we focus on the *infinite mobility* (or large-scale mobility) scenario in the rest of this section. Also, the optimal caching probabilities for all levels of mobility are seen to be shifted toward the left, allowing the network to cache the files in a more balanced way compared to the static case.

10.5.2 Effect of the Number of Attempts

The number of attempts n made to access the file of interest plays a significant role in determining the optimal cache for a given library of files. As observed from Figure 10.3, the optimal strategy for a $K = 2$ file library varies from caching only the most popular file in the network ($n = 1$) to caching both files in the library with an equal probability for a large number of attempts ($n \geq 20$). In the general case with K files in the library, it is seen that the optimal caching probability of each file in the library approaches $1/K$ asymptotically. Therefore, mobility (in particular the flexibility of accessing the file at multiple unique locations) allows the network to have a more balanced cache compared to the static case, where all the SCBSs tend to cache only the most popular files.

10.5.3 Comparison of $\mathcal{P}1$ and $\mathcal{P}2$

So far, we have focused only on $\mathcal{P}1$ in Section 10.5. We now compare the two policies in terms of the HP for different caching probabilities of file $\mathcal{F}_1$ in Figire 10.4. The results are plotted for several different numbers of attempts n. The figure depicts the optimal caching probability of file $\mathcal{F}_1$ that maximizes the HP while obtaining the file of interest from a library of 2 files. It can be observed that the optimal caching probabilities are slightly lower (while resulting in slightly higher HP) in case of $\mathcal{P}2$ compared to $\mathcal{P}1$. This asserts that it is not required to cache the most popular file (file $\mathcal{F}_1$) as frequently under this policy and can allow the SCBSs to cache the less popular files with a relatively larger probability. This slight shift in optimal caching probabilities and a higher HP of $\mathcal{P}2$ can be attributed to the policy mechanism itself. By connecting to the closest SCBS having the file in its cache under $\mathcal{P}2$, we get a better chance of obtaining the desired file than connecting to the SCBS closest to the user and hoping it has the file of interest in its cache. Finally, it can also be seen that the HP for both $\mathcal{P}1$ and $\mathcal{P}2$ increases with the number of attempts n.

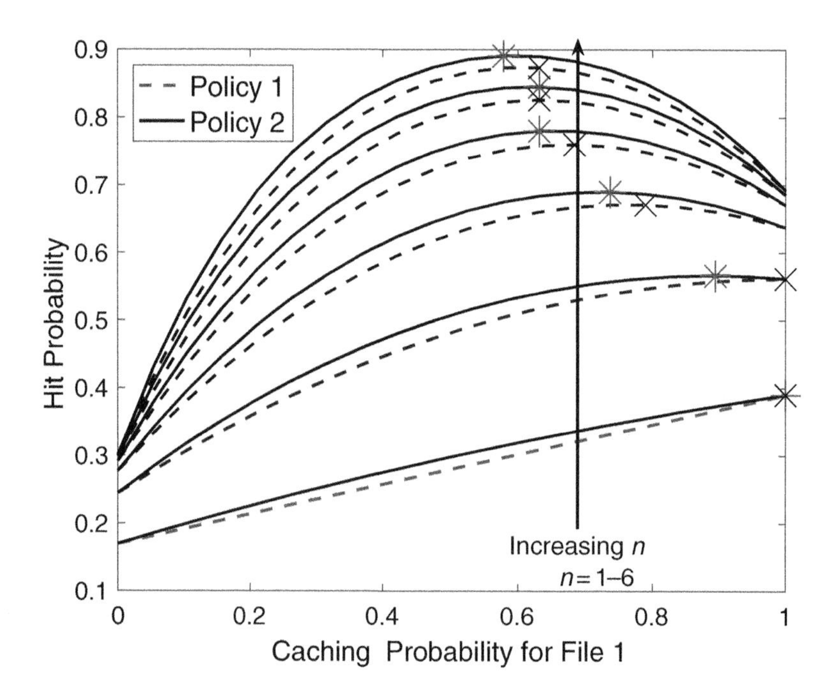

Figure 10.4 Effect of cache gathering policy (Policies 1 and 2) on the HP.

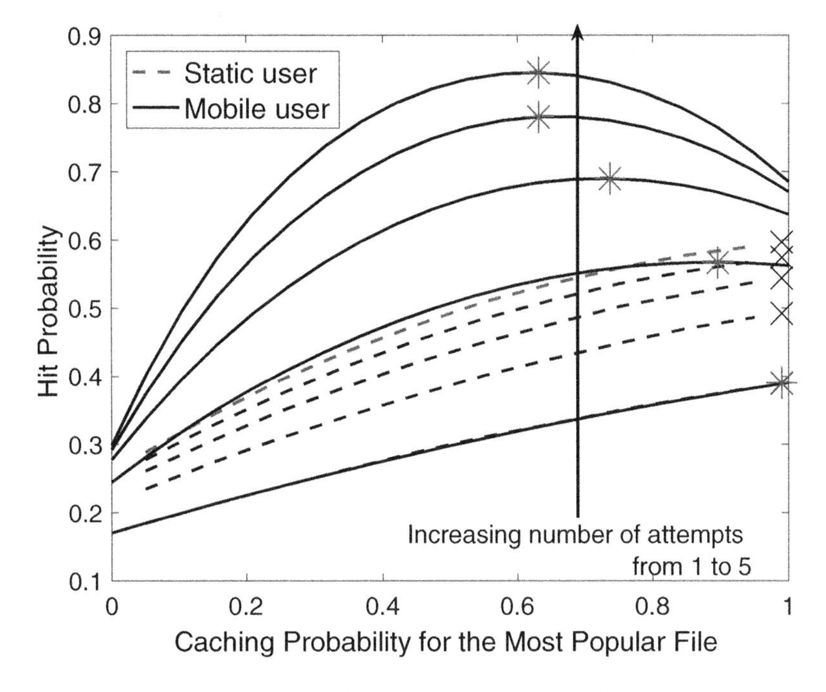

Figure 10.5 Effect of mobility on the optimal caching probabilities for varying number of attempts. User follows Policy 2.

10.5.4 Comparison of the Mobile and Static Cases as a Function of n for $\mathcal{P}2$

In the numerical result shown in Figure 10.5, we compare the HP for the mobile and static cases under the 2-library case. For this comparison, we consider that a maximum of n attempts are made by the user to receive the file of interest using Policy 2. In the mobile case, the n user attempts to obtain the file are made from different locations, whereas in the static case, all attempts are made from the same location. However, fading gains are assumed to be independent across different transmission attempts. Therefore, increasing the number of transmission attempts n also increases the HP even for the static case due to the temporal diversity. Similar observations were made earlier for $\mathcal{P}1$.

10.5.5 Effect of Library Size (K) on the Hit Probability

Figure 10.6 characterizes the optimal hit probability in a network (for a mobile user following $\mathcal{P}1$) with cache size $L = 1$ and K files in the library. It can be seen from the figure that as the number of files in the library increases, the hit probability decreases. Larger the number of files in the library, the fewer the chances of a file hit from the SCBS cache with a certain cache size. As discussed before, it can also be observed that the optimal HP increases as the number of retransmissions increases.

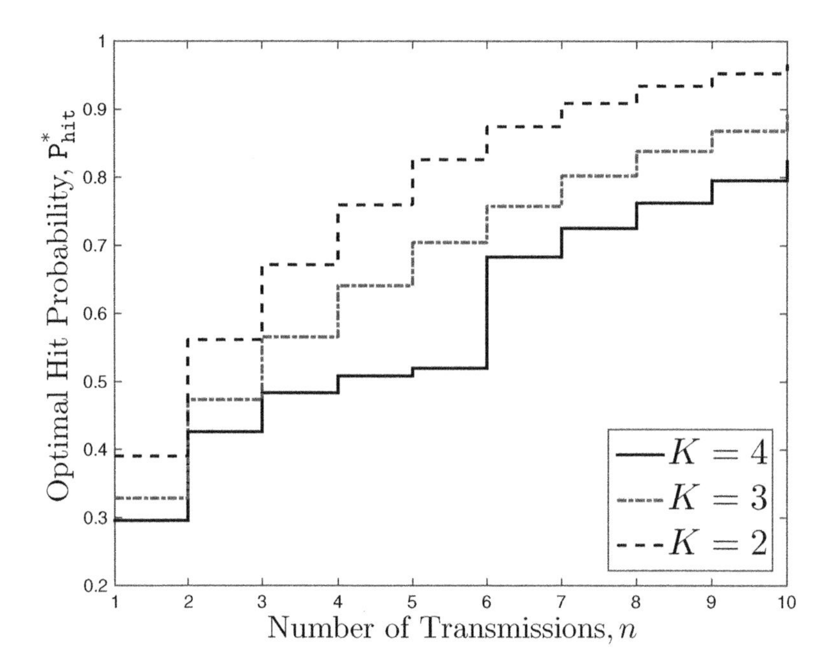

Figure 10.6 Effect of library size K on the hit probability for a mobile user following Policy 1 ($L = 1$).

10.6 Outlook

Providing reliable high-capacity backhaul to each small cell in a cellular network is prohibitively expensive. The idea of caching popular content at the SCBSs has emerged as one of the promising solutions to address this challenge. In this chapter, we focused on the problem of determining the *optimal* cache for each small cell assuming the library of popular files is known a priori. We focused on the case where the user is mobile, and can hence access its file of interest at multiple possible locations along its trajectory. Our results concretely demonstrated that the optimal cache contents are significantly different for the mobile and static cases.

In the static scenario, SCBSs need to cache the most popular files for a higher HP. In the mobile case, however, files can be cached at SCBSs more evenly compared to the static case. The overall network performance (in terms of the successful reception probability) was shown to be much better for the mobile case compared to the static case. This is due to the fact that when the user is mobile it comes across more *unique* small cells and hence more unique caches, which increases the probability of it being close to the small cell that has its file of interest.

This work has several promising extensions. First, it is important to consider *device caching* (i.e., caching content on the handheld devices) along with the *small cell caching* studied in this chapter. Jointly optimizing the cache for devices and small cells, with different constraints on their cache storage capacities, is a meaningful problem to pursue. Second, users usually form physical clusters, e.g., see [32–35], which means it is important to determine optimal caching strategies for clustered networks. Finally, the purpose of this study was to expose fundamental performance trends using simple models. It would be interesting to perform a similar analysis using actual mobility traces and actual data for the SCBS locations.

References

[1] N. Golrezaei, K. Shanmugam, A. G. Dimakis, A. F. Molisch, and G. Caire, "Femtocaching: wireless video content delivery through distributed caching helpers," in *INFOCOM, 2012 Proc. IEEE*, 2012, pp. 1107–1115.

[2] M. Afshang, H. S. Dhillon, and P. H. J. Chong, "Fundamentals of cluster-centric content placement in cache-enabled device-to-device networks," *IEEE Trans. Commun.*, vol. 64, no. 6, pp. 2511–2526, 2016.

[3] S. Krishnan, M. Afshang, and H. S. Dhillon, "Effect of retransmissions on optimal caching in cache-enabled small cell networks," *IEEE Trans. Vehicular Technol.*, vol. 66, no. 12, pp. 11 383–11 387, 2017.

[4] E. Bastug, M. Bennis, and M. Debbah, "Living on the edge: the role of proactive caching in 5G wireless networks," *IEEE Commun. Maga.*, vol. 52, no. 8, pp. 82–89, 2014.

[5] M. Cha, H. Kwak, P. Rodriguez, Y.-Y. Ahn, and S. Moon, "I tube, you tube, everybody tubes: analyzing the world's largest user generated content video system," in *Proc. 7th ACM SIGCOMM Conf. Internet Meas.*, 2007, pp. 1–14.

[6] K. Shanmugam, N. Golrezaei, A. Dimakis, A. Molisch, and G. Caire, "Femtocaching: wireless content delivery through distributed caching helpers," *IEEE Trans. Inform. Theory*, vol. 59, no. 12, pp. 8402–8413, Dec. 2013.

[7] J. Li, Y. Chen, Z. Lin, W. Chen, B. Vucetic, and L. Hanzo, "Distributed caching for data dissemination in the downlink of heterogeneous networks," *IEEE Trans. Commun.*, vol. 63, no. 10, pp. 3553–3568, Oct. 2015.

[8] M. Ji, G. Caire, and A. F. Molisch, "Wireless device-to-device caching networks: basic principles and system performance," *IEEE J. Select. Areas Commun.*, vol. 34, no. 1, pp. 176–189, Jan. 2016.

[9] B. Blaszczyszyn and A. Giovanidis, "Optimal geographic caching in cellular networks," in *Proc. IEEE Intl. Conf. Commun. (ICC)*, June 2015, 3358–3363.

[10] M. Afshang and H. S. Dhillon, "Optimal geographic caching in finite wireless networks," in *Proc. IEEE Signal Process. Adv. Wireless Commun.*, 2016, pp. 1–5.

[11] H. Tabassum, M. Salehi, and E. Hossain, "Mobility-aware analysis of 5G and B5G cellular networks: a tutorial," arXiv.org/abs/1805.02719, 2018.

[12] S. Sadr and R. S. Adve, "Handoff rate and coverage analysis in multi-tier heterogeneous networks," *IEEE Trans. Wireless Commun.*, vol. 14, no. 5, pp. 2626–2638, 2015.

[13] W. Bao and B. Liang, "Stochastic geometric analysis of user mobility in heterogeneous wireless networks," *IEEE J. Select. Areas Commun.*, vol. 33, no. 10, pp. 2212–2225, 2015.

[14] S.-Y. Hsueh and K.-H. Liu, "An equivalent analysis for handoff probability in heterogeneous cellular networks," *IEEE Commun. Lett.*, vol. 21, no. 6, pp. 1405–1408, 2017.

[15] S. Krishnan and H. S. Dhillon, "Spatio-temporal interference correlation and joint coverage in cellular networks," *IEEE Trans. Wireless Commun.*, vol. 16, no. 9, pp. 5659–5672, Sept. 2017.

[16] X. Lin, R. K. Ganti, P. J. Fleming, and J. G. Andrews, "Towards understanding the fundamentals of mobility in cellular networks," *IEEE Trans. Wireless Commun.*, vol. 12, no. 4, pp. 1686–1698, 2013.
[17] R. Wang, X. Peng, J. Zhang, and K. B. Letaief, "Mobility-aware caching for content-centric wireless networks: modeling and methodology," *IEEE Commun. Mag.*, vol. 54, no. 8, pp. 77–83, 2016.
[18] M. Chen, Y. Hao, M. Qiu, J. Song, D. Wu, and I. Humar, "Mobility-aware caching and computation offloading in 5G ultra-dense cellular networks," *Sensors*, vol. 16, no. 7, p. 974, 2016.
[19] O. Semiari, W. Saad, M. Bennis, and B. Maham, "Caching meets millimeter wave communications for enhanced mobility management in 5G networks," *IEEE Trans. Wireless Commun.*, vol. 17, no. 2, pp. 779–793, 2018.
[20] S. Krishnan and H. S. Dhillon, "Effect of user mobility on the performance of device-to-device networks with distributed caching," *IEEE Wireless Commun. Lett.*, vol. 6, no. 2, pp. 194–197, Apr. 2017.
[21] M.-H. Chiu and M. Bassiouni, "Predictive schemes for handoff prioritization in cellular networks based on mobile positioning," *IEEE J. Select. Areas Commun.*, vol. 18, no. 3, pp. 510–522, Mar. 2000.
[22] E. Bastug, M. Bennis, and M. Debbah, "Social and spatial proactive caching for mobile data offloading," in *Proc., IEEE Intl. Conf. Commun. (ICC)*, IEEE, 2014, pp. 581–586.
[23] A. Chaintreau, P. Hui, J. Crowcroft, C. Diot, R. Gass, and J. Scott, "Pocket switched networks: real-world mobility and its consequences for opportunistic forwarding," University of Cambridge, Computer Laboratory, Tech. Rep., 2005.
[24] K. Pearson, "The problem of the random walk," *Nature*, vol. 72, no. 1867, p. 342, 1905.
[25] E. Hyytia, P. Lassila, and J. Virtamo, "Spatial node distribution of the random waypoint mobility model with applications," *IEEE Trans. Mobile Comput.*, vol. 5, no. 6, pp. 680–694, 2006.
[26] H. S. Dhillon, R. K. Ganti, F. Baccelli, and J. G. Andrews, "Modeling and analysis of k-tier downlink heterogeneous cellular networks," *IEEE J. Select. Areas Commun.*, vol. 30, no. 3, pp. 550–560, 2012.
[27] M. Haenggi, "The local delay in poisson networks," *IEEE Trans. Inform. Theory*, vol. 59, no. 3, pp. 1788–1802, 2013.
[28] C. Jarray and A. Giovanidis, "The effects of mobility on the hit performance of cached D2D networks," in *Int. Symp. Model. Optim. Mobile, Ad Hoc, Wireless Networks (WiOpt)*, IEEE, 2016, pp. 1–8.
[29] J. G. Andrews, F. Baccelli, and R. K. Ganti, "A tractable approach to coverage and rate in cellular networks," *IEEE Trans. Commun.*, vol. 59, no. 11, pp. 3122–3134, 2011.
[30] S. Krishnan and H. S. Dhillon, "Distributed caching in device-to-device networks: a stochastic geometry perspective," in *Proc. IEEE Asilomar*, 2015, pp. 1280–1284.
[31] M. Haenggi, *Stochastic Geometry for Wireless Networks*, New York: Cambridge University Press, 2013.
[32] M. Afshang, H. S. Dhillon, and P. H. J. Chong, "Modeling and performance analysis of clustered device-to-device networks," *IEEE Trans. Wireless Commun.*, vol. 15, no. 7, pp. 4957–4972, 2016.
[33] C. Saha, M. Afshang, and H. S. Dhillon, "3GPP-inspired HetNet model using Poisson cluster process: sum-product functionals and downlink coverage," *IEEE Trans. Commun.*, vol. 66, no. 5, pp. 2219–2234, 2018.

[34] C. Saha, M. Afshang, and H. S. Dhillon, "Enriched k-tier HetNet model to enable the analysis of user-centric small cell deployments," *IEEE Trans. Wireless Commun.*, vol. 16, no. 3, pp. 1593–1608, Mar. 2017.

[35] M. Afshang and H. S. Dhillon, "Poisson cluster process based analysis of HetNets with correlated user and base station locations," *IEEE Trans. Wireless Commun.*, vol. 17, no. 4, pp. 2417–2431, Apr. 2018.

Part III

Cache-Aided Interference and Physical Layer Management

11 Cache-Enabled Cloud Radio Access Networks

Meixia Tao, Erkai Chen, Wei Yu, and Ya-Feng Liu

This chapter presents a content-centric framework for transmission optimization in cloud radio access networks (RANs) by leveraging wireless edge caching and physical-layer multicasting. We consider a cache-enabled cloud RAN, where each base station (BS) is connected to a central processor (CP) via a potentially capacity-limited backhaul link and equipped with a local cache to alleviate the backhaul load. We first study the caching effects on multicast-enabled access downlink, where users that request the same content form a multicast group and are served by the same BS or BS cluster using multicasting. We study the cache-aware joint design of the content-centric BS clustering and multicast beam-forming to minimize the system total power cost and backhaul cost under individual minimum transmission rate constraints for each multicast group. Through simulation results, we show that the proposed cache-aware content-centric multicast transmission is much superior to the traditional user-centric unicast transmission in terms of system total transmit power reduction and backhaul saving. We then study the caching effects on backhaul downlink with wireless multicast backhaul, where the CP delivers the requested contents to a single cluster of BSs via multicasting. Given a total cache size constraint, we study the joint cache size allocation at the BSs and the optimal multicast beam-forming transmission at the CP to minimize the expected downloading time of requested contents from the CP to the BSs. Numerical results provide some useful insights into the BS caching design and show that the optimized cache size allocation scheme outperforms the uniform allocation and other heuristic schemes.

11.1 Introduction

Cloud radio access network (cloud RAN) is a promising network architecture for the next generation of wireless cellular networks [1]. It can boost network capacity and increase energy efficiency by centralized signal processing among all the BSs that are connected to a central processor via potentially capacity-limited backhaul links. However, performing full joint processing requires the users' payload data to be shared among all the BSs, which can place a significant burden on backhaul links. As such, there is a fundamental trade-off between the access link efficiency and the backhaul link consumption in cloud RANs. This chapter presents how to exploit wireless edge caching, in conjunction with physical-layer multicasting, in cloud RAN architectures to alleviate the backhaul requirement and improve system energy efficiency.

Wireless edge caching has emerged as a promising approach that can reduce peak traffic and backhaul burden for wireless content delivery by caching some popular contents at the local BSs or pushing directly the contents at user devices during the off-peak time [2]. On the other hand, multicasting provides an efficient capacity-offloading approach to deliver a common message to multiple receivers concurrently [3, 4]. It has great potential in many applications, e.g., video streaming, mobile application updates, and public group communications. It can also be exploited in wireless backhaul to push common information from a macro BS to multiple small BSs. Caching and multicasting are thus two important enabling techniques to accelerate content delivery in wireless networks.

This chapter presents a content-centric framework for transmission optimization in cloud RANs by collectively leveraging caching and multicasting. We consider a cache-enabled cloud RAN, where each BS has a local cache with limited storage size and is connected to a CP via a dedicated or shared backhaul link. If the requested content is not cached in the local cache of a BS, it will acquire the content from the core network via the backhaul links. Users requesting a same content form a group and are served by the same BS cluster via multicast transmission. This chapter shows that caching can improve the system-level performance of cloud RAN in two different ways: for both the access link and the backhaul link. The first part of the chapter studies the design of caching and multicasting in the access link. We study the cache-aware joint content-centric BS clustering and multicast beam-forming design to minimize the system total network cost subject to a minimum rate constraint for each individual multicast group. Simulation results show that the proposed cache-aware content-centric multicast transmission is superior to the traditional user-centric unicast transmission in terms of system transmit power reduction and backhaul saving.

The second part of the chapter studies the design of caching and multicasting in the backhaul link, where the BSs fetch the requested content from the CP through a shared wireless backhaul using joint cache-channel coding. Given a total cache size constraint, we study a mixed time-scale optimization for cache size allocation among all the BSs and multicast beam-forming at the CP to minimize the expected downloading time of requested contents in the backhaul phase. Numerical results provide some useful insights into the BS caching design and show that the optimized cache size allocation scheme outperforms the uniform allocation and other heuristic schemes.

The rest of this chapter is organized as follows. Section 11.2 introduces the model of the cache-enabled cloud RAN. Section 11.3 studies caching and multicasting in the access link. Section 11.4 studies the caching and multicasting in the backhaul link. Finally, we draw conclusions in Section 11.5 and outline some possible directions for future research.

11.2 Cache-Enabled Cloud RAN Model

11.2.1 Network Model

As shown in Figure 11.1, we consider the downlink transmission of a cloud RAN, where there are N BSs and K users. Each BS has a local cache and is connected to a cloud-based CP via a backhaul link. The CP has a database consisting of F files, where the size of each file is normalized as 1. Let p_f denote the request probability (i.e., popularity distribution) of the fth file, which satisfies $0 \leq p_f \leq 1$ and $\sum_{f=1}^{F} p_f = 1$. Let C_n ($C_n \leq F$) denote the cache size of the nth BS. Each BS can prestore some file bits during off-peak time prior to user request. If the requested file of its serving user has been entirely stored in the local cache of this BS, the BS can access the file directly. Otherwise, it will download the requested file or the uncached part of this file from the CP via its backhaul link.

In this chapter, it is assumed that the channel state information (CSI) is perfectly known at the CP for joint signal processing and all BSs can precisely synchronize with each other for downlink cooperative transmission. Our focus is to illustrate a content-centric transmission framework in the cached-enabled cloud RAN and its baseband beam-forming design.

11.2.2 Content-Centric BS Clustering

A prominent approach for mitigating the backhaul load in traditional cloud RANs is to serve each user using an individually selected subset of neighboring BSs, referred to

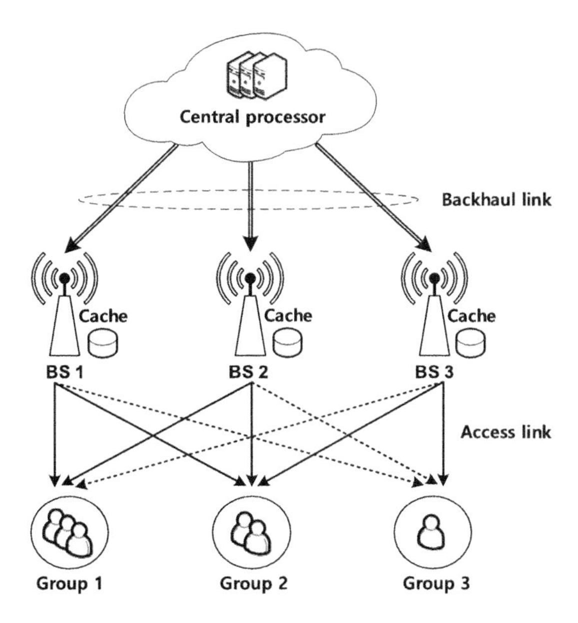

Figure 11.1 An example of cache-enabled cloud RAN downlink.

as *user-centric BS clustering*, regardless of the content each user requests. By adopting user-centric BS clustering, the CP needs only to deliver the user's payload data to its serving BSs rather than all the BSs, which can reduce the backhaul load significantly. In this case, different clusters for different users may overlap, and there are no explicit cluster boundaries [5].

Generally, the users request content according to some popularity distribution such as the Zipf distribution [6]. The more popular the content is, the more likely it will be requested and the more requests it will receive. By taking the content popularity into account, a *content-centric BS clustering* strategy is proposed in [7]. In the content-centric BS clustering, the users requesting the same content are grouped together and served by a cluster of BSs formed with respect to each content. Within each cluster, multicast transmission is then adopted to serve the users. The BS clusters for different content can overlap with each other. Compared with user-centric BS clustering, content-centric BS clustering exploits the popularity of the request contents and benefits from multicast transmission, and thus can provide efficient content delivery in the considered networks.

In the following, we present the transmission model with content-centric BS clustering in detail. We assume that each user can request content in each scheduling time slot. Denote $\mathcal{G}_m$ as the mth multicast group formed by the users requesting file f_m, for all $m = 1, \ldots, M$, where M $(1 \leq M \leq \min\{K, F\})$ is the total number of the formed multicast groups. Denote the serving BS cluster of multicast group m as $\mathcal{Q}_m$, where $\mathcal{Q}_m \subseteq \mathcal{N}$. An example with three multicast groups is illustrated in Figure 11.1, where the serving BSs of the three multicast groups are $\mathcal{Q}_1 = \{1, 2\}$, $\mathcal{Q}_2 = \{1, 2, 3\}$, and $\mathcal{Q}_3 = \{3\}$, respectively.

Define a binary matrix $\mathbf{S} \in \{0, 1\}^{M \times N}$ as the indicator of BS clustering, where $s_{m,n} = 1$ represents that BS n is within the BS cluster of multicast group m, otherwise $s_{m,n} = 0$. Denote $\mathbf{w}_m = [\mathbf{w}_{m,1}^H, \mathbf{w}_{m,2}^H, \ldots, \mathbf{w}_{m,N}^H]^H \in \mathbb{C}^{NL \times 1}$ as the network-wide beam-former for the mth group, where $\mathbf{w}_{m,n} \in \mathbb{C}^{L \times 1}$ is the beam-former of group m at BS n. Note that $\mathbf{w}_{m,n} = \mathbf{0}$ if $s_{m,n} = 0$. Therefore, $\mathbf{w}_m$ is potentially (group) sparse. For each user $k \in \mathcal{G}_m$, the received signal can be written as

$$y_k = \mathbf{h}_k^H \mathbf{w}_m x_m + \sum_{j \neq m}^{M} \mathbf{h}_k^H \mathbf{w}_j x_j + n_k, \tag{11.1}$$

where $\mathbf{h}_k = [\mathbf{h}_{k,1}^H, \mathbf{h}_{k,2}^H, \ldots, \mathbf{h}_{k,N}^H]^H \in \mathbb{C}^{NL \times 1}$ is the composite channel vector between all BSs and the kth user, $x_m \in \mathbb{C}$ is the message intended for group m, and $n_k \sim \mathcal{CN}(0, \sigma_k^2)$ is the additive white Gaussian noise. The corresponding SINR at user k can be expressed as

$$\text{SINR}_k = \frac{|\mathbf{h}_k^H \mathbf{w}_m|^2}{\sum_{j \neq m}^{M} |\mathbf{h}_k^H \mathbf{w}_j|^2 + \sigma_k^2}. \tag{11.2}$$

Accordingly, the total transmit power of the network can be expressed as

$$C_P = \sum_{m=1}^{M} \sum_{n=1}^{N} \|\mathbf{w}_{m,n}\|_2^2. \tag{11.3}$$

Compared with the traditional user-centric BS clustering, where each user is served by its nearby BSs that have good channel conditions, the content-centric BS clustering is more complicated. In the content-centric BS clustering, since the users within the same multicast group may be dispersed geographically, it is no longer feasible to determine the BS clustering simply according to the received signal strength or the physical closeness between each BS and each user. Moreover, by considering the local cache at each BS, the BS that has cached the requested file may have a higher chance to joint the cluster. As such, the content-centric BS clustering in the considered network must be aware of both channel states and cache states.

11.2.3 Caching at BSs

Caching at the BSs can enable more BSs to cooperatively transmit the same content to the users in the access link. What contents to cache at each BS is a crucial design factor in cache-enabled cloud RAN. Intuitively, in a sparse network where each user can access only one single BS, it is optimal to cache contents with the largest popularities in each BS in terms of cache-hit ratio maximization. While in a densely deployed network where each user can access to multiple BSs, finding the optimal cache placement is often intractable [2]. By allowing coded caching at each BS, one can find the optimal coded fraction of each file efficiently [8]. In this chapter, however, we restrict to uncoded caching for simplicity and consider three heuristic caching strategies as follows.

- *Popularity-aware caching (PopC):* All the storage sizes of each BS are used to cache the contents with the largest popularities. This strategy can fully exploit the benefits of full cooperation. However, when the content popularity is uniformly distributed, it may cause high backhaul load due to the low cache-hit ratio.
- *Random caching (RanC):* All the contents are cached at the BSs randomly and equally without knowing their popularity. Due to the randomness in the cache placement, it is highly probable that each user can find its requested file from the caches of the BSs without resorting to CP via backhaul. However, since different BSs tend to cache distinct contents, there is little opportunity for cooperative transmission.
- *Probabilistic caching (ProC):* Each BS randomly caches a content with a certain probability that is related to its popularity. With a higher popularity, the content is more likely to be cached at the BSs. In this caching strategy, a better trade-off between the cooperation gain and the cache-hit ratio can be made.

We will evaluate the performance of these three caching strategies via simulation in Section 11.3.2. Besides content placement, how much cache size to deploy at each BS is also an important design factor, which will be discussed in detail in Section 11.4.

11.2.4 Backhauling

The backhaul with limited capacity has become a big concern for small-cell deployment. Although the traditional fiber-based backhaul solution can provide high data rates, the prohibitive cost is high and the geographical limitations also make it impossible to deploy in many practical scenarios. Instead, with low-cost and plug-and-play installation, wireless backhauling is a promising solution. It is worth noting with wireless backhauling, the data-sharing strategy is preferred since it has the following two advantages. First, the CP can exploit the multicast transmission to deliver the user messages simultaneously to multiple BSs via the shared backhaul. Second the BSs can cache part of the user messages to further reduce the backhaul load. While for the compression strategy, since the compressed signals generated for different BSs are different and they are also adaptive to the channel conditions, it cannot exploit the benefits of multicasting or caching. In this chapter, we assume that the backhaul connections can be dedicated fiber optic cables or they can be a shared wireless link.

11.2.4.1 Dedicated Wired Backhaul

We model the cost of the dedicated backhaul link as the required transmission rate of this link. Define a binary matrix $\mathbf{C} \in \{0,1\}^{F\times N}$ to denote the cache status, where $c_{f,n} = 1$ represents that the fth file is cached in the nth BS, otherwise $c_{f,n} = 0$. For each BS, if the requested file is not cached in its local storage, it should fetch the file from the CP with the backhaul transmission rate as large as the content-delivery rate R_m. Therefore, we model the backhaul cost as the transmission rate of the corresponding file. Then the overall backhaul cost is

$$C_B = \sum_{m=1}^{M}\sum_{n=1}^{N} s_{m,n}(1 - c_{f_m,n})R_m, \tag{11.4}$$

where f_m denotes the file requested by multicast group m and R_m is the transmission rate for group m.

11.2.4.2 Shared Wireless Backhaul

Compared to the dedicated wired backhaul, shared wireless backhaul not only is much easier to deploy (when wireline infrastructure is not available) but also enjoys the crucial wireless multicast advantage that allows for efficient content delivery to multiple BSs using a same resource block. Wireless multicast is ideally suited for enabling the cooperative transmission benefit of C-RAN; but it also brings in the challenge of pathloss, fading, and shadowing effect of the wireless medium. In particular, because of the different locations of the BSs, there may be considerable disparity in the quality of their respective channels. Deploying caching at BSs [9] (i.e., BSs can pre-store contents of popular files) can handle the channel disparity issue in wireless multicast to aid the BSs with weak channels. For wireless backhauling, the backhaul efficiency is often modeled as the (expected) downloading time. In this chapter, we consider a cluster of N BSs cooperatively serving users. The CP delivers the user's message to all the BSs via multicasting. Suppose that each file has normalized size of 1 and each BS n has a local storage of size C_n that can cache some of the files. In other words, given cache size allocation C_n, each BS n can cache C_n fraction of the file. We assume that the channel coherent time is large enough such that the file delivery can be completed within one coherent time. According to [10, lemma 1], by adopting the joint cache-channel coding strategy [11], the file delivery rate R can be written as

$$R = \min_{n} \left\{ \frac{I(\mathbf{x}; y_n)}{1 - C_n} \right\}, \tag{11.5}$$

and the downloading time thus can be expressed as

$$T = \frac{1}{R} = \max_{n} \left\{ \frac{1 - C_n}{I(\mathbf{x}; y_n)} \right\}. \tag{11.6}$$

Here, $I(\mathbf{x}; y_n)$ denotes the mutual information between the transmit signal $\mathbf{x}$ and the received signal y_n. If the file size is S, then the real downloading time should be $S \times T$.

Notice that the $\{I(\mathbf{x}; y_n)\}$ depend on the channel realizations and the beam-forming vectors at the CP and hence change quickly in different fading blocks; while the cache size $\{C_n\}$ should be allocated based on the long-term statistics of the backhaul channel. Therefore, the BS cache size allocation and the beam-forming design occur in different time scales. Later in the chapter, we will focus on the downloading time T in (11.6).

11.3 Caching at BSs for Cooperation in Access Link

We now treat the optimization of caching and multicasting in the access link of cloud RAN. It is worth mentioning that the cache placement and content delivery occur in different timescales. Specifically, cache placement often happens in days or hours, while content delivery happens in a much shorter timescale [2, 12]. In the shorter timescale of each transmission slot, the cache placement is usually fixed according to some strategy. We can then optimize the content delivery scheme, which should be adaptive to the instantaneous channel realization and the cache placement. In the larger timescale, the cache placement can be optimized by taking into account the content popularity distribution as well as the long-term statistics of the wireless channel. In this section, we mainly focus on the short timescale problem in the access link, i.e., the joint optimization of content-centric BS clustering and multicast beam-forming with given cache placement. The large timescale problem, i.e., the design of cache placement will be briefly addressed via numerical results.

11.3.1 Joint BS Clustering and Beam-Forming Design

In this section, given the BS caching, we study the joint content-centric BS clustering and multicast beam-forming design in access link to seek the minimum network cost. Specifically, in the considered network architecture, the network cost is modeled as the weighted sum of the backhaul cost and the transmission power:

$$C_N = C_B + \eta C_P, \tag{11.7}$$

where $\eta > 0$ is a weighting parameter.

The total network cost minimization problem with given cache placement can be formulated as:

$$\begin{aligned} \mathcal{P}_0: \quad & \min_{\{\mathbf{w}_{m,n}\},\{s_{m,n}\}} \sum_{m=1}^{M}\sum_{n=1}^{N} s_{m,n}(1-c_{f_m,n})R_m + \eta \sum_{m=1}^{M}\sum_{n=1}^{N} \|\mathbf{w}_{m,n}\|_2^2 && (11.8a) \\ & \text{s.t.} \quad \mathrm{SINR}_k \geq \gamma_m, \ \forall\, k \in \mathcal{G}_m, \ \forall\, m && (11.8b) \\ & \qquad s_{m,n} \in \{0,1\}, \ \forall\, m,n && (11.8c) \\ & \qquad (1-s_{m,n})\mathbf{w}_{m,n} = \mathbf{0}, \ \forall\, m,n, && (11.8d) \end{aligned}$$

where $R_m = B\log(1+\gamma_m)$ is the transmission rate for group m, B is the channel bandwidth, and γ_m is the target SINR for group m.

Note that constraint (11.8d) indicates that if BS n is not in the BS clustering of group m, i.e., $s_{m,n} = 0$, then the beam-former $\mathbf{w}_{m,n}$ should be zero. We also note that problem $\mathcal{P}_0$ can be infeasible due to the QoS constraint (11.8b). In general, determining the feasibility of this problem is very difficult. Therefore, in this section, we discuss $\mathcal{P}_0$ only when it is feasible.

Problem $\mathcal{P}_0$ is a nonconvex mixed-integer nonlinear programming (MINLP) problem and is combinatorial in nature; it is in general challenging to find its global optimum solution. However, an exhaustive search can be adopted to find the global optimum BS clusters. Specifically, there are a total of 2^{MN} candidate BS clustering matrices $\{\mathbf{S}\}$. For each given $\mathbf{S}$, we can solve the following power minimization problem to obtain the power cost:

$$\begin{aligned} \mathcal{P}(\mathcal{Z}_{\mathbf{S}}): \quad & \min_{\{\mathbf{w}_{m,n}\}} \sum_{m=1}^{M}\sum_{n=1}^{N} \|\mathbf{w}_{m,n}\|_2^2 && (11.9a) \\ & \text{s.t.} \quad (11.8\text{b}), \\ & \qquad \mathbf{w}_{m,n} = \mathbf{0}, \ \forall (m,n) \in \mathcal{Z}_{\mathbf{S}}, && (11.9b) \end{aligned}$$

where $\mathcal{Z}_{\mathbf{S}} = \{(m,n) \mid s_{m,n} = 0\}$ denotes the set of inactive BS-content pairs. While the backhaul cost C_B reduces to a constant.

Similar to the traditional multicast beam-forming problems [13, 14], $\mathcal{P}(\mathcal{Z}_{\mathbf{S}})$ is a nonconvex quadratically constrained quadratic programming (QCQP) problem, which is different from the unicast beam-forming problem, which can be equivalently transformed into a second-order cone programming (SOCP) problem and hence solved efficiently. The multicast beam-forming problem is generally NP-hard. A semi-definite relaxation (SDR) method is developed in [14] to obtain a near-optimal solution. After solving $\mathcal{P}(\mathcal{Z}_{\mathbf{S}})$ with all possible matrices $\mathbf{S}$s, we can find the one with the minimum objective.

Another approach to deal with problem $\mathcal{P}_0$ is to reformulate it as a more tractable sparse multicast beam-forming (SBF) problem. Specifically, when $\mathbf{w}_{m,n} = \mathbf{0}$, we have:

$$s_{m,n} = \begin{cases} 0, & \text{if } c_{f_m,n} = 0, \\ 0 \text{ or } 1, & \text{if } c_{f_m,n} = 1. \end{cases} \qquad (11.10)$$

Otherwise, according to constraint (11.8d), there holds $s_{m,n} = 1$. Therefore, we have the following relationship between the BS cluster and the beam-former:

$$s_{m,n} = \left\| \|\mathbf{w}_{m,n}\|_2^2 \right\|_0. \tag{11.11}$$

Note that the ℓ_0-norm is defined as the number of non-zero elements of a vector. It reduces to the indicator function in the scalar case. By substituting (11.11) into the objective function (11.8a), $\mathcal{P}_0$ can be equivalently transformed into the following problem:

$$\begin{aligned} \mathcal{P}_{\text{SBF}}: \quad & \min_{\{\mathbf{w}_{m,n}\}} \quad \sum_{m=1}^{M}\sum_{n=1}^{N} \left\| \|\mathbf{w}_{m,n}\|_2^2 \right\|_0 (1 - c_{f_m,n}) R_m + \eta \sum_{m=1}^{M}\sum_{n=1}^{N} \|\mathbf{w}_{m,n}\|_2^2 \\ & \text{s.t.} \quad (11.8\text{b}). \end{aligned} \tag{11.12}$$

With ℓ_0-norm in the objective function, problem $\mathcal{P}_{\text{SBF}}$ is a sparse multicast beamforming problem. It considers the adaptive content-centric BS clustering inexplicitly, since by solving this problem, a sparse beam-former for each multicast group may be obtained, whose non-zero entries correspond to its serving BSs. The equivalent problem $\mathcal{P}_{\text{SBF}}$ is still difficult due to the nonconvex discontinuous ℓ_0-norm in the objective and the nonconvex QoS constraint (11.8b).

One way to tackle this issue is to first adopt a smoothed ℓ_0-norm approximation to replace the discontinuous ℓ_0-norm with a concave smooth function. The problem after approximation then can be represented as a general form of difference of convex (DC) programming problem, for which the convex–concave procedure (CCP) [15] based algorithm can be adopted to find a stationary solution with convergence guarantee. The main idea behind CCP is to convexify the DC problem by approximating its concave parts with their first-order Taylor expansions and then solve the approximated convex subproblems successively until convergence. The details of such an approach can be found in [7].

11.3.2 Performance Evaluation

This section provides numerical results to demonstrate the superiority of the proposed content-centric transmission framework. A hexagonal multi-cell cloud RAN consisting of $N = 7$ BSs is considered, where each BS has $L = 4$ antennas. The distance between BSs is 500 m. There are $K = 30$ users uniformly distributed within the network. The total number of contents is $F = 100$. The cache size of BS n is set to $C_n = C$ for all n. The channel bandwidth is 10 MHz. The BS antenna gain is 10 dBi. The noise power σ_k^2 is set to be -102 dBm for all users. The pathloss is modeled as $PL(\text{dB}) = 148.1 + 37.6\log_{10}(d)$, where d is the distance in km. The shadowing follows the log-normal distribution with parameter being 8 dB. The small-scale fading is modeled as the Rayleigh fading. The SINR target is $\gamma_m = 10$ dB for all multicast groups. All the results are averaged over 100 independent simulation trials.

In this section, we assume the following unequal content popularity distribution: there is one popular content accounting for 0.5 of the request probability, while the rest $F - 1$

contents follows a Zipf distribution with a skewness parameter α and the sum probability of 0.5. In the following simulation, the skewness parameter is set to $\alpha = 1$. Each BS can caches up to $C = 10$ contents. More results with different setups can be found in [7].

11.3.2.1 Effects of Caching

We first evaluate the caching effects and compare the performance of different caching strategies in Figure 11.2. We consider two scenarios with the number of users being $K = 30$ and $K = 7$, respectively. The skewness parameter α is set to $\alpha = 1$. It can be seen that by carefully designing the caching strategy, the proposed heuristic caching strategy can significantly reduce the backhaul cost and hence improve the trade-off performance between backhaul and power. In addition, it is observed that PopC is superior to ProC for most of the trade-off parameter η, except the extreme case when $\eta \to 0$. Intuitively, in PopC, the most popular content is cached in all BSs, the cooperative transmission gain can then be fully exploited. This is very helpful when the network does not care about the backhaul overhead. However, when backhaul is the main concern of the network cost (i.e., $\eta \to 0$), ProC can outperform PopC. We can also see that all the caching strategies have the minimum transmit power. This is because the minimum transmit power depends only on the target SINRs of the multicast groups.

We also illustrate the performance comparison of multicast transmission and unicast transmission with different numbers of active users in Figure 11.3. For unicast transmission, we design an individual beam-former for each of the users regardless of their requested contents. In order to ensure fairness of the backhaul link overhead, if multiple users that request the same content are served by the same BS, the BS only needs to fetch a copy of the content from the CP with the maximum requested rate if it does not cache the content. We adopt the iterative reweighted ℓ_1-norm based on the algorithm proposed in [16] to solve the sparse unicast beam-forming problem.

From Figure 11.3, it is seen that when $K = 30$, the unicast transmission performs very poorly. This is mainly due to the fact that the number of transmit antennas is less than

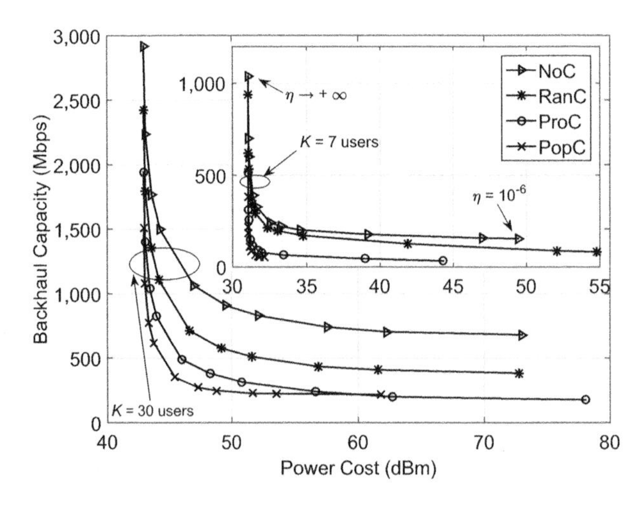

Figure 11.2 Performance comparison of different caching strategies for unequal content popularity with $\alpha = 1$.

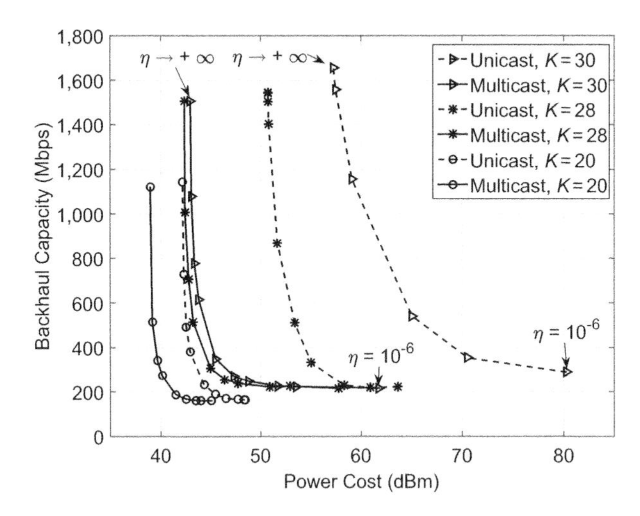

Figure 11.3 Performance comparison between multicast transmission and unicast transmission.

the number of users, and hence there are not enough design dimensions for the unicast beam-forming. On the other hand, the performance of multicast transmission is much better since it can exploit the content reuse feature among different users and hence fewer beam-formers are required. With the number of users decreasing, the performance of unicast transmission becomes better, but still far inferior to multicast transmission. Specifically, in the extreme case when $\eta \rightarrow +\infty$, which means only power cost is concerned, we can see that multicast transmission can save 3 dB power compared with unicast transmission when $K = 20$.

11.4 Caching at BSs for Multicasting in Backhaul Link

11.4.1 Joint BS Cache Allocation and Beam-Forming Design

Next, we study the effect of caching to improve the wireless backhauling of cloud RAN. We consider the downlink transmission with wireless multicast backhaul, where each user is cooperatively served by a single cluster of BSs. The CP delivers the user's message to these BSs via multicasting. The BSs can also pre-store some fraction of the popular content during the off-peak hours. The rest of the contents will be fetched from the CP using coded delivery via the wireless multicast backhaul. Assuming that the CP is equipped with multiple antennas and given a total cache size constraint, we study the joint design of cache size allocation at the BSs and the multicast beam-forming transmission at the CP so that the expected downloading time of requested files in (11.6) from the CP to the BSs is minimized. It is worthwhile emphasizing that the designs of cache size allocation and the beam-forming strategy occur in two different time scales. The cache size allocation is optimized in a much large timescale, which is adaptive to the long-term statistics of the wireless backhaul channel, while the beam-forming design is performed under a given cache size allocation and adapts to the instantaneous channel conditions.

11.4.1.1 Single-File Case

We consider the single file case of normalized size and formulate a mixed-time scale problem for a joint design of cache size allocation and multicast beam-forming. We first focus on the beam-forming design in the shorter time scale with fixed cache size allocation and given content placement. Suppose that $\mathbf{w}$ is the beam-forming vector used by the CP, and $\mathbf{h}_n$ is the channel between the nth BS and the CP. The mutual information can be expressed as

$$I(\mathbf{x}; y_n) = \log\left(1 + \frac{\mathrm{Tr}\left(\mathbf{H}_n \mathbf{W}\right)}{\sigma^2}\right),$$

where σ^2 is the variance of the complex Gaussian noise, $\mathbf{H}_n = \mathbf{h}_n \mathbf{h}_n^H$ is the channel covariance matrix, $\mathbf{W} = \mathbf{w}\mathbf{w}^H$ is the covariance matrix for the transmit signal $\mathbf{x}$, where $\{\mathbf{W} \succeq \mathbf{0} \mid \mathrm{Tr}\left(\mathbf{W}\right) \leq P, \mathrm{rank}(\mathbf{W}) = 1\}$, and P is the peak power of the CP. We shall drop the rank-one constraint in this set and define

$$\mathbb{W} = \{\mathbf{W} \succeq \mathbf{0} \mid \mathrm{Tr}\left(\mathbf{W}\right) \leq P\}.$$

With the given cache allocation $\{C_n\}$, the file downloading time (11.6) can be expressed as

$$T^* = \min_{\mathbf{W} \in \mathbb{W}} \max_n \left\{ \frac{1 - C_n}{\log\left(1 + \frac{\mathrm{Tr}(\mathbf{H}_n \mathbf{W})}{\sigma^2}\right)} \right\}. \tag{11.13}$$

Suppose that all $\mathbf{H}_n$ remain constant within a coherent block but change according to certain channel distribution in different coherent blocks, then T^* in (11.13) is a random variable. In this section, our aim is to find the optimal cache size allocation such that the long-term expected file downloading time is minimized. The problem can be mathematically formulated as [10]:

$$\min_{\{C_n\}} \quad \mathbb{E}_{\{\mathbf{H}_n\}}\left[T^*\right] \tag{11.14a}$$

$$\text{s.t.} \quad \sum_{n \in \mathcal{N}} C_n \leq C,\ 0 \leq C_n \leq 1,\ n \in \mathcal{N}, \tag{11.14b}$$

where $C (\leq N)$ is the total cache size across all the BSs.

This problem is difficult mainly due to expectation in the objective function (11.14a), which has no closed-form expression. A popular approach to handling this difficulty is to approximate the expectation in (11.14a) with its sample average [17]. By adopting the sample average approximation, the problem can be approximated as:

$$\min_{\{C_n, \mathbf{W}^m\}} \quad \frac{1}{M_s} \sum_{m=1}^{M_s} \max_n \left\{ \frac{1 - C_n}{\log\left(1 + \frac{\mathrm{Tr}(\mathbf{H}_n^m \mathbf{W}^m)}{\sigma^2}\right)} \right\} \tag{11.15a}$$

$$\text{s.t.} \quad \sum_n C_n \leq C,\ 0 \leq C_n \leq 1,\ n \in \mathcal{N}, \tag{11.15b}$$

$$\mathrm{Tr}\left(\mathbf{W}^m\right) \leq P,\ \mathbf{W}^m \succeq \mathbf{0},\ m \in \mathcal{M}_s, \tag{11.15c}$$

where M_s is the sample size, $\mathcal{M}_s := \{1, 2, \ldots, M_s\}$, $\{\mathbf{H}_n^m\}_{m \in \mathcal{M}_s}$ are the samples of $\mathbf{H}_n$, and $\mathbf{W}^m$ is the covariance matrix corresponding to the samples $\{\mathbf{H}_n^m\}_{n \in \mathcal{N}}$. Furthermore, dropping the constant $1/M_s$ in (11.15a) and introducing the auxiliary variable $\{\xi^m\}$, problem (11.15) can be reformulated as

$$\min_{\{C_n, \mathbf{W}^m, \xi^m\}} \quad \sum_{m=1}^{M_s} \frac{1}{\xi^m} \tag{11.16a}$$

$$\text{s.t.} \quad \log\left(1 + \frac{\mathrm{Tr}\left(\mathbf{H}_n^m \mathbf{W}^m\right)}{\sigma^2}\right) \geq \xi^m (1 - C_n),\ n \in \mathcal{N},\ m \in \mathcal{M}_s, \tag{11.16b}$$

$$\text{(11.15b) and (11.15c).}$$

Problem (11.16) can be efficiently solved by the trust region method [18], where the nonconvex term $\xi^m(1 - C_n)$ in (11.16b) is iteratively approximated by its first-order Taylor expansion and the approximation subproblem at each iteration is convex and can be solved by the alernating direction method of multipliers algorithm (ADMM) approach [19]. For more details of solving problem (11.14), please refer to [10].

11.4.1.2 Multi-file Case

We now study the cache size allocation problem in the general case with multiple files and different popularities. We assume that the user requests file f with probability p_f, $f \in \mathcal{F} := \{1, 2, \ldots, F\}$, where $\sum_f p_f = 1$. The fraction of file f cached in BS n is C_{nf}. Therefore, we have the total cache size constraint $\sum_n \sum_f C_{nf} \leq C$, where $C \leq NF$. If file f is requested, the downloading time, denoted as T_f^*, can be expressed as

$$T_f^* = \min_{\mathbf{W}_f \in \mathbb{W}} \max_n \left\{ \frac{1 - C_{nf}}{\log\left(1 + \frac{\mathrm{Tr}(\mathbf{H}_n \mathbf{W}_f)}{\sigma^2}\right)} \right\}. \tag{11.17}$$

Different from the downloading time (11.13) in the single-file case, the downloading time T_f^* here depends on both the channel conditions and the requested file. We then formulate the cache size allocation problem with multiple files as [10]

$$\min_{\{C_{nf}\}} \quad \sum_f p_f \mathbb{E}_{\{\mathbf{H}_n\}}\left[T_f^*\right] \tag{11.18a}$$

$$\text{s.t.} \quad \sum_n \sum_f C_{nf} \leq C,\ 0 \leq C_{nf} \leq 1,\ n \in \mathcal{N},\ f \in \mathcal{F}. \tag{11.18b}$$

This problem can be solved using the same sample approximation approach as in the single file case. Please see [10] for more details.

11.4.2 Performance Evaluation

In this section, we demonstrate the performance of the proposed cache size allocation scheme via simulations. As shown in Figure 11.4, we consider a C-RAN with

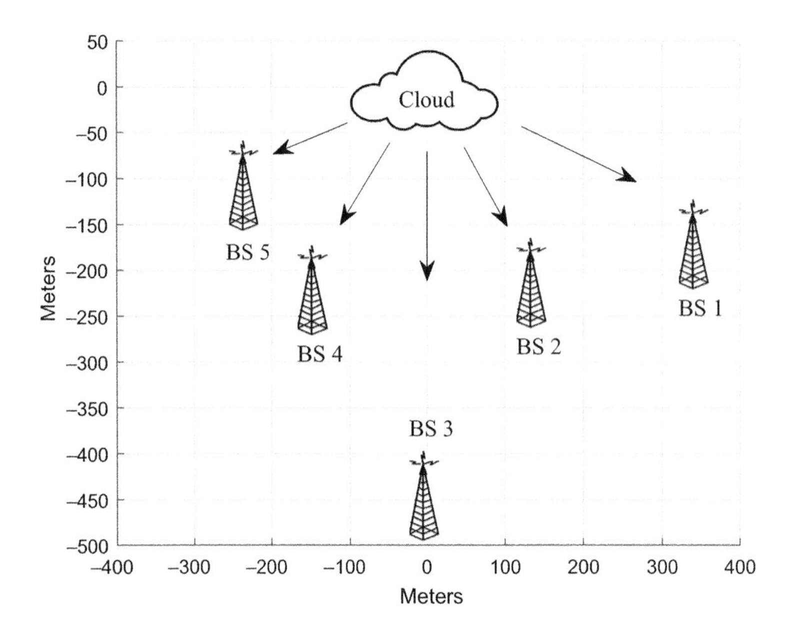

Figure 11.4 A downlink C-RAN setup with 5 BSs. The distances from the CP to the 5 BSs are 398, 278, 473, 286, and 267 m, respectively.

$N = 5$ BSs, where the BSs are randomly distributed on one side of the CP. The distances between the CP and the BSs are (398, 278, 473, 286, and 267 m, respectively. We generate 1,000 channel realizations of $\mathbf{h}_n$ according to the distribution $\mathbf{h}_n = \mathbf{K}_n^{1/2}\mathbf{v}_n$, where $\mathbf{K}_n$ denotes the large-scale pathloss component and $\mathbf{v}_n$ is the small-scale fading. The pathloss is modeled as $128.1 + 37.6\log_{10}(d)$ dB, where d is the distance in kilometers. The small-scale fading is modeled as a random vector following the independently and identically Gaussian distribution, i.e., $\mathbf{v}_n \sim \mathcal{CN}(0, 1)$. We use the first 100 samples for the sample average approximation method to optimize the cache allocation and use the remaining 900 samples to evaluate the performance with the obtained cache allocation. More parameters settings can be found in [10, table I].

11.4.2.1 Cache Allocation for BSs with Varying Channel Strengths

In this section, the superiority of the proposed scheme is demonstrated when caching a single file across multiple BSs with different channel strengths. The following schemes are considered as benchmarks:

- *Uniform cache allocation*: Each BS has the same cache size of $C_n = C/N$.
- *Proportional cache allocation*: The allocated cache sizes among the BSs satisfy that $(F - C_n)/\log\left(1 + \frac{P\mathrm{Tr}(\mathbf{K}_n)}{N\sigma^2}\right)$ are equalized for all n.
- *Lower bound*: We solve problem (11.13) to obtain the cache sizes by treating $\{C_n\}$ as the optimization variables for each channel realization. This is not practical, but can serve as a lower bound for the minimum expected file downloading time.

Table 11.1 Cache allocation for Different Schemes under Normalized Total Cache Size $C = 1$

	Schemes		
BSs	Uniform	Proportional	Optimized
BS1	0.2	0.232	0.222
BS2	0.2	0.170	0.071
BS3	0.2	0.261	0.588
BS4	0.2	0.175	0.101
BS5	0.2	0.163	0.019

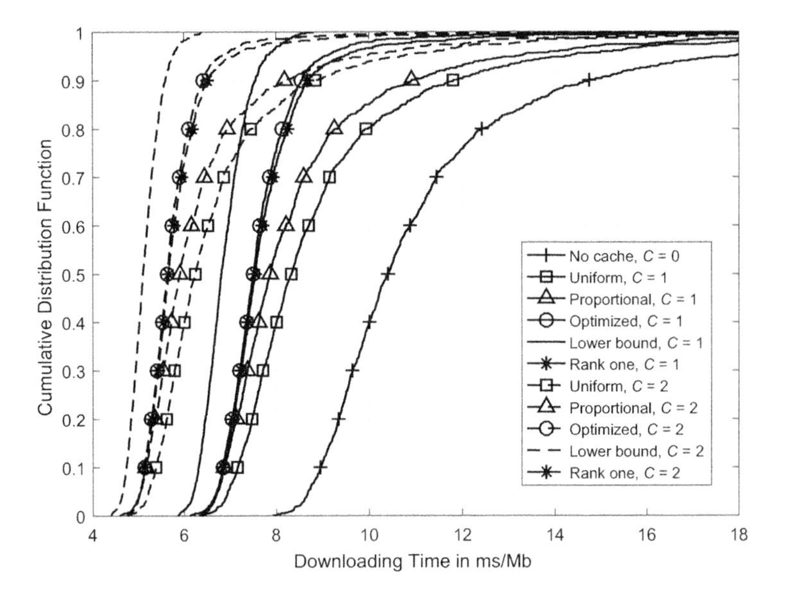

Figure 11.5 CDF of downloading time under different caching schemes.

- *Rank-one multicast beam-former*: The cache sizes are the same as the optimized scheme, but with the multicast beam-former being rank-one obtained using eigenvector decomposition.

In Table 11.1, we show the cache size allocation obtained by different schemes under normalized total cache size constraint $C = 1$. It can be seen that the proposed caching scheme and the proportional caching scheme allocate more cache size to the weaker BS 3 comparing with the uniform caching scheme, but our scheme is more aggressive. In Figure 11.5, we compare the cumulative distribution function (CDF) of the downloading time between different caching schemes. From Figure 11.5, we first see that the proposed caching scheme is superior to all the benchmark schemes in the high downloading time regime. It is also seen that the performance loss of the rank-one multicast beam-former is negligible compared to the solution obtained by solving (11.13).

Table 11.2 Optimized Cache Allocation for a 2-File Case with Different File Popularities under $C = 1$

	File Popularity (p_1, p_2)				
BSs	(0.5, 0.5)	(0.6, 0.4)	(0.7, 0.3)	(0.8, 0.2)	(0.9, 0.1)
BS1	(0.082, 0.082)	(0.132, 0.027)	(0.168, 0)	(0.202, 0)	(0.222, 0)
BS2	(0, 0)	(0, 0)	(0, 0)	(0.046, 0)	(0.071, 0)
BS3	(0.418, 0.418)	(0.482, 0.359)	(0.536, 0.27)	(0.568, 0.109)	(0.588, 0)
BS4	(0, 0)	(0, 0)	(0.026, 0)	(0.075, 0)	(0.101, 0)
BS5	(0, 0)	(0, 0)	(0, 0)	(0, 0)	(0.018, 0)
Total	(0.5, 0.5)	(0.614, 0.386)	(0.73, 0.27)	(0.891, 0.109)	(1, 0)

11.4.2.2 Cache Allocation for Files of Varying Popularity

In this part, we show simulation results for the cache size allocation schemes with multiple files and different popularities. We first consider only two files with the request probabilities being (p_1, p_2) shown in the first row of Table 11.2. Each column denotes the cache size allocation of the BSs under different file popularities given in the first row. From Table 11.2 we first see that for different file popularities, the cache size of the weakest BS 3 is always the largest, as in the single-file case shown in Table 11.1. We also see that the more popular a file is, the more cache size it will be allocated. For example, when $(p_1, p_2) = (0.9, 0.1)$, file 1 occupies all the cache space without caching any fraction of file 2.

In Figure 11.6, we compare the file downloading time of the optimized cache scheme with the following benchmarks:

- *Uniform cache allocation*: All the files have the same cache size of $C_{nf} = C/NF$ at all the BSs.
- *Proportional cache allocation*: The total allocated cache size of file f is first set as $p_f C$. The cache size of this file is then obtained according to the proportional cache allocation scheme in the single-file case.
- *Caching the most popular file*: We cache the most popular file in its entirety first, followed by caching of the second most popular file, and so on. When the remaining cache space is not enough for caching a whole file, we allocate the remaining cache space according to the proportional cache allocation scheme.

In Figure 11.6, we consider $F = 4$ files and assume the file popularity follows the the Zipf distribution [20], i.e., $p_f = \frac{f^{-\alpha}}{\sum_{i=1}^{F} i^{-\alpha}}, \forall f$. We compare the average downloading time of all the schemes with different α. Note that when α increases, the differences among the file popularities also increase. From Figure 11.6, it can be seen that for all schemes, except the uniform scheme, the average downloading time decreases when α increases. This is expected, since in a uniform cache allocation scheme, the cache sizes of all files are the same and the downloading time is the same for all files, whereas in the other three schemes, more cache size is allocated to files with greater popularity. We can also see that the proposed caching scheme outperforms the other three schemes for different α, and it converges to the scheme of caching the most popular file when $\alpha = 1.5$.

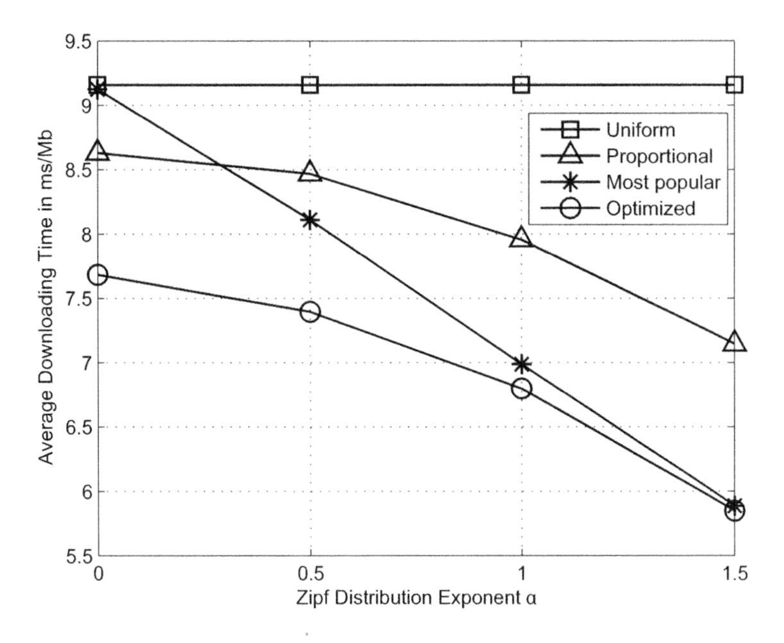

Figure 11.6 Average downloading time for different Zipf file distributions under the same number of files $F = 4$ and the total normalized cache size $C = 4$.

To sum up, from the simulation results and discussion, it is beneficial to allocate more cache sizes to the files with greater popularity, and our proposed cache allocation scheme can provide a better cache allocation solution compared to the heuristic schemes.

11.5 Conclusions and Open Issues

This chapter presented a content-centric framework for transmission optimization in cloud RANs by leveraging caching and multicasting. We first studied the effects of caching and multicasting on the access link in a cloud RAN with dedicated backhaul through the joint design of the content-centric BS clustering and multicast beamforming under different but given BS caching strategies. Simulation results showed that our proposed content-centric multicast transmission is much superior to the traditional user-centric unicast transmission in terms of system total transmit power reduction and backhaul saving. We then studied the effects of caching and multicasting on a backhaul link in a cloud RAN with wireless backhaul through the joint design of cache size allocation at the BSs and the multicast beam-forming at the CP. Numerical results showed the optimized cache size allocation scheme can greatly improve the network performance compared to other heuristic schemes.

To exploit the full potential of cache-enabled cloud RAN, it is worth investigating the joint design of access and backhaul links in the future. It is also of practical importance to seek a scalable solution for caching and multicasting in a large cache-enabled cloud RAN.

References

[1] P. Rost, C. Bernardos, A. Domenico, M. Girolamo, M. Lalam, A. Maeder, D. Sabella, and D. Wübben, "Cloud technologies for flexible 5G radio access networks," *IEEE Commun. Mag.*, vol. 52, no. 5, pp. 68–76, May 2014.

[2] K. Shanmugam, N. Golrezaei, A. Dimakis, A. Molisch, and G. Caire, "Femtocaching: wireless content delivery through distributed caching helpers," *IEEE Trans. Inf. Theory*, vol. 59, no. 12, pp. 8402–8413, Dec. 2013.

[3] D. Lecompte and F. Gabin, "Evolved multimedia broadcast/multicast service (eMBMS) in LTE-advanced: overview and rel-11 enhancements," *IEEE Commun. Mag.*, vol. 50, no. 11, pp. 68–74, Nov. 2012.

[4] N. D. Sidiropoulos, T. N. Davidson, and Z.-Q. Luo, "Transmit beamforming for physical-layer multicastings," *IEEE Trans. Signal Process.*, vol. 54, no. 6, pp. 2239–2251, June 2006.

[5] B. Dai and W. Yu, "Sparse beamforming and user-centric clustering for downlink cloud radio access network," *IEEE Access*, vol. 2, pp. 1326–1339, 2014.

[6] L. Breslau, P. Cao, L. Fan, G. Phillips, and S. Shenker, "Web caching and Zipf-like distributions: evidence and implications," in *Proc. IEEE INFOCOM*, 1999, pp. 126–134.

[7] M. Tao, E. Chen, H. Zhou, and W. Yu, "Content-centric sparse multicast beamforming for cache-enabled cloud RAN," *IEEE Trans. Wireless Commun.*, vol. 15, no. 9, pp. 6118–6131, Sep. 2016.

[8] X. Xu and M. Tao, "Modeling, analysis, and optimization of coded caching in small-cell networks," *IEEE Trans. Commun.*, vol. 65, no. 8, pp. 3415–3428, Aug. 2017.

[9] M. A. Maddah-Ali and U. Niesen, "Fundamental limits of caching," *IEEE Trans. Inf. Theory*, vol. 60, no. 5, pp. 2856–2867, May 2014.

[10] B. Dai, Y.-F. Liu, and W. Yu, "Optimized base-station cache allocation for cloud radio access network with multicast backhaul," *IEEE J. Sel. Areas Commun.*, vol. 36, no. 8, 2018.

[11] S. S. Bidokhti, M. A. Wigger, and R. Timo, "Noisy broadcast networks with receiver caching," *IEEE Trans. Inf. Theory*, 2018. [Online]. Available: http://arxiv.org/abs/1605.02317

[12] A. F. Molisch, G. Caire, D. Ott, J. R. Foerster, D. Bethanabhotla, and M. Ji, "Caching eliminates the wireless bottleneck in video aware wireless networks," *Adv. Elect. Eng.*, vol. 2014, pp. 1–13, 2014.

[13] Z. Xiang, M. Tao, and X. Wang, "Coordinated multicast beamforming in multicell networks," *IEEE Trans. Wireless Commun.*, vol. 12, no. 1, pp. 12–21, Jan. 2013.

[14] E. Karipidis, N. Sidiropoulos, and Z.-Q. Luo, "Quality of service and max-min fair transmit beamforming to multiple cochannel multicast groups," *IEEE Trans. Signal Process.*, vol. 56, no. 3, pp. 1268–1279, Mar. 2008.

[15] A. L. Yuille and A. Rangarajan, "The concave-convex procedure," *Neural Comput.*, vol. 15, no. 4, pp. 915–936, 2003.

[16] E. Chen and M. Tao, "User-centric base station clustering and sparse beamforming for cache-enabled cloud RAN," in *Proc. IEEE/CIC ICCC*, Nov. 2015, pp. 1–6.

[17] J. R. Birge and F. Louveaux, *Introduction to Stochastic Programming*, 2nd ed., New York: Springer, 2011.

[18] A. R. Conn, N. I. Gould, and P. L. Toint, *Trust Region Methods*, Philadelphia: Society for Industrial and Applied Mathematics (SIAM), 2000.

[19] S. Boyd, N. Parikh, E. Chu, B. Peleato, and J. Eckstein, "Distributed optimization and statistical learning via the alternating direction method of multipliers," *Found. Trends Mach. Learn.*, vol. 3, no. 1, pp. 1–122, 2011.

[20] M. Zink, K. Suh, Y. Gu, and J. Kurose, "Characteristics of YouTube network traffic at a campus network—measurements, models, and implications," *Comput. Netw.*, vol. 53, no. 4, pp. 501–514, 2009.

12 Fundamentals of Coded Caching for Interference Management

Meixia Tao, Fan Xu, Youlong Cao, and Kangqi Liu

12.1 Introduction

Over the last decades, there has been increasing interest in characterizing the approximate capacity of wireless networks to understand their performance limits. In particular, to provide the fundamental insights into the capacity characterization, there have been many optimal interference management schemes for several kinds of interference networks in the high signal-to-noise ratio (SNR) regime, where the local additive white Gaussian noise (AWGN) at receivers is deemphasized relative to the signal and interference powers. In the high SNR regime, the degrees of freedom (DoF) provides an approximation of the network capacity with high accuracy. A network with a sum DoF of d indicates that the sum capacity of the network can be expressed as $d\log(\mathrm{SNR}) + o(\log(\mathrm{SNR}))$. Since the dimension of each noninterfering signal stream contributes a rate of $\log(\mathrm{SNR})+o(\log(\mathrm{SNR}))$, the DoF can be interpreted as the number of resolvable signal space dimensions. There have been many works characterizing the optimal DoF of various interference networks, such as the interference channel [1], X channel [2], two-user multiple-input multiple-output (MIMO) interference channel [3], three-user MIMO interference channel [4], and two-user MIMO X channel [5].

Recently, caching has been playing an important role in wireless networks to reduce the network traffic load and improve the user perceived experience. Therefore, it is natural and timely to investigate the impact of caching in wireless interference networks. In [6], Naderializadeh et al. studied the impact of caching in the interference networks with caches at both transmitter and receiver sides. They showed that transmitter caches can achieve load balancing gain as well as transmitter cooperation gain and that receiver caches can be exploited for known interference elimination. By using a one-shot linear transmission scheme, they obtained an achievable sum DoF that is within a constant factor of 2 to the optimum. In [7], Xu et al. proposed a generic file splitting and caching scheme and transformed the interference network topology to the cooperative X-multicast channels to leverage coded multicasting gain and transmitter cooperation gain, as well as receiver local caching gain. By adopting normalized delivery time (NDT) as the performance metric, this scheme achieves the optimality in certain transmitter and receiver cache size regions and is within a bounded multiplicative gap to the optimum in the rest regions. Hachem et al. in [8] proposed a caching and delivery scheme that separates the system into a network layer and a physical layer. By exploiting receiver-coded multicasting gain in the network layer and by characterizing the optimal per-user DoF in the formed X-multicast channel via transmitter coordination in the physical layer, they provided a constant-factor characterization of the optimal system DoF. There are also other works studying caching in the interference network. For example, Roig et al. in [9] proposed a new achievable scheme where the combination of zero-forcing, interference alignment, and interference neutralization is leveraged by transmitter caches in the delivery phase, in addition to the coded multicasting gain leveraged by receiver caches; and Cao et al. in [10] considered the 3×3 MIMO interference network and obtained the spatial multiplexing gain apart from the cache-aided transmitter cooperation gain and cache-aided coded multicasting gain.

This chapter takes a close look at the impact of caching in the interference networks. Section 12.2 briefly reviews the basics of some classic interference networks and the corresponding interference management techniques. Then an interference network where all transmitters and receivers have local caches, termed as *cache-aided interference network*, is introduced in Section 12.3, where the information-theoretic metric NDT is introduced to characterize the system performance. Section 12.4 characterizes the NDT in the cache-aided interference network, for both single-antenna and multiple-antenna cases. It is shown that with different cache sizes, the network topology is changed to different classic interference networks opportunistically, where coded multicasting gain and transmitter cooperation gain (via interference alignment and interference neutralization) can be leveraged in the delivery phase, apart from local caching gain. Then, the NDT results are extended to the partially connected interference network in Section 12.5. Finally, Section 12.6 summarizes this chapter and outlines some possible directions for future research.

12.2 Preliminaries of Interference Networks and Interference Management

Before we present the fundamental limits of the cache-aided interference network in the next section, let us first briefly review some classic interference networks, including interference channel, X channel, X-multicast channel, and cooperative X-multicast channel. We will also introduce the corresponding interference management techniques.

12.2.1 Interference Channel

Consider a $K \times K$ interference channel with K receivers and K transmitters, where each transmitter k needs to send an independent message W_k to receiver k. In this channel, Cadambe and Jafar characterized the optimal per-user DoF of $\frac{1}{2}$ in [1] by using interference alignment. In what follows, we take $K = 3$ as an example to review the idea of the delivery strategy in [1].

In the 3×3 interference channel, transmitter 1, 2, 3 needs to send message W_1, W_2, W_3 to receiver 1, 2, 3, respectively. Denote x_k as the transmitted symbol for message W_k (for $k \in \{1, 2, 3\}$) and $\mathbf{v}_k$ as the $n \times 1$ beamforming vector of x_k at transmitter k, by using an

n-symbol extension. Here parameter n will be determined later. Consider an arbitrary receiver k. Its received signal $\mathbf{y}_k$ is given by

$$\mathbf{y}_k = \mathbf{H}_{k,1}\mathbf{v}_1 x_1 + \mathbf{H}_{k,2}\mathbf{v}_2 x_2 + \mathbf{H}_{k,3}\mathbf{v}_3 x_3, \tag{12.1}$$

where $\mathbf{H}_{k,1}, \mathbf{H}_{k,2}, \mathbf{H}_{k,3}$ is the $n \times n$ diagonal channel matrix between receiver k and transmitter 1, 2, 3, respectively, and their diagonal entries are independently and identically distributed (i.i.d.) as some continuous distribution. Here, we ignore the additive noise since we aim to characterize the system DoF which is defined in the high SNR regime.

If the beam-forming vectors $\mathbf{v}_1, \mathbf{v}_2, \mathbf{v}_3$ are chosen to be arbitrary non-zero vectors, each receiver k can decode its desired symbol x_k from received signal $\mathbf{y}_k$ as long as $n \geq 3$, given that the channel matrices are randomly and independently distributed. For each receiver, since its desired symbol occupies one dimension in its received signal space while the interferences occupy the rest two dimensions, a per-user DoF of $\frac{1}{3}$ is achieved.

On the other hand, if we design $\{\mathbf{v}_1, \mathbf{v}_2, \mathbf{v}_3\}$ so as to align the two undesired symbols into the same subspace at each receiver, we can leave more space for the desired symbol and achieve a higher per-user DoF. This is the key idea of *interference alignment*, where the beam-forming vectors $\mathbf{v}_1, \mathbf{v}_2, \mathbf{v}_3$ need to satisfy

$$\begin{cases} \text{span}(\mathbf{H}_{1,2}\mathbf{v}_2) = \text{span}(\mathbf{H}_{1,3}\mathbf{v}_3) & \text{at Rx 1,} \\ \text{span}(\mathbf{H}_{2,1}\mathbf{v}_1) = \text{span}(\mathbf{H}_{2,3}\mathbf{v}_3) & \text{at Rx 2,} \\ \text{span}(\mathbf{H}_{3,1}\mathbf{v}_1) = \text{span}(\mathbf{H}_{3,2}\mathbf{v}_2) & \text{at Rx 3.} \end{cases} \tag{12.2}$$

Besides these three equations, the beam-forming vectors also need to guarantee that the desired symbol and interferences lie in different directions in the received signal space so that each receiver can decode its desired symbol. The detailed design for $\mathbf{v}_1, \mathbf{v}_2, \mathbf{v}_3$ is given in [1], where asymptotic interference alignment is adopted. As a result, for each receiver, all the interferences occupy only one dimension in its received signal space while the desired symbol occupies another dimension, thus a DoF of $\frac{1}{2}$ per user is achieved.

12.2.2 X Channel

In this section, let us review another classic interference network, X channel, by using interference alignment to achieve its optimal per-user DoF. A general $N_T \times N_R$ X channel has N_T transmitters and N_R receivers. Each transmitter j (for $j \in \{1, 2, \ldots, N_T\}$) needs to send an independent message W_j^i to each receiver i (for $i \in \{1, 2, \ldots, N_R\}$). The optimal per-user DoF of $\frac{N_T}{N_T+N_R-1}$ is characterized in [2] by using asymptotic interference alignment. In what follows, we use $N_T = N_R = 3$ as an example to review the idea of the delivery strategy in [2].

In the 3×3 X channel, transmitter j (for $j \in \{1, 2, 3\}$) should deliver three independent messages, W_j^1, W_j^2, W_j^3, to receiver 1, 2, 3, respectively. Denote x_j^i (for $j, i \in \{1, 2, 3\}$) as the transmitted symbol of message W_j^i, and $\mathbf{v}_j^i$ as the $n \times 1$ beam-forming vector for

x_j^i at transmitter j, by using an n-symbol extension. Consider an arbitrary receiver i. Its received signal $\mathbf{y}_i$ is given by

$$\mathbf{y}_i = \mathbf{H}_{i,1}\left(\mathbf{v}_1^1 x_1^1 + \mathbf{v}_1^2 x_1^2 + \mathbf{v}_1^3 x_1^3\right) + \mathbf{H}_{i,2}\left(\mathbf{v}_2^1 x_2^1 + \mathbf{v}_2^2 x_2^2 + \mathbf{v}_2^3 x_2^3\right)$$
$$+ \mathbf{H}_{i,3}\left(\mathbf{v}_3^1 x_3^1 + \mathbf{v}_3^2 x_3^2 + \mathbf{v}_3^3 x_3^3\right). \quad (12.3)$$

Similar to the interference channel, to increase per-user DoF at each receiver, we aim to align the interferences into the same subspace at each receiver. For example, at receiver 1, the beam-forming vectors should satisfy

$$\begin{cases} \text{span}(\mathbf{H}_{1,1}\mathbf{v}_1^2) = \text{span}(\mathbf{H}_{1,2}\mathbf{v}_2^2) = \text{span}(\mathbf{H}_{1,3}\mathbf{v}_3^2), \\ \text{span}(\mathbf{H}_{1,1}\mathbf{v}_1^3) = \text{span}(\mathbf{H}_{1,2}\mathbf{v}_2^3) = \text{span}(\mathbf{H}_{1,3}\mathbf{v}_3^3). \end{cases} \quad (12.4)$$

Therefore, the unwanted symbols x_1^2, x_2^2, x_3^2 but intended for receiver 2 are aligned into the same subspace, and the unwanted symbols x_1^3, x_2^3, x_3^3 but intended for receiver 3 are aligned into the same subspace. Similarly, we align the unwanted symbols $\{x_1^1, x_2^1, x_3^1\}$ and $\{x_1^3, x_2^3, x_3^3\}$ by receiver 2 into the same subspaces, respectively, i.e.,

$$\begin{cases} \text{span}(\mathbf{H}_{2,1}\mathbf{v}_1^1) = \text{span}(\mathbf{H}_{2,2}\mathbf{v}_2^1) = \text{span}(\mathbf{H}_{2,3}\mathbf{v}_3^1), \\ \text{span}(\mathbf{H}_{2,1}\mathbf{v}_1^3) = \text{span}(\mathbf{H}_{2,2}\mathbf{v}_2^3) = \text{span}(\mathbf{H}_{2,3}\mathbf{v}_3^3), \end{cases} \quad (12.5)$$

and align the unwanted symbols $\{x_1^1, x_2^1, x_3^1\}$ and $\{x_1^2, x_2^2, x_3^2\}$ by receiver 3 into the same subspaces, respectively, i.e.,

$$\begin{cases} \text{span}(\mathbf{H}_{3,1}\mathbf{v}_1^1) = \text{span}(\mathbf{H}_{3,2}\mathbf{v}_2^1) = \text{span}(\mathbf{H}_{3,3}\mathbf{v}_3^1), \\ \text{span}(\mathbf{H}_{3,1}\mathbf{v}_1^2) = \text{span}(\mathbf{H}_{3,2}\mathbf{v}_2^2) = \text{span}(\mathbf{H}_{3,3}\mathbf{v}_3^2). \end{cases} \quad (12.6)$$

In addition, the beam-forming vectors should ensure that the desired symbols and interferences lie in different directions in the received signal space so that each receiver can decode its three desired symbols. The detailed design of the beam-forming vectors is given in [2].

As a result, the received signal space for an arbitrary receiver, e.g., receiver 1, is given by

$$\left[\text{span}(\mathbf{H}_{1,1}\mathbf{v}_1^1)\ \text{span}(\mathbf{H}_{1,2}\mathbf{v}_2^1)\ \text{span}(\mathbf{H}_{1,3}\mathbf{v}_3^1),\ \text{span}(\mathbf{H}_{1,1}\mathbf{v}_1^2)\ \text{span}(\mathbf{H}_{1,1}\mathbf{v}_1^3)\right] \quad (12.7)$$

where the received signal vector of desired symbols x_1^1, x_2^1, x_3^1 lies on $\text{span}(\mathbf{H}_{1,1}\mathbf{v}_1^1)$, $\text{span}(\mathbf{H}_{1,2}\mathbf{v}_2^1)$, $\text{span}(\mathbf{H}_{1,3}\mathbf{v}_3^1)$, respectively, and the received signal vectors of interferences lie on $\text{span}(\mathbf{H}_{1,1}\mathbf{v}_1^2)$ and $\text{span}(\mathbf{H}_{1,1}\mathbf{v}_1^3)$. Therefore, the desired symbols of each receiver occupy three dimensions out of the five dimensions of the received signal space, and thus an achievable per-user DoF of $\frac{3}{5}$ is obtained by interference alignment.

12.2.3 Cooperative X-Multicast Channel

The *cooperative X-multicast channel* was first proposed and analyzed in [7] when studying caching and delivery problems in the interference network. Based on their proposed parametric file splitting and caching scheme, the original interference network is transformed into the cooperative X-multicast channel. The authors obtained an achievable per-user DoF in this channel via a collective use of interference alignment and interference neutralization techniques. The general $\binom{N_T}{t}\times\binom{N_R}{\sigma}$ cooperative X-multicast channel is defined as follows.

DEFINITION 12.1 *[7, definition 2] The channel characterized as follows is called the $\binom{N_T}{t}\times\binom{N_R}{\sigma}$ cooperative X-multicast channel:*

1. *There are N_R receivers and N_T transmitters.*
2. *Each set of σ ($\sigma \leq N_R$) receivers forms a receiver multicast group.*
3. *Each set of t ($t \leq N_T$) transmitters forms a transmitter cooperation group.*
4. *Each transmitter cooperation group has an independent message for each receiver multicast group.*

In can be seen that the X-multicast channel introduced in [8] is a special case of the cooperative X-multicast channel when $t = 1$. The achievable per-user DoF of the $\binom{N_T}{t}\times\binom{N_R}{\sigma}$ cooperative X-multicast channel obtained in [7] is given by

$$d_{\sigma,t} = \begin{cases} 1, & \sigma+t-1 \geq N_R \\ \frac{\binom{N_R-1}{\sigma-1}\binom{N_T}{t}t}{\binom{N_R-1}{\sigma-1}\binom{N_T}{t}t+1}, & \sigma+t-1 = N_R-1 \\ \max\left\{d'_{\sigma,t}, \frac{\sigma+t-1}{N_R}\right\}, & \sigma+t-1 \leq N_R-2 \end{cases}, \tag{12.8}$$

where

$$d'_{\sigma,t} \triangleq \max_{1\leq t'\leq t}\left\{\frac{\binom{N_R-1}{\sigma-1}\binom{N_T}{t'}\binom{N_R-\sigma}{t'-1}t'}{\binom{N_R-1}{\sigma-1}\binom{N_T}{t'}\binom{N_R-\sigma}{t'-1}t' + \binom{N_R-1}{\sigma}\binom{N_R-\sigma-1}{t'-1}\binom{N_T}{t'-1}}\right\}. \tag{12.9}$$

We now take the $\binom{3}{2}\times\binom{3}{2}$ cooperative X-multicast channel as an example to review the idea of the delivery scheme in [7].

In the considered cooperative X-multicast channel, there are three transmitter cooperation groups: transmitters $\{1,2\}$, transmitters $\{1,3\}$, and transmitters $\{2,3\}$; and three receiver multicast groups: receivers $\{1,2\}$, receivers $\{1,3\}$, and receivers $\{2,3\}$. Denote $W_{j,l}^{i,k}$ (for $j,l,i,k \in \{1,2,3\}, j \neq l, i \neq k$) as the message transmitted by transmitters $\{j,l\}$ and desired by receivers $\{i,k\}$, which is encoded into the transmitted symbol $x_{j,l}^{i,k}$. By an n-symbol extension, the $n\times 1$ beam-forming vector of $x_{j,l}^{i,k}$ at transmitter j is given by $\mathbf{v}_{j,l}^{i,k}(j)$, and the $n\times 1$ beam-forming vector of $x_{j,l}^{i,k}$ at transmitter l is given by $\mathbf{v}_{j,l}^{i,k}(l)$. Consider an arbitrary receiver i. Its received signal $\mathbf{y}_i$ is given by

$$\begin{aligned} \mathbf{y}_i = \mathbf{H}_{i,1}&\left(\mathbf{v}_{1,2}^{1,2}(1)x_{1,2}^{1,2} + \mathbf{v}_{1,2}^{1,3}(1)x_{1,2}^{1,3} + \mathbf{v}_{1,2}^{2,3}(1)x_{1,2}^{2,3}\right) \\ + \mathbf{H}_{i,1}&\left(\mathbf{v}_{1,3}^{1,2}(1)x_{1,3}^{1,2} + \mathbf{v}_{1,3}^{1,3}(1)x_{1,3}^{1,3} + \mathbf{v}_{1,3}^{2,3}(1)x_{1,3}^{2,3}\right) \\ + \mathbf{H}_{i,2}&\left(\mathbf{v}_{1,2}^{1,2}(2)x_{1,2}^{1,2} + \mathbf{v}_{1,2}^{1,3}(2)x_{1,2}^{1,3} + \mathbf{v}_{1,2}^{2,3}(2)x_{1,2}^{2,3}\right) \\ + \mathbf{H}_{i,2}&\left(\mathbf{v}_{2,3}^{1,2}(2)x_{2,3}^{1,2} + \mathbf{v}_{2,3}^{1,3}(2)x_{2,3}^{1,3} + \mathbf{v}_{2,3}^{2,3}(2)x_{2,3}^{2,3}\right) \end{aligned}$$

$$+ \mathbf{H}_{i,3}\left(\mathbf{v}_{1,3}^{1,2}(3)x_{1,3}^{1,2} + \mathbf{v}_{1,3}^{1,3}(3)x_{1,3}^{1,3} + \mathbf{v}_{1,3}^{2,3}(3)x_{1,3}^{2,3}\right)$$
$$+ \mathbf{H}_{i,3}\left(\mathbf{v}_{2,3}^{1,2}(3)x_{2,3}^{1,2} + \mathbf{v}_{2,3}^{1,3}(3)x_{2,3}^{1,3} + \mathbf{v}_{2,3}^{2,3}(3)x_{2,3}^{2,3}\right). \quad (12.10)$$

Since each symbol causes interference to only one receiver, it can be neutralized out at this receiver by carefully designing its beam-forming vectors between the two transmitters in the corresponding cooperation group. In specific, take symbol $x_{1,2}^{1,2}$, for example. It can be neutralized out at receiver 3, by designing the beam-forming vectors $\{\mathbf{v}_{1,2}^{1,2}(1), \mathbf{v}_{1,2}^{1,2}(2)\}$ to satisfy

$$\mathbf{H}_{3,1}\mathbf{v}_{1,2}^{1,2}(1) + \mathbf{H}_{3,2}\mathbf{v}_{1,2}^{1,2}(2) = 0. \quad (12.11)$$

Since the channel matrices $\{\mathbf{H}_{i,j}\}$ are diagonal and their diagonal entries are i.i.d. as some continuous distributions, these matrices are invertible with probability 1. Then, $\mathbf{v}_{1,2}^{1,2}(2)$ can be determined by

$$\mathbf{v}_{1,2}^{1,2}(2) = -\mathbf{H}_{3,2}^{-1}\mathbf{H}_{3,1}\mathbf{v}_{1,2}^{1,2}(1), \quad (12.12)$$

and $\mathbf{v}_{1,2}^{1,2}(1)$ can be chosen as a random non-zero vector.

The interference neutralization can be applied to the rest of the eight symbols so that each symbol will be neutralized out at its undesired receiver. Then, each receiver receives only its six desired symbols, and can successfully decode all six symbols as proven in [7]. Thus each receiver can achieve a per-user DoF of 1.

In the general $\binom{N_T}{t} \times \binom{N_R}{\sigma}$ cooperative X-multicast channel, each symbol is cooperatively transmitted by t transmitters and neutralized at a maximum of $t-1$ undesired receivers. Note that for each message, when the number of receivers that do not want this message, $N_R - \sigma$, is larger than $t-1$, the interference neutralization can cancel the interference at only $t-1$ of them; the rest of the receivers that do not want this message still suffer interference from this message. Then, we should further partition the interfering messages into different groups and use asymptotic interference alignment to align interferences in a same group into a same subspace at each interfered receiver, so as to obtain the achievable per-user DoF given in (12.8), as detailed in [7].

Note that the per-user DoF (12.8) in the cooperative X-multicast channel reduces to the per-user DoF in the X channel when $\sigma = t = 1$, and reduces to the per-user DoF in the X-multicast channel when $t = 1$. In fact, (12.8) is the largest DoF among the per-user DoFs in the four considered channels, because receiver multicast, interference neutralization, and interference alignment are considered and utilized jointly.

12.3 System Model and Performance Metric

In this section, we formally define the cache-aided interference network model. Then, we use a new latency-oriented metric, namely *normalized delivery time*, to characterize the performance limit.

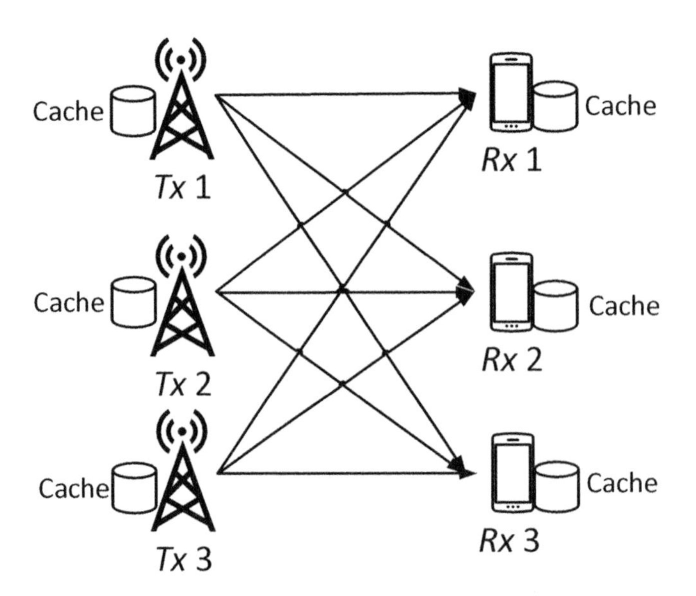

Figure 12.1 Cache-aided wireless interference network with three transmitters and three receivers.

12.3.1 Network Model

We consider an arbitrary interference network consisting of N_R (≥ 2) receivers and N_T (≥ 2) transmitters. Each node has a local cache memory with finite size. Figure 12.1 shows an example with $N_T = N_R = 3$. The channel from each transmitter to each receiver is assumed to experience both fading and additive noise.

The database consists of N files, denoted by $\{W_1, W_2, \ldots, W_N\}$, each with F bits. We assume that $N \geq N_R$ so that each receiver can demand a distinct file. Each transmitter has a storage that is able to store $\mu_T NF$ bits from the database, and each receiver has a storage that is able to store $\mu_R NF$ bits from the database. Here, we refer to μ_T and μ_R as the *normalized cache sizes* at each transmitter and receiver, respectively.

12.3.2 Two-Phase Operation Model

The cache-aided communication consists of two operation phases, a *cache placement phase* and a *content delivery phase*. In the cache placement phase, there is a caching function for each transmitter j, mapping the database into its local cached content U_j. Each receiver i also has a caching function mapping the database into its local cached content V_i. The caching functions should satisfy the cache memory constraints at nodes, i.e., $H(U_j) \leq \mu_T NF, H(V_i) \leq \mu_R NF$, and we assume that these functions are globally known.

Now let us consider the delivery phase. We assume that each receiver i requests a file W_{q_i} from the database and use $\mathbf{q} \triangleq (q_i)_{i=1}^{N_R} \in [N]^{N_R}$ to denote the demand vector. After the receiver demand is revealed, each transmitter j uses an encoding function to map its local cached content U_j, receiver demand vector $\mathbf{q}$, and channel realization to the codeword $(X_j[t])_{t=1}^T$, where T is defined as its block length. The codeword $(X_j[t])_{t=1}^T$ should satisfy the average transmit power constraint P. After receiving signals from the interference network, each receiver i uses a decoding function to decode $\hat{W}_{q_i}$ of its desired file W_{q_i} with the help of the channel realization and its local cached content V_i. We define the worst-case error probability as

$$P_\epsilon = \max_{\mathbf{q}\in[N]^{N_R}} \max_{i\in[N_R]} \mathbb{P}(\hat{W}_{q_i} \neq W_{q_i}). \tag{12.13}$$

The given caching and coding scheme is feasible if $P_\epsilon \to 0$ when $F \to \infty$ for almost all channel realizations.

In the considered network model, each receiver can obtain its desired file only from either its local cache or any of the N_T transmitters. Thus we must have $\mu_R NF + N_T \mu_T NF \geq NF$ for the cache size constraint. This is equivalent to the following region:

$$\begin{cases} 0 \leq \mu_R, \mu_T \leq 1 \\ \mu_R + N_T \mu_T \geq 1 \end{cases}, \tag{12.14}$$

which is referred to as the feasible region of normalized cache sizes.

12.3.3 Performance Metric

We characterize the performance of the cache-aided interference network using a latency-oriented metric defined as follows.

DEFINITION 12.2 *[11, definition 3] For a feasible caching and coding scheme at normalized cache size tuple (μ_R, μ_T), the normalized delivery time is defined as*

$$\tau(\mu_R, \mu_T) \triangleq \lim_{P\to\infty} \lim_{F\to\infty} \sup \frac{\max_{\mathbf{q}} T}{F/\log P}. \tag{12.15}$$

The minimum NDT is defined as

$$\tau^*(\mu_R, \mu_T) = \inf\{\tau(\mu_R, \mu_T) \colon \tau(\mu_R, \mu_T) \text{ is achievable}\}. \tag{12.16}$$

By Definition 12.2, if a feasible caching and coding scheme can achieve an NDT of τ, it means its worst-case delivery time to meet any possible use demand is τ times of the reference time required to transmit a single file of F bits in a Gaussian baseline system at high SNR regime.

The per-user capacity of this network is approximately $(d \cdot \log P + o(\log P))$ in the high SNR regime, with d being the per-user DoF. Denote R as the per-user worst-case traffic load normalized by the file size F. Then, we can rewrite the worst-case transmission time as $\max_{\mathbf{q}} T = \frac{RF}{d\cdot\log P + o(\log P)}$. By Definition 12.2, we can express NDT as

$$\tau = R/d. \tag{12.17}$$

Therefore, NDT characterizes the asymptotic time, i.e., when $P \to \infty$ and $F \to \infty$, to use a transmission rate characterized by d to deliver a per-user traffic *load* R. In the following, we will use (12.17) to calculate the NDT.

12.4 NDT Analysis in Wireless Interference Networks

In this section, we first introduce the parametric caching scheme proposed in [7]. We then present the content delivery scheme by treating the cache-aided network as the cooperative X-multicast channel. Given this delivery scheme, we obtain an achievable NDT by optimizing the file splitting ratios through a linear programming (LP) problem. Compared to the lower bound derived in [7], this achievable NDT is optimal in some transmitter and receiver cache size regions. We also prove that the multiplicative gap between the achievable NDT and this lower bound is bounded in the entire region. We then extend these results to the multiple-antenna interference networks.

12.4.1 Parametric Caching Scheme

We first present a parametric caching scheme for arbitrary transmitter and receiver node numbers N_T and N_R and any normalized cache sizes μ_R and μ_T, proposed in [7]. This scheme splits and caches each file in the database in the same manner. Specifically, consider an arbitrary file W_n (for $n \in \{1, 2, \ldots, N\}$). It is partitioned into $\sum_{r=0}^{N_R} \sum_{t=1}^{N_T} \binom{N_R}{r}\binom{N_T}{t} + 1$ subfiles exclusively. Each subfile, denoted by $W_{n,\mathrm{R}_\Phi,\mathrm{T}_\Psi}$, is cached exclusively in transmitter subset Ψ and receiver subset Φ, and with possibly different length. For example, $W_{n,\mathrm{R}_1,\mathrm{T}_{12}}$ is the subfile cached in receiver 1 and transmitters 1 and 2. Given the independence of all files as well as the symmetry of all the nodes, we also assume that subfiles cached in the same number of receivers and transmitters are of the equal length. Under this assumption, denote the size of $W_{n,\mathrm{R}_\Phi,\mathrm{T}_\Psi}$ by $a_{r,t}F$, with $r = |\Phi|, t = |\Psi|$. Here we refer to $a_{r,t}$ as the file splitting ratio, which satisfies $a_{r,t} \in [0, 1]$ and should be determined as in the following sections. For example, subfile $W_{n,\mathrm{R}_1,\mathrm{T}_{12}}$ is of $a_{1,2}F$ bits. The file splitting ratios $\{a_{r,t}\}$ must satisfy three constraints, given as follows:

$$
\begin{cases}
\sum_{r=0}^{N_R} \sum_{t=1}^{N_T} \binom{N_R}{r}\binom{N_T}{t} a_{r,t} + a_{N_R,0} = 1, & (12.18) \\
\sum_{r=1}^{N_R} \sum_{t=1}^{N_T} \binom{N_R - 1}{r - 1}\binom{N_T}{t} a_{r,t} + a_{N_R,0} \le \mu_R, & (12.19) \\
\sum_{r=0}^{N_R} \sum_{t=1}^{N_T} \binom{N_R}{r}\binom{N_T - 1}{t - 1} a_{r,t} \le \mu_T. & (12.20)
\end{cases}
$$

Constraint (12.18) results from the file size constraint, because there are $\binom{N_R}{r}\binom{N_T}{t}$ subfiles for each file that are exclusively cached in t transmitters and r receivers in the network and each of these subfiles has $a_{r,t}F$ bits, for $r \in \{0,1,\ldots,N_R\}$ and $t \in \{1,2,\ldots,N_T\}$ or $(r = N_R, t = 0)$. Similarly, constraint (12.19) and constraint (12.20) result from the cache size constraints in receivers and transmitters, respectively, which will not be explained in detail.

This scheme is called as the *parametric caching scheme* because it does not specify the values of the file splitting ratios $\{a_{r,t}\}$ once they satisfy constraints (12.18), (12.19), and (12.20). These ratios will be determined later to minimize the sum NDT in the delivery phase.

Alternatively, the authors in [6] and [9] propose another caching scheme, called *symmetric file splitting and caching* scheme, when the normalized cache sizes satisfy $N_R\mu_R = m$ and $N_T\mu_T = n$ with $m \in [0, N_R] \cap \mathbb{Z}$ and $n \in [0, N_T] \cap \mathbb{Z}$. We refer to these values as *integer point* cache size pairs, since each bit of every file is cached in n transmitters and m receivers simultaneously *on average*. More specifically, we split each file into $\binom{N_R}{m}\binom{N_T}{n}$ subfiles with equal sizes. Each subfile is cached in a unique receiver subset with m receivers and a unique transmitter subset with n transmitters. Then, each receiver and transmitter caches

$$N \cdot \binom{N_R - 1}{m - 1}\binom{N_T}{n} \cdot \frac{F}{\binom{N_R}{m}\binom{N_T}{n}} = \frac{m}{N_R} NF = \mu_R NF, \tag{12.21}$$

$$N \cdot \binom{N_R}{m}\binom{N_T - 1}{n - 1} \cdot \frac{F}{\binom{N_R}{m}\binom{N_T}{n}} = \frac{n}{N_T} NF = \mu_T NF \tag{12.22}$$

bits, respectively, and satisfies the local cache sizes.

It can be seen that the parametric caching scheme degenerates to the symmetric caching scheme by letting $a_{m,n} = \frac{1}{\binom{N_R}{m}\binom{N_T}{n}}$ and other file splitting ratios be 0. Since we can select suitable file splitting ratios in the parametric caching scheme for NDT optimization, this scheme is more general than the symmetric caching scheme.

12.4.2 Content Delivery Strategy

In this section, we aim to introduce the content delivery scheme. We focus on the worst demand that each receiver demands a different file. Note that we can still apply our scheme if one file is requested by two or more receivers, by seeing the demands as different demands. We assume that receiver i needs file W_i. Given local caches, receiver i requires only subfiles $\{W_{i,\mathrm{R}_\Phi,\mathrm{T}_\Psi} : \Phi \not\ni i\}$.

We first divide these subfiles into $N_T N_R$ groups, where the subfiles in each group, denoted as (r,t), are cached exactly at r receivers and t transmitters. By this grouping method, group (r,t) contains all the subfiles with length $a_{r,t}F$ bits. In this group, there are in total $N_R\binom{N_R-1}{r}\binom{N_T}{t}$ subfiles, among which each receiver needs $\binom{N_R-1}{r}\binom{N_T}{t}$ subfiles. We deliver each group individually in the time division manner. In this section, we illustrate the transmission scheme for an arbitrary group (r,t), for $r \in [0, N_R - 1] \cap \mathbb{Z}$ and $t \in [0, N_T] \cap \mathcal{Z}$.

In this group, each subfile is cached at t transmitters and r receivers. We can thus use coded multicasting via bit-wise XOR to combine these subfiles, similar to [12].

More specifically, for any transmitter set Ψ with $|\Psi| = t$ and any receiver set Φ^+ with $|\Phi^+| = r + 1$, message $\bigoplus_{i \in \Phi^+} W_{i,\mathrm{R}_{\Phi^+ \setminus \{i\}},\mathrm{T}_\Psi}$ can be generated by each transmitter in Ψ, and all receivers in Φ^+ desire this message. To illustrate it, consider an arbitrary receiver $i \in \Phi^+$. Since it caches subfiles $\{W_{i',\mathrm{R}_{\Phi^+ \setminus \{i'\}},\mathrm{T}_\Psi} : i' \in \Phi^+ \setminus \{i\}$, it is able to decode the required subfile $W_{i,\mathrm{R}_{\Phi^+ \setminus \{i\}},\mathrm{T}_\Psi}$ from $\bigoplus_{i \in \Phi^+} W_{i,\mathrm{R}_{\Phi^+ \setminus \{i\}},\mathrm{T}_\Psi}$. By applying this method, a coded message is generated by combining $r + 1$ subfiles via bit-wise XOR. Transmitters thus need to deliver only $\binom{N_R}{r+1}\binom{N_T}{t}$ messages to receivers, each available at t transmitters and desired by $r + 1$ receivers. The network is transformed into the $\binom{N_T}{t} \times \binom{N_R}{r+1}$ cooperative X-multicast channel, with per-user DoF $d_{r+1,t}$ achieved in (12.8) where $\sigma = r + 1$.

Since each receiver requires $\binom{N_R-1}{r}\binom{N_T}{t}$ messages and each message has $a_{r,t}F$ bits, the NDT for the considered group is $\tau_{r,t} = \frac{\binom{N_R-1}{r}\binom{N_T}{t}}{d_{r+1,t}} a_{r,t}$. Combining the delivery for all $N_T N_R$ groups, the sum NDT is given by

$$\tau = \sum_{r=0}^{N_R-1} \sum_{t=1}^{N_T} \frac{\binom{N_R-1}{r}\binom{N_T}{t}}{d_{r+1,t}} a_{r,t}. \tag{12.23}$$

12.4.3 Achievable NDT

Now, we aim to minimize the total NDT (12.23) by optimizing the file splitting ratios $\{a_{r,t}\}$ subject to constraints (12.18), (12.19), and (12.20). This results in the following achievable upper bound of NDT [7].

THEOREM 12.3 (Achievable NDT) [7, theorem 1] *For the considered network consisting of $N_T \geq 2$ transmitters, $N_R \geq 2$ receivers, and a database with $N \geq N_R$ files, where each transmitter and each receiver have normalized cache sizes μ_T and μ_R, respectively, the minimum NDT is upper-bounded by the optimal solution of the LP problem given as follows:*

$$\mathcal{P}_1 : \tau_U(\mu_R, \mu_T) \triangleq \min_{\{a_{r,t}:(r,t) \in \mathcal{A}\}} \sum_{r=0}^{N_R-1} \sum_{t=1}^{N_T} \frac{\binom{N_R-1}{r}\binom{N_T}{t}}{d_{r+1,t}} a_{r,t}, \tag{12.24}$$

$$\text{s.t.} \quad 0 \leq a_{r,t} \leq 1, \forall (r,t) \in \mathcal{A} \tag{12.25}$$

$$(12.18)(12.19)(12.20) \tag{12.26}$$

where $\mathcal{A} \triangleq \{(r,t) : r + N_R t \geq N_R, 0 \leq r \leq N_R, 0 \leq t \leq N_T, r,t \in \mathbb{Z}\}$, $\{a_{r,t}\}$ are the file splitting ratios, and $d_{r+1,t}$ is the achievable per-user DoF in the $\binom{N_T}{t} \times \binom{N_R}{r+1}$ cooperative X-multicast channel in (12.8) *where $\sigma = r + 1$.*

Compared to the theoretical lower bound in [7, theorem 2] where interfile coding is not permitted in the caching strategy, the optimality of the achievable NDT and its gap to this lower bound are given by two corollaries as follows, whose proofs are in [7, appendix B] and [7, appendix C], respectively.

COROLLARY (Optimality) [7, corollary 1] *When interfile coding is not allowed in the cache placement strategy, the achievable NDT in Theorem 12.3 is optimal if* (μ_R, μ_T) *satisfies any of the following conditions:*

1. $N_R\mu_R + N_T\mu_T \geq N_R$: *the optimal NDT is* $\tau^* = 1 - \mu_R$.
2. $(\mu_R, \mu_T) = (0, 1)$: *the optimal NDT is* $\tau^* = \frac{N_R}{\min\{N_T, N_R\}}$.
3. $(\mu_R, \mu_T) = (0, 1/N_T)$: *the optimal NDT is* $\tau^* = \frac{N_T+N_R-1}{N_T}$.
4. $\mu_R + N_T\mu_T = 1$ *if intrafile coding is not permitted in the cache placement: the optimal NDT is* $\tau^* = \frac{N_T+N_R-1}{N_T}(1-\mu_R)$.

COROLLARY (Gap of NDT) [7, corollary 2] *If interfile coding is not permitted in the caching strategy, the multiplicative gap between the achievable NDT in 12.3 and the lower bound in [7, theorem 2] is within 2 when* $N_T \geq N_R$, *within 12 when* $N_T < N_R, \mu_T \geq \frac{1}{N_T}$, *and within* $\frac{N_T+N_R-1}{N_T}$ *when* $N_T < N_R, \mu_T < \frac{1}{N_T}$.

The optimality results of the achievable NDT in Theorem 12.3 are based on the assumption that there is no interfile coding in the cache placement strategy. The authors in [8] derived a theoretical lower bound of NDT without any assumptions on the caching functions. By using this lower bound, they found that the multiplicative gap between their achievable NDT and the optimum is no larger than 13.5. The authors in [6] also proposed an achievable scheme by using interference neutralization in a one-shot linear transmission scheme, and found that the multiplicative gap between their achievable NDT and the optimum is no larger than 2 under the assumptions of uncoded prefetching in the cache placement and one-shot linear transmission schemes.

In the following, we focus on the achievable NDT in Theorem 12.3, and present some discussions.

12.4.3.1 On Caching at Integer Point Cache Size Pairs

In the achievable scheme presented earlier, caching gains are reflected by file splitting ratios in the LP problem. Here, we consider only the symmetric file splitting and caching scheme at any integer point cache size pair $(\mu_R = m/N_R, \mu_T = n/N_T)$, with $m \in [0, N_R] \cap \mathbb{Z}$ and $n \in [0, N_T] \cap \mathbb{Z}$, while the caching gains exploited at general tuple (μ_R, μ_T) is detailed in [7].

In the symmetric file splitting and caching scheme, each file is split to $\binom{N_R}{m}\binom{N_T}{n}$ subfiles with equal sizes. Each subfile is cached at n transmitters and m receivers, corresponding to the file splitting ratio $a_{m,n} = \frac{1}{\binom{N_R}{m}\binom{N_T}{n}}$, with the rest being 0. In the delivery phase, the network topology is transformed into the $\binom{N_T}{n} \times \binom{N_R}{m+1}$ cooperative X-multicast channel by the coded generation method, whose achievable NDT is given by

$$\tau_{m,n} = \frac{1-\mu_R}{d_{m+1,n}}, \tag{12.27}$$

where $d_{m+1,n}$ is the achievable per-user DoF in (12.8).

We can rewrite (12.27) as

$$\tau_{m,n} = \frac{N_R(1-\mu_R)}{(m+1)d_{\text{sum}}}, \tag{12.28}$$

where $d_{\text{sum}} = \frac{N_R d_{m+1,n}}{m+1}$ is the sum DoF derived from per-user DoF (12.8) in the cooperative X-multicast channel. In d_{sum}, the numerator $N_R d_{m+1,n}$ comes from the fact that there are N_R receivers in total, each with per-user DoF of $d_{m+1,n}$; and the dominator $m+1$ comes from the fact that each message in the cooperative X-multicast channel is desired by $m+1$ receivers.

The expression in (12.28) reveals the caching gains more explicitly. Specifically, the local caching gain is reflected by $(1-\mu_R)$, because each receiver caches $\mu_R F$ bits of its desired file in advance; the coded multicasting gain is reflected by $(m+1)$, which is exactly the number of desired receivers of each transmitted message; the transmitter cooperation gain is reflected by d_{sum}, which is achieved through joint interference alignment and interference neutralization techniques.

12.4.3.2 On the Optimal File Splitting Ratios

We can use equation substitutions and other suitable manipulations to solve the LP problem in Theorem 12.3. In general, it does not have unique solutions, which implies that we may have multiple solutions for optimal file splitting ratios of the LP problem in Theorem 12.3. The different optimal file splitting ratios reflect different caching and delivery schemes, even though the same NDT is achieved. In the following, we will use the 3×3 network as an example to show the impact of different solutions of file splitting ratios. The following corollary presents the optimal solution of the LP problem in Theorem 12.3 for this network.

COROLLARY *For the cache-aided 3×3 interference network, the minimum NDT is upper bounded by*

$$\tau^*(\mu_R,\mu_T) \le \tau_U = \begin{cases} 1-\mu_R, & (\mu_R,\mu_T)\in\mathcal{R}_{33}^1 \\ \frac{4}{3}-\frac{4}{3}\mu_R-\frac{1}{3}\mu_T, & (\mu_R,\mu_T)\in\mathcal{R}_{33}^2 \\ \frac{3}{2}-\frac{5}{3}\mu_R-\frac{1}{2}\mu_T, & (\mu_R,\mu_T)\in\mathcal{R}_{33}^3 \\ \frac{13}{6}-\frac{8}{3}\mu_R-\frac{3}{2}\mu_T, & (\mu_R,\mu_T)\in\mathcal{R}_{33}^4 \\ \frac{8}{3}-\frac{8}{3}\mu_R-3\mu_T, & (\mu_R,\mu_T)\in\mathcal{R}_{33}^5 \end{cases}, \tag{12.29}$$

where $\{\mathcal{R}_{33}^i\}_{i=1}^5$ are given as follows and sketched in Figure 12.2.

$$\begin{cases} \mathcal{R}_{33}^1 = \{(\mu_R,\mu_T)\colon \mu_R+\mu_T \ge 1, \mu_R \le 1, \mu_T \le 1\} \\ \mathcal{R}_{33}^2 = \{(\mu_R,\mu_T)\colon \mu_R+\mu_T < 1, 2\mu_R+\mu_T \ge 1, \mu_R+2\mu_T > 1\} \\ \mathcal{R}_{33}^3 = \{(\mu_R,\mu_T)\colon 3\mu_R+3\mu_T \ge 2, 2\mu_R+\mu_T < 1, \mu_R \ge 0\} \\ \mathcal{R}_{33}^4 = \{(\mu_R,\mu_T)\colon 3\mu_R+3\mu_T < 2, \mu_R \ge 0, 3\mu_T > 1\} \\ \mathcal{R}_{33}^5 = \{(\mu_R,\mu_T)\colon 3\mu_T \le 1, \mu_R+2\mu_T \le 1, \mu_R+3\mu_T \ge 1\} \end{cases}. \tag{12.30}$$

Let us take the integer point cache size pair $(\mu_R = \frac{1}{3}, \mu_T = \frac{2}{3})$ as an example. We have two solutions of file splitting ratios to achieve (12.29). The first solution is $a_{0,3}^* = \frac{2}{3}$ and $a_{3,0}^* = \frac{1}{3}$ while the rest ratios being 0. It implies that we should split

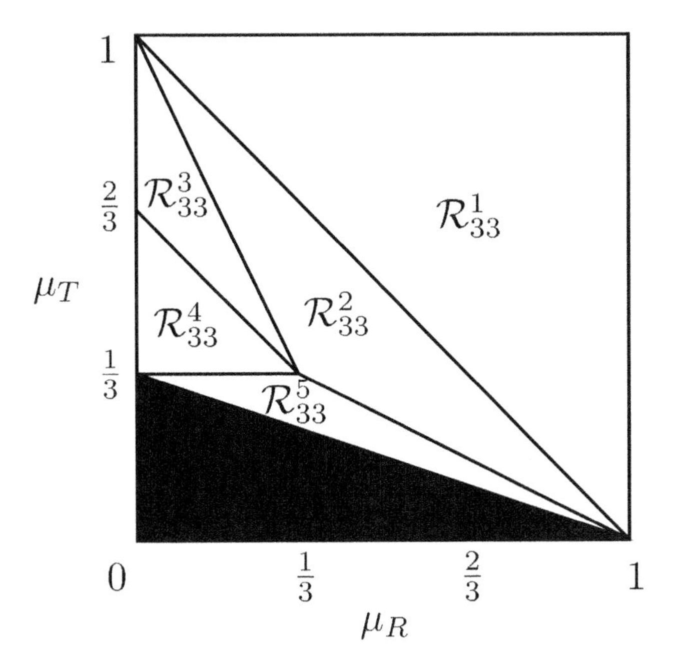

Figure 12.2 Cache size regions in the 3×3 network.

each file into two subfiles. The first subfile has $\frac{2}{3}F$ bits, and we let all three transmitters but not any receivers cache this subfile. The second subfile has $\frac{1}{3}F$ bits and we let all receivers but not any transmitters cache this subfile. By transforming the network into a multiple-input single-output (MISO) broadcast channel in the delivery phase, this solution exploits transmitter cooperation gain as well as local caching gain.

We can also let $a_{1,2}^* = \frac{1}{9}$ and the rest ratios be 0. Note that symmetric file splitting and caching scheme is adopted in this solution, i.e., each file is split into 9 subfiles, each with the same $\frac{1}{9}F$ bits. Each subfile is exclusively cached at two transmitters and one receiver. By transforming the interference network into a partially cooperative X-multicast channel in the delivery phase, this solution exploits transmitter cooperation gain as well as coded multicasting gain, besides local caching gain. Since both solutions achieve the NDT $\tau^* = \frac{2}{3}$, they are both optimal when ($\mu_R = \frac{1}{3}, \mu_T = \frac{2}{3}$).

In general, we find that the caching and transmission scheme is not unique at integer point cache size pairs ($\mu_R = \frac{m}{N_R}, \mu_T = \frac{n}{N_T}$) with $m + n \geq N_R$. In specific, when $N_T \geq N_R$, besides the symmetric file splitting and caching strategy, we can also let $a_{N_R,0}^* = \mu_R, a_{0,N_R}^* = \frac{1-\mu_R}{\binom{N_T}{N_R}}$, which implies that we split each file into $1 + \binom{N_T}{N_R}$ subfiles. The first subfile has $\mu_R F$ bits and we let all N_R receivers but not any transmitters cache this subfile, while each of the other subfiles has $\frac{1-\mu_R}{\binom{N_T}{N_R}}F$ bits and we let N_R out of N_T distinct transmitters but not any receivers cache it. In the delivery phase, we can transform the network to the $\binom{N_T}{N_R} \times \binom{N_R}{1}$ cooperative X-multicast channel with $d_{1,N_R} = 1$ and achieve the following NDT

$$\tau = \frac{1-\mu_R}{d} = 1-\mu_R. \tag{12.31}$$

In this scheme, we do not use bit-wise XOR; therefore, we exploit only transmitter cooperation gain but not coded multicasting gain here. Since (12.31) achieved by this scheme is the same as (12.27) achieved by symmetric file splitting and caching scheme when $m+n \geq N_R$ and $N_T \geq N_R$, it implies that transmitter cooperation gain can have the same contribution as the combined transmitter cooperation and coded-multicasting gain in this case.

If $N_T < N_R$, the system performance is limited by the number of transmitters. Thus, unlike the previous case, we should use bit-wise XOR here. Specifically, we let $a^*_{N_R,0} = 1 - \frac{N_R}{N_T}(1-\mu_R), a^*_{N_R-N_T,N_T} = \frac{\frac{N_R}{N_T}(1-\mu_R)}{\binom{N_R}{N_R-N_T}}$, which implies that we split each file into $1+\binom{N_R}{N_R-N_T}$ subfiles. The first subfile has $1-\frac{N_R}{N_T}(1-\mu_R)F$ bits and we let all N_R receivers but not any transmitters cache this subfile, while each of the rest of the subfiles is of $\frac{\frac{N_R}{N_T}(1-\mu_R)}{\binom{N_R}{N_R-N_T}}F$ bits and we let all N_T transmitters and $N_R - N_T$ out of N_R distinct receivers cache it. Given this caching strategy, only the subfiles with size $\frac{\frac{N_R}{N_T}(1-\mu_R)}{\binom{N_R}{N_R-N_T}}F$ bits should be transmitted, and transmitter cooperation gain, coded-multicasting gain, and local caching gain can be exploited in the delivery phase.

Given the various choices to select suitable file splitting ratios, we should choose a proper scheme to improve the system performance in practical systems with other constraints, such as file splitting number constraints or computation limitations at nodes.

12.4.4 MIMO Interference Network

In this section, we extend our scheme to the multiple-antenna case, where each transmitter and each receiver have M_T and M_R antennas, respectively.

We still adopt the parametric caching scheme presented in Section 12.4.1. The delivery phase is similar to the previous case, where the network is transformed to the cooperative X-multicast channels with multiple antennas. However, unlike the single antenna case, it is much more difficult to obtain the achievable DoF in the multiple antenna case. In [10], interference management techniques in finite symbol extensions, such as zero forcing, interference neutralization, and interference alignment, are proposed by using linear precoding to obtain the achievable DoFs in the cooperative X-multicast channels when $N_T = N_R = 3$. Substituting this DoF into problem $\mathcal{P}_1$ and solving it, we obtain an achievable upper bound of NDT in the 3×3 MIMO network in [10, theorem 1]. It is seen that each additive term in the achievable NDT is inversely proportional to M_T or M_R. This means that spatial multiplexing gain is exploited here. When interfile coding is not permitted in the caching strategy, we prove that this achievable NDT achieves the optimality for some cache size regions and some antenna configurations, and the multiplicative gap between the achievable NDT and the optimum is no larger than 3 in the rest cases.

In the special case when $M_T = M_R = 1$, the MIMO interference network degenerates to the single-antenna case, and the achievable NDT in [10] is no larger than 1.2 times of the NDT obtained in [7] with single antenna. Compared to [7], the increase is from the fact that interference management schemes in [10] are restricted to linear precoding with finite symbol extensions, while [7] applies infinite symbol extensions.

Unlike the single-antenna case, the symmetric file splitting and caching scheme in the 3×3 MIMO interference network is not always optimal at integer point cache size pairs. To illustrate, let us consider the case $(\mu_R = 0, \mu_T = \frac{2}{3})$ with antenna configuration $(M_R = 5, M_T = 3)$. By using symmetric file splitting and caching scheme, we split each file into three subfiles with equal size. Each subfile is cached in two different transmitters. In the delivery phase, by transforming the network into the partially cooperative MIMO X channel, a per-user DoF of 2 is achieved. Thus the achievable NDT is given by $\frac{1}{2}$. On the other hand, by applying the parametric file splitting, each file can be split into four subfiles. The first subfile has $\frac{1}{2}F$ bits and is cached at all the three transmitters, while each of the rest three subfiles has $\frac{1}{6}F$ bits and is cached at a distinct transmitter. In the delivery phase, for the transmission of subfiles cached at all the three transmitters, the network is transformed into the MIMO broadcast channel, and we can achieve a per-user DoF of 3; for the transmission of subfiles cached at one transmitter, the network is transformed into the MIMO interference channel, and we can achieve a per-user DoF of 2. The achievable NDT is thus $\frac{5}{12}$, which is less than the NDT of $\frac{1}{2}$ obtained by symmetric file splitting and caching scheme.

12.5 Partially Connected Interference Network

Section 12.4 focuses on a fully connected interference network where all the channel realizations are assumed to be i.i.d. However, in practice, there always exist some links weaker than the rest, due to the signal attenuation resulting from blocking objects or the natural pathloss resulting from radio propagation. We refer to this scenario as the partially connected interference network, where each receiver is locally connected to a part of the transmitters, and extend the previous discussion to this network in the following.

12.5.1 Network Model

We focus on a $(K + L - 1) \times K$ partially connected linear interference network. There are K receivers, given by $\{0, 1, \ldots, K - 1\}$, and $K + L - 1$ transmitters, given by $\{0, 1, \ldots, K + L - 2\}$. Each receiver i can locally communicate with L continuous transmitters $\{i, i + 1, \ldots, i + L - 1\}$, and we assume $L \leq K$ where L is referred to as the receiver connectivity. Figure 12.3 plots the 6×4 network with $L = 3$ and $K = 4$. We use set $\mathcal{T}_i \triangleq \{i, i + 1, \ldots, i + L - 1\}$ to denote the L transmitters communicated with receiver i, and use set $\mathcal{R}_j \triangleq \{j, j - 1, \ldots, j - L + 1\} \cap \{0, 1, \ldots, K - 1\}$, with $|\mathcal{R}_j| \leq L$, to denote the receivers communicated with transmitter j. Each transmitter and receiver has a local storage and has a single antenna. We assume that the normalized

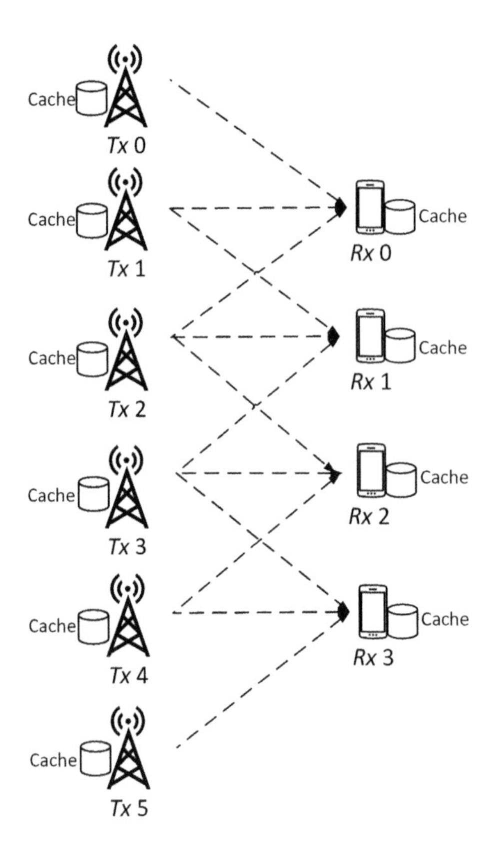

Figure 12.3 6×4 partially connected linear interference network with receiver connectivity $L = 3$.

cache sizes should satisfy $\mu_T \geq \frac{1}{L}$ and $\mu_R \geq 0$, which implies that the entire database can be stored at any L transmitters.

Note that our considered model is an extension of the conventional Wyner's model [13] and the $K \times K$ partially connected network studied in [14]. There are $L - 1$ more transmitters in this model, indexed by $\{K, K+1, \ldots, K+L-2\}$, to ensure that each receiver is connected to L transmitters. Note that this constant receiver connectivity is critical in this work so that the following achievable scheme proposed in [15] is tractable in the linear network topology.

The linear network has a direct relation with the $K \times K$ partially connected circular interference network. In the circular network, each receiver i communicates with L circulant transmitters, given by $\mathcal{T}_i^c \triangleq \{i, i+1, \ldots, i+L-1\} \bmod K$. Denote $\mathcal{R}_j^c \triangleq [j-L+1 : j] \bmod K$ as the L receivers communicated with transmitter j in the circular network. Note that we have $\mathcal{R}_j^c = \mathcal{R}_j \cup \mathcal{R}_{K+j}$ for all $j \in [0, L-2]$. This implies that, we can transform the linear network into the circular network by combining transmitters $K + j$ and j in the linear network, for all $j \in [0, L-2]$, into transmitter j in the circular network. If $L = K$, the circular network degenerates to the fully connected network. In the final section, our study on the linear network will be extended to the circular network, and we will also compare the results to the fully connected network.

12.5.2 Achievable Scheme

12.5.2.1 Cache Placement

Each file W_n is first split into L subfiles with equal size, given as $\{W_{n,T_p}\}_{p=0}^{L-1}$. Subfiles $\{W_{n,T_p}\}_{n=0}^{N-1}$ are cached at transmitters $\{j: j \bmod L = p\}$. By using this caching strategy, each receiver is able to obtain the entire database from the L transmitters it communicates with. Note that each transmitter caches only $\frac{NF}{L}$ bits by this cache placement strategy, which satisfies its local cache capacity.

To fill the receiver caches, we further split each subfile W_{n,T_p} into 2^L subfiles, denoted by $\{W_{n,T_p,R_Q}\}$, where $Q \subseteq \{0, 1, \ldots, L-1\}$. Each subfile W_{n,T_p,R_Q} is stored in receivers $\tilde{\mathcal{R}}_Q \triangleq \{i : (i \bmod L) \in Q\}$. It can be seen that the same subfiles are cached at transmitters congruent modulo L, given as $\{j, j+L, j+2L, \ldots\}$, and the same subfiles are cached at receivers congruent modulo L, given as $\{i, i+L, i+2L, \ldots\}$. Taking the 6×4 network (see Fig. 12.3) as an example, we have $U_0 = U_3$, $U_1 = U_4$, $U_2 = U_5$, and $V_0 = V_3$. Similar to the parametric caching in fully connected network, we assume that the size of each subfile W_{n,T_p,R_Q} satisfying $|Q| = r$ is $a_r F$ bits, where a_r is referred to as the file splitting ratio. Similar to (12.18) and (12.19), $\{a_r\}$ should satisfy two constraints given as follows:

$$\begin{cases} L\sum_{r=0}^{L}\binom{L}{r}a_r = 1, & (12.32) \\ L\sum_{r=1}^{L}\binom{L-1}{r-1}a_r \leq \mu_R, & (12.33) \end{cases}$$

where constraints (12.32) and (12.33) come from the file size constraint and receiver cache constraint, respectively.

12.5.2.2 Content Delivery

Each receiver i is assumed to request file W_{i+1}, and it only needs subfiles:

$$W_{i+1}^{\text{need}} \triangleq \left\{W_{i+1,T_p,R_Q} : p \in \{0, \ldots, L-1\}, Q \subseteq \{0, 1, \ldots, L-1\} \setminus \{i \bmod L\}\right\}, \tag{12.34}$$

given its local cache. To deliver these subfiles, we partition them into L groups according to the cardinality of Q. Subfiles in the same group are with the same $|Q|$. There are $KL\binom{L-1}{r}$ subfiles in each group r, for $r = |Q| \in \{0, \ldots, L-1\}$. We deliver the groups in time division manner. In this section, we focus on an arbitrary group r to present the transmission scheme.

Since each subfile is cached at r of any L consecutive receivers, similar to the fully connected network, coded multicasting opportunities can be exploited by combining $r+1$ transmitted subfiles via XOR. To do this, we first add $2L-2$ virtual receivers to expand the considered linear network, as shown in [15], so that each transmitter needs to send an independent message to any $r+1$ of its connected L receivers via coded multicasting. Then, the original network is transformed into the expanded $(K+L-1) \times K$

partially connected X-multicast channel, whose achievable DoF is $d = \frac{L}{L+\frac{L-r-1}{r+1}}$, via asymptotic interference alignment.

Given that each receiver needs $L\binom{L-1}{r}$ subfiles and each subfile has $a_r F$ bits, the NDT for this group is

$$\tau_r = \frac{L\binom{L-1}{r}a_r}{d} = \left[L\binom{L-1}{r} + \binom{L-1}{r+1}\right]a_r. \tag{12.35}$$

12.5.3 Achievable NDT

Minimizing the sum NDT for the delivery of all groups subject to the constraints (12.32) and (12.33), an achievable NDT is given in the following theorem.

THEOREM 12.4 [15, theorem 1] *(Achievable NDT) For the considered network consisting of K receivers and $K + L - 1$ transmitters with receiver connectivity L, an achievable NDT is obtained by solving the LP problem given as follows:*

$$\tau^*(\mu_R, \mu_T) \leq \tau_{ub} \triangleq \min_{\{a_r\}} \sum_{r=0}^{L-1} \left[L\binom{L-1}{r} + \binom{L-1}{r+1}\right] a_r \tag{12.36}$$

$$\text{s.t.} \quad \textit{Constraints } (12.32), (12.33). \tag{12.37}$$

It is shown in [15] that the multiplicative gap between the achievable NDT in Theorem 12.4 and the optimal NDT is no larger than 2.

In Theorem 12.4, the caching gains can be reflected more explicitly at $\mu_R = \frac{l}{L}$, for $l \in \{0, 1, \ldots, L\}$. In specific, a valid solution when $\mu_R = \frac{l}{L}$ is given by $a_l^* = \frac{1}{L\binom{L}{l}}$ while the rest $a_l^* = 0$. This solution achieves the following NDT

$$\tau_{ub} = \frac{L\binom{L-1}{l} + \binom{L-1}{l+1}}{L\binom{L}{l}} = \left(1 - \frac{1}{L} + \frac{1}{1 + L\mu_R}\right) \cdot (1 - \mu_R). \tag{12.38}$$

In (12.38), $(1 - \mu_R)$ reflects receiver local caching gain, and $(1 - \frac{1}{L} + \frac{1}{1+L\mu_R})$ reflects a combined transmitter coordination and coded multicasting gain.

12.5.4 Application to Circular Network

As mentioned before, the linear network can be transformed into the circular network by combining transmitters j and $K + j$ together, for $j \in \{0, \ldots, L - 2\}$. To apply the scheme into the circular network, we need to ensure that these two transmitters cache the same contents, which is automatically satisfied when L divides K. As a result, we can extend Theorem 12.4 directly to the $K \times K$ circular network if L divides K.

If $K = L$, the circular network reduces to the fully connected network, and the achievable DoF in the considered circular channel in [15] is the same as the DoF in the X-multicast channel.

12.6 Conclusion and Open Issues

This chapter presents interference management techniques in the cache-aided wireless networks. We first reviewed some basic interference networks and their achievable DoF results. Then we introduced the network model for the considered cache-aided interference network. Specifically, we adopted normalized delivery time as the performance metric. The achievable caching and delivery schemes for both single-antenna and multiple-antenna cases thus transform the interference network into cooperative X-multicast channels. We can thus opportunistically leverage transmitter cooperation gain and coded multicasting gain, apart from local caching gain in the delivery phase. The achievable NDTs are optimal in some cases, while the multiplicative gap between this achievable NDT and the optimal NDT is bounded in all cases. Finally, we extended our results to the partially connected networks and present a caching and delivery scheme which is tailor-made for the linear and circular network models. We prove that the the multiplicative gap between the achievable NDT and the optimal NDT in this network is no larger than 2.

There are several open issues in the cache-aided interference networks. First, the globally optimal NDT remains unknown in some cases. To approach the optimal NDT, one possible solution is to adopt coded cache placement so as to achieve a higher DoF in the cooperative X-multicast channel.

Second, while this chapter focuses only on the cache size region where all the requested files can be retrieved from either the receiver local cache or the transmitter caches, it still needs to investigate the caching and delivery scheme when the cache size is not large enough so that some file bits need to be fetched from a cloud through backhaul or fronthaul links.

Last, but not least, it is of practical importance to study the performance of a cache-aided interference network at the finite SNR regime, where the exact delivery time rather than NDT is more valued.

References

[1] V. R. Cadambe and S. A. Jafar, "Interference alignment and degrees of freedom of the k-user interference channel," *IEEE Transactions on Information Theory*, vol. 54, no. 8, pp. 3425–3441, Aug. 2008.

[2] V. Cadambe and S. Jafar, "Interference alignment and the degrees of freedom of wireless X networks," *IEEE Transactions on Information Theory*, vol. 55, no. 9, pp. 3893–3908, Sept. 2009.

[3] S. A. Jafar and M. J. Fakhereddin, "Degrees of freedom for the MIMO interference channel," *IEEE Transactions on Information Theory*, vol. 53, no. 7, pp. 2637–2642, July 2007.

[4] C. Wang, T. Gou, and S. A. Jafar, "Subspace alignment chains and the degrees of freedom of the three-user MIMO interference channel," *IEEE Transactions on Information Theory*, vol. 60, no. 5, pp. 2432–2479, May 2014.

[5] S. A. Jafar and S. Shamai, "Degrees of freedom region of the MIMO X channel," *IEEE Transactions on Information Theory*, vol. 54, no. 1, pp. 151–170, Jan. 2008.

[6] N. Naderializadeh, M. A. Maddah-Ali, and A. S. Avestimehr, "Fundamental limits of cache-aided interference management," *IEEE Transactions on Information Theory*, vol. 63, no. 5, pp. 3092–3107, May 2017.
[7] F. Xu, M. Tao, and K. Liu, "Fundamental tradeoff between storage and latency in cache-aided wireless interference networks," *IEEE Transactions on Information Theory*, vol. 63, no. 11, pp. 7464–7491, Nov. 2017.
[8] J. Hachem, U. Niesen, and S. N. Diggavi, "Degrees of freedom of cache-aided wireless interference networks," *IEEE Transactions on Information Theory*, vol. 64, no. 7, pp. 5359–5380, July 2018.
[9] J. S. P. Roig, D. Gündüz, and F. Tosato, "Interference networks with caches at both ends," in *IEEE International Conference on Communications (ICC)*, May 2017.
[10] Y. Cao, M. Tao, F. Xu, and K. Liu, "Fundamental storage-latency tradeoff in cache-aided MIMO interference networks," *IEEE Transactions on Wireless Communications*, vol. 16, no. 8, pp. 5061–5076, Aug. 2017.
[11] A. Sengupta, R. Tandon, and O. Simeone, "Cache aided wireless networks: tradeoffs between storage and latency," in *Annual Conference on Information Science and Systems (CISS)*, Mar. 2016.
[12] M. A. Maddah-Ali and U. Niesen, "Fundamental limits of caching," *IEEE Transactions on Information Theory*, vol. 60, no. 5, pp. 2856–2867, May 2014.
[13] A. D. Wyner, "Shannon-theoretic approach to a Gaussian cellular multiple-access channel," *IEEE Transactions on Information Theory*, vol. 40, no. 6, pp. 1713–1727, Nov. 1994.
[14] A. E. Gamal, V. S. Annapureddy, and V. V. Veeravalli, "Interference channels with coordinated multipoint transmission: degrees of freedom, message assignment, and fractional reuse," *IEEE Transactions on Information Theory*, vol. 60, no. 6, pp. 3483–3498, June 2014.
[15] F. Xu and M. Tao, "Cache-aided interference management in partially connected wireless networks," in *IEEE Global Communications Conference (GLOBECOM)*, Dec. 2017.

13 Full-Duplex Radios for Edge Caching

Italo Atzeni and Marco Maso

Recent studies have shown that edge caching may have a beneficial effect on the sustainability of future wireless networks. While its positive impact at the network level is rather clear (in terms of, e.g., access delay and backhaul load), assessing its potential benefits at the physical layer is less straightforward. This chapter builds upon this observation and focuses on the performance enhancement brought by the addition of caching capabilities to full-duplex (FD) radios in the context of ultra-dense networks (UDNs). More specifically, we aim at showing that the interference footprint of such networks, i.e., the major bottleneck to overcome to observe the theoretical FD throughput doubling at the network level, can be significantly reduced thanks to edge caching. As a matter of fact, fundamental results available in the literature show that most of the gain, as compared to their half-duplex (HD) counterparts, can be achieved by such networks only if costly modifications to their infrastructure are performed and/or if high-rate signaling is exchanged between user equipments (UEs) over suitable control links. Therefore, we aim at proposing a viable and cost-effective alternative to these solutions based on prefetching locally popular content at the network edge. We start by considering an interference-rich scenario such as an ultra-dense FD small-cell network, in which several noncooperative FD base stations (BSs) serve their associated UE while communicating with a wireless backhaul node (BN) to retrieve the content to deliver. We then describe a geographical caching policy aiming at capturing local files popularity and compute the corresponding cache-hit probability. Thereupon, we calculate the probability of successful transmission of a file requested by a UE, either directly by its serving small cell base station (SBS) or by the corresponding BN: this quantity is then used to lower-bound the throughput of the considered network. Our approach leverages tools from stochastic geometry in order to guarantee both analytical tractability of the problem and generality of the results. A set of suitable numerical simulations is finally performed to confirm the correctness of the theoretical findings and characterize the performance enhancement brought by the adoption of edge caching. The most striking result in this sense is the remarkable performance improvement observed when shifting from cache-free to cache-aided FD small-cell networks.

The work of Italo Atzeni was supported by the European Research Council under the Horizon 2020 Programme (ERC 670896 PERFUME).

13.1 Introduction

The last decade has witnessed the progressive introduction of the 4G cellular network technology and the concurrent adoption of increasingly competitive pricing strategies by device manufacturers and telcos. As a consequence, devices that are able to offer reliable broadband data connections to their users, i.e., smartphones, ceased to be premium products and became a commodity. Their market penetration is already massive and keeps progressing at steady pace. Recent studies forecast that smartphones will represent 86% of the total mobile data traffic by 2021, compared to 81% in 2016, and that monthly mobile data traffic will reach 49 exabytes worldwide (or, equivalently, a run rate of 587 exabytes annually) [1].

The amount of network resources needed to support these trends is ever-increasing. Telcos already anticipate that current mobile networks will have to be restructured to cope with both future service demands and the multitude of novel mobile broadband applications constantly introduced in the market. Many important requirements have been identified in this context, such as the need for higher spectral and energy efficiency, lower end-to-end delays, better coverage, large scalability, and lower capital expense (CAPEX) and operating expenses (OPEX), just to name a few [2]. As a consequence, one of the strongest drivers in the last years for several research groups in both industry and academia has been the need to define a more advanced and flexible network technology as compared to 4G, i.e., the so-called 5G. The remarkable results of such activities have already yielded significant outcomes within standardization development organizations like the 3GPP, who have already published the first version of the standard that will guide the deployment of future 5G wireless networks, i.e., 3GPP Release 15 [3, 4].

From a practical point of view, NR deployments will be characterized by the introduction, or further development, of several key solutions expected to bring the sought performance enhancement as compared to existing networks. Interestingly, only some strategies and network configurations have been and are subject to standardization, whereas some others are considered as part of the implementation aspects. Noteworthy and representative examples of these two categories are [3, 4]:

- *Massive multiple-input multiple-output (MIMO)*: this natural candidate for the physical layer of NR has imposed a revision of the reference sequences and channel state information (CSI) feedback mechanisms [5, 6]; and
- *Advanced MIMO precoding*: the adoption of such strategies at the BS should be completely transparent to the UE, i.e., precoding solutions are implementation aspects that are not specified in the standard.

As a matter of fact, the relevance and impact of many other technologies and approaches will increase in future 5G networks as compared to their current role in mobile and fixed networks, regardless of their 3GPP standardization status (i.e., specified or not). In this chapter, we specifically focus on two of these approaches to study and discuss the potential brought by their mutual interactions:

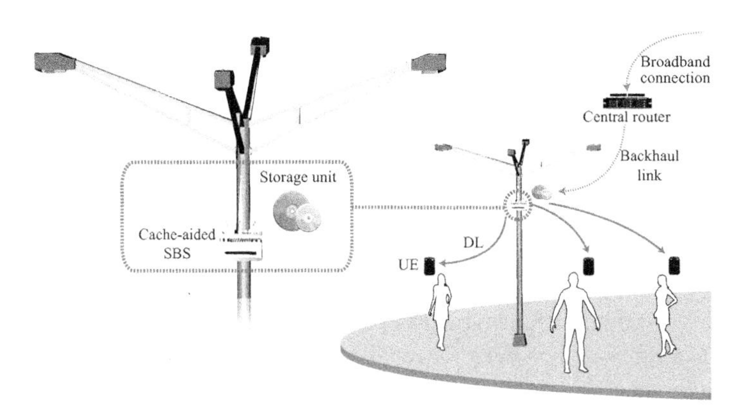

Figure 13.1 Cache-aided SBS. The SBS is equipped with a storage unit to prefetch popular content.

- The *proactive caching* at the network edge, by means of which contents (e.g., videos, images, and news) are brought closer to the users and intelligently cached at SBS equipped with high-capacity storage units, as illustrated in Figure 13.1. As a result, the end-to-end access delay is significantly reduced, the mobile infrastructure is offloaded, and the impact of limited-capacity backhaul on the network performance is mitigated [7–12]. The role of edge caching becomes particularly crucial in case of UDN deployments, i.e., massively populated (and possibly heterogeneous) networks in which the distance between BS and served UE is reduced as compared to classic macro-cell networks [13–16]. Such UDNs may comprise several layers, each of them including different categories of cells (i.e., femto, pico, micro, and macro cells) [17–19]. In general, this layered architecture allows us to design efficient strategies to offload the preexisting macro-cell infrastructure and enhance the network capacity, especially when several nodes provide caching support [20].
- The transition from HD to *FD operations* at radio terminals also promises to offer many benefits, although subject to some peculiar limitations [21–25]. An FD device does not require separate time/frequency resources to be able to support data transmission and reception. In other words, it can simultaneously transmit and receive data over the same bandwidth, thus having the potential to achieve a theoretical throughput doubling and energy efficiency enhancement in comparison to HD radios. In particular, equipping network nodes with FD capabilities can simplify the adoption of flexible duplexing strategies such as dynamic time division duplex (TDD) and enable readjustments to frame structures on the fly. Additionally, FD transmission offers advantages in terms of operation, cost, and efficiency as compared to traditional HD operating mode [26].

The aforementioned approaches certainly have significant potential if taken individually. Nevertheless, assessing the extent of their interoperability is not straightforward.

This is mostly due to the interference footprint of the FD links [24], which may complicate a seamless integration of caching capabilities at each network node. At this stage, a brief introduction of such technology is in order, to better characterize its features and issues, before studying the impact of edge caching on the performance of FD radios and networks.

13.1.1 Full-Duplex Communications

The majority of current wireless radios operate in HD mode. In practice, these devices perform data transmission and reception over separate time/frequency resources. Depending on the way such resources are used, we can have either TDD or frequency division duplex (FDD) operations, i.e., uplink (UL) and downlink (DL) transmissions occur over two different time or frequency resources, respectively. This approach has several advantages in terms of both ease of implementation and rather straightforward network operations to perform multi-cell transmissions. As a matter of fact, it can be argued that this implicitly sets a hard constraint on the spectral efficiency of the system. For this reason, many research efforts have been performed lately to investigate the potential and the feasibility of FD communications, in which the same time/frequency resource is used to perform the UL and DL data transmissions. However, the strong self-interference (SI) observed by the FD radio during the signal reception enforces a crucial obstacle to the feasibility of such approach. In other terms, a nonnegligible portion of the transmitted signal is always received by the device's receive chain, in turn reducing the signal-to-interference-plus-noise ratio (SINR) of the incoming useful signals [27–29]. This situation is depicted in Figure 13.2. Many different transceiver designs and self-interference cancellation (SIC) algorithms have been devised to ensure the feasibility of FD operations [28–38]. These solutions can be classified into two major categories based on passive or active cancellation. In the former case, SIC is achieved in the propagation domain by physical separation of the transmit and receive antennas. Conversely, active SIC solutions exploit the FD node's knowledge of its own transmitted signal to subtract it from the receive signal after appropriate manipulations and processing.

Unfortunately, the SI is not the only problem that system designers must face when dealing with FD radios. The major obstacle to their practical adoption in future 5G networks is arguably the aggregated interference footprint resulting from multiple and concurrent FD communications within the network. Let us provide an example to highlight this issue. Consider a simple network composed of several FD nodes arranged in BS/UE pairs and take an active BS/UE pair as reference. During UE-to-BS UL operations, every neighboring BS engaging in DL transmission strongly interferes with the considered BS, inducing the so called *BS-to-BS interference*. Similarly, during BS-to-UE DL operations, all the UEs performing UL transmission heavily interfere with the considered UE, creating the so called *UE-to-UE interference*, also referred to as INI [39, 40]. In practice, the FD throughput gain tends to 2 in case of very sparse deployment of nodes. Nevertheless, such gain saturates quickly as the network density

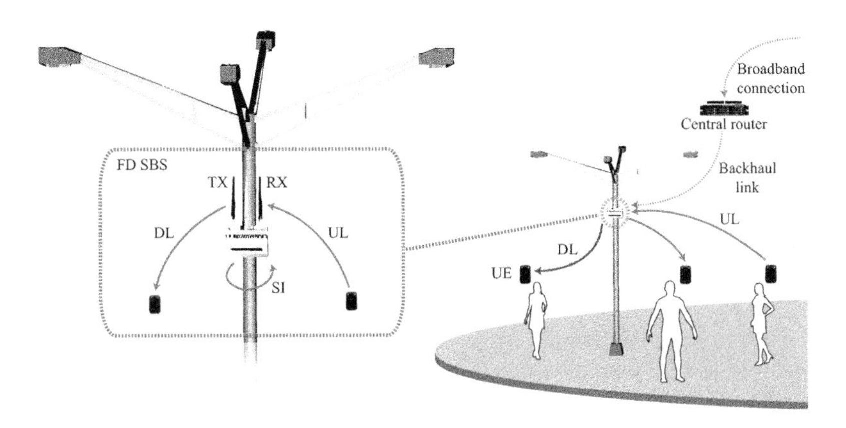

Figure 13.2 FD SBS. Data transmission and reception occur over the same time/frequency resource.

increases, the fundamental reason being that the number of interfering nodes also doubles with respect to the HD case. This becomes more significant when either the link distance decreases or the node density increases [41]. In other words, the theoretical throughput doubling brought by FD at the device level does not seem to materialize straightforwardly at the network level (regardless of the effectiveness of the adopted SIC algorithms), unless specific and possibly costly countermeasures are taken. As a result, the aggressive spatial frequency reuse inherent to dense network deployments may not be feasible due to the presence of a multitude of FD links mutually interfering at all times.

Studies and analysis of the FD interference footprint have been recently carried out to identify viable strategies to reduce it and improve the scalability of the FD throughput enhancement. Myriad approaches have been proposed to address this problem; indeed they range from user scheduling algorithms to advanced interference management and power control techniques. A common feature shared by such solutions is that they require the adoption of additional signaling among nodes or the implementation of heavy infrastructural changes [24, 25, 38–40, 42, 43]. In this context, two fundamental results can be highlighted: on the one hand, it is shown in [24] that most of the theoretical network throughput gain is achievable if only the BSs operate in FD while the UEs operate in HD, and centralized scheduling decisions are taken by a central unit enjoying full access to global system information; on the other hand, it is shown in [43] that, in case of distributed network control and operations, FD gains can be observed only if the UEs can exchange suitable information about INI over one or more orthogonal control links. Hence, the relevance of the aforementioned results is mostly theoretical, as the advocated infrastructural changes at the network level are extremely expensive. One of the goals of this chapter is to investigate the feasibility of a constructive alternative to such approaches: this considers the interactions between FD operations and smart caching strategies, and avoids any substantial changes to the network infrastructure and to the signaling exchange among nodes.

13.2 System Model

13.2.1 Network Model

Consider a UDN made up of (1) a tier of macro-cell BNs equipped with internet access, (2) a tier of SBSs providing network coverage, and (3) a set of mobile UEs. Each SBS communicates with only one BN in the UL direction and transmits contents to only one UE in the DL direction, functioning as a relay between the two. The SBSs operate in FD, whereas both BNs and UEs operate in HD mode; the same time/frequency resource is used for the communications in both directions. In the following, and focusing our attention on the SBSs, the BNs and the UEs are referred to as *UL nodes* and *DL nodes*, respectively.

Spatial random models allow us to seize the randomness of realistic ultra-dense small-cell deployments and, in addition, to derive tractable and accurate expressions for system-level performance analysis [44]. Therefore, we model the spatial distribution of the network nodes (i.e., SBSs and UL/DL nodes) using the homogeneous, independently marked PPP $\Phi \triangleq \{(x, u(x), d(x))\} \subset \mathbb{R}^2 \times \mathbb{R}^2 \times \mathbb{R}^2$. Here, we let $\Phi \triangleq \{x\}$ denote the ground PPP of the SBSs with spatial density λ (measured in [SBSs/m^2]), whereas the isotropic marks $\Phi_{\mathrm{UL}} \triangleq u(\Phi) = \{u(x)\}_{x \in \Phi}$ and $\Phi_{\mathrm{DL}} \triangleq d(\Phi) = \{d(x)\}_{x \in \Phi}$ denote the PPPs of the UL and DL nodes, respectively. Furthermore, let $r_{y,z} \triangleq \|y - z\|$ be the distance between nodes $y, z \in \Phi$; the distances of the UL and DL nodes from their associated SBSs are assumed fixed and are denoted by $R_{\mathrm{UL}} \triangleq r_{u(x),x}$ and $R_{\mathrm{DL}} \triangleq r_{x,d(x)}$, $\forall x \in \Phi$, respectively. It is thus evident that, according to these definitions, the PPPs Φ_{UL} and Φ_{DL} are dependent on the ground PPP Φ and have the same spatial density of the latter. Last, since the SBSs cover small areas compared with the BNs, one can reasonably assume that $R_{\mathrm{UL}} \gg R_{\mathrm{DL}}$. A snapshot of the considered two-tier network is given in Figure 13.3.

13.2.2 Cache-Aided Network Nodes

Let us assume that the UL nodes have direct access to the *global file catalog* $\mathcal{F} \triangleq \{f_1, f_2, \ldots, f_F\}$, with $|\mathcal{F}| = F$, which can be interpreted as a subset of all the contents available on the internet. Without loss of generality, we assume that all files have identical length, as files with different lengths can be always split into chunks of equal size. In this context, whenever a DL node sends a request for a content in $\mathcal{F}$, its serving SBS, operating in FD mode, fetches the corresponding file from the associated UL node and delivers it to the DL node. In UDN scenarios, however, the reliability of the content transmission may be reduced by the aggressive spatial frequency reuse, which may sensibly diminish the throughput with respect to an equivalent HD network.

Assume that SBS $x \in \Phi$ is equipped with a *storage unit* Δ_x with size $S < F$ files and that DL node $d(x)$ sends a request for file $f_i \in \mathcal{F}$. Let $\mathcal{P} \triangleq \{p_1, p_2, \ldots, p_F\}$, with $\sum_{i=1}^{F} p_i = 1$, be the set of request probabilities of each file, which depends on the files popularity over the whole network. Now, a *cache-hit* event occurs whenever $f_i \in \Delta_x$, i.e., if f_i is cached at SBS x. In this case, DL node $d(x)$ is served directly

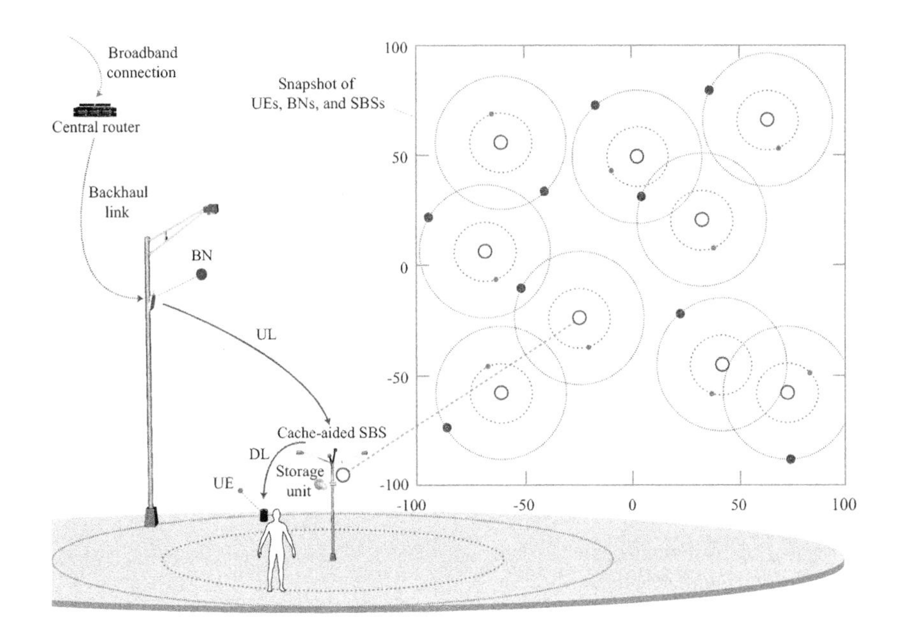

Figure 13.3 Snapshot of the marked Poison point process (PPP) modeling the SBSs, UL nodes, and DL nodes.

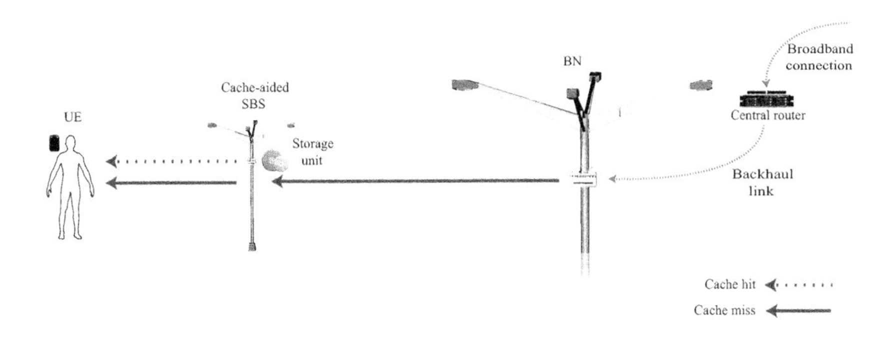

Figure 13.4 Communication links in case of cache hit and cache miss.

by SBS x without any communication between SBS x and UL node $u(x)$; alternatively, a *cache-miss* event occurs whenever $f_i \notin \Delta_x$, i.e., if f_i is not available in the cache, and SBS x must fetch the file from UL node $u(x)$ and deliver it to $d(x)$ in FD mode (see Figure 13.4). Thus a cache-hit event allows us to offload the overlaying macro-cell infrastructure and, since the UL becomes inactive, removes the need for the SBS to operate in FD mode. As a consequence, two major advantages can be observed in terms of reduced interference: (1) at the single-cell level, both the SI (at the SBS) and the INI (at the DL node) disappear, and (2) at the network level, the inter-cell interference is substantially reduced. Figure 13.5 provides a representation of the so-obtained scenario, whose interference terms are described in Section 13.2.4.

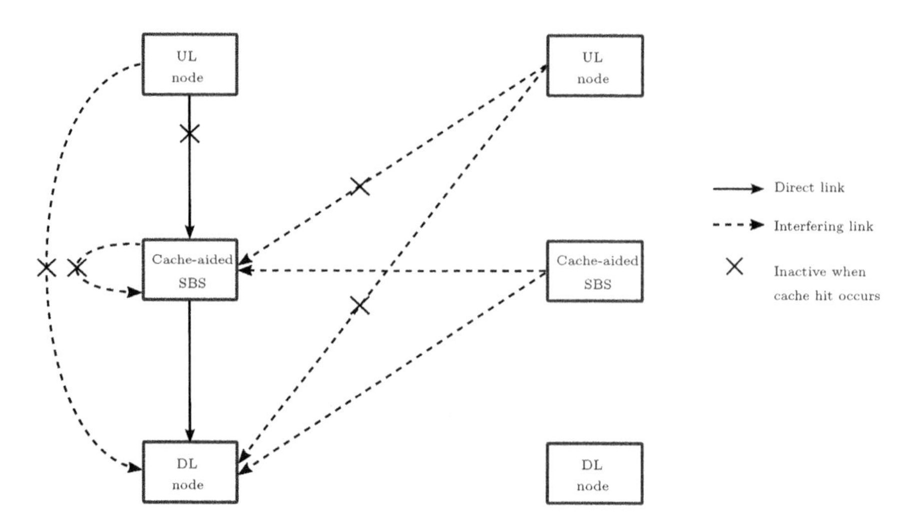

Figure 13.5 System model with cache-aided SBSs, UL nodes, and DL nodes, with corresponding direct and interfering links.

A key parameter to assess the effectiveness of the considered cache-aided approach is the *cache-hit probability*, denoted by $\mathsf{P}_{\mathrm{hit}}$, which is the probability that any file requested by a given DL node is cached at its associated SBS. The framework adopted in this chapter to model such probability is presented in Section 13.3. In particular, such framework is designed to capture the local files popularity in noncooperative random networks. Subsequently, we investigate the system-level performance gain brought by the deployment of cache-aided SBSs in FD networks for a given $\mathsf{P}_{\mathrm{hit}}$ in Section 13.4.

13.2.3 Channel Model

In the considered system model, we assume that all nodes are single-antenna devices; the extension of our study to multi-antenna settings goes beyond the scope of this chapter and can be accomplished using the analytical framework presented in [25, 45]. In addition, it is assumed that the UL nodes and the SBSs transmit with powers ρ_{UL} and ρ_{DL}, respectively.

The wireless channel propagation is characterized as the combination of two main parameters, i.e., large-scale pathloss attenuation and small-scale fading. Let $\ell(y,z) \triangleq r_{y,z}^{-\alpha}$ be the pathloss function between nodes y and z. We base our model upon the ITU-R urban micro-cellular (UMi) pathloss model described in [46], where different attenuations are specified for the links between different types of nodes. Accordingly, we let $\alpha = \alpha_2$ if $y \in \Phi_{\mathrm{UL}} \wedge z \in \Phi_{\mathrm{DL}}$ (i.e., between BNs and UEs) and $\alpha = \alpha_1$ otherwise (i.e., between BNs and SBSs as well as between SBSs and UEs). In this respect, we assume non-line-of-sight propagation between UL and DL nodes, which results in stronger pathloss attenuation as compared to the other links, and set $\alpha_2 \geq \alpha_1 > 2$. Switching the focus to the small-scale fading, let $h_{y,z}$ denote the channel power fading gain between nodes y and z. We assume that the SI channel is subject to Rician fading [27], whereas all the other channels are subject to Rayleigh fading. In other terms, we have that $h_{y,z} \sim \exp(1)$ if $y \neq z$ and $h_{y,y} \sim \Gamma(a,b)$. In particular, the shape parameter a and scale parameter b of the SI distribution can be computed in closed form from the Rician K-factor K and the SI attenuation Ω measured at the SBS when communicating in FD, as detailed in [25, lemma 1].

13.2.4 Signal-to-Interference Ratio

Massive and dense small-cell deployments, such as the one considered in this chapter, are often characterized by heavy inter-cell interference as a result of the very short inter-site distance [25]. As a consequence, it is meaningful to specifically focus on the interference-limited regime, where the noise is overwhelmed by interference. In this context, the definition of an appropriate metric, such as the measured SIR at the SBSs and at the DL nodes, is paramount to be able to capture the essential features of the interference-limited regime. We start by denoting a cache-miss event at SBS x with the notation $\not\Delta_x$ and accordingly define the indicator function

$$\mathbb{1}_{\not\Delta_x} \triangleq \begin{cases} 1, & \text{if } \not\Delta_x, \\ 0, & \text{otherwise.} \end{cases} \tag{13.1}$$

The SIR at SBS x may be written as

$$\mathtt{SIR}_x \triangleq \frac{\rho_{\mathrm{UL}} R_{\mathrm{UL}}^{-\alpha_1} h_{u(x),x}}{I_x}, \tag{13.2}$$

with aggregate interference given by

$$I_x \triangleq \sum_{y \in \Phi \setminus \{x\}} \left(\rho_{\mathrm{DL}} r_{y,x}^{-\alpha_1} h_{y,x} + \rho_{\mathrm{UL}} r_{u(y),x}^{-\alpha_1} h_{u(y),x} \mathbb{1}_{\not\Delta_y} \right) + h_{x,x} \mathbb{1}_{\not\Delta_x} \tag{13.3}$$

as the interference term. Similarly, the SIR at DL node $d(x)$ may be written as

$$\mathtt{SIR}_{d(x)} \triangleq \frac{\rho_{\mathrm{DL}} R_{\mathrm{DL}}^{-\alpha_1} h_{x,d(x)}}{I_{d(x)}}, \tag{13.4}$$

with aggregate interference given by

$$\begin{aligned} I_{d(x)} \triangleq & \sum_{y \in \Phi \setminus \{x\}} \left(\rho_{\mathrm{UL}} r_{y,d(x)}^{-\alpha_1} h_{y,d(x)} + \rho_{\mathrm{DL}} r_{u(y),d(x)}^{-\alpha_1} h_{u(y),d(x)} \mathbb{1}_{\not\Delta_y} \right) \\ & + \rho_{\mathrm{UL}} r_{u(x),d(x)}^{-\alpha_2} h_{u(x),d(x)} \mathbb{1}_{\not\Delta_x}. \end{aligned} \tag{13.5}$$

The effect of equipping the SBSs with storage capabilities and shifting from a cache-free to a cache-aided scenario is rather evident upon observing (13.3) and (13.5). More precisely, a cache-hit event induces a reduction of the following major interference components, at both the network and the device level:

- Aggregate network interference
- SI at the SBSs [27]
- INI at the DL nodes [39, 40]

We recall that the last two interference terms are the two main causes hindering the practical feasibility of FD technology at the network level.

13.3 Caching Model

A necessary step when performing studies on the performance of cache-aided networks is the definition of a caching model, whose role is to establish how files are requested and cached by DL nodes and SBSs, respectively [7]. Accordingly, this section introduces the noncooperative, static caching model used throughout this chapter, which aims at mimicking a geographical caching policy based on local files popularity.[1] In this regard, it is important to note that existing literature typically does not consider geographical aspects of the files popularity of the UEs when defining caching models (we refer to [12] for an overview on content request models).

Here, the spatial distribution of the contents from the global file catalog $\mathcal{F}$ is modeled by means of the homogeneous, independently marked PPP $\Psi \triangleq \big\{(y, f(y))\big\} \subset \mathbb{R}^2 \times \mathcal{F}$, where $\Psi_{\mathcal{F}} \triangleq \{y\}$ is the PPP of the files with spatial density η (measured in [files/m^2]). In this context, each file $f_i \in \mathcal{F}$ corresponds to a thinned PPP with spatial density $p_i \eta$. Moreover, we assume that the files in $\mathcal{F}$ are ordered by decreasing popularity, i.e., $p_1 \geq p_2 \geq \cdots \geq p_F$. The considered caching model consists of two core concepts, i.e., the *request region* and the *caching policy*, which describe how DL nodes request files and how SBSs cache files, respectively. The introduction of the last notation is in order to be able to explicitly add a geographical dimension to these two concepts. Accordingly, we let $\mathcal{B}(z, v)$ denote the ball of radius v (measured in [m]) centered at node $z \in \Phi \cup \Phi_{\mathrm{DL}}$.

DEFINITION 13.1 (Request region) *Assume that DL node $d(x) \in \Phi_{\mathrm{DL}}$ is interested in requesting locally popular files. Then the request region of DL node $d(x)$ is defined as*

$$\mathcal{R}_{d(x)} \triangleq \big\{\Psi_{\mathcal{F}} \cap \mathcal{B}(d(x), R_{\mathrm{R}})\big\}, \tag{13.6}$$

with R_{R} defined as the radius of the request region.

REMARK 1 *From a qualitative point of view, R_{R} is related to the local interests of the UEs with respect to globally requested files. In other terms, if DL node $d(x)$ is interested in requesting all possible files in the global file catalog $\mathcal{F}$, then $R_{\mathrm{R}} \to \infty$ (provided that $\{p_i > 0\}_{i=1}^{F}$).*

DEFINITION 13.2 (Caching policy) *Assume that SBS $x \in \Phi$ is interested in caching locally popular files. Then the potential cache region is defined as*

$$\mathcal{C}_x \triangleq \big\{\Psi_{\mathcal{F}} \cap \mathcal{B}(x, R_{\mathrm{C}})\big\}, \tag{13.7}$$

[1] More complex cooperative caching policies can be devised. However, this goes beyond the scope of this chapter.

with R_{C} defined as the radius of the potential cache region. The caching policy of SBS $x \in \Phi$ is defined as

$$\Delta_x \triangleq \{f_i : f_i \in \mathcal{C}_x \wedge i \leq S\}. \tag{13.8}$$

REMARK 2 *SBSs operating according to such caching policy will cache only geographically close (and, therefore, popular) files, in turn aiming at reducing the overhead associated with pre-fetching files from the BNs.*

REMARK 3 *Similarly to what has been previously observed for the request region, as $R_{\mathrm{C}} \to \infty$, we note that such caching policy will always converge to storing globally popular files as in* [7].

Finally, the following lemma formalizes the cache-hit probability under the described caching model.

LEMMA 13.3 *The cache-hit probability is given by*

$$\mathsf{P}_{\mathrm{hit}} = \frac{1}{F} \sum_{i=1}^{S} \left(1 - e^{-p_i \eta \pi R_{\mathrm{R}}^2}\right)\left(1 - e^{-p_i \eta \pi R_{\mathrm{C}}^2}\right). \tag{13.9}$$

Proof By definition, each file f_i is distributed according to a thinned PPP with spatial density $p_i \eta$. Hence we can straightforwardly infer that the probabilities of f_i falling independently into the request region $\mathcal{R}_{d(x)}$ and into the potential cache region $\mathcal{C}_x$ are $1 - e^{-p_i \eta \pi R_{\mathrm{R}}^2}$ and $1 - e^{-p_i \eta \pi R_{\mathrm{C}}^2}$, respectively. Assume now that $S \to \infty$, i.e., the SBSs are equipped with unlimited storage. Then, in this case, the cache-hit probability of file f_i can be derived as the probability of file f_i falling into both the request region and the potential cache region, which is readily given by $\left(1 - e^{-p_i \eta \pi R_{\mathrm{R}}^2}\right)\left(1 - e^{-p_i \eta \pi R_{\mathrm{C}}^2}\right)$. Finally, considering the totality of the contents included in the global file catalog $\mathcal{F}$ and imposing storage constrains from Definition 13.2 yields the expression in (13.9). □

REMARK 4 *Note that, in noncooperative caching settings, the maximization of* $\mathsf{P}_{\mathrm{hit}}$ *is straightforwardly achieved by caching the S most popular files at the SBSs.*

13.4 Performance Analysis

In this section, we use tools from stochastic geometry to analyze the system-level performance enhancements brought by the considered cache-aided FD network over its cache-free counterpart. This choice provides analytical tractability of the problem and is crucial to guarantee the generality of our results. As a main performance metric, we study the probability that a DL node successfully receives a requested content, either through a direct transmission from its associated SBS or with the aid of the corresponding UL node. We term this metric as *probability of successful transmission*, which is denoted by $\mathsf{P}_{\mathrm{suc}}(\cdot)$. In this context, it is convenient to recall that the delivery of a requested file will be performed over different links depending on the occurrence of a cache-hit event. In particular:

- *Cache-hit event*: the transmission involves one hop, i.e., from the SBS to the DL node.
- *Cache-miss event*: the transmission requires two hops, i.e., first from the UL node to the SBS and then from the latter to the DL node through the SBS (which introduces additional interference).

In our analysis, we focus on a *typical SBS*, indexed by x, and its marks $u(x)$ and $d(x)$, referred to as *typical UL node* and *typical DL node*, respectively. Building on Slivnyak's theorem [44, chapter 8.5] and on the stationarity of Φ (resp. of Φ_{DL}), the statistics of the typical SBS's (resp. of the typical DL node's) signal reception are representative of the statistics seen by any SBS (resp. by any DL node) in the system.

Switching our focus back to $\mathsf{P}_{\mathrm{suc}}(\cdot)$, we consider that a requested file is successfully received by the typical DL node (i.e., through the two-hop communication link involving the typical UL node, the typical SBS, and the typical DL node) if $\mathtt{SIR}_x > \theta \wedge \mathtt{SIR}_{d(x)} > \theta$, with θ defined as a target SIR threshold. Additionally, we consider that the correct reception of the requested file over one hop uniquely depends on the SIR experienced at the receiver, regardless of the considered hop. For simplicity, and without loss of generality, we assume the same SIR threshold for both UL and DL directions.

Now, thanks to the caching capabilities at the typical SBS, we can state that the UL communication does not occur with probability $\mathsf{P}_{\mathrm{hit}}$. We can then express the probability of successful transmission as

$$\mathsf{P}_{\mathrm{suc}}(\theta) \triangleq \mathsf{P}_{\mathrm{hit}}\mathbb{P}(\mathtt{SIR}_{d(x)} > \theta) + (1 - \mathsf{P}_{\mathrm{hit}})\mathbb{P}(\mathtt{SIR}_x > \theta, \mathtt{SIR}_{d(x)} > \theta). \tag{13.10}$$

Building upon this definition, other useful performance metrics can be expressed in terms of probability of successful transmission. Noteworthy examples are the outage probability, given by $\mathsf{P}_{\mathrm{out}}(\theta) \triangleq 1 - \mathsf{P}_{\mathrm{suc}}(\theta)$, and the achievable areal spectral efficiency (ASE), defined as $\mathsf{ASE}(\theta) \triangleq \lambda \mathsf{P}_{\mathrm{suc}}(\theta)\log_2(1+\theta)$ (measured in [bps/Hz/m^2]).

Before proceeding with our analysis, we provide some useful preliminary definitions for the sake of notational simplicity in the remainder of the section:

$$\widehat{\Upsilon}(s) \triangleq \frac{\pi(s\rho_{\mathrm{DL}})^{\frac{2}{\alpha_1}} \csc\left(\frac{2\pi}{\alpha_1}\right)}{\alpha_1}, \tag{13.11}$$

$$\widetilde{\Upsilon}(s) \triangleq \int_0^\infty \left(1 - \frac{1}{1 + s\rho_{\mathrm{DL}} r^{-\alpha_1}} \Xi(s,r)\right) r \mathrm{d}r, \tag{13.12}$$

$$\Xi(s,r) \triangleq \frac{1}{2\pi}\int_0^{2\pi} \frac{\mathrm{d}\varphi}{1 + s\rho_{\mathrm{UL}}(R_{\mathrm{UL}}^2 + r^2 + 2R_{\mathrm{UL}} r \cos\varphi)^{-\frac{\alpha_2}{2}}}. \tag{13.13}$$

Recalling the expressions of I_x and $I_{d(x)}$ in (13.3) and (13.5), respectively, a tight analytical lower bound on $\mathsf{P}_{\mathrm{suc}}(\theta)$ is provided next in Theorem 13.4, with additional properties given in Corollary 13.4.

THEOREM 13.4 *The probability of successful transmission in* (13.10) *is bounded as*$\mathsf{P}_{\mathrm{suc}}(\theta) \geq \underline{\mathsf{P}}_{\mathrm{suc}}(\theta)$, *with*

$$\underline{\mathsf{P}}_{\mathrm{suc}}(\theta) \triangleq \mathsf{P}_{\mathrm{hit}}\mathcal{L}_{I_{d(x)}}(\theta\rho_{\mathrm{DL}}^{-1}R_{\mathrm{DL}}^{\alpha_1}) + (1-\mathsf{P}_{\mathrm{hit}})\mathcal{L}_{I_x}^{\mathbb{A}_x}(\theta\rho_{\mathrm{UL}}^{-1}R_{\mathrm{UL}}^{\alpha_1})\mathcal{L}_{I_{d(x)}}^{\mathbb{A}_x}(\theta\rho_{\mathrm{DL}}^{-1}R_{\mathrm{DL}}^{\alpha_1}), \tag{13.14}$$

where $\mathcal{L}_{I_{d(x)}}(s)$ is the Laplace transform of the interference observed at DL node $d(x)$ in case of cache hit, whereas $\mathcal{L}_{I_x}^{\mathbb{A}_x}(s)$ and $\mathcal{L}_{I_{d(x)}}^{\mathbb{A}_x}(s)$ are the Laplace transforms of the interference observed at SBS x and at DL node $d(x)$, respectively, in case of cache miss:

$$\mathcal{L}_{I_{d(x)}}(s) \triangleq \exp\big(-2\pi\lambda\mathsf{P}_{\mathrm{hit}}\widehat{\Upsilon}(s)\big)\exp\big(-2\pi\lambda(1-\mathsf{P}_{\mathrm{hit}})\widetilde{\Upsilon}(s)\big), \tag{13.15}$$

$$\mathcal{L}_{I_x}^{\mathbb{A}_x}(s) \triangleq \frac{1}{(1+s\rho_{\mathrm{DL}}b)^a}\mathcal{L}_{I_{d(x)}}(s), \tag{13.16}$$

$$\mathcal{L}_{I_{d(x)}}^{\mathbb{A}_x}(s) \triangleq \Xi(s,R_{\mathrm{DL}})\mathcal{L}_{I_{d(x)}}(s). \tag{13.17}$$

Proof The construction of (13.14) relies on the assumption of uncorrelated locations of the UL and DL nodes in presence of a cache-miss event. As a matter of fact, according to the Fortuin–Kasteleyn–Ginibre (FKG) inequality [44, chapter 10.4.2], such an uncorrelated case yields a lower bound on the network performance for the correlated case (we refer to [25] for further details). Therefore, given $\mathsf{P}_{\mathrm{suc}}(\theta)$ in (13.10), we can write

$$\mathsf{P}_{\mathrm{suc}}(\theta) \geq \mathsf{P}_{\mathrm{hit}}\mathsf{P}_{\mathrm{suc},2}(\theta) + (1-\mathsf{P}_{\mathrm{hit}})\mathsf{P}_{\mathrm{suc},1}^{\mathbb{A}_x}(\theta)\mathsf{P}_{\mathrm{suc},2}^{\mathbb{A}_x}(\theta), \tag{13.18}$$

where $\mathsf{P}_{\mathrm{suc},2}(\theta)$ represents the probability of successfully transmitting a requested file from the typical SBS to the typical DL node in case of a cache-hit event, and $\mathsf{P}_{\mathrm{suc},1}^{\mathbb{A}_x}(\theta)$ (resp. $\mathsf{P}_{\mathrm{suc},2}^{\mathbb{A}_x}(\theta)$) denotes the probability of successfully transmitting a requested file from the typical UL node to the typical SBS (resp. from the typical SBS to the typical DL node) in case of a cache-miss event.

We begin by focusing on the latter components, i.e., $\mathsf{P}_{\mathrm{suc},1}^{\mathbb{A}_x}(\theta)$ and $\mathsf{P}_{\mathrm{suc},2}^{\mathbb{A}_x}(\theta)$. In particular, $\mathsf{P}_{\mathrm{suc},1}^{\mathbb{A}_x}(\theta)$ is obtained as the Laplace transform of I_x in (13.3) in presence of SI [25, theorem 1], which is given by $\mathcal{L}_{I_x}^{\mathbb{A}_x}(s)$ in (13.16). Likewise, $\mathsf{P}_{\mathrm{suc},2}^{\mathbb{A}_x}(\theta)$ is obtained as the Laplace transform of $I_{d(x)}$ in (13.5) in presence of INI, which is given by $\mathcal{L}_{I_{d(x)}}^{\mathbb{A}_x}(s)$ in (13.17). Following a similar approach, $\mathsf{P}_{\mathrm{suc},2}(\theta)$ can be obtained as the Laplace transform of $I_{d(x)}$ in (13.5) in absence of INI, which is given by $\mathcal{L}_{I_{d(x)}}(s)$. Now, let us define $\widetilde{\Phi} \triangleq \{x \in \Phi : \mathbb{A}_x\}$ and $\widehat{\Phi} \triangleq \Phi\backslash\widetilde{\Phi}$, which are, by definition, independent PPPs with spatial densities $\mathsf{P}_{\mathrm{hit}}\lambda$ and $(1-\mathsf{P}_{\mathrm{hit}})\lambda$, respectively. As a consequence, $\mathcal{L}_{I_{d(x)}}(s)$ in (13.15) can be derived as

$$\mathcal{L}_{I_{d(x)}}(s) = E[e^{-sI_{d(x)}}] \tag{13.19}$$

$$= E\Bigg[\exp\Bigg(-s\sum_{y\in\Phi\backslash\{x\}}\big(\rho_{\mathrm{DL}}r_{yd(x)}^{-\alpha_1}h_{yd(x)} + \rho_{\mathrm{UL}}r_{u(y)d(x)}^{-\alpha_1}h_{u(y)d(x)}\mathbb{1}_{\mathbb{A}_y}\big)\Bigg)\Bigg] \tag{13.20}$$

$$= E\Bigg[\prod_{y\in\Phi\backslash\{x\}}\exp\Big(-s\big(\rho_{\mathrm{DL}}r_{yd(x)}^{-\alpha_1}h_{yd(x)} + \rho_{\mathrm{UL}}r_{u(y)d(x)}^{-\alpha_1}h_{u(y)d(x)}\mathbb{1}_{\mathbb{A}_x}\big)\Big)\Bigg] \tag{13.21}$$

$$= E\Big[\prod_{y\in\widehat{\Phi}\setminus\{x\}} \exp\big(-s\rho_{\mathrm{DL}} r_{yd(x)}^{-\alpha_1} h_{yd(x)}\big)\Big]$$
$$\times E\Big[\prod_{y\in\widetilde{\Phi}\setminus\{x\}} \exp\Big(-s\big(\rho_{\mathrm{DL}} r_{yd(x)}^{-\alpha_1} h_{yd(x)} + \rho_{\mathrm{UL}} r_{u(y)d(x)}^{-\alpha_1} h_{u(y)d(x)}\big)\Big)\Big], \tag{13.22}$$

and, using the moment-generating function of the exponential distribution, we obtain

$$\mathcal{L}_{I_{d(x)}}(s) = E_{\widehat{\Phi}}\Big[\prod_{y\in\widehat{\Phi}\setminus\{x\}} \frac{1}{1+s\rho_{\mathrm{DL}} r_{yd(x)}^{-\alpha_1}}\Big]$$
$$\times E_{\widetilde{\Phi}}\Big[\prod_{y\in\widetilde{\Phi}\setminus\{x\}} \frac{1}{1+s\rho_{\mathrm{DL}} r_{yd(x)}^{-\alpha_1}} \frac{1}{1+\rho_{\mathrm{UL}} r_{u(y)d(x)}^{-\alpha_1}}\Big]. \tag{13.23}$$

Then applying the probability generating functional of a PPP [44, chapter 4.3] yields

$$\mathcal{L}_{I_{d(x)}}(s) = \exp\Big(-2\pi\lambda \mathsf{P}_{\mathrm{hit}} \int_0^\infty \Big(1-\frac{1}{1+s\rho_{\mathrm{DL}} r^{-\alpha_1}}\Big) r\,\mathrm{d}r\Big)$$
$$\times \exp\Big(-2\pi\lambda(1-\mathsf{P}_{\mathrm{hit}}) \int_0^\infty \Big(1-\frac{1}{1+s\rho_{\mathrm{DL}} r^{-\alpha_1}}\,\Xi(s,r)\Big) r\,\mathrm{d}r\Big). \tag{13.24}$$

Finally, the integral appearing in the first exponential of (13.24) has a closed-form solution given by $\widehat{\Upsilon}(s)$ in (13.11). This concludes the proof. □

COROLLARY *The lower bound on the probability of successful transmission in* (13.14) *is characterized by the following properties:*

(1) $\underline{\mathsf{P}}_{\mathrm{suc}}(\theta) \to \mathsf{P}_{\mathrm{suc}}(\theta)$ *as* $\mathsf{P}_{\mathrm{hit}} \to 1$.
(2) $\underline{\mathsf{P}}_{\mathrm{suc}}(\theta) = \mathsf{P}_{\mathrm{suc}}(\theta)$ *in case of uncorrelated locations of the nodes between UL and DL communications.*

Proof The proof follows directly from Theorem 13.4. □

Last, we introduce the *FD throughput gain*, which will be used as a performance metric in Section 13.5, defined as

$$\mathsf{TG}_{\mathrm{FD}}(\theta) \triangleq 2\mathsf{P}_{\mathrm{suc}}(\theta) \exp\Bigg(2\pi\lambda \frac{\pi\theta^{\frac{2}{\alpha_1}}\big(R_{\mathrm{UL}}^2 + R_{\mathrm{DL}}^2\big)\csc\big(\frac{2\pi}{\alpha_1}\big)}{\alpha_1}\Bigg). \tag{13.25}$$

This performance metric quantifies the throughput gain of a cache-aided small-cell network operating in FD mode as compared to its cache-free HD counterpart by relating the probability of successful transmission $\mathsf{P}_{\mathrm{suc}}(\theta)$ in the two settings and taking into account the theoretical FD throughput doubling (we refer to [25] for details). In particular, we note that the FD setting outperforms its HD counterpart when $\mathsf{TG}_{\mathrm{FD}}(\theta) > 1$.

13.5 Numerical Results and Discussion

In this section, we present and discuss numerical results obtained by means of suitable Monte Carlo simulations in order to assess the validity of our theoretical findings. We specifically focus on the analytical expressions obtained in Sections 13.3 and 13.4.

As commonly assumed in the literature (see, e.g., [12]), the global file catalog follows a Zipf popularity distribution such that the request probability $p_i \in \mathcal{P}$ of each file $f_i \in \mathcal{F}$ can be written as

$$p_i = \left(i^{\gamma} \sum_{j=1}^{F} \frac{1}{j^{\gamma}} \right)^{-1} \tag{13.26}$$

for a certain catalog shape parameter γ. The SBSs cache contents from the global file catalog (depending on the policy defined in Definition 13.2) serve the corresponding UEs accordingly. The corresponding storage-to-catalog ratio is defined as $\kappa \triangleq \frac{S}{F} \leq 1$. The values of the most relevant parameters adopted for the simulations are listed in Table 13.1; furthermore, the shape parameter a and the scale parameter b of the SI distribution, which appear in (13.16), are computed from the Rician K-factor K and the SI attenuation Ω measured at the FD SBSs as in [25, lemma 1].

The probability of successful transmission $\mathsf{P}_{\text{suc}}(\theta)$ in (13.10) and its analytical lower bound $\underline{\mathsf{P}}_{\text{suc}}(\theta)$ in (13.14) are illustrated in Figure 13.6 as functions of the file request density η for a fixed SBS density $\lambda = 5 \times 10^{-4}$ SBSs/m^2. With reference to Corollary 13.4, we recall that $\underline{\mathsf{P}}_{\text{suc}}(\theta)$ gives the exact expression of $\mathsf{P}_{\text{suc}}(\theta)$ for uncorrelated locations of the nodes between UL and DL phases. Moreover, note that the curves for $\kappa = 0$ are related to the cache-free scenario analyzed in [25] (see also Figures 13.7 and 13.8). Qualitatively, it is evident from Figure 13.6 that the probability of successful transmission grows with both the file request density η and the storage-to-catalog ratio κ. On the one hand, we observe

Table 13.1 System Parameters Used in the Simulations

System Parameter	Symbol	Value
Radius of request region	R_{R}	8 m
Radius of potential cache region	R_{C}	40 m
Catalog shape parameter	γ	0.7
Storage-to-catalog ratio	κ	$\{0.1, 0.35, 0.6\}$
Distance UL node–SBS	R_{UL}	20 m
Distance SBS–DL node	R_{DL}	5 m
Transmit power of UL nodes	ρ_{UL}	30 dBm
Transmit power of DL nodes	ρ_{DL}	24 dBm
Pathloss exponent UL nodes–SBSs/SBSs–DL nodes	α_1	3
Pathloss exponent UL nodes–DL nodes	α_2	4
Target SIR	θ	0 dB
Rician K-factor	K	1
SI attenuation	Ω	60 dB

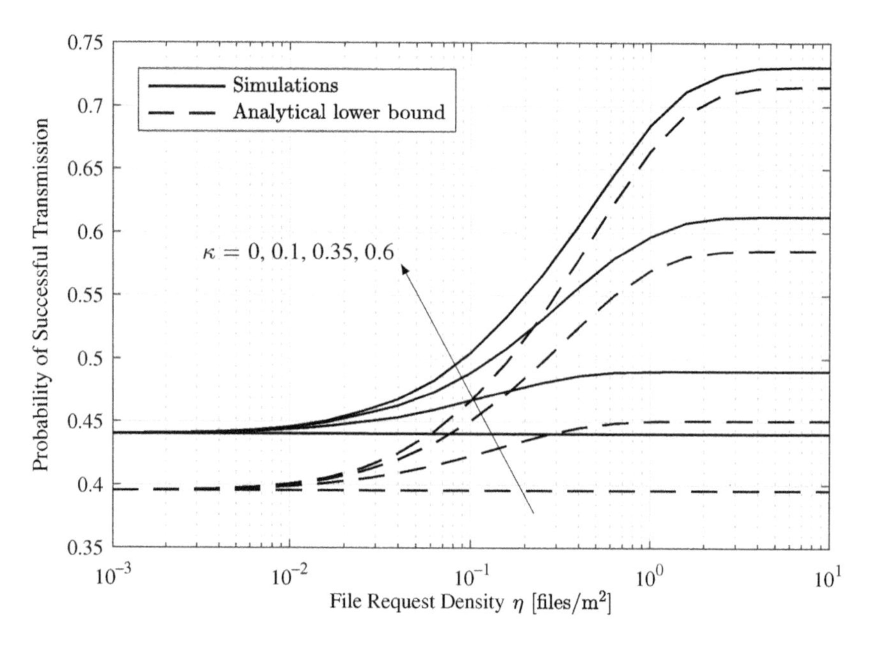

Figure 13.6 Probability of successful transmission against file request density, with SBS density $\lambda = 5 \times 10^{-4}$ SBSs/m^2.

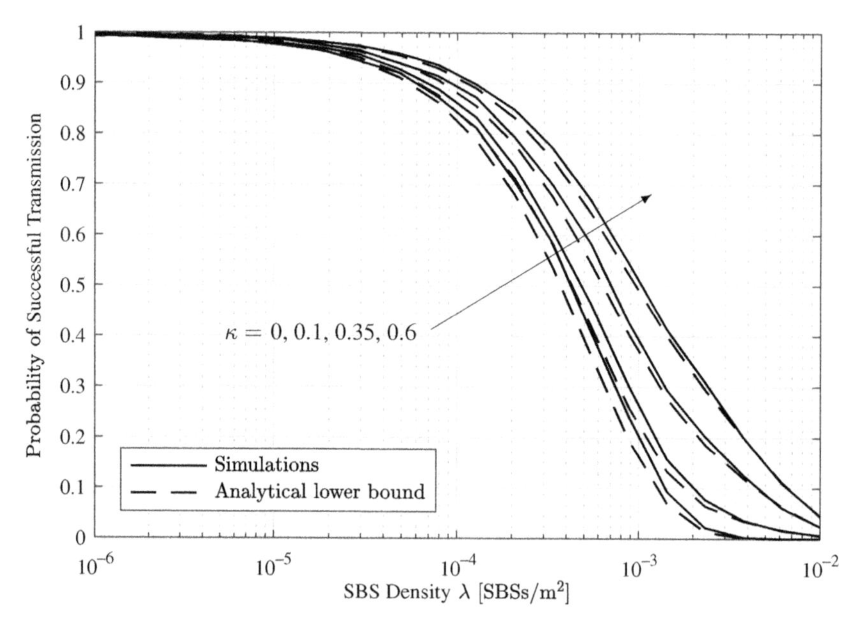

Figure 13.7 Probability of successful transmission against SBS density, with file request density $\eta = 1$ files/m^2.

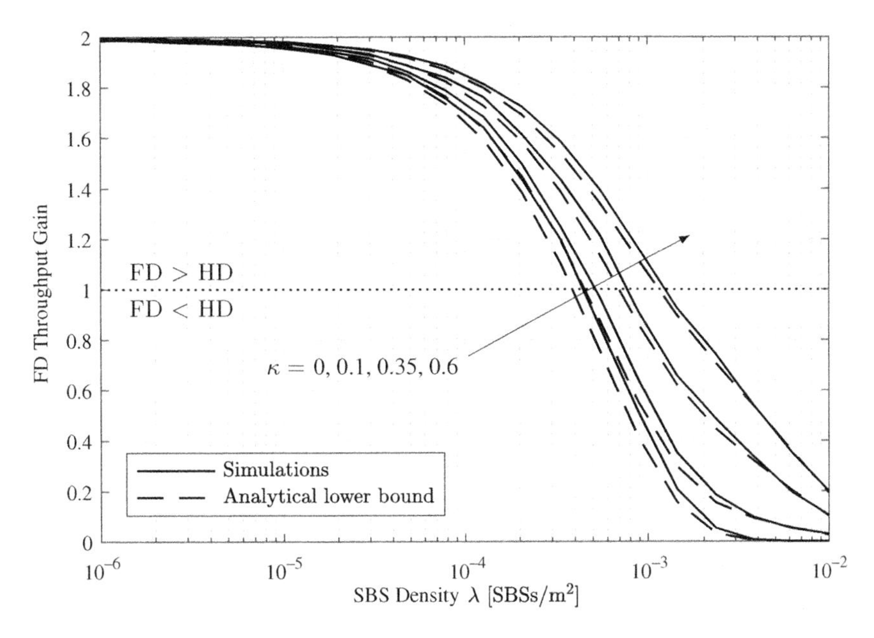

Figure 13.8 FD throughput gain against SBS density, with file request density $\eta = 1$ files/m^2.

that a higher file request density yields a larger P_{hit}, which in turn improves the efficiency of the storage use. On the other hand, the variation experienced by the probability of successful transmission over η increases with the storage capabilities at the SBSs, and so does the tightness of the analytical lower bound. Concerning this last aspect, we note that, even if such bound may look rather loose for $\kappa \leq 0.35$, its quantitative difference with the actual numerical performance never exceeds 10%.

Assume now a file request density $\eta = 1$ files/m^2. Figure 13.7 plots the probability of successful transmission $\mathsf{P}_{\text{suc}}(\theta)$ in (13.10) and its analytical lower bound $\underline{\mathsf{P}}_{\text{suc}}(\theta)$ in (13.14) as functions of the SBS density λ. The analytical lower bound is remarkably tight and, in accordance with Figure 13.6, becomes increasingly accurate as the storage-to-catalog ratio κ grows. Nonetheless, it is even more meaningful to analyze the FD throughout gain in (13.25) together with its analytical lower bound (obtained by replacing $\mathsf{P}_{\text{suc}}(\theta)$ with $\underline{\mathsf{P}}_{\text{suc}}(\theta)$ in the aforementioned expression), which are illustrated in Figure 13.8 against the SBS density λ. In practice, higher ASE can be achieved by deploying a very dense FD network in which each SBS is equipped with suitable caching capabilities. In this respect, we observe that:

- A SBS density $\lambda = 10^{-4}$ SBSs/m^2 yields $\mathsf{TG}_{\text{FD}}(\theta) = 1.7$ with $\kappa = 0$ and $\mathsf{TG}_{\text{FD}}(\theta) = 1.85$ with $\kappa = 0.6$.
- A SBS density $\lambda = 10^{-3}$ SBSs/m^2 yields $\mathsf{TG}_{\text{FD}}(\theta) = 0.42$ with $\kappa = 0$ and $\mathsf{TG}_{\text{FD}}(\theta) = 1.11$ with $\kappa = 0.6$.

It is evident that the optimal trade-off between the SBS density and the storage size installed at each SBS must be found by network planners taking into account the deployment cost of each element; the interested reader may refer to [16] for further details on this subject.

13.6 Conclusions

Several research efforts in both academic and industrial contexts have highlighted that edge caching can provide significant benefits in terms of network performance as, e.g., end-to-end access delay. Conversely, very few straightforward insights can be drawn on the benefits experienced at the physical layer when the network nodes are equipped with caching capabilities. This is mostly due to the complexity of the physical interactions occurring among devices in modern network.

This chapter takes a step forward with respect to the aforementioned position by showing that edge caching can actually offer a remarkable degree of interoperability with one of the most promising technologies for next-generation network deployments, i.e., FD communications. More specifically, we show that integrating caching capabilities at the FD SBSs is a cost-effective means of improving the scalability of the theoretical throughput doubling brought by the FD paradigm from the device to the network level.

Our study considers an interference-limited UDN setting consisting of several non-cooperative SBSs with FD capabilities, which simultaneously communicate with both their served UEs and wireless BNs. In this case, the interference footprint of the UDN, already significant by design, is further increased by the FD operations. In fact, the latter induce higher levels of inter-cell and inter-node interference as compared to the HD scenario, in turn causing a spectral efficiency bottleneck that prevents the theoretical FD throughput doubling to occur at the network level. Fundamental results available in the literature show that most of such doubling can be achieved only if the network infrastructure is subject to radical and expensive modifications or if high-rate signaling is exchanged between UEs over suitable control links.

In this context, we add file storage capabilities to SBSs and consider a geographical caching policy aiming at capturing local files popularity, whereby the SBSs intelligently store popular contents anticipating the UEs' requests. The rationale of this choice is that the presence of prefetched popular files at the SBSs reduces the need for the latter to retrieve contents from the wireless BNs upon the UEs' request. This clearly diminishes the number of transmissions performed by the BNs toward the SBSs, in turn reducing the interference footprint of the UDN. Remarkably, this low-cost solution can be implemented without the need for additional signaling between the nodes or any infrastructural change.

The performance of such a cache-aided FD network is characterized in terms of throughput gain as compared to its HD counterpart. To this end, two fundamental metrics are identified and analyzed:

- The probability that any file requested by a given DL node is cached at its serving SBS, which is termed as *cache-hit probability*.
- The probability of successful transmission of a file requested by a UE, either directly by its associated SBS (if present in its cache) or by the corresponding BN.

In particular, the second metric is used to derive an analytical lower bound on the throughput of the UDN. As a final step, we perform a set of suitable numerical simulations to assess the performance enhancement brought by the adoption of edge caching. The obtained results highlight that shifting from cache-free to cache-aided UDNs allows us to effectively operate the network in FD mode while supporting higher SBS densities, in turn improving the ASE. In other words, the deployment of cache-aided SBSs has beneficial effects on the network throughput experienced over a given area, thanks to a nonnegligible reduction of the aggregate interference observed in the FD network in comparison with the cache-free case.

The results presented in this chapter demonstrate that the interoperability between edge caching and FD communications is not only possible but also desirable from the network throughput perspective. However, from a quantitative point of view, the extent of the benefits may strongly depend on several parameters, such as:

- Adopted caching policy;
- UEs association policy; and
- Mobility of the network nodes, either in the form of moving SBSs or classic UEs' dynamics.

Therefore, future additional studies and investigations should be performed in these directions to further deepen our understanding of the benefits brought by edge caching to the physical layer of wireless communication networks, especially when the FD paradigm is adopted.

References

[1] "Cisco visual networking index: global mobile data traffic forecast, 2016–2021," Cisco, White Paper, 2017.

[2] J. G. Andrews, S. Buzzi, W. Choi, S. V. Hanly, A. Lozano, A. C. K. Soong, and J. C. Zhang, "What will 5G be?" *IEEE J. Sel. Areas Commun.*, vol. 32, no. 6, pp. 1065–1082, June 2014.

[3] "TR 38.211 (V15.1.0): NR; physical channels and modulation," 3GPP, technical report, Mar. 2018.

[4] "TR 38.214 (V15.1.0): NR; physical layer procedures for data," 3GPP, Technical Report, Mar. 2018.

[5] F. Boccardi, R. W. Heath, A. Lozano, T. L. Marzetta, and P. Popovski, "Five disruptive technology directions for 5G," *IEEE Commun. Mag.*, vol. 52, no. 2, pp. 74–80, Feb. 2014.

[6] E. Björnson, J. Hoydis, and L. Sanguinetti, "Massive MIMO networks: spectral, energy, and hardware efficiency," *Found. Trends Signal Process.*, vol. 11, no. 3–4, pp. 154–655, 2017.

[7] E. Baştuğ, M. Bennis, M. Kountouris, and M. Debbah, "Cache-enabled small cell networks: modeling and tradeoffs," *EURASIP J. Wireless Commun. Netw.*, no. 1, pp. 41–51, 2015.

[8] K. Li, C. Yang, Z. Chen, and M. Tao, "Optimization and analysis of probabilistic caching in n-tier heterogeneous networks," *IEEE Trans. Wireless Commun.*, vol. 17, no. 2, pp. 1283–1297, Feb. 2018.

[9] K. Hamidouche, W. Saad, M. Debbah, and H. V. Poor, "Mean-field games for distributed caching in ultra-dense small cell networks," in *Proc. IEEE American Control Conf. (ACC)*, July 2016.

[10] C. Yang, B. Xia, W. Xie, K. Huang, Y. Yao, and Y. Zhao, "Interference cancellation at receivers in cache-enabled wireless networks," *IEEE Trans. Veh. Technol.*, vol. 67, no. 1, pp. 842–846, Jan. 2018.

[11] S. Krishnan and H. S. Dhillon, "Effect of user mobility on the performance of device-to-device networks with distributed caching," vol. 6, no. 2, pp. 194–197, Apr. 2017.

[12] G. Paschos, E. Baştuğ, I. Land, G. Caire, and M. Debbah, "Wireless caching: technical misconceptions and business barriers," *IEEE Commun. Mag.*, vol. 54, no. 8, pp. 16–22, Aug. 2016.

[13] D. Liu and C. Yang, "Caching policy toward maximal success probability and area spectral efficiency of cache-enabled HetNets," *IEEE Trans. Commun.*, vol. 65, no. 6, pp. 2699–2714, June 2017.

[14] A. Khreishah, J. Chakareski, and A. Gharaibeh, "Joint caching, routing, and channel assignment for collaborative small-cell cellular networks," *IEEE J. Sel. Areas Commun.*, vol. 34, no. 8, pp. 2275–2284, Aug. 2016.

[15] M. Maso, I. Atzeni, I. Ghamnia, E. Baştuğ, and M. Debbah, "Cache-aided full-duplex small cells," in *Proc. Int. Symp. Modeling Optimiz. Mobile, Ad Hoc, Wireless Netw. (WiOpt)*, May 2017.

[16] I. Atzeni, M. Maso, I. Ghamnia, E. Baştuğ, and M. Debbah, "Flexible cache-aided networks with backhauling," in *Proc. IEEE Int. Workshop Signal Process. Adv. Wireless Commun. (SPAWC)*, July 2017.

[17] J. G. Andrews, F. Baccelli, and R. K. Ganti, "A tractable approach to coverage and rate in cellular networks," *IEEE Trans. Commun.*, vol. 59, no. 11, pp. 3122–3134, Nov. 2011.

[18] H. S. Dhillon, R. K. Ganti, F. Baccelli, and J. G. Andrews, "Modeling and analysis of K-tier downlink heterogeneous cellular networks," *IEEE J. Sel. Areas Commun.*, vol. 30, no. 3, pp. 550–560, Apr. 2012.

[19] S. F. Yunas, M. Valkama, and J. Niemelä, "Spectral and energy efficiency of ultra-dense networks under different deployment strategies," *IEEE Commun. Mag.*, vol. 53, no. 1, pp. 90–100, Jan. 2015.

[20] E. Baştuğ, M. Bennis, and M. Debbah, "Living on the edge: the role of proactive caching in 5G wireless networks," *IEEE Commun. Mag.*, vol. 52, no. 8, pp. 82–89, Aug. 2014.

[21] A. Sabharwal, P. Schniter, D. Guo, D. W. Bliss, S. Rangarajan, and R. Wichman, "In-band full-duplex wireless: challenges and opportunities," *IEEE J. Sel. Areas Commun.*, vol. 32, no. 9, pp. 1637–1652, Sept. 2014.

[22] I. Atzeni and M. Kountouris, "Full-duplex MIMO small-cell networks: performance analysis," in *Proc. IEEE Global Commun. Conf. (GLOBECOM)*, Dec. 2015.

[23] Z. Tong and M. Haenggi, "Throughput analysis for full-duplex wireless networks with imperfect self-interference cancellation," *IEEE Trans. Commun.*, vol. 63, no. 11, pp. 4490–4500, Nov. 2015.

[24] S. Goyal, P. Liu, S. S. Panwar, R. A. Difazio, R. Yang, and E. Bala, "Full duplex cellular systems: will doubling interference prevent doubling capacity?" *IEEE Commun. Mag.*, vol. 53, no. 5, pp. 121–127, May 2015.

[25] I. Atzeni and M. Kountouris, "Full-duplex MIMO small-cell networks with interference cancellation," *IEEE Trans. Wireless Commun.*, vol. 16, no. 12, pp. 8362–8376, Dec. 2017.

[26] M. Maso, C.-F. Liu, C.-H. Lee, T. Q. S. Quek, and L. S. Cardoso, "Energy-recycling full-duplex radios for next generation networks," *IEEE J. Sel. Areas Commun.*, vol. 33, no. 12, pp. 2948–2962, Dec. 2015.

[27] M. Duarte, C. Dick, and A. Sabharwal, "Experiment-driven characterization of full-duplex wireless systems," *IEEE Trans. Wireless Commun.*, vol. 11, no. 12, pp. 4296–4307, Dec. 2012.

[28] D. Bharadia, E. McMilin, and S. Katti, "Full duplex radios," *ACM SIGCOMM Comput. Commun. Review*, vol. 43, no. 4, pp. 375–386, Aug. 2013.

[29] D. Bharadia and S. Katti, "Full duplex MIMO radios," in *Proc. USENIX Symp. Netw. Syst. Design and Implementation (NSDI)*, Apr. 2014.

[30] J. I. Choi, M. Jain, K. Srinivasan, P. Levis, and S. Katti, "Achieving single channel, full duplex wireless communication," in *Proc. Ann. Int. Conf. Mobile Comput. Netw. (MOBICOM)*, Sept. 2010.

[31] M. E. Knox, "Single antenna full duplex communications using a common carrier," in *Proc. IEEE Ann. Wireless Microw. Technol. Conf. (WAMICON)*, Apr. 2012.

[32] D. Bharadia, K. Joshi, and S. Katti, "Robust full duplex radio link," *ACM SIGCOMM Comput. Commun. Review*, vol. 44, no. 4, pp. 9147–148, Aug. 2014.

[33] N. Phungamngern, P. Uthansakul, and M. Uthansakul, "Digital and RF interference cancellation for single-channel full-duplex transceiver using a single antenna," in *Int. Conf. Elect. Eng./Electron., Comput., Telecommun. Inf. Technol. (ECTI-CON)*, May 2013.

[34] M. Jain, J. I. Choi, T. Kim, D. Bharadia, S. Seth, K. Srinivasan, P. Levis, S. Katti, and P. Sinha, "Practical, real-time, full duplex wireless," in *Proc. Ann. Int. Conf. Mobile Comput. Netw. (MOBICOM)*, Sept. 2011.

[35] N. Li, W. Zhu, and H. Han, "Digital interference cancellation in single channel, full duplex wireless communication," in *Proc. Int. Conf. Wireless Commun., Netw. Mobile Comput. (WiCOM)*, Sept. 2012.

[36] E. Ahmed and A. Eltawil, "All-digital self-interference cancellation technique for full-duplex systems," *IEEE Trans. Wireless Commun.*, vol. 14, no. 7, pp. 3519–3532, July 2015.

[37] D. Korpi, L. Anttila, and M. Valkama, "Reference receiver based digital self-interference cancellation in MIMO full-duplex transceivers," in *Proc. IEEE Global Commun. Conf. (GLOBECOM)*, Dec. 2014.

[38] I. Atzeni, M. Maso, and M. Kountouris, "Optimal low-complexity self-interference cancellation for full-duplex MIMO small cells," in *Proc. IEEE Int. Conf. Commun. (ICC)*, May 2016.

[39] G. C. Alexandropoulos, M. Kountouris, and I. Atzeni, "User scheduling and optimal power allocation for full-duplex cellular networks," in *Proc. IEEE Int. Workshop Signal Process. Adv. Wireless Commun. (SPAWC)*, July 2016.

[40] I. Atzeni, M. Kountouris, and G. C. Alexandropoulos, "Performance evaluation of user scheduling for full-duplex small cells in ultra-dense networks," in *European Wireless (EW) Conf.*, May 2016.

[41] S. Wang, V. Venkateswaran, and X. Zhang, "Fundamental analysis of full-duplex gains in wireless networks," *IEEE/ACM Trans. Netw.*, vol. 25, no. 3, pp. 1401–1416, June 2017.

[42] L. Wang, F. Tian, T. Svensson, D. Feng, M. Song, and S. Li, "Exploiting full duplex for device-to-device communications in heterogeneous networks," *IEEE Commun. Mag.*, vol. 53, no. 5, pp. 146–152, May 2015.

[43] J. Bai and A. Sabharwal, "Distributed full-duplex via wireless side-channels: bounds and protocols," *IEEE Trans. Wireless Commun.*, vol. 12, no. 8, pp. 4162–4173, Aug. 2013.

[44] M. Haenggi, *Stochastic Geometry for Wireless Networks*, New York: Cambridge University Press, 2012.

[45] I. Atzeni, J. Arnau, and M. Kountouris, "Downlink cellular network analysis with LOS/NLOS propagation and elevated base stations," *IEEE Trans. Wireless Commun.*, vol. 17, no. 1, pp. 142–156, Jan. 2018.

[46] "TR 36.828 (V11.0.0): Further enhancements to LTE time division duplex (TDD) for downlink-uplink (DL-UL) interference management and traffic adaptation," 3GPP, Technical Report, June 2012.

14 Caching in Mobile Millimeter Wave: Sub-6 GHz Networks

Omid Semiari, Walid Saad, and Mehdi Bennis

14.1 Background, Related Works, and Summary of Contributions

Emerging applications with high data rate requirements such as uncompressed video streaming, high-definition maps for autonomous vehicles [1], and mobile TV have substantially increased the traffic in cellular networks. To manage the high traffic demand, new promising techniques are introduced to boost the capacity of wireless cellular networks, including but not limited to (1) operating at high-frequency millimeter wave (mmW) bands that can yield extremely large throughput by using large beam-forming gains from antenna arrays with many elements and transmitting over large GHz bandwidth, (2) dense deployment of small base stations (SBSs) to remove coverage holes and improve spectral efficiency, and (3) using the storage capabilities of use equipment (UE) to cache the data for *mobile UE (MUE)*. Via caching, MUE can store the content proactively and use the cached content without requesting wireless resources from the network during peak traffic hours.

Nonetheless, dense deployment of SBSs within the coverage of macro-cell base stations (MBSs) will pose several critical issues in cellular networks. (1) As an MUE travels across densely deployed SBSs, frequent handovers (HOs) are inevitable. Frequent HOs increase signaling overhead for MUE. Also, with more HOs, an MUE will naturally experience greater handover failure (HOF), particularly if the MUE moves very fast [2]. Indeed, the MUE cannot finish the HO by the time it travels across a target SBS. (2) The discovery target SBSs needs inter-frequency measurements, which are power consuming. (3) Given that sub-6 GHz μW frequencies are highly congested, frequency resources needed for frequent HOs of MUE will strain the capacity of the network for static users due to limiting available resources. Hence transferring traffic from congested μW frequency bands to mmW bands can play a key role in reducing traffic at the μW network.

14.1.1 Related Works

In [3], various decentralized protocols are presented for mobility management in emerging heterogeneous networks (HetNets). The work in [4] introduces a new energy-efficient SBS discovery method. The work in [5] studies HO methods for enhancing HO across LTE-Advanced and femtocell systems. In [6], existing vertical HO approaches are discussed. The work in [7] shows that the ping-pong effect (i.e., making an MUE perform HOs repeatedly between two base stations) is impacted by the HO sampling period. The work in [8] develops a new HO policy that accounts for the speed of MUE to prevent frequent HOs. The authors in [9] introduce an HO policy that supports soft HOs by enabling MUE to connect simultaneously to both an MBS and SBSs. Additionally, the work in [10] proposes a decentralized framework that leverages different bands to separate the data and control traffic for MUEs.

The works in [3–11], despite being interesting, do not take into account leveraging caching capabilities at mmW frequencies and consider only μW networks. The work in [12] proposes an HO policy at mmW frequencies that allows an MBS to help mmW SBSs at a control plane. Nonetheless, [12] considers only line of sight (LoS) mmW links and does not analyze the stochastic and unreliable communications at mmW frequencies. In [13], a joint mmW-μW HO scheme is developed that uses mmW links to buffer videos. Nonetheless, mobility challenges are not addressed in [13]. Furthermore, the work in [14] does not account for dynamic energy and mobility management in mobile scenarios.

The works in [15–18] study proactive caching for optimizing mobility management. The authors in [15] overview different aspects of caching to offload the core network traffic. In addition, the work in [16] incorporates the mobility of an MUE to cache the content proactively and transfer it across different data centers along the trajectory of MUE. In [17], the authors develop a strategy that places different portions of a file at multiple SBSs. This enables MUE to opportunistically download different cached parts of the original content as it moves within different cells. Moreover, in [18], a proactive caching scheme is introduced that exploits the MUE's trajectory information to improve mobility management. However, the body of work in [15–18] focuses on developing protocols at higher network layers. Moreover, caching directly at the MUE is overlooked. However, we will demonstrate how exploiting mmW frequencies complements content caching at MUE and how it will provide opportunities to effectively reduce the number of necessary HOs.

14.1.2 Summary of Contributions

This chapter presents a new framework that addresses challenging mobility management problems, such as substantial energy consumption and high rate of handover failure in dense wireless networks. In particular, we develop a model that exploits broadband mmW connectivity whenever available to cache content that the MUE is interested in. Thus it will enable the MUE to skip unnecessary HOs to small-sized SBSs. To this end, tractable expressions will be derived to analyze the performance of caching over mmW links. In addition, performance improvements (such as reduction of handover failure rate) achievable via caching in mmW–μW will be studied. Furthermore, we will propose a decentralized mobility management solution via dynamic matching. Then, we discuss why conventional matching algorithms such as those proposed in [19] and [20] cannot be applied to solve the proposed problem. Thereby, a new decentralized solution will be developed that yields a dynamically stable HO policy in mobile networks. Finally, we will discuss the complexity of our developed method and show the

Table 14.1 Description of Notation

Notation	Description	Notation	Description
$\mathcal{U}$	MUE set	$\mathcal{K}$	SBSs set
θ_u	Moving angle of MUE	v_u	Speed of MUE
p_k	Transmit power	B	Segment size of video (bits)
Ω_u	Cache size of MUE u	$\Omega_u^{\mathtt{max}}$	Maximum cache size
t_u^c	Caching duration of MUE u	Q	Video play rate
$\bar{R}^c(u,k)$	Average achievable caching rate	d^c	Traversed distance using cached content
ΔT	Time-to-trigger (TTT)	r^c	Traversed distance in caching duration
T_s	Interfrequency cell scanning interval	t_{MTS}	Minimum time-of-stay (ToS)
θ_k	Beamwidth for SBS k	E^s	Energy consumption for each cell search
$t_{u,k}$	Time-of-stay for MUE u at SBS k	t_{MTS}	Minimum required time-of-stay

achievable performance gains via extensive simulations. See Table 14.1 for notation used in this chapter.

14.2 System Model

Consider a set $\mathcal{K}$ of K uniformly distributed SBSs along with an MBS in a heterogeneous cellular network. In addition, consider U randomly distributed MUE, in a set $\mathcal{U}$. $\theta_u \in [0, 2\pi]$ is a random direction of an MUE u, during the considered time T. The average speed of an MUE u is $v_u \in [v_{\mathtt{min}}, v_{\mathtt{max}}]$. Let p_k denote the transmit power of each SBS (either picocell or femtocell) $k \in \mathcal{K}$. No interference exists between the MBS and SBSs since the MBS operates at a different μW frequency [4, 21]. The SBSs can also serve MUE over mmW frequencies [14, 22–26]. We note that the MAC layer integration of μW and mmW RATs can reduce the time and signaling overhead associated with vertical HOs [27].

14.2.1 Channel Model

The pathloss (in dB) of a mmW link is

$$L(u,k) = 20\log_{10}\left(\frac{4\pi r_0}{\lambda}\right) + 10\alpha\log_{10}\left(\frac{r_{u,k}}{r_0}\right) + \chi, \tag{14.1}$$

where (14.1) holds for $r_{u,k} \geq r_{\mathtt{ref}}$, and $r_{\mathtt{ref}}$ and $r_{u,k}$ represent the reference distance and the MUE u to SBS k distance, respectively. Moreover, λ is the wavelength at $f_c = 73$ GHz, α is the pathloss exponent, and χ is a Gaussian random variable with

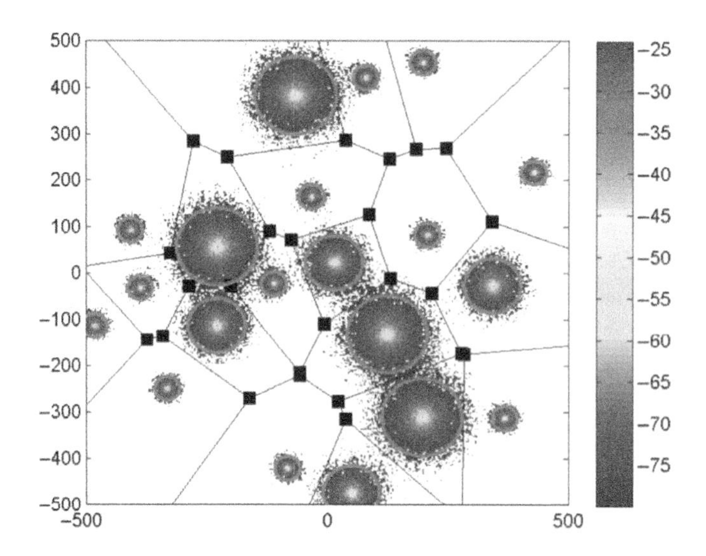

Figure 14.1 SBSs coverage (RSS > −80 dB) and the simplified cell coverage (shown by circles).

a zero mean and variance ξ^2. Depending on the state of the link, i.e, whether the link is LoS or non–line of sight (NLoS), pathloss parameters will be different. Similar pathloss models can be used for the μW links with different values for the parameters.

Figure 14.1 shows the considered HetNet along with the coverage area of each μW SBS according to the maximum receive signal strength (max-RSS) criteria. In Figure 14.1, white areas are covered by the MBS. Because of the shadowing, an MUE may have to perform frequent HOs repeatedly between two base stations (known as the ping-pong effect).

14.2.2 Antenna Gain Pattern

Due to the sever pathloss over mmW frequencies, it is necessary to perform beamforming to exploit antenna gains and extend the communication range. At the MUE, the simplified antenna gain model is [28]:

$$G(\theta) = \begin{cases} G_{\mathtt{max}}, & \mathtt{if} \quad \theta < |\theta_m|, \\ G_{\mathtt{min}}, & \mathtt{otherwise}. \end{cases} \tag{14.2}$$

In this model, $G_{\mathtt{max}}$ and $G_{\mathtt{min}}$ represent the main lobe and side lobes gains, respectively. Moreover, θ_m denotes the beam width of the antenna main lobe. A similar sectorized pattern is also considered at SBSs; nonetheless, every SBS k can have N_k beams. Figure 14.2 shows the beam pattern configuration of an SBS k with $N_k = 3$ beams formed uniformly in $\theta \in [0, 2\pi]$. Due to the mobility of MUEs, dynamic beam training will result in significant control overhead to establish mmW links. Hence, in this work, we consider fixed mmW beams at each SBS and let the MUE access the mmW link opportunistically when the MUE traverses across the coverage of a mmW beam. We assume that the total precoding and combining gains is $\psi_{u,k} = G^2_{\max}$.

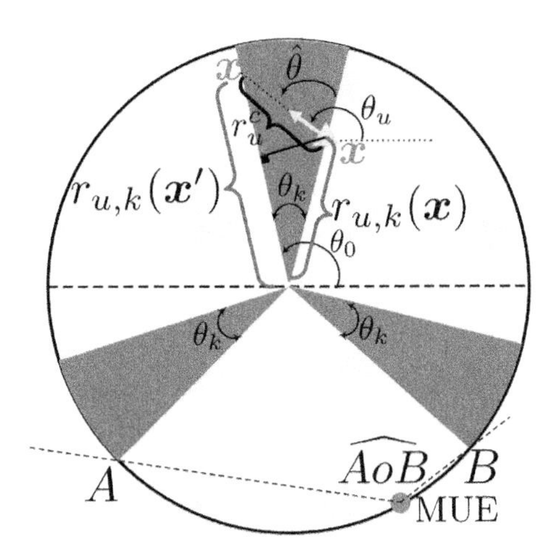

Figure 14.2 Example of a dual-mode SBS with $N_k = 3$ mmW beams (shaded areas).

14.2.3 Traffic Model

High-definition video transmission is one of the prominent wireless services with very high quality of service (QoS) requirements. In mobile HetNets, frequent HOs as well as HOFs will degrade the QoS for such applications. Therefore, the focus of this work is addressing the challenges of HOs for MUE, particularly with streaming service requests. The key idea is to use caching to download incoming B bits of video segments to an MUE, once it is feasible to be served over a mmW link. With this approach, fast caching of the video can be done within a limited time. The number of segments that can be cached is given by:

$$\Omega_u(k) = \min\left\{\left\lfloor \frac{\bar{R}^c(u,k)t_u^c}{B} \right\rfloor, \Omega_u^{\max}\right\}, \tag{14.3}$$

where $\Omega_u^{\max}$ is the maximum cache size. Moreover, the time it takes for an MUE u to traverse across its serving SBS's mmW beam is defined as the *caching duration* and denoted by t_u^c. In fact, as shown in Figure 14.2, $t_u^c = r_u^c/v_u$, where r_u^c is the traversed distance across the mmW beam. The notations $\lfloor.\rfloor$ and $\min\{.,.\}$ represent the floor and minimum operations, respectively. In addition, the *average achievable rate* within the caching duration is denoted by $\bar{R}^c(u,k)$. Accordingly, the distance that an MUE can travel and use the cached content will be

$$d^c(u,k) = \frac{\Omega_u(k)}{Q} v_u, \tag{14.4}$$

where Q is the video play rate. In fact, during the time that an MUE travels the distance $d^c(u,k)$, there is no need to search for an SBS to perform HO and the control information can be managed solely via the MBS. As we show later, this caching mechanism results in a more efficient and reliable HO in small cell networks.

14.2.4 Handover Process and Relevant Parameters

The HO procedure introduced by 3GPP includes the following steps: First, a cell search will be performed by the MUE periodically (every T_s seconds). Second, if the value of the received signal strength exceeds that of the current serving cell, the average RSS will be calculated by the MUE during the time to trigger (TTT) of ΔT seconds. Then, if the serving SBS has a smaller average RSS than that of the target SBS during TTT, the MUE will trigger the HO. Finally, relevant HO data will be sent to the target SBS and the HO will be executed.

In this work, the described HO process will be changed to accommodate caching in mobile scenarios. In particular, instead of a fixed value for T_s, determining the search period will depend on the caching metrics such as Q and Ω_u. During the time period at which the MUE can play the cached video segments, HO search will be muted. The HO search will be resumed once Ω_u/Q is less than the ΔT.

Given that in small cell networks, the time of stay (ToS) is generally small, the MUE may not have sufficient time to successfully execute the HO process. This results in an HOF and can reduce the quality of mobile services. Hence we characterize the HOF as a main performance indicator. The HOF can be formulated as:

$$\gamma_{\text{HOF}}(u,k) = \begin{cases} 1, & \text{if } t_{u,k} < t_{\text{MTS}}, \\ 0, & \text{otherwise,} \end{cases} \tag{14.5}$$

where t_{MTS} is the minimum ToS (MTS) required to successfully perform the HO. In addition, $t_{u,k}$ represents the ToS for MUE u when attempting HO to the SBS k. Although limited available time for HO is not the only cause for HOFs, it can be a major factor in an ultra-dense HetNet that encompasses high-speed MUE [29].

Additionally, let E^s be the energy required per each cell search. Therefore, the cell search energy consumption during time T will be

$$E^s_{\text{total}} = E^s \frac{T}{T_s}. \tag{14.6}$$

Here, we note that this energy is needed to acquire primary and secondary synchronization signals of the target SBSs [4]. We also need to note that there is a trade-off between energy efficiency and the amount of traffic that can be managed by SBSs over the mmW band, rather than the μW MBS. In fact, although increasing T_s is clearly more energy efficient, it will result in fewer HOs to SBSs, and the MUE will be primarily served by the MBS. The proposed caching-based HO strategy, on the other hand, will allow us to increase T_s without reducing offloading traffic from the MBS.

14.3 Caching-Enabled Mobility Management

As mentioned earlier, we consider SBSs with both mmW and μW radio interfaces (dual-mode). Here, we focus on analyzing the performance in terms of the probability for caching content at an MUE via the mmW radio interface.

14.3.1 Probability of Caching via mmW Links

The small circle in Figure 14.2 shows the point at which the MUE u's path intersects with the cell k. From Figure 14.2, we can observe that caching via a mmW link is feasible only if the MUE's moving path falls within $\widehat{AoB}$ in order to pass across the coverage area of a mmW link. Thus if we denote the probability of caching via a mmW link by $\mathbb{P}_k^c(N_k, \theta_k)$, we can make the following observation directly from geometry:

THEOREM 14.1 *If there are $N_k \geq 2$ main lobes for the mmW radio interface of an SBS, each with a beamwidth $\theta_k > 0$, then:*

$$\mathbb{P}_k^c(N_k, \theta_k) = \left[\frac{N_k \theta_k}{2\pi}\right] + \left[1 - \frac{N_k \theta_k}{2\pi}\right]\left[\frac{1}{2}\left(1 - \frac{1}{N_k}\right) + \frac{\theta_k}{4\pi}\right]. \tag{14.7}$$

Proof See appendix A in [22]. □

We note that (14.7) can be verified for an example scenario with $N_k = 3$ and $\theta_k = \frac{2\pi}{3}$. For this scenario, (14.7) results in $\mathbb{P}_k^c(N_k, \theta_k) = 1$ which is valid since for this scenario, the mmW interface will provide coverage for the entire cell.

14.3.2 Statistics of the Caching Duration

Given the random direction and speed of an MUE, the time duration available for caching a content will not be deterministic. Hence it is important to derive the statistics of the caching duration that then will be used for the performance analysis.

To this end, let the triangle in Figure 14.2 show the location of an arbitrary MUE u, $\boldsymbol{x_u} = (x_u, y_u) \in \mathbb{R}^2$, crossing a mmW beam. In addition, let the SBS of interest be located at $\boldsymbol{x}_k = (0,0)$. Accordingly, we can characterize the geometry of a mmW beam by representing its two sides as follows:

$$y = x \tan(\theta_0 - \theta_k), y = x \tan(\theta_0), \quad x > 0. \tag{14.8}$$

From Figure 14.2, the MUE (shown as a triangle) is considered to be located on the side $x = y\cos(\theta_0 - \theta_k)$. Hence θ_0 in (14.8) is $\theta_0 = \arccos\left(\frac{x_u}{r_{u,k}(\boldsymbol{x_u})}\right) + \theta_k$, where $r_{u,k}(\boldsymbol{x}) = \sqrt{x_u^2 + y_u^2}$. The statistics of the caching duration t^c is captured by

$$F_{t_u^c}(t_0) = \mathbb{P}(t_u^c \leq t_0) = \mathbb{P}(r_u^c \leq v_u t_0), \tag{14.9}$$

where r_u^c is the distance that MUE u will traverse across the mmW beam, as shown in Figure 14.2, and $F_{t^c}(.)$ is the cumulative distribution function (CDF) of t^c. Based on $\boldsymbol{x}_u$, the minimum distance will be:

$$r_u^{\min} = \frac{\left|x_u \tan\theta_0 - y_u\right|}{\sqrt{1 + \tan^2\theta_0}}. \tag{14.10}$$

If $r_u^{\min} > v_u t_0$, then $F_{t_u^c}(t_0) = 0$. Thus we let $r_u^{\min} \leq v_u t_0$. Moreover, we can easily observe that $\boldsymbol{x}'_u = \left(x_u + r_u^c \cos\theta_u, y_u + r_u^c \sin\theta_u\right)$ where $\boldsymbol{x}'_u$ is the point at which the MUE's trajectory intersects with the line $y = x\tan(\theta_0)$. Therefore, $y_u + r_u^c \sin\theta_u = \left[x_u + r_u^c \cos\theta_u\right]\tan\theta_0$, and

$$r_u^c = v_u t_u^c = \frac{y_u - x_u \tan\theta_0}{\tan\theta_0 \cos\theta_u - \sin\theta_u}, \tag{14.11}$$

where r_u^c clearly represents the traversed distance during the caching duration t^c. Next, from (14.9) and (14.11), the CDF can be written as

$$F_{t_u^c}(t_0) = \mathbb{P}\left(\frac{y_u - x_u \tan\theta_0}{\tan\theta_0 \cos\theta_u - \sin\theta_u} \leq v_u t_0\right). \tag{14.12}$$

LEMMA 14.2 *Considering an arbitrary MUE u with a random speed v_u, the CDF of the caching duration is*

$$F_{t^c}(t_0) = \frac{1}{\pi - \theta_k}\left(\arccos\left(\frac{r_u^{min}}{v_u t_0}\right) + \min\left\{\arccos\left(\frac{r_u^{min}}{r_{u,k}(\boldsymbol{x})}\right), \arccos\left(\frac{r_u^{min}}{v_u t_0}\right)\right\}\right). \tag{14.13}$$

Proof See appendix B in [22]. □

Figure 14.3 shows the CDF of t^c for different MUE locations. It can be observed that with smaller distances between the MUE and the SBS, it is more probable to have less time for caching the content over a mmW link. On the other hand, the achievable data rate will be higher if the MUE is close to the SBS. That being said, in the next section, we present comprehensive performance analysis while considering both data rates and caching duration.

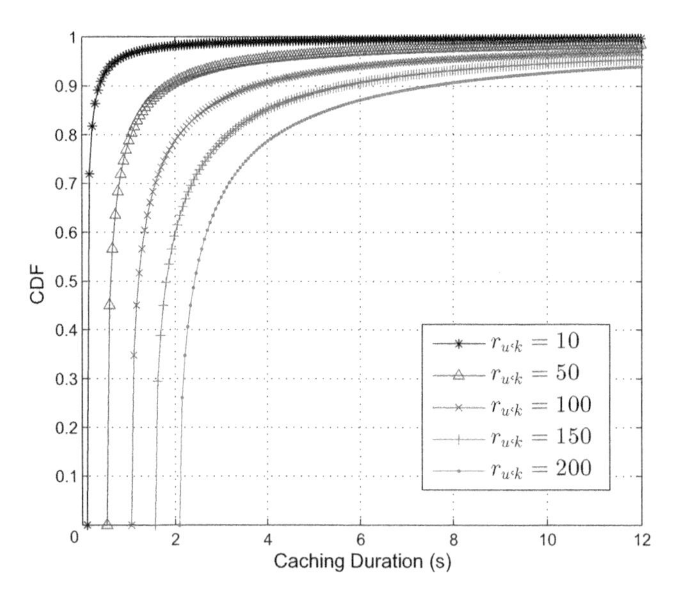

Figure 14.3 Distribution of caching duration t^c.

14.4 Performance Analysis of the Proposed Cache-Enabled Mobility Management Scheme

In this section, we consider only free space pathloss and disregard the shadowing effect to derive tractable expressions for the average achievable rate for content caching over the mmW radio interface.

14.4.1 Average Caching Data Rate

Considering $r_{u,k}(\boldsymbol{x})$ as the initial distance between the SBS and the MUE, and if the MUE travels across a mmW beam, the achievable data rate for caching is

$$R^c(u,k) = \frac{1}{v_u t_u^c} \int_{r_{u,k}(\boldsymbol{x})}^{r_{u,k}(\boldsymbol{x}')} w \log\left(1 + \frac{\beta P_t \psi r_{u,k}^{-\alpha}}{wN_0}\right) dr_{u,k}, \tag{14.14}$$

where $\beta = (\frac{\lambda}{4\pi r_0})^2 r_0^\alpha$. Using this rate, we can find the following result.

THEOREM 14.3 *The average caching data rate for an MUE u is*

$$\bar{R}^c(u,k) = \mathbb{P}_k^c(N_k,\theta_k) R^c(u,k), \tag{14.15}$$

$$= \delta_2 \int_{f(\theta_k)}^{f(0)} \frac{1}{f^2(\theta)} \log\left(1 + \delta_1 f^\alpha(\theta)\right) df(\theta), \tag{14.16}$$

$$\stackrel{(a)}{=} \frac{\delta_2}{\ln(2)}\Bigg[2\sqrt{\delta_1}\arctan(\sqrt{\delta_1} f(\theta_k)) - \frac{\ln(\delta_1 f^2(\theta_k)+1)}{f(\theta_k)}, - 2\sqrt{\delta_1}\arctan(\sqrt{\delta_1} f(0)) + \frac{\ln(\delta_1 f^2(0)+1)}{f(0)}\Bigg], \tag{14.17}$$

where $\delta_1 = \frac{\beta P_t \psi}{wN_0}\left[r_{u,k}(\boldsymbol{x})\sin\hat{\theta}\right]^{-\alpha}$. In addition, $\delta_2 = w r_{u,k}(\boldsymbol{x})\sin\hat{\theta}\mathbb{P}_k^c(N_k,\theta_k)/v_u t^c$, and $\hat{\theta} = \theta_u - \theta_0 + \theta_k$. We note that (a) results from setting $\alpha = 2$, which is a valid assumption for LoS links [30].

Proof See appendix C in [22]. □

14.4.2 Analysis of Performance Gains from the Proposed Caching-Based Mobility Management

Using the previous results in (14.3), (14.4), and (14.17), we can find the performance gains of the proposed caching-based mobility management approach, based on the number cell searches that can be avoided by using the cached content. In fact,

$$\eta \approx \left\lfloor \frac{\mathbb{E}\left[d^c(u,k)\right]}{l} \right\rfloor, \tag{14.18}$$

where $d^c(u,k)$, as defined earlier, represents the distance that the MUE u can travel while playing the cached video content. In (14.18), the expected value is due to the fact that $d^c(u,k)$ is a random variable that depends on θ_u and can be obtained directly from (14.4) and $F_{t^c}(.)$ in Lemma 14.2. Moreover, l represents the average cell size.

Remark 14.1 Considering η defined in (14.18), the proposed method will minimize the average energy consumption E^s for interfrequency cell search by a factor of $1/\eta$, if the energy consumption linearly increases with the number of scans.

In addition, from the definition of γ_{HOF} in (14.5), we define the HOF probability as $\mathbb{P}(D_{u,k} < v_u t_{\text{MTS}})$ [31], where $D_{u,k} = t_{u,k} v_u$, and $t_{u,k}$ is the ToS. The HOF probability is given by:

$$\mathbb{P}(D_{u,k} < v_u t_{\text{MTS}}) = \int_0^{v_u t_{\text{MTS}}} \frac{2}{\pi\sqrt{4a_k^2 - D^2}} dD = \frac{2}{\pi} \arcsin\left(\frac{v_u t_{\text{MTS}}}{2a_k}\right). \tag{14.19}$$

where (14.19) considers a_k as the radius of the cell and the MUE's trajectory as a random cord with length $D_{u,k} = t_{u,k} v_u$ within the cell. In fact,

$$f_D(D) = \frac{2}{\pi\sqrt{4a^2 - D^2}}, \tag{14.20}$$

where $f_D(D)$ is the PDF of the chord's length. This analysis clearly demonstrates the gains of caching-based mobility management for a single mobile users. In the next section, we focus on developing a framework in multi-user scenarios that leverages caching to enhance the mobility management.

14.5 Proposed Cache-Enabled Mobility Management Based on Dynamic Matching

Our framework allows a single MUE to use caching in order to avoid unnecessary HOs to SBSs. Therefore, the proposed scheme provides this flexibility for an MUE to decide whether to perform an HO to an SBS or remain connected to the MBS and receive data content over μW band or use the cached content without performing any HO. Similar options are available for an MUE that is already served by an SBS. At the network level, the goal can be maximizing the traffic offloads from the congested μW frequencies to the mmW links provided by SBSs. In this network performance optimization we need to account for other described challenges of mobility management, such as guaranteeing bounds for the HOF, total MUEs that a single SBS can serve, and the MUEs capacity constraints to cache video contents. Given these constraints, we seek to find an optimal HO policy as follows:

$$\arg\min_{\zeta} \sum_{u\in\mathcal{U}} \zeta(u, k_0), \tag{14.21a}$$

$$\text{s.t. } \mathbb{P}\left(\sum_{k\in\mathcal{K}} \zeta(u,k) D_{u,k} < v_u t_{\text{MTS}}\right) \leq P_u^{\text{th}}, \tag{14.21b}$$

$$\left[1 - \sum_{k \in \mathcal{K}'} \zeta(u,k)\right] T_s \leq \frac{\Omega_u}{Q}, \tag{14.21c}$$

$$\sum_{k \in \mathcal{K}'} \zeta(u,k) \leq 1, \tag{14.21d}$$

$$\sum_{u \in \mathcal{U}} \zeta(u,k) \leq U_k^{\text{th}}, \forall k \in \mathcal{K}, \tag{14.21e}$$

$$\zeta(u,k) \in \{0,1\}, \tag{14.21f}$$

where $\mathcal{K}' = \mathcal{K} \cup \{k_0\}$ and ζ includes decision variables $\zeta(u,k) \in \{0,1\}$. Indeed, $\zeta(u,k) = 1$, if HO is done by an MUE u to the target cell k and $\zeta(u,k) = 0$, otherwise. The objective function in (14.21a) aims to maximize the traffic offloads from the MBS to SBSs. Constraint (14.21b) ensures that the probability of HOF for an MUE that is assigned to an SBS does not exceed a threshold P_u^{th} (selected based on the QoS demands); (14.21c) guarantees that MUE u is assigned to a BS, unless enough content is cached for the next T_s seconds. In addition, (14.21d) and (14.21e) represent the quotas for MUEs and BSs.

From (14.19), (14.21b) can be represented as a linear constraint

$$\sum_{k \in \mathcal{K}} \frac{2}{\pi} \arcsin\left(\frac{v_u t_{\text{MTS}}}{2a_k}\right) \zeta(u,k) \leq P_u^{\text{th}}.$$

Therefore, the proposed problem in (14.21a) to (14.21f) is an integer linear program (ILP) problem, and hence it is NP-hard. Although approximation methods can be adopted to solve (14.21a) to (14.21f), centralized approaches are not desirable in real-time mobile scenarios, due to the scalability issues and the additional latency and overhead. Furthermore, in dynamic mobile networks, it is imperative to account for future situations to properly optimize the network performance. We elaborate on this by explaining two examples in Figure 14.4. As scenario 1, let's assume that a feasible solution for (14.21a) to (14.21f) does not assign an MUE u to the target SBS k. Instead, as shown in scenario 1, the solution suggests that the MUE uses the cached content for the next T_s seconds. Given that the goal is to maximize the traffic offloads from the MBS to the SBSs, the given solution in scenario 1 is not efficient since the MUE has to eventually perform an HO to the MBS. That is because the solution does not take into account the future HO instance after the cached content runs out. Alternatively, the SBS could be assigned to the target SBS, fill out the cache, and then start using the cached content. With this strategy, the MUE could reach to the coverage of the next target SBS without the need to perform HO to the MBS.

As the second scenario in Figure 14.4, let a feasible solution for (14.21a) to (14.21f) assign an arbitrary MUE 1 to a target SBS k_1. Again, the target is to avoid performing HO to the MBS as much as possible. Now, if HO fails and the MUE 1 does not have sufficient cached content, the MUE has to perform an HO to the MBS. This could be avoided if the MUE 2 with a large cache size was assigned to the SBS k_1. In this way, even with an HOF, the MUE 2 could still reach the next target SBS k_2 by playing the cached video segments.

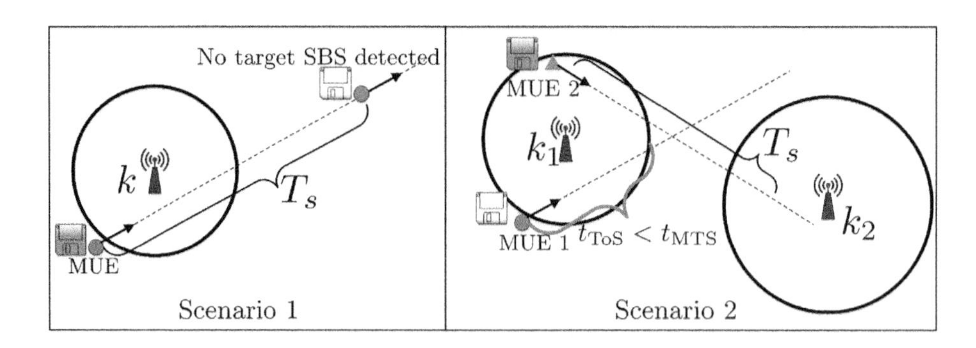

Figure 14.4 Example scenarios for dynamic HO.

These scenarios demonstrate that an effective HO strategy must account for future instances. Hence we develop an HO strategy based on *dynamic matching theory* [32] to solve the problem in (14.21a) to (14.21f).

14.5.1 Mobility Management as a Matching Game

As shown in prior works [33–37], matching theory is an effective method for developing efficient solutions for complex problems such as (14.21a) to (14.21f). The matching problem can be defined as follow:

DEFINITION 14.4 *Considering the MUEs and BSs sets, $\mathcal{U}$ and $\mathcal{K}' = \mathcal{K} \cup \{k_0\}$, a single-period* HO matching *is defined as a many-to-one mapping $\mu: \mathcal{U} \cup \mathcal{K}' \to \mathcal{U} \cup \mathcal{K}'$ that*

1. $\forall u \in \mathcal{U}$, $\mu(u) \in \mathcal{K}' \cup \{u\}$.
2. $\forall k \in \mathcal{K}'$, $\mu(k) \subseteq \mathcal{U} \cup \{k\}$, *and* $\forall k \in \mathcal{K}$, $|\mu(k)| \leq U_k^{th}$.
3. $\mu(u) = k$, *if and only if* $u \in \mu(k)$.

Here, $\mu(u) = k$ indicates that u is matched to k, and $\mu(u) = u$ means that the MUE u is not assigned to any BS, and hence it uses the cached content. Similarly, $\mu(k) = k$ means that the BS k has no MUE to serve.

By definition, the matching meets the HO conditions in (14.21d) to (14.21f). The utility of MUE $u \in \mathcal{U}$ for an SBS $k \in \mathcal{K}$ is

$$\Phi(u,k) = P_u^{\text{th}} - \mathbb{P}\left(\sum_{k \in \mathcal{K}} \zeta(u,k) D_{u,k} < v_u t_{\text{MTS}}\right) = P_u^{\text{th}} - \frac{2}{\pi} \arcsin\left(\frac{v_u t_{\text{MTS}}}{2a_k}\right). \tag{14.22}$$

We observe that the defined utiltiy in (14.22) increases with the cell radius a_k. The SBS k's utility for an MUE u is

$$\Gamma(u,k) = T_s - \frac{\Omega_u}{Q}. \tag{14.23}$$

Clearly, the utility increases as the available cache size for the next time duration T_s decreases. This gives higher priority to MUEs with urgent need to be served by an SBS before their run out of cached content. According to the defined utilities, the preference profile of an MUE u, $\succ_u$, is

$$k \succ_u k' \quad \Leftrightarrow \quad \Phi(u,k) > \Phi(u,k'), \tag{14.24a}$$

$$u \succ_u k \quad \Leftrightarrow \quad \Phi(u,k) < 0, \tag{14.24b}$$

where $k \succ_u k'$ means that SBS k is strictly preferred to SBS k'. In addition, $u \succ_u k$ indicates that if the utility is negative, MUE u does not accept SBS k. In fact, (14.24b) is in line with meeting (14.21b). We also define the preference profile of an SBS k, $\succ_k$, as

$$u \succ_k u' \quad \Leftrightarrow \quad \Gamma(u,k) > \Gamma(u',k), \tag{14.25a}$$

$$k \succ_k u \quad \Leftrightarrow \quad \Gamma(u,k) < 0, \tag{14.25b}$$

where (14.25b) is equivalent to meeting the condition in (14.21c). The tuple $\Pi \triangleq (\mathcal{U} \cup \mathcal{K},\ \succ_{\boldsymbol{u}},\ \succ_{\boldsymbol{k}})$ formally defines the proposed matching problem where $\succ_{\boldsymbol{u}} = \{\succ_u\}_{u \in \mathcal{U}}$ and $\succ_{\boldsymbol{k}} = \{\succ_k\}_{k \in \mathcal{K}}$.

Within this framework, we are interested in finding the *two-sided stable matching*, μ^*, defined as follow [38]:

DEFINITION 14.5 *Matching μ is stable, $\mu \equiv \mu^*$, if and only if there is no (blocking) pair $(u,k) \notin \mu$ such that $k \succ_u \{\mu(u), u\}$ and $u \succ_k \{\mu(k), k\}$.*

It is easy to observe that the notion of stable matching ensures two-sided fairness in assignment of MUEs to SBSs.

Remark 14.2 The *deferred acceptance (DA)* algorithm (see Algorithm 5) always converges to a stable matching in a single-period HO matching game [19].

Algorithm 5 Greedy DA Algorithm for Single-Period HO

Inputs: $\Pi \triangleq (\mathcal{U} \cup \mathcal{K},\ \succ_{\boldsymbol{u}},\ \succ_{\boldsymbol{k}})$.

Outputs: Stable matching μ^*.

1: Every unassigned MUE $u \in \mathcal{U}$ proposes to the most preferred SBS $k \succ_u u$. Exclude k from u's preference profile $\succ_u$.

2: Every $k \in \mathcal{K}$ collects the proposals in Step 1 and tentatively accepts the U_k^{th} most preferred proposals.

3: **repeat** Steps 1 to 2

4: **until** Every u is applied for all SBSs $k \succ_u u$, or already accepted.

5: **if** $\exists u \in \mathcal{U}, \mu(u) \notin \mathcal{K}$ and $\Omega_u / Q < T_s$, **then**

6: $\quad \mu(u) = u$,

7: **else**

8: $\quad$ Associate u with the MBS.

9: **end if**

Here we note that it is not feasible to guarantee stability in a dynamic scenario if the DA algorithm is used to associate MUEs to SBSs. That is because the DA algorithm cannot account for the scenarios that may occur after an HOF (such as the two scenarios described earlier). Hence in order to achieve a stable matching in dynamic mobile scenarios, we need to extend the notion of stability from one stage to *dynamic stability*, as described next.

14.5.2 Mobility Management Based on Dynamic Matching

In order to take into account network instances that may occur after HO for an MUE and to guarantee the dynamic stability, we extend the concept of preferences from an ordering of individual SBSs to the ranking of *association plans*. In fact, we define an association plan as two consecutive matchings for an MUE or SBS. In this framework, if an MUE is associated with the SBS k and then performs a HO to another SBS k', such an association plan will be denoted as kk'. Accordingly, $k_1k_2 \succ_u k_1'k_2'$ means that plan k_1k_2 is more preferred than $k_1'k_2'$. We also extend the notion of matching in Definition 14.4 to $\mu^\dagger : \mathcal{U} \cup \mathcal{K}' \rightarrow (\mathcal{U} \cup \mathcal{K}')^2$, such that $\mu^\dagger(u) = (\mu_1(u), \mu_2(u))$, where μ_1 and μ_2 are one-period matchings. The definition of stability can then be revisited as follow [32]:

DEFINITION 14.6 *If a pair (u,k) meets the following conditions: (1) $kk \succ_u \mu^\dagger(u)$ and $uu \succ_k \mu^\dagger(k)$, (2) $ku \succ_u \mu^\dagger(u)$ and $uk \succ_k \mu^\dagger(k)$, (3) $uk \succ_u \mu^\dagger(u)$ and $ku \succ_k \mu^\dagger(k)$, or (4) $uu \succ_u \mu^\dagger(u)$ and $kk \succ_k \mu^\dagger(k)$, then it is said that the pair (u,k) can* period-1 block *the matching. A matching that cannot be period-1 blocked by any pair is considered* ex-ante stable.

To account for scenarios in which an MUE or BS may block the matching after μ_1 is done, we define the dynamic stability as follows:

DEFINITION 14.7 *If there is an MUE u such that $(\mu_1(u), u) \succ_u \mu^\dagger(u)$, then the MUE u can* period-2 block *the matching $\mu^\dagger$. Additionally, a pair (u,k) can* period-2 block *if either of the following conditions is met: (1) $(\mu_1(u), k) \succ_u \mu^\dagger(u)$ and $(\mu_1(k), u) \succ_k \mu^\dagger(k)$ or (2) $(\mu_1(u), u) \succ_u \mu^\dagger(u)$ and $(\mu_1(k), k) \succ_k \mu^\dagger(k)$. A* dynamically stable *matching cannot be period-1 or period-2 blocked by any MUE or MUE-BS pair.*[1]

We note that although dynamic stability is not ensured if matching is ex-ante stable, a dynamically stable matching is also an ex-ante stable. With this in mind, we propose a mechanism that is guaranteed to yield a dynamically stable solution for the proposed problem.

14.5.3 Proposed Algorithm for Dynamically Stable Mobility Management

The proposed algorithm is given in Algorithm 6. We first develop an algorithm, based on [32], to find an ex-ante stable (Phase 1). Then, we design a strategy to remove all the period-2 blocking pairs (Phase 2).

[1] This notion can be extended to more than two periods.

Algorithm 6 Proposed Algorithm for Mobility Management

Inputs: κ for all users and base stations.
Outputs: Dynamically stable matching μ^*.

Phase 1:

1: Let $\mathcal{P}_u = \cup_{k \in \mathcal{K}}\{kk, uk, ku\}$ be the set of all plans for u. For every $u \in \mathcal{U}$, if $uu \succ_u \kappa$, for all $\kappa \in \mathcal{P}_u$, then u does not propose to base stations. Otherwise, MUE u selects the most preferred plan κ_u^* and proposes to the corresponding BS.
2: Every base station $k \in \mathcal{K}$ receives the proposals and tentatively accepts most preferred plans, while considering the quota U_k^{th}. Obviously, any accepted plan κ by SBS k meets $\kappa \succ_k kk$.
3: **repeat** Steps 1 to 2
4: **until** No plan is rejected. The resulting ex-ante stable matching is shown by $\mu^\dagger = (\mu_1^\dagger, \mu_2^\dagger)$.
Phase 2:
5: **if** $\exists u \in \mathcal{U}, \mu_2^\dagger(u) = u$, **then** run DA algorithm in Algorithm 5 to the subset of users with $\mu_2^\dagger(u) = u$ and the subset of base stations with $|\mu_2^\dagger(k)| < U_k^{\text{th}}$, while ensuring that (14.26) and (14.27) are satisfied. Return the outcome.
6: **else**
7: return $\mu^\dagger$.
8: **end if**

Proposition 1 The outcome of the Stage 1 in the proposed algorithm is an ex-ante stable assignment.

Proof See appendix D in [22]. □

The proposed algorithm follows the following strategy to avoid period-2 blockage: For any SBS that $|\mu_2^\dagger(k)| = U_k^{\text{th}}$, where U_k^{th} denotes the maximum quota,

$$\mu^\dagger \succ_k \left(\mu_1^\dagger(k), \tilde{\mu_2}^\dagger(k) \cup \{u\}\right), \tag{14.26}$$

where $\tilde{\mu_2}^\dagger(k)$ is identical to $\mu_2{}^\dagger(k)$, except one of associated MUE is replaced by the new matching with MUE u. This condition guarentees that the SBS k cannot pair with any MUE to period-2 block the matching. Moreover,

$$(\mu_1(k_0), u) \succ_{k_0} \mu^\dagger \iff P^{\text{th}}_{\mu_1^\dagger(u)} - \frac{2}{\pi}\arcsin\left(\frac{v_u t_{\text{MTS}}}{2a_{\mu_1^\dagger(u)}}\right) < \epsilon, \tag{14.27}$$

where $\epsilon \geq 0$. This constraint ensures that if the HOF criteria is not met for an MUE in the period-1 association, the MUE must be assigned to the MBS in the second period.

THEOREM 14.8 *Algorithm 6 always converges to a dynamically stable MUE association in heterogeneous networks.*

Proof See appendix E in [22]. □

To consider the overhead of implementing the proposed algorithm in practice, it is interesting to consider the number of HO request signals transmitted by MUEs. We note that other control messages sent by SBSs can be handled over broadcast channels and do not significantly contribute to the signaling overhead. As discussed in [22], for a scenario in which the MUEs have the same preference profile, the signaling overhead

Table 14.2 Simulation Parameters

Notation	Parameter	Value
f_c	Carrier frequency	73 GHz
$P_{t,k}$	Total transmit power of SBSs	dBm [20; 27; 30]
K	Total number of SBSs	50
w	Available bandwidth	5 GHz
$(\alpha_{\text{LoS}}, \alpha_{\text{NLoS}})$	Pathloss exponent	(2, 3.5) [30]
d_0	Pathloss reference distance	1 m [30]
$G_{\max}$	Antenna main lobe gain	18 dB [28]
$G_{\min}$	Antenna side lobe gain	−2 dB [28]
N_k	Number of mmW beams	3
θ_m, θ_k	beam width	10° [28]
N_0	Noise power spectral density	−174 dBm/Hz
t_{MTS}	Minimum time of stay	1s [29]
Q	Play rate	1k segments per second
B	Size of video segments	1 Mbits
$(v_{\min}, v_{\max})$	Minimum and maximum MUE speeds	(1, 16) m/s
E^s	Energy per interfrequency scan	3 mJ [21]

of the proposed scheme will be $\mathcal{O}(UK)$. In the next section, we demonstrate via simulations the merits of the proposed caching-based mobility management in reducing the signaling overhead.

14.6 Simulation Results

An area with a radius of 500 m is considered with $K = 50$ small base stations distributed uniformly around the MBS located at the center. The distance between SBSs is at least 30 m. Other key parameters are listed in Table 14.2.

14.6.1 Performance Analysis for Single-User Scenarios

Figure 14.5 shows a performance comparison between the proposed scheme and a HO policy with no caching. The considered performance metric is the average handover failure rate. The results show that the proposed approach significantly improves the performance in mobile small cell networks. Figure 14.5 demonstrates that using mmW link to cache content can effectively minimize the handover failure. For example, the gain can reach up to 45% at $v_u = 60$ km/h speed.

The achievable rate of caching is shown in Figure 14.6 for $v_u = 60$ km/h, versus $r_{u,k}(\boldsymbol{x})$ for different θ_u. Figure 14.6 demonstrates that the caching rate is significant, even at high speeds for all θ_u. However, the performance is noticeably degraded by blockage.

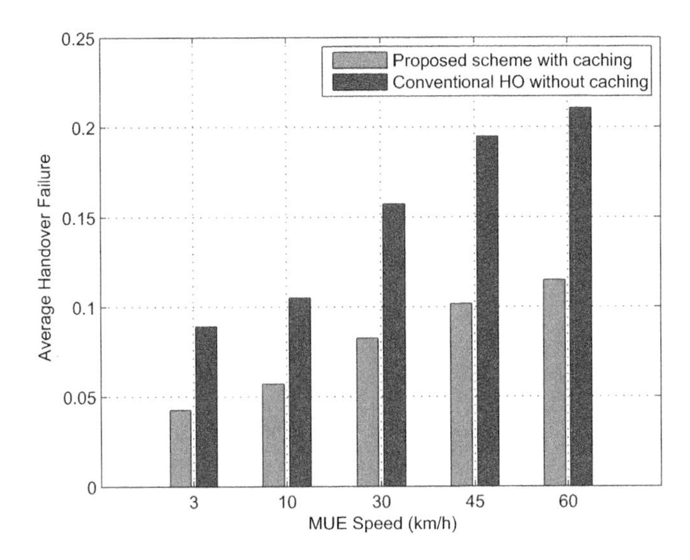

Figure 14.5 HOF versus different MUE speeds.

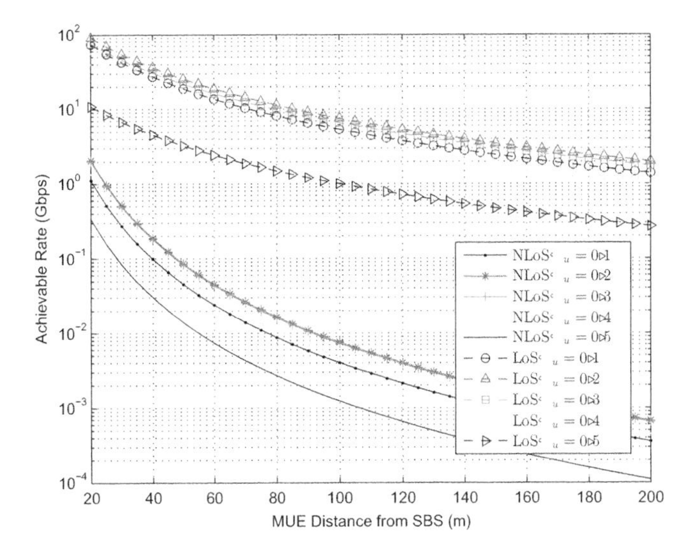

Figure 14.6 Achievable rate of caching versus $r_{u,k}(\boldsymbol{x})$ for different θ_u.

14.6.2 Performance Analysis of the Developed Algorithm

A set of mobile UEs is considered, each with a randomly selected speed and direction entering the coverage of a target cell. In addition, $\Omega_u = 10^4$ segments for all MUEs and $U_k^{\text{th}} = 10$ for each SBS.

Figure 14.7 compares the average probability of handover failure for the proposed algorithm as compared to the baseline scheme. The results confirm that as the speed increases, the ToS reduces, which results in a higher HOF probability. Figure 14.7 also

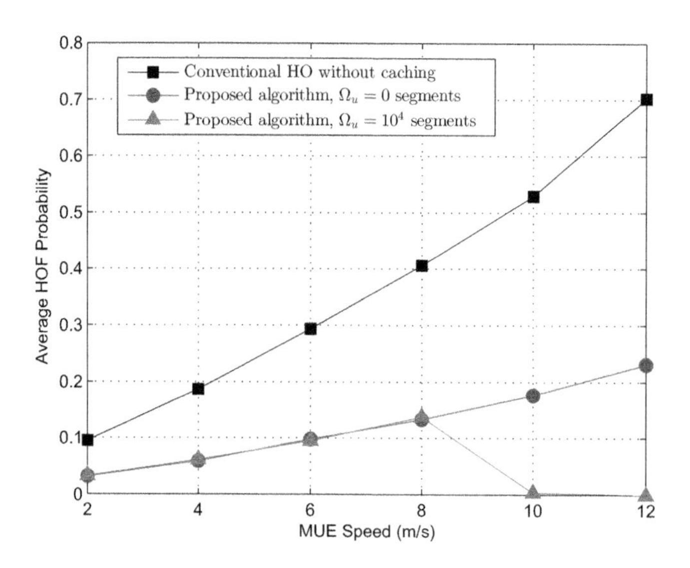

Figure 14.7 Average HOF probability versus MUEs speeds.

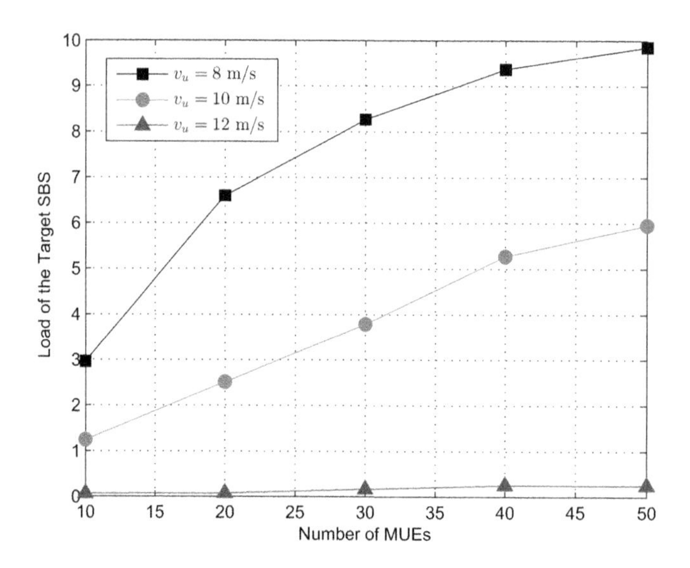

Figure 14.8 Number of assigned MUEs to the target SBS versus the network size.

demonstrates that at speeds larger than $v_u = 8$ m/s, the MUE can use caching to travel a larger distance, which results in lower HOF.

Figure 14.8 shows the number of users assigned to the target cell versus the number of MUEs. The speed of MUEs are selected from $v_u = 8, 10, 12$ m/s, $U_k^{\text{th}} = 10$, and $\Omega_u = 10^4$. At higher speeds, the load of the target cell decreases since the ToS is small and HOF is more likely. Additionally, as the speed increases, it is more feasible for the MUE to reach the next cell by only using the cached content. As shown in Figure 14.8, there is an up to 45% load reduction once v_u changes from 8 to 10 m/s with $U = 40$.

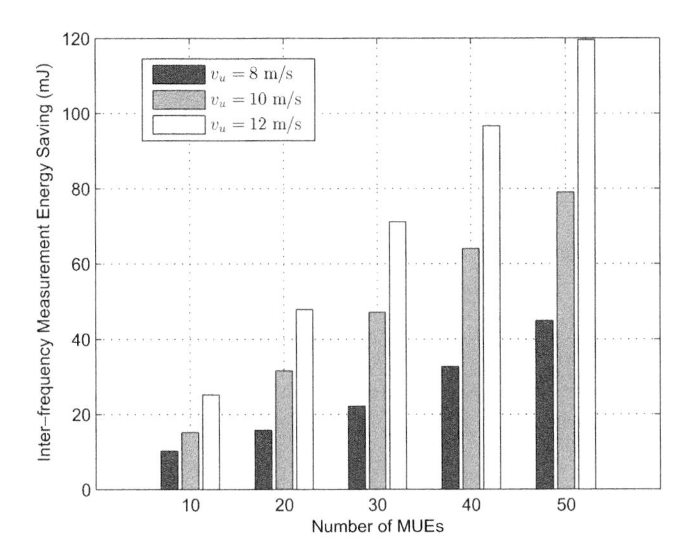

Figure 14.9 Energy savings versus the network size.

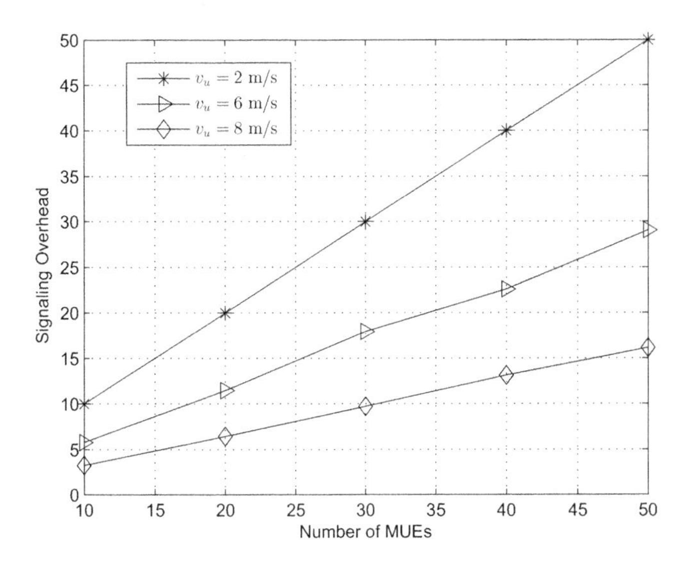

Figure 14.10 Signaling overhead versus the network size.

Figure 14.9 shows the advantages of the proposed scheme in terms of energy savings. With $U = 50$, Figure 14.9 demonstrates that the energy saving gains are up to 80%, 52%, and 29%, respectively, for $v_u = 8, 10,$ and 12 m/s achieved via skipping unnecessary cell searches when cached content is available.

The overhead of the proposed scheme is shown in Figure 14.10 for $\Omega_u = 10^4$ and various speeds. The overhead is defined as the number of HO requests that MUEs send to the target base station. We can observe that at low speeds (e.g., $v_u = 2$ m/s), the overhead is larger. However, it is clear that the overhead is not very large and does not exceed 17 signals for $U = 50$ with $v_u = 8$ m/s. Figure 14.10 shows a key advantage of the proposed scheme, which is less overhead at higher speeds. In fact, although handover is overall more challenging at higher speeds, the proposed algorithm allows users at higher speeds to use caching more effectively to reduce the complexity and overhead of mobility management.

14.7 Summary

We have developed a new cache-enabled mobility management framework in integrated mmW–μW small cell networks. We have demonstrated the key merits of caching via fast mmW links to optimize the performance in mobile scenarios. We have presented fundamental analysis for caching data rate, caching duration, and the probability of caching. Moreover, we have proposed a novel handover policy, based on dynamic matching games and have proposed a new algorithm that guarantees dynamically stable handover. Furthermore, both the theoretical results and simulations have shown substantial performance gains achievable via the proposed scheme, in terms of reducing the handover failure and energy consumption, as compared with a baseline scheme with no caching mechanism. Future extensions of this work may consider more complex mobility patterns for mobile users, variable speeds, and multi-stage stable mobility management.

References

[1] O. Semiari, M. K. Tareq, M. A. Salehi, and W. Saad, "Ultra reliable, low latency vehicle-to infrastructure wireless communications with edge computing," in *Proceedings of IEEE Global Communications Conference (GLOBECOM) Mobile and Wireless Networks Symposium*, Dec. 2018, pp. 1–7.

[2] D. Lopez-Perez, I. Guvenc, and X. Chu, "Mobility management challenges in 3GPP heterogeneous networks," *IEEE Communications Magazine*, vol. 50, no. 12, pp. 70–78, Dec. 2012.

[3] F. Giust, L. Cominardi, and C. J. Bernardos, "Distributed mobility management for future 5G networks: overview and analysis of existing approaches," *IEEE Communications Magazine*, vol. 53, no. 1, pp. 142–149, Jan. 2015.

[4] A. Prasad, O. Tirkkonen, P. Lundén, O. N. C. Yilmaz, L. Dalsgaard, and C. Wijting, "Energy-efficient inter-frequency small cell discovery techniques for LTE-advanced heterogeneous network deployments," *IEEE Communications Magazine*, vol. 51, no. 5, pp. 72–81, May 2013.

[5] D. Xenakis, N. Passas, L. Merakos, and C. Verikoukis, "Mobility management for femtocells in LTE-advanced: key aspects and survey of handover decision algorithms," *IEEE Communications Surveys Tutorials*, vol. 16, no. 1, pp. 64–91, 2014.

[6] A. Ahmed, L. M. Boulahia, and D. Gaiti, "Enabling vertical handover decisions in heterogeneous wireless networks: a state-of-the-art and a classification," *IEEE Communications Surveys Tutorials*, vol. 16, no. 2, pp. 776–811, 2014.

[7] K. Vasudeva, M. Simsek, D. Lopez-Perez, and I. Guvenc, "Impact of channel fading on mobility management in heterogeneous networks," in *2015 IEEE International Conference on Communication Workshop*, June 2015, pp. 2206–2211.

[8] M. Khan and K. Han, "An optimized network selection and handover triggering scheme for heterogeneous self-organized wireless networks," *Mathematical Problems in Engineering*, vol. 16, pp. 1–11, 2014.

[9] H. Zhang, N. Meng, Y. Liu, and X. Zhang, "Performance evaluation for local anchor-based dual connectivity in 5G user-centric network," *IEEE Access*, vol. 4, pp. 5721–5729, Sept. 2016.

[10] I. Elgendi, K. S. Munasinghe, and A. Jamalipour, "Mobility management in three-tier SDN architecture for densenets," in *2016 IEEE Wireless Communications and Networking Conference*, Apr. 2016, pp. 1–6.

[11] I. F. Akyildiz, J. Xie, and S. Mohanty, "A survey of mobility management in next-generation all-IP-based wireless systems," *IEEE Wireless Communications*, vol. 11, no. 4, pp. 16–28, Aug. 2004.

[12] S. G. Park and Y. S. Choi, "Mobility enhancement in centralized mmwave-based multi-spot beam cellular system," in *2015 International Conference on Information and Communication Technology Convergence*, Oct. 2015, pp. 200–205.

[13] J. Qiao, X. S. Shen, J. W. Mark, and L. Lei, "Video quality provisioning for millimeter wave 5G cellular networks with link outage," *IEEE Transactions on Wireless Communications*, vol. 14, no. 10, pp. 5692–5703, Oct. 2015.

[14] O. Semiari, W. Saad, M. Bennis, and B. Maham, "Mobility management for heterogeneous networks: leveraging millimeter wave for seamless handover," in *Proceedings of IEEE Global Communications Conference (GLOBECOM) Mobile and Wireless Networks Symposium*, Dec. 2017, pp. 1–6.

[15] X. Wang, M. Chen, T. Taleb, A. Ksentini, and V. C. M. Leung, "Cache in the air: exploiting content caching and delivery techniques for 5G systems," *IEEE Communications Magazine*, vol. 52, no. 2, pp. 131–139, Feb. 2014.

[16] Y. Rao, H. Zhou, D. Gao, H. Luo, and Y. Liu, "Proactive caching for enhancing user-side mobility support in named data networking," in *2013 Seventh International Conference on Innovative Mobile and Internet Services in Ubiquitous Computing*, July 2013, pp. 37–42.

[17] K. Poularakis and L. Tassiulas, "Exploiting user mobility for wireless content delivery," in *2013 IEEE International Symposium on Information Theory*, July 2013, pp. 1017–1021.

[18] A. S. Gomes, B. Sousa, D. Palma, V. Fonseca, Z. Zhao, E. Monteiro, T. Braun, P. Simoes, and L. Cordeiro, "Edge caching with mobility prediction in virtualized LTE mobile networks," *Future Generation Computer Systems*, 2016, www.sciencedirect.com/science/article/pii/S0167739X16302072.

[19] D. Gale and L. Shapley, "College admissions and the stability of marriage," *American Mathematical Monthly*, vol. 69, no. 1, pp. 9–15, Jan. 1962.

[20] E. Jorswieck, "Stable matchings for resource allocation in wireless networks," in *Proceedings of 17th International Conference on Digital Signal Processing (DSP)*, July 2011, pp. 1–8.

[21] A. Prasad, O. Tirkkonen, P. Lunden, O. N. Yilmaz, L. Dalsgaard, and C. Wijting, "Energy-efficient inter-frequency small cell discovery techniques for LTE-advanced heterogeneous network deployments," in *IEEE Communications Magazine*, May 2013, pp. 72–81.

[22] O. Semiari, W. Saad, M. Bennis, and B. Maham, "Caching meets millimeter wave communications for enhanced mobility management in 5G networks," *IEEE Transactions on Wireless Communications*, vol. 17, no. 2, pp. 779–793, Feb. 2018.

[23] O. Semiari, W. Saad, M. Bennis, and Z. Dawy, "Inter-operator resource management for millimeter wave multi-hop backhaul networks," *IEEE Transactions on Wireless Communications*, vol. 16, no. 8, pp. 5258–5272, Aug. 2017.

[24] O. Semiari, W. Saad, and M. Bennis, "Joint millimeter wave and microwave resources allocation in cellular networks with dual-mode base stations," *IEEE Transactions on Wireless Communications*, vol. 16, no. 7, pp. 4802–4816, July 2017.

[25] O. Semiari, W. Saad, M. Bennis, and M. Debbah, "Performance analysis of integrated sub-6 GHz-millimeter wave wireless local area networks," in *Proceedings of IEEE Global Communications Conference (GLOBECOM) Mobile and Wireless Networks Symposium*, Dec. 2017 pp. 1–7.

[26] O. Semiari, W. Saad, M. Bennis, and M. Debbah, "Integrated millimeter wave and sub-6 ghz wireless networks: a roadmap for joint mobile broadband and ultra-reliable low-latency communications," *IEEE Wireless Communications Magazine*, arXiv:1802.03837v2, Oct. 2018.

[27] A. Ravanshid, P. Rost, D. S. Michalopoulos, V. V. Phan, H. Bakker, D. Aziz, S. Tayade, H. D. Schotten, S. Wong, and O. Holland, "Multi-connectivity functional architectures in 5G," in *2016 IEEE International Conference on Communications Workshops*, May 2016, pp. 187–192.

[28] S. Singh, M. N. Kulkarni, A. Ghosh, and J. G. Andrews, "Tractable model for rate in self-backhauled millimeter wave cellular networks," *IEEE Journal on Selected Areas in Communications*, vol. 33, no. 10, pp. 2196–2211, Oct. 2015.

[29] 3GPP, "E-UTRA: mobility enhancements in heterogeneous networks," 3rd Generation Partnership Project, vol. Rel 11, Sept. 2012.

[30] A. Ghosh, R. Ratasuk, P. Moorut, T. S. Rappaport, and S. Sun, "Millimeter-wave enhanced local area systems: a high-data-rate approach for future wireless networks," *IEEE Journal on Selected Areas in Communications*, vol. 32, no. 6, pp. 1152 –1163, June 2014.

[31] C. H. M. de Lima, M. Bennis, and M. Latva-aho, "Modeling and analysis of handover failure probability in small cell networks," in *Proceedings of IEEE Conference on Computer Communications Workshops*, Apr. 2014, pp. 736–741.

[32] S. Kadam and M. H. Kotowski, "Multi-period matching," http://scholar.harvard.edu/kadam/publications/multi-period-matching, Apr. 2016.

[33] O. Semiari, W. Saad, S. Valentin, M. Bennis, and B. Maham, "Matching theory for priority-based cell association in the downlink of wireless small cell networks," in *2014 IEEE International Conference on Acoustics, Speech and Signal Processing (ICASSP)*, May 2014, pp. 444–448.

[34] O. Semiari, W. Saad, Z. Dawy, and M. Bennis, "Matching theory for backhaul management in small cell networks with mmwave capabilities," in *2015 IEEE International Conference on Communications (ICC)*, June 2015, pp. 1–6.

[35] O. Semiari, W. Saad, and M. Bennis, "Context-aware scheduling of joint millimeter wave and microwave resources for dual-mode base stations," in *2016 IEEE International Conference on Communications (ICC)*, May 2016, pp. 1–6.

[36] O. Semiari, W. Saad, and M. Bennis, "Downlink cell association and load balancing for joint millimeter wave-microwave cellular networks," in *2016 IEEE Global Communications Conference (GLOBECOM)*, Dec. 2016.

[37] O. Semiari, W. Saad, S. Valentin, M. Bennis, and H. V. Poor, "Context-aware small cell networks: how social metrics improve wireless resource allocation," *IEEE Transactions on Wireless Communications*, vol. 14, no. 11, pp. 5927 – 5940, Nov. 2015.

[38] A. E. Roth and M. A. O. Sotomayor, *Two-Sided Matching: A Study in Game-Theoretic Modeling and Analysis*. Cambridge: Cambridge University Press, 1992.

Part IV

Energy-Efficiency, Security, Economic, and Deployment

15 Energy-Efficient Deployment in Wireless Edge Caching

Thang X. Vu, Symeon Chatzinotas, and Björn Ottersten

In this chapter, we investigate the performance of edge caching wireless networks by taking into account the caching capability when designing the signal transmission. We consider hierarchical caching systems in which the contents can be prefetched at both user terminals or the base station (BS) and investigate the energy performance under two notable uncoded and coded caching strategies. The backhaul and access throughputs are derived for both caching policies for arbitrary values of base station and user cache sizes from which closed-form expressions for the corresponding system energy efficiency (EE) are obtained. Furthermore, we propose two optimization problems to maximize the system EE and minimize the content delivery time subject to some given quality of service requirements.

15.1 Introduction

Future wireless networks will have to address stringent requirements of delivering content at high speed and low latency due to the proliferation of mobile devices and data-hungry applications. Toward this goal, there have proposed various network architectures to improve the system throughput such as cloud radio access networks (C-RANs) [1, 2] and heterogeneous networks (HetNets). Unfortunately, traffic congestion might still occur during peak-traffic hours, due to the uneven popularity distribution of the convent. A promising solution to overcome this challenge is to distribute the content in advance close to end users via distributed storages in the network, which is known as caching [3]. In general, caching consists of two phases: placement and delivery. The placement phase is implemented during off-peak hours when the network resources are abundant, which prefetch some popular contents in the caches. In the latter, which usually occurs during peak-traffic times, the users request desirable files. The requested files will be served immediately if they have been prefetched in the local caches. In this manner, caching improves load balancing and significantly reduces the backhaul's load during peak-traffic times [3, 4].

Existing research on caching focuses on two methods—namely uncoded and coded caching [3, 5–7]. In the uncoded caching strategy, the placement phase is designed

This work is supported by the Luxembourg National Research Fund under project FNR CORE ProCAST, grant R-AGR-3415-10.

to maximize the local caching gain that linearly increases with the number of file parts available in the local storage. On the other hand, the coded caching can further improve the caching gain by allowing multicast transmission of coded messages during the delivery phase [4, 8, 9]. Thanks to the careful design in coded caching placement phase, all users can recover the requested files via a multicast stream [4]. The trade-off between the memory and the transmission latency is studied in [10], which is normalized delivery time. The rate-memory trade-off of multi-layer coded caching networks is studied in [11, 12]. It is worthy noting that in order to achieve the global gain in the coded caching, the data center requires to know the number of users in order to construct the coded messages.

The performance of caching systems can be further improved via joint design of content caching and signal transmission. Such improvements come from the consideration of the cached content at the edge nodes when designing the signal transmission. It is shown in [13] that transmission costs (power and bandwidth) can be reduced, thanks to a joint design of multicast beam-forming and power allocation. In [14], the authors study the impact of wireless backhaul on the energy efficiency in edge caching wireless networks. A joint optimization of caching, routing, and channel assignment is proposed in [15]. The benefit of caching in wireless networks are analysed in [16–20] for device-to-device (D2D) networks and in [21–23] under energy efficiency perspective. It is worth noting that the above mentioned works investigate either only the uncoded caching policy or separate caching at higher layers without considering the signal transmission.

In this chapter, we investigate edge caching wireless networks in which caching can be implemented at both the BS and user levels. The contributions of this chapter are summarized as follows:

- First, we investigate the performance of two popular *uncoded* and (inter-file) *coded* caching policies and consider arbitrary storage capacities when calculating the backhaul and access throughputs.
- Second, we analyze the system EE via closed-form expressions for both caching strategies, which give insights of the key system parameters. The system EE is further optimized subject to some given quality-of-service (QoS) constraints. Furthermore, we derive a closed-form expression for the maximum EE under zero-forcing (ZF) precoding design.
- Third, the delivery time under both caching policies are analyzed and optimized via the joint design of the beam-forming vectors and power allocation. The effectiveness of the proposed designs are validated via numerical results. In particular, we show that the uncoded caching can surpass the coded caching in terms of EE in some cases.

The rest of this chapter is organized as follows. Section 15.2 describes the system model and the caching strategies. Section 15.3 provides analyses for EE. Section 15.4 provides details of our proposed optimizations. Section 15.5 minimizes the delivery time. Section 15.6 derives the EE for general content popularity. Section 15.7 shows numerical results. Finally, Section 15.8 concludes the chapter.

Notation: $(.)^H$,$(x)^+$ and Tr(.) denote the Hermitian transpose, $\max(0,x)$ and the trace(.) function, respectively. $\lfloor x \rfloor$ denotes the largest integer not exceeding x.

15.2 Signal Transmission and Caching Model

A system under consideration consists of K users, denoted by $\mathcal{K} = \{1,\ldots,K\}$, served by a data center via a common BS, shown in Figure 15.1. The considered model can find application in many practical use cases in which the users can play the role of various cache-assisted edge nodes such as small-cell BSs in the HetNet. Let L denote the antenna number at the BS, where $L \geq K$, which serves all users via a shared wireless medium. The BS has full access to the contents at the data center. We consider block Rayleigh fading channels, in which the channel fading coefficients are fixed within a block and are mutually independent across the users. It is assumed that the block duration is sufficiently long such that all the user requests can be served. Without loss of generality, the data center contains N files of equal size of Q bits and is denoted by $\mathcal{F} = \{F_1,\ldots,F_N\}$.

15.2.1 Caching Model

A multi-layer caching network is considered in which popular contents can be stored at both the BS and users. Denote M_b and M_u with $0 \leq M_b, M_u \leq N$ as the cache size at the BS and users, respectively. In this chapter, offline caching policy is employed, which comprise two separate *placement* and *delivery phases*. The placement phase is executed during off-peak hours [4]. For robustness, we employ distributed cache placement in which the BS is unaware of user cache's content. As such, the BS's cache is feed with $\frac{M_b Q}{N}$ (non-overlapping) bits of every file, which are randomly chosen. In the same

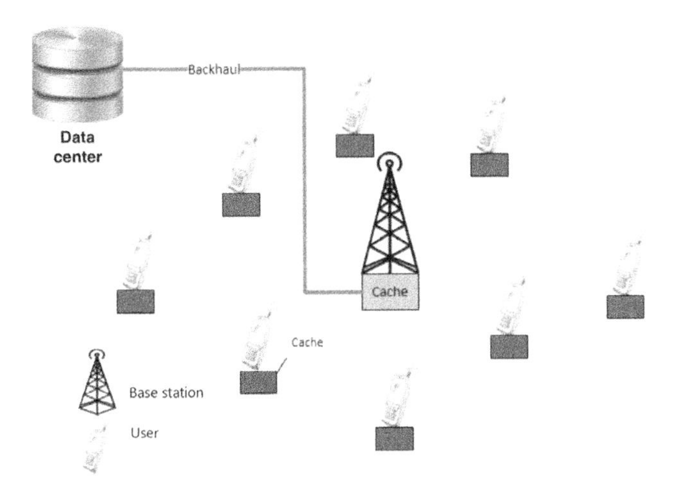

Figure 15.1 Cache-assisted wireless networks. Caching is available at the BS or users.

manner, each user keeps $\frac{M_u Q}{N}$ bits of every file in its local storage under the uncoded caching strategy. For the coded caching policy, the placement phase at the user caches is implemented as in [4]. As a result, total number of bits stored at the BS and user caches are respectively $M_b Q$ and $M_u Q$ bits, which satisfy the memory constraints.

In the *delivery phase*, each user send a file index to the data center via the BS. Similar to [4], we consider the worst case in which the files are equally popular and one user request is different from the others. Non-uniform distributions of the content popularity, e.g., Zipf distribution, is discussed in Section 15.6. Denote $d_1, \ldots, d_K$ as the file indexes requested by user $1, \ldots, K$, respectively. We note that the requested file can be served immediately if it is available in the user cache. Otherwise, the requested file will be sent from the BS. In the following, we describe in details the delivery phase for two popular uncoded and coding caching policies.

15.2.1.1 Uncoded Caching

In the uncoded caching policy, the user requests are served independently. The benefits of this policy lay in the fact that the users do not know the cache content of each other and hence do not require any cooperation among the users. Let $Q_{\text{unc, BH}}$ and $Q_{\text{unc, AC}}$ denote the aggregated number of bits transmitted through the backhaul and access links, respectively.

Proposition 2 The total number of bits transmitted through the backhaul and access links under the uncoded caching policy are $Q_{\text{unc, BH}} = KQ\left(1 - \frac{M_u}{N}\right) \times \left(1 - \frac{M_b}{N}\right)$ and $Q_{\text{unc, AC}} = KQ\left(1 - \frac{M_u}{N}\right)$, respectively.

The proof of Proposition 2 is given in [8, section II].

15.2.1.2 Coded Caching

In order for the coded caching to exploit the multicast benefit of the shared wireless medium, the BS needs to know the number of users in both placement and delivery phases in order to intelligently encodes the requested files.

Proposition 3 Denote $m = \lfloor \frac{KM_u}{N} \rfloor \in \mathbb{Z}^\star$ and $\delta = \frac{KM_u}{N} - m$ with $0 \leq \delta < 1$. The access throughput under the coded caching is $Q_{\text{cod, AC}} = (1-\delta)\frac{Q(K-m)}{m+1} + \delta\frac{Q(K-m-1)}{m+2}$, and the backhaul throughput is $Q_{\text{cod, BH}} = (1-\delta)\left(1 - \left(\frac{M_b}{N}\right)^{m+1}\right) \frac{Q(K-m)}{m+1} + \delta\left(1 - \left(\frac{M_b}{N}\right)^{m+2}\right)\frac{Q(K-m-1)}{m+2}$.

Proof The proof of Proposition 3 is given in [24]. □

The results in Proposition 3 provide the close-form expressions of the backhaul and access throughputs under the two caching policies for arbitrary values $M_u \in [0, N]$. Specially if $\frac{KM_u}{N} \in \mathbb{Z}^\star$ then $\delta = 0$, the results are simplified as $\left(1 - \left(\frac{M_b}{N}\right)^{m+1}\right)\frac{Q(K-m)}{m+1}$ and $\frac{KQ(1-M_u/N)}{1+KM_u/N}$, respectively, which can also be found in [4]. It is worth highlighting that the results in [4] apply only to integer $\frac{KM_u}{N}$. Meanwhile Proposition 3 derives general results for arbitrary values of the user cache size.

15.2.2 Transmission Model

The requested files, which are not available in the user's local storage, will be transmitted from the BS. Denote $\mathbf{h}_k \in \mathbb{C}^{L\times 1}$ as the channel coefficients from the BS antennas to the kth user, whose elements follows a complex Gaussian distribution with zero-mean and variance $\sigma^2_{h_k}$, which accounts for the pathloss from the BS antennas to user k. We assume that the BS has full knowledge of all channels thanks to pilot-assisted channel estimation.

15.2.2.1 Signal Transmission for Uncoded Caching Strategy

In uncoded caching, the BS serves the user requests separately. Let F_{d_k} denote the file requested by user k, where d_k indicates the index of the requested file. Let $\bar{F}_{d_k}$ denote part of the requested file that is not at the user k's cache. Upon receiving the user requests, the BS first modulates $\bar{F}_{d_k}$ in to a corresponding modulated signal x_k and then sends the precoded signal through the shared access channels. In order to mitigate interuser interference, the BS applies a precoding vector $\mathbf{w}_k \in \mathbb{C}^{L\times 1}$ for user k. Denote by n_k the noise at user k, which is a Gaussian random variable with zero-mean and variance σ^2. User k will receive a signal given as $y_k = \mathbf{h}_k^H \mathbf{w}_k x_k + \sum_{l\neq k} \mathbf{h}_k^H \mathbf{w}_l x_l + n_k$. The achievable information rate of user k is

$$R_{\text{unc},k} = B\log_2(1+\text{SINR}_k), 1 \le k \le K, \tag{15.1}$$

where $\text{SINR}_k = \frac{|\mathbf{h}_k^H \mathbf{w}_k|^2}{\sum_{l\neq k} |\mathbf{h}_k^H \mathbf{w}_l|^2 + \sigma^2}$ is the signal-to-interference-plus-noise ratio at user k, and B is the access channel bandwidth.

The total transmit power in this case is $P_{\text{unc}} = \sum_{k=1}^{K} \| \mathbf{w}_k \|^2$.

15.2.2.2 Signal Transmission for Coded Caching Strategy

It is observed in the policy that one coded message is needed by a group of users. Thus in order to exploit the coded-caching architecture, physical-layer multicasting [25] is employed to multicast the coded messages to the corresponding users.

In the coded caching strategy, the BS transmits the total number of C_K^{m+1} coded messages (of length $\frac{Q}{C_K^m}$ bits), each of which is received by a subset of $m+1$ users [4]. Let $S \subset \mathcal{K}$ denote an arbitrary subset of $m+1$ users, and let $\mathcal{S} = \{S \mid |S| = m+1\}$ denote all possible subsets. It is straightforward to have $|\mathcal{S}| = C_K^{m+1}$. For ease of presentation, let $X_{\mathcal{S}}$ denote the coded message interested by the users in $\mathcal{S}$. The coded message is first modulated into $x_{\mathcal{S}}$ before being multiplied by a common precoding vector $\mathbf{w}_{\mathcal{S}}$. The achievable rate for the users within $\mathcal{S}$ is given as

$$R_{\text{cod},\mathcal{S}} = \min_{k\in\mathcal{S}} \left\{ B\log_2\left(1 + \frac{|\mathbf{h}_k^H \mathbf{w}_{\mathcal{S}}|^2}{\sigma^2}\right)\right\}. \tag{15.2}$$

The transmit power in the coded caching is $P_{\text{cod}} = \| \mathbf{w}_{\mathcal{S}} \|^2$.

15.3 Energy-Efficiency Analysis

In this section, we analyze EE performance of the two considered caching policies.

DEFINITION 15.1 (Energy efficiency) *The EE measured in bit/Joule is defined as:*

$$\mathrm{EE} = \frac{KQ}{E_{\Sigma}},$$

where KQ stands for the total bits requested by K users and E_{Σ} is the amount of energy spent to serve these bits.

Remember that E_{Σ} in the EE represents the energy cost during the delivery phase since the cost in the cache placement phase is negligible because it occurs rarely compared with the delivery phase [4, 13].

15.3.1 EE Analysis for Uncoded Caching Strategy

In uncoded caching, the total energy consumption is computed as $E_{\mathrm{unc},\Sigma} = E_{\mathrm{unc,BH}} + E_{\mathrm{unc,AC}}$, where $E_{\mathrm{unc,BH}}$ and $E_{\mathrm{unc,AC}}$ represents the energy consumption on the backhaul and access channels, respectively.[1] It is noted that each user requests $\frac{Q_{\mathrm{unc,AC}}}{K}$ bits, which are sent to the users independently via unicast mode. With the required rate $R_{\mathrm{unc},k}$, it costs $\frac{Q_{\mathrm{unc,AC}}}{KR_{\mathrm{unc},k}}$ seconds to serve user k's request. As a result, the access links consume a total amount of energy given as

$$E_{\mathrm{unc,AC}} = \frac{Q_{\mathrm{unc,AC}}}{KR_{\mathrm{unc},k}} P_{\mathrm{unc}} = Q\left(1 - \frac{M_u}{N}\right)\sum\nolimits_{k=1}^{K} \frac{\| \mathbf{w}_k \|^2}{R_{\mathrm{unc},k}}.$$

Because the backhaul link has sufficient capacity to serve the access network, the backhaul's energy consumption can be calculated as a linear function of the backhaul throughput, given as

$$E_{\mathrm{unc,BH}} = \eta Q_{\mathrm{unc,BH}} = \eta KQ\left(1 - \frac{M_u}{N}\right)\left(1 - \frac{M_b}{N}\right),$$

where η is a constant, which plays a role as the pricing factor used to trade energy for delivered bits [13]. We note that the practical value of η is determined by the backhaul technology.

As a result, the EE under the uncoded caching strategy is given as

$$\mathrm{EE}_{\mathrm{unc}} = \frac{K}{\left(1 - \frac{M_u}{N}\right)\left(\eta K\left(1 - \frac{M_b}{N}\right) + \sum_{k=1}^{K} \frac{\|\mathbf{w}_k\|^2}{R_{\mathrm{unc},k}}\right)}. \tag{15.3}$$

From (15.3) we can see that $\mathrm{EE}_{\mathrm{unc}}$ is jointly determined by the cache sizes M_u and M_b and the access transmitted power.

[1] E_{Σ} may also include a static energy consumption factor in practice.

15.3.2 EE Analysis for Coded Caching Strategy

Similar to the uncoded caching, the backhaul energy cost is given as $E_{\text{cod,BH}} = \eta Q_{\text{cod,BH}}$, where η is the pricing factor. To compute the access energy cost $E_{\text{cod,AC}}$, we note that the BS transmits to all users in $\mathcal{S}$ simultaneously. With the rate $R_{\text{cod},\mathcal{S}}$, it takes $\frac{Q_{\text{cod,AC}}}{C_K^{m+1} R_{\text{cod},\mathcal{S}}}$ seconds to send $X_{\mathcal{S}}$. As a result, the total energy cost in this case is calculated as $E_{\text{cod,AC}} = \frac{Q_{\text{cod,AC}}}{C_K^{m+1}} \sum_{\mathcal{S}\in\mathcal{S}} \frac{P_{\text{cod},\mathcal{S}}}{R_{\text{cod},\mathcal{S}}}$. Thus the EE under the coded caching strategy is given as

$$\text{EE}_{\text{cod}} = \frac{KQ}{E_{\text{cod},\Sigma}} = \frac{KQ}{\eta Q_{\text{cod,BH}} + \frac{Q_{\text{cod,AC}}}{C_K^{m+1}} \sum_{\mathcal{S}\in\mathcal{S}} \frac{P_{\text{cod},\mathcal{S}}}{R_{\text{cod},\mathcal{S}}}}. \tag{15.4}$$

From Proposition 3 we obtain

$$\text{EE}_{\text{cod}} = \frac{1+KM_u/N}{\left(1-\frac{M_u}{N}\right)\left(\eta\left(1-\left(\frac{M_b}{N}\right)^{\frac{KM_u}{N}+1}\right)+\frac{1}{C_K^{m+1}}\sum_{\mathcal{S}\in\mathcal{S}}\frac{\|\mathbf{w}_{\mathcal{S}}\|^2}{R_{\text{cod},\mathcal{S}}}\right)}.$$

It is observed that the EE in the coded-caching is driven by the BS and user cache size as well as the BS's transmit power.

15.3.3 Comparison between the Two Strategies

In this subsection, we give insights on the comparison between the two caching strategies in some specific scenarios, e.g., the sate quality of service requirements $R_{\text{unc},k} = R_{\text{cod},\mathcal{S}} = \gamma, \forall k, \mathcal{S}$.

15.3.3.1 Free-Cost Backhaul Link

When the backhaul capacity is sufficiently large, we can assume that all the requested files are available at either the BS or user storages. Such a case is equivalent to $M_b = N$ or $\eta = 0$. As a result, we have

$$\text{EE}_{\text{unc}} = \frac{K}{\left(1-\frac{M_u}{N}\right)\frac{P_{\text{unc}}}{\gamma}}, \quad \text{EE}_{\text{cod}} = \frac{1+KM_u/N}{\left(1-\frac{M_u}{N}\right)\frac{P_{\text{cod}}}{\gamma}}.$$

Consider the case that the BS operates at the same transmit power, i.e., $P_{\text{unc}} = P_{\text{cod}}$, we have $\text{EE}_{\text{unc}} > \text{EE}_{\text{cod}}$. As such the coded caching method obtains a higher EE than the uncoded caching strategy when $M_u > \left(\frac{P_{\text{cod}}}{P_{\text{unc}}} - \frac{1}{K}\right)N$.

No caching at the BS: $M_b = 0$ In this case, all the un-cached files at the user storage will be transmitted from the data center, and thus

$$\text{EE}_{\text{unc}} = \frac{1}{\left(1-\frac{M_u}{N}\right)\left(\eta+\frac{P_{\text{unc}}}{\gamma K}\right)}, \quad \text{EE}_{\text{cod}} = \frac{1+KM_u/N}{\left(1-\frac{M_u}{N}\right)\left(\eta+\frac{P_{\text{cod}}}{\gamma}\right)}.$$

It is clearly shown that the coded-caching strategy will surpass the uncoded caching for the same transmit power because $\frac{KM_u}{N} > 0$ and $\frac{P_{\text{unc}}}{K} < P_{\text{cod}}$.

15.4 Energy-Efficiency Maximization in Edge Caching Wireless Networks

In this section, we maximize the system EE for the two considered caching policies via joint design of the precoding vectors. In general, the optimization can be stated as

$$\underset{\{\mathbf{w}_k\}_{k=1}^K, \mathbf{w}}{\texttt{Maximize}}\ \text{EE} \quad \texttt{s.t.}\ \text{QoS constraint}, \tag{15.5}$$

where $\text{EE} \in \{\text{EE}_{\text{unc}}, \text{EE}_{\text{cod}}\}$.

15.4.1 EE Maximization for Uncoded Caching Strategy

In the uncoded caching method, the BS serves the users separately. If the requested file has not been cached, it takes $t_k = \frac{Q}{\gamma_k}$ seconds to deliver the requested file to user k, where γ_k is the rate requirement of user k (bits per second). In case that part of the requested file is available at user's local cache, the BS will send only noncached parts to the users. The equivalent QoS constraint in this case becomes $\bar{\gamma}_k = (1 - \frac{M_u}{N})Q/t_k = (1 - \frac{M_u}{N})\gamma_k$. An important observation from (15.3) is that the network topology and the cache sizes are usually fixed in practice. Thus problem (15.5) is equivalent to the minimization of the transmit energy, stated as

$$\underset{\{\mathbf{w}_k \in \mathbb{C}^L\}_{k=1}^K}{\texttt{Minimize}} \sum_{k=1}^K \frac{\| \mathbf{w}_k \|^2}{R_{\text{unc},k}}, \ \texttt{s.t.}\ \frac{|\mathbf{h}_k^H \mathbf{w}_k|^2}{\sum_{l \neq k} |\mathbf{h}_k^H \mathbf{w}_l|^2 + \sigma^2} \geq \zeta_k, \forall k, \tag{15.6}$$

where the rate constraint is transferred into a SINR requirement $\zeta_k = 2^{\frac{\bar{\gamma}_k}{B}} - 1$.

In the following section, we optimize the EE based on two beamforming vector designs.

15.4.1.1 Cost Minimization by Zero-Forcing Precoding

The ZF design is selected thanks to its fast implementation. In this design, the direction of the beam-forming vectors are chosen to fully mitigate interuser interference. Thus the main goal is to optimize transmit power on each beam. Denote $p_k, 1 \leq k \leq K$, as the transmit power designed for user k. The precoding vector for user k under the ZF design is $\mathbf{w}_k = \sqrt{p_k}\tilde{\mathbf{h}}_k$, where $\tilde{\mathbf{h}}_k$ the kth column of the ZF matrix $\mathbf{H}^H(\mathbf{H}\mathbf{H}^H)^{-1}$, with $\mathbf{H} = [\mathbf{h}_1, \ldots, \mathbf{h}_K]^T$.

THEOREM 15.2 *Under the ZF design, the uncoded caching strategy achieves the maximum EE*

$$\text{EE}_{\text{unc}}^{\text{ZF}} = \frac{K}{\left(1 - \frac{M_u}{N}\right)\left(\eta K\left(1 - \frac{M_b}{N}\right) + \sigma^2 \sum_{k=1}^K \frac{\zeta_k \|\tilde{\mathbf{h}}_k\|^2}{\bar{\gamma}_k}\right)}.$$

Proof By design, $|\mathbf{h}_l^H \mathbf{w}_k|^2 = p_k \delta_{lk}$, where δ_{ij} is equal to 1 if $i = j$ and 0 otherwise. As a result, we can reformulate problem (15.6) as

$$\underset{\{p_k:p_k\geq 0\}_{k=1}^K}{\texttt{Minimize}} \sum_{k=1}^{K} \frac{a_k p_k}{\log_2(1+p_k/\sigma^2)}, \quad \texttt{s.t.} \;\; p_k \geq \zeta_k\sigma^2, \forall k, \tag{15.7}$$

where $a_k = \| \tilde{\mathbf{h}}_k \|^2$.

Now we consider a function $f(x) = \frac{ax}{\log_2(1+bx)}$ with $a,b \geq 0$ in $\mathbb{R}^+$. It is straightforward to see that this function strictly increases in its domain. As a result, the optimal solution of (15.7) is achieved at $p_k^\star = \zeta_k\sigma^2$, and the minimum transmit power is $\sigma^2\sum_{k=1}^{K}\zeta_k \;\| \tilde{\mathbf{h}}_k \|^2$. We complete the proof of Theorem 15.2 by substituting the minimum transmit power into $\mathrm{EE}_{\mathrm{unc}}$. □

15.4.1.2 Cost Minimization by Semi-definite Relaxation

The EE can be further improved via the optimization of both the direction and magnitude of the precoding vectors. Since problem (15.6) is difficult to solve due to the nonconvexity of its objective function, we instead solve (15.6) suboptimally by using the upper bound of the objective function. Obviously, we have $\frac{\|\mathbf{w}_k\|^2}{R_{\mathrm{unc},k}} \leq \frac{\|\mathbf{w}_k\|^2}{\bar{\gamma}_k}$. Let $\mathcal{K}_t \triangleq \left\{k \mid \bar{\gamma}_k \leq \frac{Q}{t}, \forall t \in \left[0, \frac{Q}{\min_k(\bar{\gamma}_k)}\right]\right\}$ denote a group of users that are active at time instance t. Then we can approximate problem (15.6) as follows:

$$\underset{\mathbf{w}_k\in\mathbb{C}^L}{\texttt{Minimize}} \sum_{k\in\mathcal{K}_t} \frac{\| \mathbf{w}_k \|^2}{\bar{\gamma}_k}, \;\texttt{s.t.}\; \frac{|\mathbf{h}_k^H\mathbf{w}_k|^2}{\sum_{k\neq l\in\mathcal{K}_t} |\mathbf{h}_k^H\mathbf{w}_l|^2 + \sigma^2} \geq \zeta_k, \forall k. \tag{15.8}$$

In the next step, we employ new variables $\mathbf{X}_k = \mathbf{w}_k\mathbf{w}_k^H \in \mathbb{C}^{L\times L}$ and denote parameters $\mathbf{A}_k = \mathbf{h}_k\mathbf{h}_k^H \in \mathbb{C}^{L\times L}$. It is straightforward to verify that $|\mathbf{h}_l^H\mathbf{w}_k|^2 = \mathbf{h}_l^H\mathbf{w}_k\mathbf{w}_k^H\mathbf{h}_l = \mathrm{Tr}(\mathbf{h}_l\mathbf{h}_l^H\mathbf{w}_k\mathbf{w}_k^H) = \mathrm{Tr}(\mathbf{A}_l\mathbf{X}_k)$. Thus problem (15.8) can be reformulated as follows:

$$\underset{\mathbf{X}_k\in\mathbb{C}^{L\times L}}{\texttt{Minimize}} \sum_{k\in\mathcal{K}_t} \mathrm{Tr}(\mathbf{X}_k), \;\texttt{s.t.}\; \mathrm{Tr}(\mathbf{A}_k\mathbf{X}_k) \geq \zeta_k \sum_{k\neq l\in\mathcal{K}_t} \mathrm{Tr}(\mathbf{A}_l\mathbf{X}_k) + \zeta_k\sigma^2, \forall k, \tag{15.9}$$
$$\mathbf{X}_k \succeq \mathbf{0}, \mathrm{rank}(\mathbf{X}_k) = 1, \forall k.$$

Because the trace function is linear, the objective and the two first constraints of (15.9) are convex. The challenge to solve (15.9) lays in the nonconvex rank-one constraint. Fortunately, this constraint can be handled efficiently by the semi-definite relaxation (SDR) method [26]. In order to improve the performance of the SDR, a Gaussian randomization procedure is used if the rank-one constraint is violated. It has been shown in [27] that the SDR method can approach the performance of the optimal solution.

15.4.2 EE Maximization for Coded Caching Strategy

In order to satisfy user k with a QoS γ_k, the BS must serve the requested file within $t_k = \frac{Q}{\gamma_k}$ seconds. In the coded caching method, each user receives only C_{K-1}^m coded messages in a total of C_K^{m+1}. Therefore, user k is actually active in $\frac{C_{K-1}^m}{C_K^{m+1}} t_k = \frac{(m+1)Q}{K\gamma_k}$. Consequently, the equivalent QoS for user k is $\bar{\gamma}_k = \left(\frac{Q*C_{K-1}^m}{C_K^m}\right)/\left(\frac{(m+1)Q}{K\gamma_k}\right) = \frac{K-m}{m+1}\gamma_k$, where $\frac{Q*C_{K-1}^m}{C_K^m}$ is total number of bits sent to user k. It is observed from (15.4) that the EE maximization is equivalent to minimizing $\frac{P_{\text{cod}}}{R_{\text{cod},\mathcal{S}}}$. The optimization problem is formulated as

$$\underset{\mathbf{w}_{\mathcal{S}} \in \mathbb{C}^{L\times 1}}{\texttt{Minimize}} \ \frac{\| \mathbf{w}_{\mathcal{S}} \|^2}{R_{\text{cod},\mathcal{S}}}, \quad \texttt{s.t.} \ R_{\text{cod},\mathcal{S}} \geq \bar{\gamma}_k, \forall k \in \mathcal{S}, \tag{15.10}$$

where $\mathcal{S}$ denote the subset of active users.

Solving problem (15.10) is challenging due to the nonconvexity of the objective function. Thus we instead minimize the upper bound of the objective function, i.e., $\frac{\text{Tr}(\mathbf{X})}{\bar{\gamma}_{\min,\mathcal{S}}}$, where $\bar{\gamma}_{\min,\mathcal{S}} = \min_{k\in\mathcal{S}} \bar{\gamma}_k$. Let $\mathbf{X} = \mathbf{w}_{\mathcal{S}}^H \mathbf{w}_{\mathcal{S}} \in \mathbb{C}^{L\times L}$ denote a new variable. We then reformulate the problem as

$$\underset{\mathbf{X} \in \mathbb{C}^{L\times L}}{\texttt{Minimize}} \frac{\text{Tr}(\mathbf{X})}{\bar{\gamma}_{\min,\mathcal{S}}}, \quad \texttt{s.t.} \ \mathbf{X} \succeq \mathbf{0}; \ \text{rank}(\mathbf{X}) = 1; \tag{15.11}$$
$$\text{Tr}(\mathbf{A}_k\mathbf{X}) \geq \sigma^2 \left(2^{\frac{\bar{\gamma}_{\min,\mathcal{S}}}{B}} - 1 \right), \forall k \in \mathcal{S}.$$

A similar observation is that the objective function of (15.11) is convex since function trace is linear. In addition, the constraints are also convex, except the rank-one constraint. Therefore, we employ the SDR method to approximately solve (15.11) by avoiding the rank-one constraint. Since the SDR does not always provide a rank-one solution, we employ Gaussian randomization to improve the SDR solution [27].

15.5 Minimization of Content Delivery Time

One important performance metric is the average delivery time it takes to serve all the user requests. In general, there are two components that contribute to the delivery time: backhaul and access. Because the backhaul capacity is generally much larger than the access capacity, the delivery time on the backhaul link is negligible. In addition, we assume that the processing time at the BS is fixed and negligible. Thus the access links mainly contribute to the total delivery time.

15.5.1 Minimization of Delivery Time for Uncoded Caching Strategy

Denote by t_k a time that the BS needs to send all user k's requested bits. With a serving rate $R_{\text{unc},k}$, we have $t_k = \frac{Q_{\text{unc,AC}}}{K R_{\text{unc},k}}$. Therefore, the uncoded caching policy spends an average delivery time

$$\tau_{\text{unc}} = \frac{1}{K} \sum\nolimits_{k=1}^{K} t_k = \frac{Q}{K} \left(1 - \frac{M_u}{N} \right) \sum\nolimits_{k=1}^{K} \frac{1}{R_{\text{unc},k}}.$$

The average delivery time minimization problem can be formulated as follows:

$$\underset{\{\mathbf{w}_k \in \mathbb{C}^L\}_{k=1}^K}{\texttt{Minimize}} \ \frac{Q}{K} \left(1 - \frac{M_u}{N} \right) \sum\nolimits_{k=1}^{K} \frac{1}{R_{\text{unc},k}} \tag{15.12}$$
$$\texttt{s.t.} \ R_{\text{unc},k} \geq \bar{\gamma}_k, \forall k; \ \sum\nolimits_{k=1}^{K} \| \mathbf{w}_k \|^2 \leq P_{\Sigma}.$$

In (15.12), the first constraint is to satisfy the QoS requirement and the second constraint is to not exceed for the total transmit power budget.

15.5.1.1 Zero-Forcing Precoding Design

Similar to the previous section, we consider ZF-based design to reduce the computational complexity. The precoding vectors in ZF design is given as $\mathbf{w}_k = \sqrt{p_k}\tilde{\mathbf{h}}_k$, where p_k is the scaling power for user k and $\tilde{\mathbf{h}}_k$ is the the ZF beam-forming vector that is the kth column of $\mathbf{H}^H(\mathbf{H}\mathbf{H}^H)^{-1}$. Since $\mathbf{h}_l^H\tilde{\mathbf{h}}_k = \delta_{lk}$, we thus obtain $R^{ZF}_{\text{unc},k} = \log_2\left(1 + \frac{p_k}{\sigma^2}\right)$. Therefore, we can equivalently reformulate (15.12) as

$$\begin{aligned} &\underset{\{p_k: p_k \geq 0\}_{k=1}^K}{\texttt{Minimize}} \ \frac{Q}{K}\left(1 - \frac{M_u}{N}\right)\sum\nolimits_{k=1}^{K}\frac{1}{\log_2(1 + p_k/\sigma^2)} \\ &\texttt{s.t.} \ \frac{p_k}{\sigma^2} \geq \zeta_k, \forall k; \ \sum\nolimits_k p_k \parallel \tilde{\mathbf{h}}_k \parallel^2 \leq P_\Sigma. \end{aligned} \tag{15.13}$$

Proposition 4 Given the total power P_Σ satisfying $P_\Sigma \geq \sigma^2 \sum_{k=1}^K \zeta_k \parallel \tilde{\mathbf{h}}_k \parallel^2$, the problem (15.13) is convex and feasible.

The proof of Proposition 4 can be found in [24, proposition 3]

15.5.1.2 General Beam-Forming Design

In a general beam-forming design, we jointly optimize the direction and magnitude for every beamforming vector. Since the original problem is difficult to solve as its objective function is nonconvex, we instead minimize the upper bound of the objective function of (15.12), i.e., $\max\{t_1, \ldots, t_K\} = \frac{Q\left(1 - \frac{M_u}{N}\right)}{\min\{R_{\text{unc},1}, \ldots, R_{\text{unc},K}\}}$. The suboptimal optimization of (15.12) is given as

$$\begin{aligned} &\underset{\{\mathbf{w}_k \in \mathbb{C}^L\}_{k=1}^K}{\texttt{Maximize}} \ \min\{R_{\text{unc},1}, \ldots, R_{\text{unc},K}\} \\ &\texttt{s.t.} \ R_{\text{unc},k} \geq \bar{\gamma}_k, \forall k; \ \sum\nolimits_k \parallel \mathbf{w}_k \parallel^2 \leq P_\Sigma. \end{aligned} \tag{15.14}$$

In the next step, we introduce a positive variable x and express the first constraint via SINR representation. Then problem (15.14) can be equivalently reformulated as

$$\begin{aligned} &\underset{x>0, \{\mathbf{w}_k \in \mathbb{C}^L\}_{k=1}^K}{\texttt{Maximize}} \ x, \ \texttt{s.t.} \ \frac{|\mathbf{h}_k^H \mathbf{w}_k|^2}{\sum_{l \neq k} |\mathbf{h}_k^H \mathbf{w}_l|^2 + \sigma^2} \geq x, \forall k, \\ &x \geq \zeta_k; \ \sum\nolimits_k \parallel \mathbf{w}_k \parallel^2 \leq P_\Sigma. \end{aligned} \tag{15.15}$$

Next, we introduce new variables $\mathbf{X}_k = \mathbf{w}_k\mathbf{w}_k^H$ and remember that $\mathbf{A}_k = \mathbf{h}_k\mathbf{h}_k^H$. The problem (15.15) can be reformulated to

$$\begin{aligned} &\underset{\{\mathbf{X}_k \in \mathbb{C}^{L\times L}\}_{k=1}^K, x}{\texttt{Maximize}} \ x, \ \texttt{s.t.} \ x \geq \zeta_k; \ \sum_k \text{Tr}(\mathbf{X}_k) \leq P_\Sigma; \ \text{rank}(\mathbf{X}_k) = 1 \\ &\text{Tr}(\mathbf{A}_k\mathbf{X}_k) - x\sum_{l \neq k}\text{Tr}(\mathbf{A}_k\mathbf{X}_l) \geq x\sigma^2, \forall k; \ \mathbf{X}_k \succeq \mathbf{0}. \end{aligned} \tag{15.16}$$

Table 15.1 Algorithm to Solve (15.16)

1.	Initialize $A_H, A_L = \zeta$, and the accuracy ϵ.
2.	$A_M = (A_H + A_L)/2$.
3.	Given A_M, if (15.17) is feasible, then $A_L := A_M$. Otherwise $A_H := A_M$.
4.	Repeat Steps 2 and 3 until $\lvert A_H - A_L \rvert \le \epsilon$.

Table 15.2 Algorithm to Solve (15.19)

1.	Initialize $A_H, A_L = 2^{\bar{\gamma}_{\min,\mathcal{S}}} - 1, \epsilon$.
2.	$A_M = (A_H + A_L)/2$.
3.	Given A_M, if (15.20) is feasible, then $A_L := A_M$. Otherwise $A_H := A_M$.
4.	Repeat Steps 2 and 3 until $\lvert A_H - A_L \rvert \le \epsilon$.

We observe that for a given x, all the constraints of (15.16) are convex except the rank-one constraint. This motivates us to employ the SDR method to suboptimally solve (15.16) via bisection, the details of which are provided in Table 15.1.

$$\texttt{find } \{\mathbf{X}_k \in \mathbb{C}^{L\times L}\}_{k=1}^{K}, \texttt{ s.t. } \sum\nolimits_k \mathrm{Tr}(\mathbf{X}_k) \le P_\Sigma; \ \mathbf{X}_k \succeq \mathbf{0}, \forall k$$
$$\mathrm{Tr}(\mathbf{A}_k\mathbf{X}_k) - A_M\Big(\sum\nolimits_{l\neq k}\mathrm{Tr}(\mathbf{A}_k\mathbf{X}_l) + \sigma^2\Big) \ge 0, \forall k. \tag{15.17}$$

15.5.2 Minimization of Delivery Time for Coded Caching Strategy

Remember that the BS under coded caching policy multicasts a common (coded) message $\mathbf{X}_\mathcal{S}$ to all users in $\mathcal{S}$. The delivery time of coded messages, taking into account the fact that each $\mathbf{X}_\mathcal{S}$ has $\frac{Q_{\text{cod, AC}}}{C_K^{m+1}}$ bits, is calculated as $\tau_{\text{cod}} = \frac{Q_{\text{cod, AC}}}{C_K^{m+1}} \sum_{\mathcal{S}\in\mathcal{S}} \frac{1}{R_{\text{cod},\mathcal{S}}}$, where $R_{\text{cod},\mathcal{S}}$ is given in (15.2). Because the BS sends the coded messages in a consecutive manner, minimizing the average delivery time τ_{cod} is equivalent to minimizing the transmission time of each coded message. Thus the optimization is formulated as follows:

$$\underset{\mathbf{w}_\mathcal{S}\in\mathbb{C}^L}{\texttt{Minimize}} \ \frac{1}{R_{\text{cod},\mathcal{S}}}, \ \texttt{s.t.} \ R_{\text{cod},\mathcal{S}} \ge \bar{\gamma}_{\min,\mathcal{S}}; \ \|\mathbf{w}_\mathcal{S}\|^2 \le P_\Sigma. \tag{15.18}$$

With the help of a slack variable $x > 0$ and $\mathbf{X} = \mathbf{w}_\mathcal{S}\mathbf{w}_\mathcal{S}^H \in \mathbb{C}^{L\times L}$, problem (15.18) can be equivalently reformulated as

$$\underset{x,\mathbf{X}\in\mathbb{C}^{L\times L}}{\texttt{Maximize}} \ x, \ \texttt{s.t.} \ \mathrm{Tr}(\mathbf{A}_k\mathbf{X}) \ge x\sigma^2, \forall k \in \mathcal{S}; \ \mathbf{X} \succeq \mathbf{0};$$
$$x \ge 2^{\bar{\gamma}_{\min,\mathcal{S}}} - 1; \ \mathrm{Tr}(\mathbf{X}) \le P_\Sigma; \ \mathrm{rank}(\mathbf{X}) = 1. \tag{15.19}$$

It is observed that for a given x, the first constraint in (15.19) is convex. Problem (15.19) can thus be solved via SDR and the bisection methods, whose steps are provided in Table 15.2.

$$\texttt{find } \mathbf{X} \in \mathbb{C}^{L \times L}, \ \texttt{s.t. } \operatorname{Tr}(\mathbf{A}_k \mathbf{X}) - A_M \sigma^2 \geq 0, \forall k \in \mathcal{S} \qquad (15.20)$$
$$\operatorname{Tr}(\mathbf{X}) \leq P_\Sigma; \ \mathbf{X} \succeq \mathbf{0}.$$

15.6 Non-uniform File Popularity Distribution

In practice, the requested files are not always equally popular. In fact, there is a small number of files that are more frequently requested by a large number of users. To capture such scenarios, we assume generic distribution of content popularity and focus on uncoded caching in this section. Denote by $\mathbf{p}_k = \{q_{k,1}, \ldots, q_{k,N}\}$ the content popularity of user k, in which $q_{k,n}$ represents the average request rate of user k to file n. Obviously, $\sum_{n=1}^{N} q_{k,n} = 1$.

The BS collects all the user requests, and hence sees the global popularity as

$$q_{G,n} = \frac{1}{K} \sum_{k=1}^{K} q_{k,n}. \qquad (15.21)$$

For robustness, we consider arbitrary user cache sizes. Denote by M_0 (files) the BS's cache size and by M_k (files) the cache size at user k. The distributed cache placement phase is implemented in which each user prefetches its cache with the most locally popular files until full. In addition, we denote $\tilde{\mathbf{q}}_k = \Pi(\mathbf{q}_k)$ and $\tilde{\mathbf{q}}_G = \Pi(\mathbf{q}_G)$ as the ranked version in decreasing order of $\mathbf{q}_k$ and $\mathbf{q}_G$, respectively. Then user k stores the first $n_k = M_k$ files in $\tilde{\mathbf{q}}_k$. In a similar way, the BS stores the first $n_G = M_0$ files in $\tilde{\mathbf{q}}_G$.

Proposition 5 Denote by $\mathbf{D} = \{d_1, \ldots, d_K\}$ the requested file indexes, e.g., user k requests file d_k. The total throughput on the backhaul and access links are $Q_{\text{BH}}(\mathbf{D}) = Q\sum_{k=1}^{K} \mathbb{I}_{n_G}(\Pi_G(d_k))$ and $Q_{\text{AC}}(\mathbf{D}) = Q\sum_{k=1}^{K} \mathbb{I}_{n_k}(\Pi_k(d_k))$, respectively, where $\Pi_k(d_k)$ denotes the sorted position of file d_k, and $\mathbb{I}_n(i) = 1$ if $i > n$ and 0 otherwise.

The proof of Proposition 5 can be found by considering the user's and BS's cached contents.

It is worth noting that if a file is cached, the whole content is available in the cache. As such, only a subset of users $\tilde{\mathcal{K}}(\mathbf{D}) = \{k \mid \Pi_k(d_k) > n_k\}$ who do not have the requested files receive data from the BS. A following transmission procedure is applied to serve the users:

$$\underset{\mathbf{w}_{k \in \tilde{\mathcal{K}}(\mathbf{D})} \in \mathbb{C}^L}{\texttt{Minimize}} \sum\nolimits_{k \in \tilde{\mathcal{K}}(\mathbf{D})} \frac{\| \mathbf{w}_k \|^2}{\tilde{R}_{\text{unc},k}}, \ \texttt{s.t. } \tilde{R}_{\text{unc},k} \geq \gamma, \forall k \in \tilde{\mathcal{K}}(\mathbf{D}), \qquad (15.22)$$

where $\tilde{R}_{\text{unc},k} = B \log_2 \left(1 + \frac{|\mathbf{h}_k^H \mathbf{w}_k|^2}{\sum_{k \neq l \in \tilde{\mathcal{K}}(\mathbf{D})} |\mathbf{h}_k^H \mathbf{w}_l|^2 + \sigma^2}\right)$.

The delivery time minimization problem is formulated as:

$$\underset{\mathbf{w}_{k \in \tilde{\mathcal{K}}(\mathbf{D})} \in \mathbb{C}^L}{\texttt{Minimize}} \sum\nolimits_{k \in \tilde{\mathcal{K}}(\mathbf{D})} \frac{Q}{\tilde{R}_{\text{unc},k}}, \ \texttt{s.t. } \tilde{R}_{\text{unc},k} \geq \gamma, \forall k \in \tilde{\mathcal{K}}(\mathbf{D}). \qquad (15.23)$$

By using a similar technique as that in in Sections 15.4.1 and 15.5.1, we can obtain the solutions of problems (15.22) and (15.23), respectively.

15.7 Numerical Results

The effectiveness of the proposed designs are demonstrated via numerical results, which are measured over 500 channel realizations. In the figures, we use *SDR* to refer to the SDR-based precoding design for the uncoded caching and *ZF* to refer to the zero-forcing design in Section 15.4.1.1. In addition, the system parameters are as follows: $N = 1{,}000$ files, $L = 10$ antennas, $K = 8$ users, $B = 1$ MHz, $\eta = 10^{-6}$ bits/Joule [13], $\sigma^2_{h_k} = 1, \forall k$, $Q = 10$ Mb, $\gamma_k = 2$ Mbps , $\forall k$.

15.7.1 Energy Efficiency Performance

Figure 15.2a plots the EE performance of the two caching strategies as a function of $\frac{M_u}{N}$, the normalized cache size, and no cost on the backhaul. In general, it is observed that the SDR-based uncoded caching design is more efficient than the coded caching for small cache sizes. This interesting observation suggests an important guideline for employing the uncoded caching in practice because user storage capacity is usually small. For user storage capacity larger than 20% of the library size, i.e., N, it is better to switch to the coded caching. Furthermore, it is shown that SDR-based design always surpasses the ZF for all cache sizes, which confirms the efficiency of SDR over the ZF.

Figure 15.2b shows the EE performance for various user cache sizes with $M_b = 0.7N$ (backhaul's cost is counted). In this case, the coded caching strategy performs better than its counterpart uncoded caching when the users are able to cache certain files. The larger the cache size, the better the coded caching outperforms the uncoded method. Focusing on the uncoded caching strategy, the SDR design achieves slightly better EE than the ZF design, at the cost of higher complexity (Table 15.3). The relationship between the EE version and the user cache size is presented in Figure 15.2c for small BS and user storage capacities. An interesting observation is that that the uncoded caching surpasses the coded caching method, which is in agreement with Figure 15.2b. Figure 15.2d plots the EE versus the BS cache size with $M_u = 0.5N$. It is shown that caching at the BS affects both caching strategies. Furthermore, coded caching is more efficient than uncoded caching regardless M_b.

The EE under non-uniform Zipf distribution is presented in Figure 15.3a as a function of the normalized user cache size. Note that only the uncoded caching algorithm is shown. Under the Zipf content popularity distribution, we have $q_{k,n} = \frac{n^{-\alpha}}{\sum_{i=1}^{N} i^{-\alpha}}, \forall k$. The figure shows that the SDR-based design achieves a significantly larger EE than ZF. Specifically, the SDR achieves an EE of almost 3 times higher than the ZF design when the user is able to store up to 40% the library. It is also observed that a larger Zipf

Table 15.3 Simulation Time in Seconds, $m = K - 1$

K	Coded	Uncoded-SDR	Uncoded-ZF
4	0.197	0.384	8.7e-5
8	0.204	1.131	10e-5

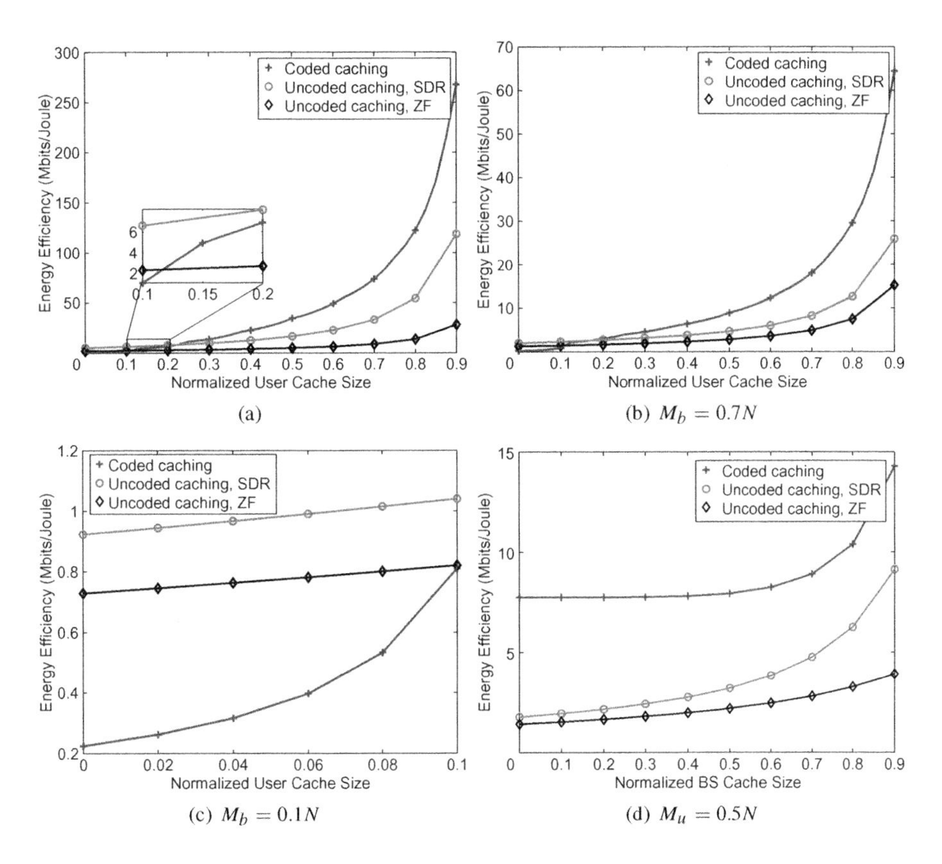

Figure 15.2 Energy efficiency of the two caching methods. (a) EE versus normalized user cache size, cost-free on backhaul; (b) EE versus normalized user cache size; (c) EE versus normalized user cache size with small values (d) EE versus normalized BS cache size.

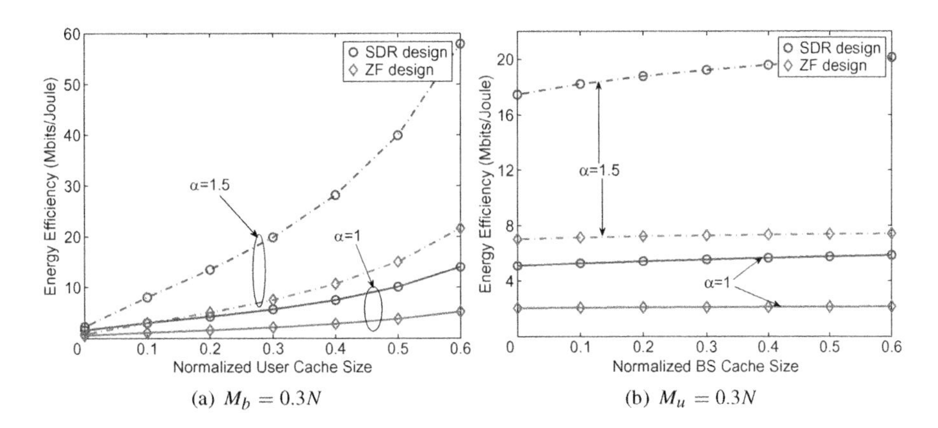

Figure 15.3 Energy efficiency of the uncoded caching algorithm with Zipf content popularity distribution with different Zipf exponents.

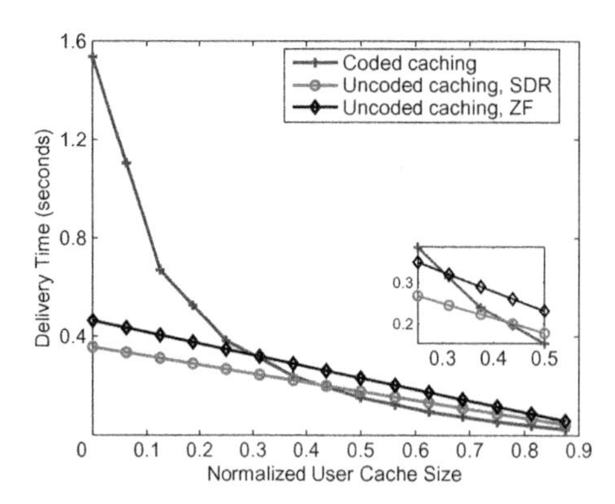

Figure 15.4 Delivery time of the two caching methods versus the normalized user memory M_u. Average transmit power is 10 dB.

exponent factor provides higher EE. In such cases, the requests are more concentrated to the most popular files. Figure 15.3b shows the EE performance for various BS cache sizes. Similar conclusions are obtained, i.e., the SDR design is more efficient than the ZF design.

15.7.2 Delivery Time Performance

Figure 15.4 plots the delivery times for various user cache sizes with $K = 8$ and $P_\Sigma = 10$ dB. When the user is equipped with a small storage memory, employing uncoded caching is more efficient than the coded caching policy. As the cache size increases, the coded-caching policy performs better than the uncoded caching. From practical point of view, employing which strategy depends on the memory availability in order to exploit the benefit of both caching strategies. We also observe that the uncoded caching performance is linearly proportional to the cache size, which is in line with Proposition 2.

Figure 15.5 presents the delivery times as a function of the BS transmit power. Spending more transmit power will significantly decrease the delivery times in both caching policies. It is shown that in the small user cache size regime (Figure 15.5a), the uncoded caching strategy spends less time serving the requests than the coded caching method, which is also observed from Figure 15.4. In the large cache size regime, (Figure 15.5b), the coded caching strategy is more efficient than the uncoded caching. In addition, the SDR and ZF-based designs obtain the same performance for large transmit power, which is because a transmission with high transmit power can accommodate both SDR and ZF.

Figure 15.6 presents the effect of the number of users on the delivery times for a given BS transmit power. When there are less active users, e.g., small K, the uncoded caching slightly outperforms the coded caching method. When there are more active

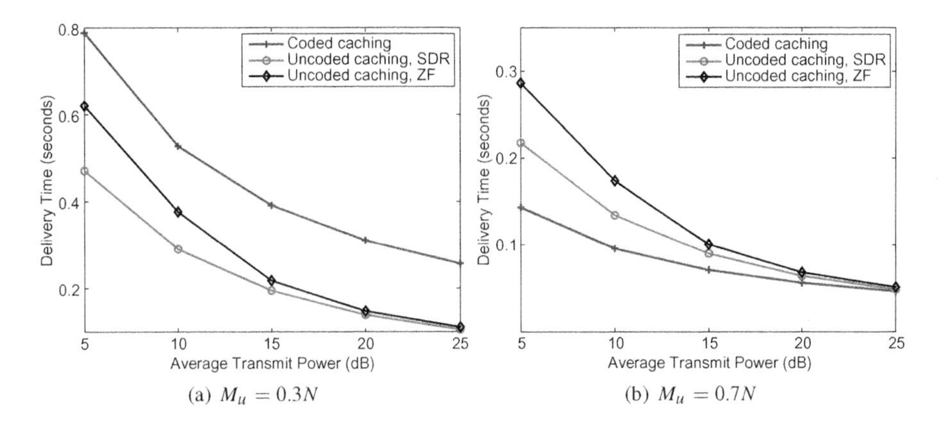

Figure 15.5 Delivery time of the two caching methods versus the average transmit power.

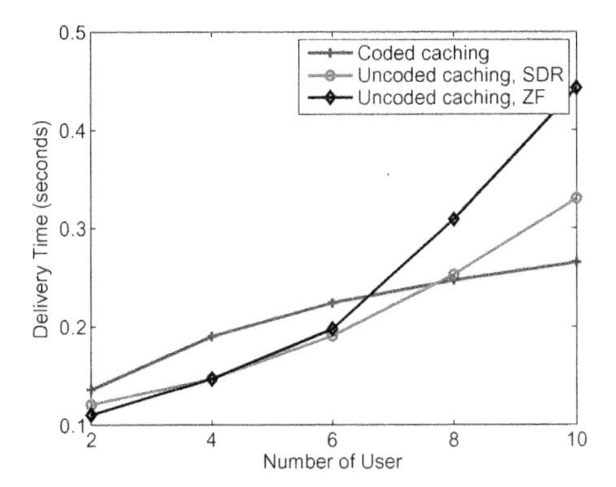

Figure 15.6 Delivery time of the two caching methods versus the number of users. Average transmit power is 10 dB, $M_u = 0.4N$.

users, e.g., large K, the uncoded caching is defeated by the coded caching. This is reasonable because larger K is equivalent to having larger total cache size, thus the coded caching algorithm is more efficient.

15.8 Conclusions

There have been two popular coded and uncoded caching strategies in the edge caching wireless networks whose performance in terms of energy efficiency and delivery time depend on key system parameters, e.g., cache size, number of users, and transmit power. By jointly designing the signal transmission while taking into consideration the cached contents, we can significantly improve the performance metrics of interests. In general, the coded caching achieves better performance than the uncoded caching if the user cache size is sufficient large. This result provides a useful guideline on the implementation of which caching policy to use in practice.

References

[1] T. X. Vu, H. D. Nguyen, T. Q. S. Quek, and S. Sun, "Adaptive cloud radio access networks: compression and optimization," *IEEE Trans. Signal Process*, vol. 65, no. 1, pp. 228–241, Jan. 2017.

[2] T. X. Tran and D. Pompili, "Dynamic radio cooperation for user-centric cloud-RAN with computing resource sharing," *IEEE Trans. Wireless Commun.*, vol. 16, no. 4, pp. 2379–2393, Apr. 2017.

[3] S. Borst, V. Gupta, and A. Walid, "Distributed caching algorithms for content distribution networks," in *Proc. IEEE Int. Conf. Comput. Commun.*, Mar. 2010, pp. 1–9.

[4] M. A. Maddah-Ali and U. Niesen, "Fundamental limits of caching," *IEEE Trans. Inf. Theory*, vol. 60, no. 5, pp. 2856–2867, May 2014.

[5] K. C. Almeroth and M. H. Ammar, "The use of multicast delivery to provide a scalable and interactive video-on-demand service," *IEEE J. Sel. Areas Commun.*, vol. 14, no. 6, pp. 1110–1122.

[6] D. Christopoulos, S. Chatzinotas, and B. Ottersten, "Cellular-broadcast service convergence through caching for COMP cloud RAN," in *Proc. IEEE Symp. Commun. Veh. Tech. Benelux*, 2015, pp. 1–6.

[7] T. X. Vu, S. Chatzinotas, and B. Ottersten, "Blockchain-based content delivery networks: Content transparency meets user privacy," in *Proc. IEEE Wireless Commun. Netw. Conf.*, 2019, pp. 1–6.

[8] T. X. Vu, S. Chatzinotas, and B. Ottersten, "Coded caching and storage allocation in heterogeneous networks," in *Proc. IEEE Wireless Commun. Netw. Conf.*, 2017, pp. 1–5.

[9] T. X. Vu, L. Lei, S. Chatzinotas, B. Ottersten, and T. A. Vu, "Energy efficient design for coded caching delivery phase," in *Proc. Int. Conf. Recent Adv. Sig. Process. Telecom. Comput.*, 2019, pp. 1–5.

[10] S. H. Park, O. Simeone, W. Lee, and S. Shamai, "Coded multicast fronthauling and edge caching for multi-connectivity transmission in fog radio access networks," in *Proc. IEEE Int. Workshop Signal Process. Adv. Wireless Commun.*, 2017, pp. 1–5.

[11] N. Karamchandani, U. Niesen, M. A. Maddah-Ali, and S. N. Diggavi, "Hierarchical coded caching," *IEEE Trans. Inf. Theory*, vol. 62, no. 6, pp. 3212–3229, June 2016.

[12] L. Tang and A. Ramamoorthy, "Coded caching for networks with the resolvability property," in *Proc. IEEE Int. Symp. Inf. Theory*, July 2016, pp. 420–424.

[13] M. Tao, E. Chen, H. Zhou, and W. Yu, "Content-centric sparse multicast beamforming for cache-enabled cloud RAN," *IEEE Trans. Wireless Commun.*, vol. 15, no. 9, pp. 6118–6131, Sept. 2016.

[14] T. X. Vu, S. Chatzinotas, and B. Ottersten "Energy minimization for cache-assisted content delivery networks with wireless backhaul," *IEEE Wireless Commun. Lett.*, vol. 7, no. 3, pp. 332–335, June 2018.

[15] A. Khreishah, J. Chakareski, and A. Gharaibeh, "Joint caching, routing, and channel assignment for collaborative small-cell cellular networks," *IEEE J. Sel. Areas Commun.*, vol. 34, no. 8, pp. 2275–2284, IEEE Trans. Inf. Theory. 2016.

[16] L. Zhang, M. Xiao, G. Wu, and S. Li, "Efficient scheduling and power allocation for D2D-assisted wireless caching networks," *IEEE Trans. Commun.*, vol. 64, no. 6, pp. 2438–2452, June 2016.

[17] M. Gregori, J. Gómez-Vilardebó, J. Matamoros, and D. Gündüz, "Wireless content caching for small cell and D2D networks," *IEEE J. Sel. Areas Commun.*, vol. 34, no. 5, pp. 1222–1234, May 2016.

[18] M. Ji, G. Caire, and A. F. Molisch, "Wireless device-to-device caching networks: basic principles and system performance," *IEEE J. Sel. Areas Commun.*, vol. 34, no. 1, pp. 176–189, Jan. 2016.

[19] C. Yang, Y. Yao, Z. Chen, and B. Xia, "Analysis on cache-enabled wireless heterogeneous networks," *IEEE Trans. Wireless Commun.*, vol. 15, no. 1, pp. 131–145, Jan. 2016.

[20] Z. Chen, J. Lee, T. Q. Quek, and M. Kountouris, "Cooperative caching and transmission design in cluster-centric small cell networks," *IEEE Trans. Wireless Commun.*, vol. 16, no. 5, pp. 3401–3415, May 2016.

[21] G. Alfano, M. Garetto, and E. Leonardi, "Content-centric wireless networks with limited buffers: when mobility hurts," *IEEE/ACM Trans. Netw.*, vol. 24, no. 1, pp. 299–311, Jan. 2016.

[22] F. Gabry, V. Bioglio, and I.Land, "On energy-efficient edge caching in heterogeneous networks," *IEEE J. Sel. Areas Commun.*, vol. 34, no. 12, pp. 3288–3298, Dec. 2016.

[23] D. Liu and C. Yang, "Energy efficiency of downlink networks with caching at base stations," *IEEE J. Sel. Areas Commun.*, vol. 34, no. 4, pp. 907–922, Apr. 2016.

[24] T. X. Vu, S. Chatzinotas, and B. Ottersten, "Edge-caching wireless networks: performance analysis and optimization," *IEEE Trans. Wireless Commun.*, vol. 17, no. 4, pp. 2827–2839, Apr. 2018.

[25] N. D. Sidiropoulos, T. N. Davidson, and Z.-Q. Luo, "Transmit beamforming for physical-layer multicasting," *IEEE Trans. Signal Process*, vol. 54, no. 6, pp. 2239–2251, June 2006.

[26] S. Boyd and L. Vandenberghe, *Convex Optimization*. Cambridge: Cambridge University Press, 2004.

[27] Z.-Q. Luo, W. K. Ma, A. M. C. So, Y. Ye, and S. Zhang, "Semidefinite relaxation of quadratic optimization problems," *IEEE Signal Process. Mag.*, vol. 27, no. 3, pp. 20–34, Mar. 2010.

[28] T. X. Vu, S. Chatzinotas, and B. Ottersten, "Energy-efficient design for edge-caching wireless networks: when is coded-caching beneficial?" in *Proc. IEEE Int. Workshop Signal Process. Wireless Commun.*, 2017, pp. 1–5.

16 Cache-Enabled UAVs in Wireless Networks

Mingzhe Chen, Walid Saad, and Changchuan Yin

16.1 Introduction

Drones, small aircrafts, and tethered balloons are being considered to provide service to ground wireless users that cannot be served by the ground communication system. Compared to traditional ground communication systems, an unmanned aerial vehicle (UAV) based wireless system is flexible to deploy and can provide line-of-sight links for wireless communications so as to improve the data rates of the ground users. However, enabling the UAVs to service wireless ground users requires a high data rate and reliable wireless connections to ground base stations (BSs) or to the core network. Meeting these requirements is very challenging due to the inherent bandwidth and spectrum limitations of wireless transmission links. To address this challenge, one promising solution is to use content caching techniques at the level of UAVs. In such *cache-enabled UAV networks*, UAVs can cache content locally (e.g., at off peak hours) and then directly send the stored content to ground users so as to reduce the traffic over UAV-BS links and UAV-core network links.

Caching at ground BSs has been studied in many existing works such as in [1–7]. A content pushing scheme is developed in [1] to decrease the traffic load. The authors in [2] used echo state networks (ESNs) for the predictions of the users' content request distribution and mobility. In [3], the authors developed a cache-enabled framework, during which BSs can store multimedia content in advance to efficiently serve users. The authors in [4] proposed a hierarchical framework in cloud radio access networks (C-RANs) for caching. The work in [5] used a centralizer to make decisions for all BSs so as to determine the cached contents. In [6], the authors proposed a caching strategy to reduce the traffic load over backhaul. The authors in [7] proposed a caching strategy to offload the fronthaul traffic. However, mobile users may not be effectively served as the cache is placed at static ground BSs in cases of frequent handovers. In particular, as a given ground user connects to a new ground BS, the contents that the ground user requests may not be stored at the new BSs. In consequence, the new BSs may not efficiently service the ground users that need to connect to the new BSs. In order to serve ground users effectively, one can store the contents that the users request at multiple

This work was supported in part by the National Natural Science Foundation of China under Grant 61671086 and Grant 61629101, the 111 Project under Grant B17007, the Director Funds of Beijing Key Laboratory of Network System Architecture and Convergence under Grant 2017BKL-NSAC-ZJ-04, the BUPT Excellent Ph.D. Students Foundation, and in part by the U.S. National Science Foundation under Grants IIS-1633363.

ground BSs. However, this caching scheme will waste caching storage and increase signaling overheads. In consequence, to improve caching efficiency, it is necessary to deploy cache-enabled UAVs that can change their deployment according to the users' mobility so as to effectively transmit the contents that are requested by the ground users.

The cache-enabled UAVs can be used for air BSs, relays, and user equipment [8]. In particular, when UAVs act as BSs or relays, they can cache the contents requested by the users and send these contents to the ground users so as to reduce delay. When the UAVs act as user equipment, they can cache data files collected via cameras and sensors, which are used for generating videos. Recently, research [8–14] has investigated a number of cache-enabled UAV problems. In [9] and [10], the authors proposed a scheme for UAV-based networks by using proactive caching techniques. The authors in [11] optimized the resource allocation in a cache-based UAV network. A UAV-based transmission scheme is proposed in [12] for scalable videos in hyperdense networks. The work in [8] and [13] review the opportunities and challenges for the cache-enabled cellular connected UAVs. The authors in [14] introduced four representative scenarios to explain the applications of UAV-based networks in caching, energy transfer, and communications. However, a number of opportunities and challenges in cache-enabled UAV networks still needs to be solved. First, since the communication links of UAVs are line of sight (LoS) and the mobility of UAVs is flexible, one can study the cooperative caching for UAVs so as to improve the caching hit ratio and thus efficiently service the users on the ground. Moreover, it is of interest to consider content delivery with caching for cache-enabled UAVs. In particular, one can study how to change the cache-enabled UAV deployment so as to maximize the content delivery rate of the links from core network to the cache-enabled UAVs and from the UAVs to users. In addition, due to the capacity-limited battery installed at UAVs, it is natural to investigate how to use caching technology to improve the energy efficiency of cache-enabled UAVs. In addition, it is imperative to investigate the machine learning–based algorithm to investigate the behaviors of mobile users so as to optimize the routing path, cached content, and deployment of cache-enabled UAVs. Finally, to enable the cache-enabled UAVs to service the emerging wireless services, such as virtual reality applications, it is necessary to consider cache-enabled UAVs with other optimizing metrics such as delay and computation.

In this chapter, we study one application of cache-enabled UAVs in C-RANs so as to maximize the users' quality of experience (QoE).

16.2 Cache-Enabled UAVs for Users' QoE Maximization

In this section, we study how a network can deploy cache-enabled UAVs that can change their deployment based on the users' locations and service the mobile users so as to maximize the users' QoE in a C-RAN. In particular, we develop a novel prediction algorithm based on the conceptor-based echo state networks [15] for the predictions of the users' content request distributions and mobility patterns. Based on these predictions, the locations of cache-enabled UAVs and the contents stored at the UAV cache can be determined and the users' QoE can be maximized. First, we motivate the problem and develop a basic system model. Then, we present a novel prediction algorithm based on an echo state network to predict the users' content request distributions and mobility patterns. Based on these predictions, the optimal deployment of cache-enabled UAVs and optimal cached contents can be determined. We conclude by evaluating the performance of cache-enabled UAVs via simulations.

16.2.1 Motivation

In a C-RAN, various wireless users are served by a large amount of remote radio heads (RRHs) connected with cloud-based baseband units (BBUs) via fronthaul links [16]. In order to improve the spectral efficiency, one can implement cooperative signal processing techniques at the BBUs [17]. However, a C-RAN performance will be limited by the capacity-limited backhaul (C-RAN to core) and fronthaul links [17]. In fact, due to the C-RAN feature, it it impossible to rely on backhaul and fiber fronthaul links. Hence it is necessary to find methods to overcome the problems related to the capacity limited backhaul and fronthaul links. Caching techniques [18–22] in which the ground users can receive content from cache placed at the RRH or cloud level can be utilized to reduce the data traffic load over backhaul and fronthaul links in C-RANs. However, deploying cache in a C-RAN faces multiple challenges, such as determination of cache placement, accurate content request distribution predictions, and optimal cache update.

Recently, a number of existing works have studied the a number of problems related to caching such as [1–7, 23–25]. However, the existing literature, such as [1–7, 23–25], can be used only for a static network during which the users cannot move. This is due to the fact that the cache in the existing works [1–7, 23–25] is placed at the static terrestrial BSs. However, the cache-enabled ground static BSs cannot satisfy the high data rate requirements of the users in an area with high-rise buildings (i.e., hotspots or stadiums) and ultradense users. Moreover, cache-enabled ground static BSs cannot service the mobile users efficiently since they may move outside the BSs' coverage range. Hence it is necessary to investigate a base station that can move and can track the mobility patterns of the users so as to improve the caching efficiency. In such a case, UAVs must be used as flying base stations to track the ground users' mobility patterns, store the popular content, and then effectively service them. Due to the UAVs' flexible deployment, they can build reliable transmission links between the UAVs and users by the mitigation of the blockage effect.

Using UAVs for improving wireless communications was studied in [26–33]. However, the existing works such as in [26–33] focused on the performance analysis of deploying UAVs over wireless networks. They completely ignored the user behavior predictions and did not investigate deploying cache at UAVs. In fact, the user behavior prediction such as mobility pattern prediction enables the UAVs to move effectively, thus immediately servicing the ground users. Moreover, deploying UAVs into a C-RAN must consider the fact that the UAV-cloud transmission links will be capacity limited. This is because the transmission links between UAVs and the cloud are wireless, and the bandwidth of the UAV-cloud transmission links is limited. To solve this challenge,

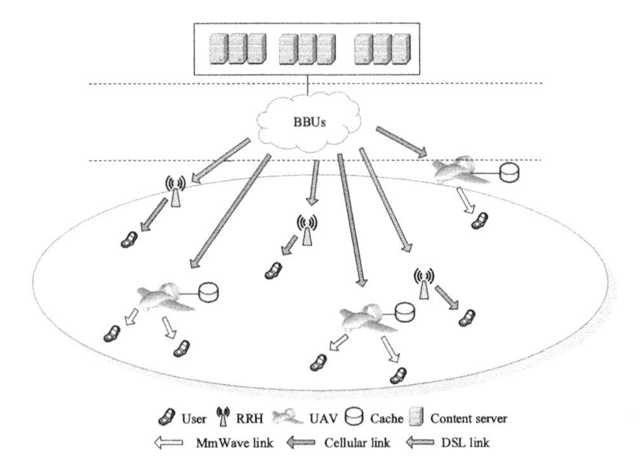

Figure 16.1 A C-RAN with cache-enabled UAVs.

the network can use caching techniques to store the contents at the UAVs. Cache-enabled UAVs can transmit the contents to its requested users and offload the traffic over fronthaul links.

16.2.2 Basic Problem

Consider a downlink C-RAN network that consists of a set $\mathcal{U}$ ground users and a set $\mathcal{R}$ RRHs. We assume that the RRHs are grouped into E clusters via *K-mean clustering* approach [34], as shown in Figure 16.1. In consequence, the RRHs and UAVs use zero-forcing beam-forming (ZFBF) [35] to serve the ground mobile users. In this model, a set $\mathcal{K}$ cache-enabled UAVs and ground RRHs jointly service the ground mobile users. For the UAV–user transmission links, since UAVs are located at high altitude and the blocking effect due to obstacles can be reduced, we use the millimeter wave (mmWave) frequency spectrum for air-to-ground transmissions from UAVs to the ground users. Meanwhile, the cellular band is used for the transmission from ground RRHs to the users. The RRHs are associated with the BBUs using capacity-limited fronthaul links. Moreover, the content servers are associated with the BBUs using fiber backhaul links. The licensed cellular band is used for the wireless fronthaul links from the UAVs to the cloud. In consequence, the wireless UAV-cloud fronthaul links will interfere with the RRH-ground mobile user transmission links.

In this model, a set $\mathcal{N}$ of N content that is requested by all of the ground mobile users is stored at the content server. We assume that each content data size is L. UAV cache is used to cache the users' requested contents. Caching the contents that are requested by the users can significantly reduce the transmission delay of the content server–UAV transmission links since cache-enabled UAVs can directly transmit their cached contents to their associated ground users.

We assume that a set $\mathcal{C}_k$ contents is stored at cache of UAV k. We also assume that, at each time slot τ, each ground mobile user will request only one content. Δ_τ denotes the duration of each time slot τ. The contents cached at the UAV will be updated every T time slots.

For pedestrian mobility patterns, we consider a realistic model in which each user will periodically visit a number of certain locations. For instance, a certain ground user will often work at the same office during weekdays. In this system, the BBUs will collect the users' locations once every H time slots. We also assume that each ground mobile user moves with a constant speed. The mobility pattern of each user can be used to decide the cached contents and the optimal locations of UAVs, thus impacting each user's QoE. In this system, the ground mobile user association will change according to the requirement of QoE. To service the ground mobile users effectively, one must adjust the deployment of the UAVs as the users' locations continuously vary.

The transmission models of BBU–UAV links, UAV–user links, and RRH–user links are introduced next. We assume that a time slot τ consists of F equally time intervals Δ_t, which is given by $\Delta_\tau = F\Delta_t$. We also assume that the duration of a time interval Δ_t is sufficiently small. In consequence, the location of each user is constant [27]. The standard log-normal shadowing model [36] is used to model the UAVs–user transmission links over mmWave. By setting specific parameters, this model can be used to model the non-line-of-sight (NLoS) and LoS links. In consequence, for UAV k that is located at $\left(x_{\tau,k}, y_{\tau,k}, h_{\tau,k}\right)$ and sending a content to user i, the NLoS and LoS pathloss at interval t is (in dB) [37]:

$$l_{t,ki}^{\mathrm{LoS}}\left(\boldsymbol{w}_{\tau,t,k}, \boldsymbol{w}_{\tau,t,i}\right) = L_{FS}\left(d_0\right) + 10\mu_{\mathrm{LoS}} \log\left(d_{t,ki}\left(\boldsymbol{w}_{\tau,t,k}, \boldsymbol{w}_{\tau,t,i}\right)\right) + \chi_{\sigma_{\mathrm{LoS}}}, \quad (16.1)$$

$$l_{t,ki}^{\mathrm{NLoS}}\left(\boldsymbol{w}_{\tau,t,k}, \boldsymbol{w}_{\tau,t,i}\right) = L_{FS}\left(d_0\right) + 10\mu_{\mathrm{NLoS}} \log\left(d_{t,ki}\left(\boldsymbol{w}_{\tau,t,k}, \boldsymbol{w}_{\tau,t,i}\right)\right) + \chi_{\sigma_{\mathrm{NLoS}}}, \quad (16.2)$$

where $\boldsymbol{w}_{\tau,t,k} = \left[x_{\tau,k}, y_{\tau,k}, h_{\tau,k}\right]$ represents UAV k's coordinate. Here, $h_{\tau,k}$ is UAV k's altitude. Meanwhile, $\boldsymbol{w}_{\tau,t,k} = \left[x_{t,i}, y_{t,i}\right]$ represents user i's coordinate at interval t. Finally, $L_{FS}\left(d_0\right)$ is the free space pathloss that is expressed by $20\log\left(d_0 f_c 4\pi/c\right)$, where d_0 represents the reference distance, c is the light speed, and f_c is the carrier frequency. The distance between user i and UAV k is $d_{t,ki}\left(\boldsymbol{w}_{\tau,t,k}, \boldsymbol{w}_{\tau,t,i}\right) = \sqrt{\left(x_{t,i} - x_{\tau,k}\right)^2 + \left(y_{t,i} - y_{\tau,k}\right)^2 + h_{\tau,k}^2}$; μ_{NLoS} and μ_{LoS}, respectively, represent the pathloss exponents of NLoS and LoS links; and $\chi_{\sigma_{\mathrm{NLoS}}}$ and $\chi_{\sigma_{\mathrm{LoS}}}$ represent the shadowing random variables, which the Gaussian distribution with $\sigma_{\mathrm{NLoS}}, \sigma_{\mathrm{LoS}}$ standard deviations, and zero mean.

The LoS connection probability relies on the height and density of buildings, the elevation angle between the user and the UAV, as well as the UAVs' and users' locations. Then, the LoS connection probability can be given by [26]:

$$\Pr\left(l_{t,ki}^{\mathrm{LoS}}\right) = \left(1 + X \exp\left(-Y\left[\phi_t - X\right]\right)\right)^{-1}, \quad (16.3)$$

where Y and X are constants and $\phi_t = \sin^{-1}\left(h_{\tau,k}/d_{t,ki}\left(\boldsymbol{w}_{\tau,t,k}, \boldsymbol{w}_{\tau,t,i}\right)\right)$ represents the elevation angle. The average pathloss between UAV k and user i can be given by [29]:

$$\bar{l}_{t,ki}\left(\boldsymbol{w}_{\tau,t,k}, \boldsymbol{w}_{\tau,t,i}\right) = \Pr\left(l_{t,ki}^{\mathrm{LoS}}\right) \times l_{t,ki}^{\mathrm{LoS}} + \Pr\left(l_{t,ki}^{\mathrm{NLoS}}\right) \times l_{t,ki}^{\mathrm{NLoS}}, \quad (16.4)$$

where $\Pr\left(l_{t,ki}^{\mathrm{NLoS}}\right) = 1 - \Pr\left(l_{t,ki}^{\mathrm{LoS}}\right)$. Based on this pathloss model, for user i that is located at $\boldsymbol{w}_{\tau,t,i}$ and associated with UAV k, the average signal-to-noise ratio (SNR) is given by:

$$\gamma_{t,ki}^{\mathrm{V}} = \frac{P_{t,ki}}{10^{\bar{l}_{t,ki}(\boldsymbol{w}_{\tau,t,k},\boldsymbol{w}_{\tau,t,i})/10}\sigma^2}, \tag{16.5}$$

where $P_{t,ki}$ indicates the transmit power of UAV k, and σ^2 represents the Gaussian noise. Let B_V be the total bandwidth of each UAV. The total bandwidth of each UAV is equally allocated to the connected users. For user i associated with UAV k, the channel capacity is given by:

$$C_{\tau,ki}^{\mathrm{V}} = \frac{1}{F_{\tau,i}} \sum_{t=1}^{F_{\tau,i}} \frac{B_V}{U_k} \log_2\left(1 + \gamma_{t,ki}^{\mathrm{V}}\right), \tag{16.6}$$

where U_k represents the number of users that are associated with UAV k. $F_{\tau,i}$ is the number of intervals for which each user i requests a content.

The probabilistic NLoS and LoS links over Sub-6 GHz band are considered for the BBU–UAV links, since the content transmission using Sub-6 GHz band will have a smaller pathloss and a more reliable transmission compared to the use of mmWave band. The NLoS links will experience higher attenuations compared to the LoS links due to the diffraction loss. The pathloss of the NLoS and LoS BBU-UAV k transmission links are [26]:

$$L_{t,k}^{\mathrm{LoS}} = d_{t,ki}\left(\boldsymbol{w}_{\tau,t,k}, \boldsymbol{w}_{\tau,t,B}\right)^{-\beta}, \tag{16.7}$$

$$L_{t,k}^{\mathrm{NLoS}} = \eta d_{t,ki}\left(\boldsymbol{w}_{\tau,t,k}, \boldsymbol{w}_{\tau,t,B}\right)^{-\beta}, \tag{16.8}$$

where $\boldsymbol{w}_{\tau,t,B} = \left[x_B, y_B\right]$ represents the BBUs' location, and β denotes the exponent of the pathloss. According to (16.3)–(16.5), for the link between UAV k and the BBUs, one can calculate the average SNR and LoS probability.

For the users connected to RRH cluster q, the received signals at interval t can be given by:

$$\boldsymbol{b}_{t,q} = \sqrt{P_R}\boldsymbol{H}_{t,q}\boldsymbol{F}_{t,q}\boldsymbol{a}_{t,q} + \boldsymbol{n}, \tag{16.9}$$

where $\boldsymbol{H}_{t,q} \in \mathbb{R}^{U_q \times R_q}$ represents the pathloss matrix. Here U_q denotes the number of users connected to RRH cluster q. R_q represents the number of antennas of each RRH. P_R represents each RRH's transmit power. $\boldsymbol{a}_{t,q} \in \mathbb{R}^{U_q \times 1}$ represents the content that needs to transmit at interval t. $\boldsymbol{n}_{t,q} \in \mathbb{R}^{U_q \times 1}$ denotes the noise power. $\boldsymbol{F}_{t,q} = \boldsymbol{H}_{t,q}^{\mathrm{H}}\left(\boldsymbol{H}_{t,q}\boldsymbol{H}_{t,q}^{\mathrm{H}}\right)^{-1} \in \mathbb{R}^{R_q \times U_q}$ represents the beam-forming matrix [38]. We also assume that each user's bandwidth is B. For user i within cluster $\mathcal{M}_q$, the received signal-to-interference-plus-noise-ratio (SINR) at interval t is

$$\gamma_{t,qi}^{\mathrm{H}} = \frac{P_R\left\|\boldsymbol{h}_{t,qi}\boldsymbol{f}_{t,qi}\right\|^2}{\underbrace{\sum_{j=1,j\neq q}^{E}\sum_{u\in\mathcal{U}_j} P_R\left\|\boldsymbol{h}_{t,ji}\boldsymbol{f}_{t,ju}\right\|^2}_{\text{other cluster RRHs interference}} + \underbrace{P_B g_{t,Bi} d_{t,Bi}^{-\beta}}_{\text{wireless fronthaul interference}} + \sigma^2},$$

where $\mathcal{M}_j$ represents a set of RRHs in cluster j, $\mathcal{U}_j$ represents a set of users associated with the RRHs in cluster j, $\boldsymbol{h}_{t,qi} \in \mathbb{R}^{1\times R_q}$ represents the channel gain from the RRHs to the users in group $\mathcal{M}_q$, where $h_{t,ki} = g_{t,ki} d_{t,ki}(x_i, y_i)^{-\beta}$. $g_{t,ki}$ represents the parameters of the Rayleigh fading at interval t. $d_{t,ki}(x_i, y_i) = \sqrt{\left(x_{t,k} - x_{t,i}\right)^2 + \left(y_{t,k} - y_{t,i}\right)^2}$ is the distance between RRH k and user i. $\boldsymbol{f}_{t,qi} \in \mathbb{R}^{R_q \times 1}$ indicates the beam-forming vector. Given each user i associated with the RRHs in cluster $\mathcal{M}_q$, the channel capacity of transmitting each content can be given by:

$$C^{\mathrm{H}}_{\tau,qi} = \frac{1}{F_{\tau,i}} \sum_{t=1}^{F_{\tau,i}} B\log_2\left(1 + \gamma^{\mathrm{H}}_{t,qi}\right). \tag{16.10}$$

Next, we introduce each user's QoE model. In our work, the QoE of each user captures the delay, data rate, and device type of each user.

In the considered system, the users can receive the contents by three types of transmission links: (1) content servers–BBU–RRH–user, (2) content servers–BBU–UAV–user, and (3) UAV cache–user. In our model, the links between the core network and the cloud is fiber. In consequence, the transmission delay between the cloud and the core network can be neglected. The capacity of the wired BBU–RRH link is assumed to be limited, and its maximum data rate for all users is v_F. In consequence, the BBU–RRH fronthaul data rate for each user can be given by $v_{FU} = v_F / N_{FR}$, with N_{FR} being the number of users that receive the content from RRHs. Hence for a user i that receives a content n, the delay is

$$D_{\tau,i,n} = \begin{cases} \frac{L}{v_{FU}} + \frac{L}{C^{\mathrm{H}}_{\tau,qi}}, & \text{link } (a), \\ \frac{L}{C^{F}_{\tau,k}} + \frac{L}{C^{\mathrm{V}}_{\tau,ki}}, & \text{link } (b), \\ \frac{L}{C^{\mathrm{V}}_{\tau,ki}}, & \text{link } (c), \end{cases} \tag{16.11}$$

where $C^{F}_{\tau,k}$ is the data rate of the transmission link between UAV k and the BBUs. The sensitivity to the delay is categorized into five levels via the mean opinion score (MOS) model [39]. The mapping between the MOS model [39] and delay can be given by:

$$\bar{D}_{\tau,i,n} = \frac{\Delta_\tau - D_{\tau,i,n}}{\Delta_\tau - \min\left\{\frac{L}{v_F}, \frac{L}{C^{\max}_{K}}\right\}}, \tag{16.12}$$

as shown in Table 16.1.

The screen size of each device will affect each user's QoE. In fact, users whose devices have larger screens may be more sensitive to QoE compared to those who have smaller devices (such as small smartphones). The screen size's impact is captured by a parameter S_i, which reflects the length of the device of each user. In particular, a device with a larger screen must display a content with a higher resolution. Therefore, the user

Table 16.1 Mean Opinion Score Model [39]

QoE	Poor	Fair	Good	Very Good	Excellent
Interval scale	0–0.2	0.2–0.4	0.4–0.6	0.6–0.8	0.8–1

that owns a larger screen requires a higher data rate. For user i that uses device S_i to receive a content n, the data rate requirement can be given $\delta_{S_i,n} = S_i \hat{C}_n$, with $\hat{C}_n$ being the data rate requirement. The mapping from the data rate requirement $\delta_{S_i,n}$ to the MOS model can be given by:

$$V_{t,i} = \begin{cases} 1, & j \geq \delta_{S_i,n}, \\ 0, & j < \delta_{S_i,n}, \end{cases} \tag{16.13}$$

where $j \in \left\{ C^{\mathrm{V}}_{t,ki}, C^{\mathrm{H}}_{t,qi} \right\}$. $V_{t,i} = 1$ means that user i's data rate meets the requirement, $V_{t,i} = 0$, otherwise. For each user i that receives content n, the QoE is [39]:

$$Q_{\tau,i,n} = \zeta_1 \bar{D}_{\tau,i,n} + \frac{\zeta_2}{F_{\tau,i}} \sum_{t=1}^{F_{\tau,i}} V_{t,i}, \tag{16.14}$$

where ζ_1 and ζ_2 are the weighting parameters with $\zeta_1 + \zeta_2 = 1$.

Next, we first determine the minimum data rate required to meet the QoE requirement. Then, we decide each UAV's minimum transmit power that can meet the QoE requirement of its connected users. Finally, we introduce the minimization problem. Table 16.1 shows that, as $0.8 \leq \bar{D}_{\tau,i,n} \leq 1$, the delay MOS will be "excellent". This indicates that a given user's delay is minimized. In such a case, the minimum value that can maximize the delay is $\bar{D}_{\min} = 0.8$. Based on (16.11), for UAV k that transmits a content n to user i, the delay requirement of user i is

$$C^R_{\tau,ki,n} = \begin{cases} \dfrac{L}{\left(\Delta_\tau - \bar{D}_{\min}\left(\Delta_\tau - \min\left\{\frac{L}{v_F}, \frac{L}{C_k^{\max}}\right\}\right) - \frac{L}{C^F_{\tau,k}}\right)}, & n \notin \mathcal{C}_k, \\ \dfrac{L}{\left(\Delta_\tau - \bar{D}_{\min}\left(\Delta_\tau - \min\left\{\frac{L}{v_F}, \frac{L}{C_k^{\max}}\right\}\right)\right)}, & n \in \mathcal{C}_k. \end{cases} \tag{16.15}$$

Eq. (16.15) shows that, as UAV k stores a content n, the requirement of the delay decreases.

We assume that user i's device data rate requirement is $\delta_{S_i,n}$. Hence, the QoE is optimized as $C^{\mathrm{V}}_{t,ki} \geq \max\left\{C^R_{\tau,ki,n}, \delta_{S_i,n}\right\}$. In consequence, the minimum data rate that is needed to optimize the QoE of each user i is given by $\delta^R_{i,n} = \max\left\{C^R_{\tau,ki,n}, \delta_{S_i,n}\right\}$. Given (16.5), the minimum transmit power that is required to meet the QoE requirement at interval t can be given by:

$$P^{\min}_{t,ki}\left(\boldsymbol{w}_{\tau,t,k}, \delta^R_{i,n}, n\right) = \left(2^{\delta^R_{i,n} U_k / B_V} - 1\right)\sigma^2 10^{\bar{l}_{t,ki}(\boldsymbol{w}_{\tau,t,k}, \boldsymbol{w}_{\tau,t,i})/10}. \tag{16.16}$$

Eq. (16.16) shows that UAV k's minimum transmit power depends on the transmitted content n, the data rate that is required to meet user i's QoE requirement, and UAV's location.

Given the proposed model, our purpose is to design an effective deployment for UAVs so as to improve each user's QoE while minimizing the UAVs' transmit power. This QoE maximization problem is given by:

$$\min_{\mathcal{C}_k, \mathcal{U}_{\tau,k}, \boldsymbol{w}_{\tau,t,k}} \sum_{\tau=1}^{T} \sum_{k \in \mathcal{K}} \sum_{i \in \mathcal{U}_{\tau,k}} \sum_{t=1}^{F_{\tau,i}} P^{\min}_{\tau,t,ki}\left(\boldsymbol{w}_{\tau,t,k}, \delta^R_{i,n}, n_{\tau,i}\right), \tag{16.17}$$

$$\text{s. t. } h_{\min} \leq h_{\tau,k}, k \in \mathcal{K}, \tag{16.17a}$$

$$m \neq j, m, j \in \mathcal{C}_k, \mathcal{C}_k \subseteq \mathcal{N}, k \in \mathcal{K}, \tag{16.17b}$$

$$0 < P_{\tau,t,ki}^{\min} \leq P_{\max}, i \in \mathcal{U}, k \in \mathcal{K}, \tag{16.17c}$$

where $P_{\tau,t,ki}^{\min}$ represents UAV k's minimum transmit power to user i. $n_{\tau,i}$ indicates the content requested by user i. $\mathcal{U}_{\tau,k}$ represents a set of users connected to UAV k. $h_{\min}$ represents each UAV's minimum altitude. Eq. (16.17b) indicates that each UAV cache can store a unique, single contrent: Eq. (16.17c) captures the fact that each UAV's transmit power must be minimized. Since the problem in (16.17) is to meet the data rate that is needed to maximize each user's QoE at future time slots, the user behavior predictions will significantly affect the solution. Eq. (16.17) shows that the mobility pattern prediction enables the BBUs to determine each UAV's optimal location. Meanwhile, the prediction related to the content requests enables the BBUs to find the optimal cached contents of UAVs. Moreover, since the transmission links impacts the UAVs' transmit power needed to satisfy the QoE requirement, the problem in (16.17) includes the optimization of cached contents.

16.2.3 Conceptor Echo State Networks for Content Request Distribution and Mobility Pattern Predictions

Next, we introduce a conceptor-based ESN prediction approach for the predictions of the content request distributions and mobility patterns of each user. These predictions will be used in Section 16.2.4 to determine the optimal cached contents at UAVs, the UAVs' optimal locations, and user-UAV association. We first specify the components of the prediction approach. Next, we introduce how the prediction approach for the predictions of the mobility patterns and content request distribution of each user.

16.2.3.1 Components of the Conceptor-Based ESN

The conceptor-based ESN prediction algorithm has four components: input, output, ESN model, and conceptor.

The components of the conceptor-based ESN learning algorithm for content request distribution prediction is given as follows:

- *Input:* The input of the conceptor-based ESN learning algorithm is a vector $\boldsymbol{x}_{t,j} = \left[x_{tj1}, \cdots, x_{tjN_x}\right]^{\mathrm{T}}$ that denotes user j's context including occupation, gender, device type (e.g., smartphone or tablet), and age. In this vector, N_x represents the number of properties of each user j's context information.
- *Output:* The output of the conceptor ESN learning algorithm is $\boldsymbol{y}_{t,j} = \left[p_{tj1}, p_{tj2}, \ldots, p_{tjN}\right]$, which denotes the content request distribution probability of user j, where p_{tjn} represents the probability that user j requests content n.
- *ESN model:* An ESN model of the conceptor ESN prediction algorithm of each user j is used to find the relationship between the output $\boldsymbol{y}_{t,j}$ and input $\boldsymbol{x}_{t,j}$ so as to build the function between the content request distribution and the user's context. Typically, an ESN model consists of the recurrent matrix

$\boldsymbol{W}_j^{\alpha} \in \mathbb{R}^{N_w \times N_w}$, the input weight matrix $\boldsymbol{W}_j^{\alpha,\text{in}} \in \mathbb{R}^{N_w \times N_x}$, and the output weight matrix $\boldsymbol{W}_j^{\alpha,\text{out}} \in \mathbb{R}^{N \times N_w}$. Here, the dynamic reservoir consists of the input weight matrix $\boldsymbol{W}_j^{\alpha,\text{in}} \in \mathbb{R}^{N_w \times N_x}$ and the input weight matrix $\boldsymbol{W}_j^{\alpha,\text{in}} \in \mathbb{R}^{N_w \times N_x}$. N_w represents the number of neurons in the dynamic reservoir.

The dynamic reservoir of each user j is used to record each user's historical context. The output weight matrix $\boldsymbol{W}_j^{\alpha,\text{out}}$ and the reservoir can be used to predict the content request distribution. The generation of the ESN model is given in [40].

- *Conceptors:* To predict the content request distribution, one must collect the content requests and each user's context information at the same time slots in different weeks so as to train the conceptor ESN model. In this model, each content request distribution is referred to one prediction pattern. The reservoir states are given by $\boldsymbol{v}_j^i = \left[\boldsymbol{v}_{1,j}^i, \ldots, \boldsymbol{v}_{t,j}^i\right]$, where $\boldsymbol{v}_{t,j}^i = \left[v_{t,j1}^i, \ldots, v_{t,jN_w}^i\right]^{\text{T}}$ represents the prediction pattern of i's reservoir state at time t and $\boldsymbol{R}_j^i = \mathbb{E}\left[\boldsymbol{v}_{t,j}^i\left(\boldsymbol{v}_{t,j}^i\right)^{\text{T}}\right]$ is the state correlation matrix. Then, prediction pattern i's conceptor can be given by [41]:

$$\boldsymbol{M}_j^i = \boldsymbol{R}_j^i\left(\boldsymbol{R}_j^i + \chi^{-2}\boldsymbol{I}\right)^{-1}, \tag{16.18}$$

  where χ is aperture, which is defined in [41]. To accurately learn several prediction patterns, the aperture χ must be set appropriately. As the aperture is small, the states of the reservoir of the conceptor ESN will slightly change to learn a new prediction pattern. However, as a aperture is large, the states of the reservoir will change significantly to learn a new prediction pattern.

The components of the conceptor ESN learning algorithm for the prediction of the mobility pattern include the following:

- *Input:* The input is a vector $\boldsymbol{m}_{t,j} = \left[m_{tj1}, \ldots, m_{tjN_x+1}\right]^{\text{T}}$, which represents user j's current location and context at time t. The input $\boldsymbol{m}_{t,j}$ is used to predict each user j's future locations.
- *Output:* The output of the conceptor ESN algorithm for the mobility pattern prediction is a vector $\boldsymbol{s}_{t,j} = \left[s_{tj1}, \ldots, s_{tjN_s}\right]^{\text{T}}$, which denotes each user j's predicted locations in the next time slots, with N_s being the number of predicted locations.
- *ESN model:* The ESN model of the conceptor ESN algorithm for mobility pattern prediction also consists of the dynamic reservoir and the output weight matrix $\boldsymbol{W}_j^{\text{out}} \in \mathbb{R}^{N_s \times N_w}$. Similar to the ESN model used for the prediction of the content request distribution, the dynamic reservoir of an ESN model consists of the recurrent matrix $\boldsymbol{W}_j \in \mathbb{R}^{N_w \times N_w}$ and the input weight matrix $\boldsymbol{W}_j^{\text{in}} \in \mathbb{R}^{N_w \times N_x+1}$. The generation of an ESN model for the mobility pattern prediction is similar to the ESN model that is used for content request distribution prediction.
- *Conceptors:* The mobility of each user in each day is referred to one prediction pattern. The equation of the conceptors for the prediction of the mobility patterns is similar to the equation given in (16.18).

16.2.3.2 Predictions of Content Request Distribution and Mobility Patterns

Next, we introduce the learning approach for the predictions of the mobility patterns and content request distribution. The prediction procedure has two stages: the training stage and the prediction stage.

For the training stage, user j's prediction pattern i's dynamic reservoir state $\boldsymbol{v}_{t,j}^{i}$ is [40]:

$$\boldsymbol{v}_{t,j}^{i} = f\left(\boldsymbol{W}_j^{\alpha}\boldsymbol{v}_{t-1,j}^{i} + \boldsymbol{W}_j^{\alpha,\text{in}}\boldsymbol{x}_{t,j}\right), \tag{16.19}$$

where $f(x) = \frac{e^x - e^{-x}}{e^x + e^{-x}}$. Here, the input and its corresponding output is training data that consist of N_{tr} contexts of each user. Using (16.19) and N_{tr} training data, for each prediction pattern j, the reservoir states before update are $\boldsymbol{v}_{\text{old},j}^{i} = \left[0, \boldsymbol{v}_{1,j}^{i}, \ldots, \boldsymbol{v}_{N_{tr}-1,j}^{i}\right]$. The reservoir states after update are $\boldsymbol{v}_{j}^{i} = \left[\boldsymbol{v}_{1,j}^{i}, \ldots, \boldsymbol{v}_{N_{tr},j}^{i}\right]$. $\boldsymbol{v}_{\text{old},j}^{i}$ will be used for training an *input simulation matrix* $\boldsymbol{D}_j \in \mathbb{R}^{N_w \times N_w}$. $\boldsymbol{v}_{j}^{i}$ with the updated reservoir states can be used to update $\boldsymbol{W}_j^{\alpha,\text{out}}$.

$\boldsymbol{D}_j$ with $\boldsymbol{W}_j^{\alpha,\text{out}}$ can be used to predict each prediction pattern. The procedure of adding prediction pattern i to user j and updating $\boldsymbol{D}_j$ can be given by [41]:

$$\boldsymbol{D}_j = \boldsymbol{D}_{\text{old},j} + \boldsymbol{D}_{\text{inc},j}^{i}, \tag{16.20}$$

where $\boldsymbol{D}_{\text{inc},j}^{i} = \left(\left(\boldsymbol{S}\boldsymbol{S}^{\text{T}} / N_{tr} + \chi^{-2}\boldsymbol{I}\right)^{\dagger} \boldsymbol{S}\boldsymbol{T}^{\text{T}} / N_{tr}\right)^{\text{T}}$, $\boldsymbol{S} = \boldsymbol{F}_j^{i-1}\boldsymbol{v}_{\text{old},j}^{i}$, and $\boldsymbol{T} = \boldsymbol{W}_j^{\alpha,\text{in}}\boldsymbol{x}_j^{i} - \boldsymbol{D}_{\text{old},j}\boldsymbol{v}_{\text{old},j}^{i}$. In (16.20), $\boldsymbol{F}_j^{i-1} = \neg \vee \left\{\boldsymbol{M}_j^{1}, \ldots, \boldsymbol{M}_j^{i-1}\right\}$ represents the reservoir's memory that has not been used to record the input data, where $\neg$ and $\vee$ represent the Boolean operators. $\boldsymbol{x}_j^{i} = \left[\boldsymbol{x}_{1,j}^{i}, \ldots, \boldsymbol{x}_{N_{tr},j}^{i}\right]$ represents prediction pattern i's input sequences. During the training stage, $\boldsymbol{M}_j^{i}$ will be updated based on (16.18).

$\boldsymbol{W}_j^{\alpha,\text{out}}$ in the conceptor ESN learning algorithm can be trained in an offline manner via the ridge regression method [40] so as to approximate the prediction function. The training procedure can be given by:

$$\boldsymbol{W}_j^{\alpha,\text{out}} = \boldsymbol{y}_j\boldsymbol{v}_j^{\text{T}}\left(\boldsymbol{v}_j^{\text{T}}\boldsymbol{v}_j + \lambda^2\mathbf{I}\right)^{-1}, \tag{16.21}$$

where $\boldsymbol{v}_j = \left[\boldsymbol{v}_j^{1}, \boldsymbol{v}_j^{2}, \ldots, \boldsymbol{v}_j^{N_M}\right]^{\text{T}}$, $\boldsymbol{v}_j^{i} = \left[\boldsymbol{v}_{1,j}^{i}, \ldots, \boldsymbol{v}_{N_{tr},j}^{i}\right]$ represents the sequence of the prediction pattern i's reservoir states, λ denotes the learning rate, and N_M is the number of the content request distribution prediction patterns of each user. $\boldsymbol{v}_j^{i}$ will also be utilized to update the conceptor $\boldsymbol{M}_j^{i}$ for user j's prediction pattern i.

After training the ESNs and conceptors, the matrices $\boldsymbol{D}_j$, $\boldsymbol{W}_j^{\alpha,\text{out}}$, and the conceptors $\boldsymbol{M}_j = \left[\boldsymbol{M}_j^{1}, \ldots, \boldsymbol{M}_j^{N_M}\right]$ can be used for the predictions of the user's mobility patterns and content request distributions. During the prediction stage, the reservoir states of user j's prediction pattern i can be given by [41]:

$$\boldsymbol{v}_{t,j}^{i} = \boldsymbol{C}_j^{i} f\left(\boldsymbol{W}_j^{\alpha}\boldsymbol{v}_{t-1,j}^{i} + \boldsymbol{D}_j\boldsymbol{v}_{t-1,j}^{i}\right). \tag{16.22}$$

Table 16.2 The Conceptor ESN Prediction Algorithm

Inputs: N_{tr} training data,

Training Stage:

for each prediction pattern i **do**.
if the reservoir memory's space $F_j^{i-1} > 0$ **do**.
(a) BBUs collect the reservoir states $v_{\text{old},j}^i$ and v_j^i to update D_j, using (16.20).
(b) BBUs use the reservoir states v_j^i to calculate C_j^i using (16.18).
else
(c) Update the reservoir matrix W_j^{α}, retrain all prediction patterns.
end if
end for
(d) BBUs collect the reservoir states for all prediction patterns v_j to train $W_j^{\alpha,\text{out}}$ using (16.21).

Prediction Stage:

(a) BBUs choose the conceptor to get the corresponding reservoir states based on (16.22).
(b) Obtain the content request distribution prediction based on (16.23).

Output: Prediction $y_{t,j}$

Eq. (16.22) shows that the conceptor of prediction pattern j, C_j^i, determines the reservoir states' update. Through changing C_j^i, the conceptor-based ESN learning algorithm can predict different prediction patterns using one ESN model. In such case, content request distribution i's prediction is given as follows:

$$y_{t,j} = W_j^{\alpha,\text{out}} v_{t,j}^i. \tag{16.23}$$

Eq. (16.22) and (16.23) show that the conceptor-based ESN prediction algorithm uses a matrix D_j to manage the ESN's reservoir memory. The process of using the proposed algorithm for the prediction of each user j's content request distribution is given in Table 16.2. The table shows that the proposed algorithm can use a unique nonlinear system to approximate each prediction pattern. Based on this property, the ESNs can predict the mobility patterns using different nonlinear systems. Moreover, using the proposed algorithm, the cloud can learn the information related to the memory of the reservoir and identify the learned patterns.

16.2.4 Optimal Content Caching and Locations for UAVs

Next, we introduce the use of the predictions to solve the problem (16.17). In the considered model, the BBUs will select a subset of users to associate with the RRHs. Then the remaining users will be grouped into K groups. Each UAV will service the users in one group. Given the predictions and the user association, the optimal cached contents and locations of UAVs can be determined. Finally, we analyze the complexity and implementation of the conceptor ESN approach.

16.2.4.1 RRH-Users Association

The RRH-users association is determined according to the predicted locations of the users. Typically, the prediction accuracy of each user's mobility patterns will directly impacts the user–RRH association. A ground user will connect to the RRHs as the requirement is satisfied:

THEOREM 16.1 *Given the device screen size S_i and the minimum data rate requirement $\bar{D}_{\min}$ of a given user i, a given user i will connect to the RRHs within cluster k if the following rate condition is satisfied:*

$$C_{\tau,qi}^{H} \geq \max\left\{\frac{L}{\left(\Delta_\tau - \bar{D}_{\min}\left(\Delta_\tau - \min\left\{\frac{L}{v_F}, \frac{L}{C_k^{\max}}\right\}\right) - \frac{L}{v_{FU}}\right)}, \delta_{S_i,n}\right\}. \tag{16.24}$$

Proof According to $\bar{D}_{\min}$ and (16.12), the delay of each user i is $D_{\tau,i,n} = \Delta_\tau - \bar{D}_{\min}\left(\Delta_\tau - \min\left\{\frac{L}{v_F}, \frac{L}{C_k^{\max}}\right\}\right)$. In consequence, the delay requirement of transmitting content n from RRHs within cluster q to user i is:

$$C_{\tau,qi}^{R} = \frac{L}{\Delta_\tau - \bar{D}_{\min}\left(\Delta_\tau - \min\left\{\frac{L}{v_F}, \frac{L}{C_k^{\max}}\right\}\right) - \frac{L}{v_{FU}}}. \tag{16.25}$$

Hence the delay requirement will be $C_{\tau,qi}^{R}$. Since $\delta_{S_i,n}$ is the device data rate requirement, the data rate of transmitting content n to user i, $C_{\tau,qi}^{\mathrm{H}}$ must follow $C_{\tau,qi}^{\mathrm{H}} \geq \max\left\{C_{\tau,qi}^{R}, \delta_{S_i,n}\right\}$. □

Theorem 16.1 shows that the RRH–user association is determined by the requirements of delay, the data rate, as well as the fronthaul rate. Eq. (16.24) shows that each user's fronthaul rate decreases when the number of users connected to the RRHs increases. This is because as the fronthaul rate decreases, the delay requirement will be improved. Here we do not consider the RRH energy consumption since the ground can continuously supply power to the RRHs. However, the UAVs are powered by limited energy batteries. Thus it is necessary to encourage the ground users to first connect to the RRHs as the RRHs can satisfy the users' QoE requirements.

16.2.4.2 Optimal Cached Contents for UAVs

As the users that are associated with the RRHs are determined, the remaining users will be serviced by the cache-enabled UAVs. In such cases, we need to determine the users–UAVs association. First, we exploit K-mean approach [34] to divide the users into K groups. Performing the K-mean algorithm enables the users that are located near each other to be clustered into one group. We assume that each UAV will serve one cluster of users and the association between UAVs and the users can be determined. According to the user-UAV association, the cached contents of UAVs can be determined. Eq. (16.15) shows that the optimal cached contents of UAVs can reduce the UAV's transmit power required to satisfy the QoE requirement of each user. This is due to the fact that the UAVs can directly transmit the cached contents to the requested users, and hence the transmission delay decreases and the UAV transmit power is reduced.

We assume that $\boldsymbol{p}_{j,i} = \left[p_{j,i1}, p_{j,i2}, \ldots, p_{j,iN}\right]$ represents user i's content request distribution at period j. The optimal cached contents of each UAV is decided by the following theorem.

THEOREM 16.2 The optimal cached contents $\mathcal{C}_k$ of UAV k during period T is given by:

$$\mathcal{C}_k = \arg\max_{\mathcal{C}_k} \sum_{j=1}^{T/H} \sum_{\tau=1}^{H} \sum_{i \in \mathcal{U}_{\tau,k}} \sum_{n \in \mathcal{C}_k} \left(p_{j,in} \Delta P_{j,\tau,ki,n}\right), \tag{16.26}$$

where $\Delta P_{j,\tau,ki,n} =$

$$\begin{cases} P^{\min}_{\tau,ki}\left(C^R_{\tau,ki}\right)_{n \notin \mathcal{C}_k} - P^{\min}_{\tau,ki}\left(C^R_{\tau,ki}\right)_{n \in \mathcal{C}_k}, C^R_{\tau,ki,n \notin \mathcal{C}_k} \geq \delta_{S_i,n}, \\ P^{\min}_{\tau,ki}\left(\delta_{S_i,n}\right)_{n \notin \mathcal{C}_k} - P^{\min}_{\tau,ki}\left(C^R_{\tau,ki}\right)_{n \in \mathcal{C}_k}, \delta_{S_i,n} > C^R_{\tau,ki,n \notin \mathcal{C}_k}, \end{cases}$$

and $P^{\min}_{\tau,ki}\left(C^R_{\tau,ki}\right)$ refers to $P^{\min}_{\tau,ki}\left(\boldsymbol{w}_{\tau,t,k}, C^R_{\tau,ki}, n\right)$.

Proof Since $C^R_{\tau,ki,n}$ is determined by the cached contents of each UAV, it can be given by $\delta^R_{i,n} = \max\left\{C^R_{\tau,ki,n(n \in \mathcal{C}_k)}, C^R_{\tau,ki,n(n \notin \mathcal{C}_k)}, \delta_{S_i,n}\right\}$. We assume that $P^{\min}_{\tau,ki} = \sum_{t=1}^{F_{\tau,i}} P^{\min}_{j,\tau,t,ki}$. Then the transmit power reduction of each UAV is given by:

$$\Delta P_{j,\tau,ki,n} = \begin{cases} P^{\min}_{\tau,ki}\left(C^R_{\tau,ki}\right)_{n \notin \mathcal{C}_k} - P^{\min}_{\tau,ki}\left(C^R_{\tau,ki}\right)_{n \in \mathcal{C}_k}, C^R_{\tau,ki,n \notin \mathcal{C}_k} \geq \delta_{S_i,n}, \\ P^{\min}_{\tau,ki}\left(\delta_{S_i,n}\right)_{n \notin \mathcal{C}_k} - P^{\min}_{\tau,ki}\left(C^R_{\tau,ki}\right)_{n \in \mathcal{C}_k}, \quad \delta_{S_i,n} > C^R_{\tau,ki,n \notin \mathcal{C}_k}. \end{cases} \tag{16.27}$$

Since the content request distribution of each user will change once a period, the power minimization problem of each UAV k can be given by:

$$\begin{aligned} \min_{\mathcal{C}_k} \sum_{\tau=1}^{T} \sum_{i \in \mathcal{U}_{\tau,k}} P^{\min}_{\tau,ki} \min_{\mathcal{C}_k} &= \sum_{j=1}^{T/H} \sum_{\tau_j=1}^{H} \sum_{i \in \mathcal{U}_{\tau,k}} P^{\min}_{\tau_j,ki} \\ &= \min_{\mathcal{C}_k} \sum_{j=1}^{T/H} \sum_{\tau=1}^{H} \sum_{i \in \mathcal{U}_{\tau,k}} P^{\min}_{j,\tau,ki} \overset{(a)}{\Leftrightarrow} \max_{\mathcal{C}_k} \sum_{j=1}^{T/H} \sum_{\tau=1}^{H} \sum_{i \in \mathcal{U}_{\tau,k}} \Delta P_{j,\tau,ki,n}, \\ &\overset{(b)}{=} \max_{\mathcal{C}_k} \sum_{j=1}^{T/H} \sum_{\tau=1}^{H} \sum_{i \in \mathcal{U}_{\tau,k}} \left(\sum_{n \in \mathcal{C}_k} \left(p_{j,in} \Delta P_{j,\tau,ki,n}\right) + \sum_{n \notin \mathcal{C}_k} \left(p_{j,in} \Delta P_{j,\tau,ki,n}\right) \right), \\ &= \max_{\mathcal{C}_k} \sum_{j=1}^{T/H} \sum_{\tau=1}^{H} \sum_{i \in \mathcal{U}_{\tau,k}} \sum_{n \in \mathcal{C}_k} \left(p_{j,in} \Delta P_{j,\tau,ki,n}\right), \end{aligned} \tag{16.28}$$

where (a) stems from the fact that the minimization of each UAV's transmit power is equal to the maximized transmit power reduction due to caching, and (b) is obtained by using each user's content request probability distribution to compute the average transmit power reduction. This completes the proof. □

Theorem 16.2 shows that as all users' fronthaul data rates are equal, $\Delta P_{j,\tau,ki,n}$ will be a constant. Hence the optimal cached contents is $\mathcal{C}_k = \arg\max_{\mathcal{C}_k} \sum_{j=1}^{T/H} \sum_{\tau=1}^{H} \sum_{i \in \mathcal{U}_{\tau,k}} \sum_{n \in \mathcal{C}_k} p_{j,in}$. Theorem 16.2 also shows that the optimal cached contents of each UAV depends on the user–UAV association and the predictions. In consequence, using the prediction results, we can decide the optimal cached contents of each UAV.

16.2.4.3 Optimal UAV Locations

Once the UAVs determined the optimal cached contents, the content transmission link and the delay requirement $C_{\tau,ki,n}^{R}$ will be determined. In consequence, the QoE requirement $\delta_{i,n}^{R}$ is also decided. Next, a closed form expression of each UAV k's optimal location within two special cases is derived.

THEOREM 16.3 *To minimize the transmit power of each UAV k, each UAV k's optimal location for the two cases—(a) UAV k is located at low altitude,* $h_{\tau,k}^2 \ll \left(x_{t,i} - x_{\tau,k}\right)^2 + \left(y_{t,i} - y_{\tau,k}\right)^2$ *and* $\mu_{NLoS} = 2$*, and (b) UAV k is located at high altitude,* $h_{\tau,k}^2 \gg \left(x_{t,i} - x_{\tau,k}\right)^2 + \left(y_{t,i} - y_{\tau,k}\right)^2$*—are given as follows:*

$$x_{\tau,k} = \frac{\sum\limits_{i \in \mathcal{U}_{\tau,k}} \sum\limits_{t=1}^{F_{\tau,i}} x_{t,i}\psi_{t,ki}}{\sum\limits_{i \in \mathcal{U}_{\tau,k}} \sum\limits_{t=1}^{F_{\tau,i}} \psi_{t,ki}}, \quad y_{\tau,k} = \frac{\sum\limits_{i \in \mathcal{U}_{\tau,k}} \sum\limits_{t=1}^{F_{\tau,i}} y_{t,i}\psi_{t,ki}}{\sum\limits_{i \in \mathcal{U}_{\tau,k}} \sum\limits_{t=1}^{F_{\tau,i}} \psi_{t,ki}}, \tag{16.29}$$

where $\psi_{t,ki} = \left(2^{\delta_{i,n}^{R}/B} - 1\right)\sigma^2 10^{(L_{FS}(d_0)+\chi_\sigma)/10}$ *and* $\sigma = \begin{cases} \sigma_{NLoS}, & \textit{for case a}), \\ \sigma_{LoS}, & \textit{for case b}). \end{cases}$

Proof As the altitude of each UAV k is very low, $h_{\tau,k}^2 \ll \left(x_{t,i} - x_{\tau,k}\right)^2 + \left(y_{t,i} - y_{\tau,k}\right)^2$, $\frac{h_{\tau,k}}{d_{t,ki}\left(\boldsymbol{w}_{\tau,t,k}, \boldsymbol{w}_{\tau,t,i}\right)} \approx 0$ results in $\phi_t = 0°$. Thus $\Pr\left(l_{t,ki}^{\text{NLoS}}\right) = 1$. As such, $\bar{l}_{t,ki}\left(\boldsymbol{w}_{\tau,t,k}, \boldsymbol{w}_{\tau,t,i}\right) = l_{t,ki}^{\text{NLoS}}$. Eq. (16.16) can be given as follows:

$$P_{\tau,t,ki}^{\min} = \left(2^{\delta_{i,n}^{R}/B} - 1\right)\sigma^2 10^{\left(L_{FS}(d_0)+\chi_{\sigma_{\text{NLoS}}}\right)/10} d_{t,ki}\left(\boldsymbol{w}_{\tau,t,k}, \boldsymbol{w}_{\tau,t,i}\right)^{\mu_{\text{NLoS}}}. \tag{16.30}$$

The derivation of $\sum\limits_{i \in \mathcal{U}_{\tau,k}} \sum\limits_{t=1}^{F_{\tau,i}} P_{\tau,t,ki}^{\min}$ with respect to $x_{\tau,k}$ can be given by:

$$\frac{\partial \sum\limits_{i \in \mathcal{U}_{\tau,k}} \sum\limits_{t=1}^{F_{\tau,i}} P_{\tau,t,ki}^{\min}}{\partial x_{\tau,k}} = \frac{\sum\limits_{i \in \mathcal{U}_{\tau,k}} \sum\limits_{t=1}^{F_{\tau,i}} \partial P_{\tau,t,ki}^{\min}}{\partial x_{\tau,k}} =$$

$$\sum_{i \in \mathcal{U}_{\tau,k}} \sum_{t=1}^{F_{\tau,i}} \mu_{\text{NLoS}}\left(x_{\tau,k} - x_{t,i}\right)\psi_{t,ki}\left(\left(x_{\tau,k} - x_{t,i}\right)^2 + \left(y_{\tau,k} - y_{t,i}\right)^2 + h_{\tau,k}^2\right)^{\frac{\mu_{\text{NLoS}}}{2} - 1}. \tag{16.31}$$

When $\mu_{\mathrm{NLoS}} = 2$, (16.31) can be simplified to $\sum_{i \in \mathcal{U}_{\tau,k}} \sum_{t=1}^{F_{\tau,i}} 2\left(x_{\tau,k} - x_{t,i}\right)\psi_{t,ki} = 0$. As a result, $x_{\tau,k} = \frac{\sum_{i \in \mathcal{U}_{\tau,k}} \sum_{t=1}^{F_{\tau,i}} x_{t,i}\psi_{t,ki}}{\sum_{i \in \mathcal{U}_{\tau,k}} \sum_{t=1}^{F_{\tau,i}} \psi_{t,ki}}$. Similarly, we can obtain that $y_{\tau,k} = \frac{\sum_{i \in \mathcal{U}_{\tau,k}} \sum_{t=1}^{F_{\tau,i}} y_{t,i}\psi_{t,ki}}{\sum_{i \in \mathcal{U}_{\tau,k}} \sum_{t=1}^{F_{\tau,i}} \psi_{t,ki}}$.

For case (*b*), $h_{\tau,k}^2 \gg \left(x_{t,i} - x_{\tau,k}\right)^2 + \left(y_{t,i} - y_{\tau,k}\right)^2$, $d_{t,ki}\left(\boldsymbol{w}_{\tau,t,k}, \boldsymbol{w}_{\tau,t,i}\right) \approx h_{\tau,k}$. Thus $\frac{h_{\tau,k}}{d_{t,ki}\left(\boldsymbol{w}_{\tau,t,k}, \boldsymbol{w}_{\tau,t,i}\right)} \approx 1 \to \phi_t = 90°$. In consequence, $\Pr\left(l_{t,ki}^{\mathrm{LoS}}\right) = 1$. Then we have $\bar{l}_{t,ki}\left(\boldsymbol{w}_{\tau,t,k}, \boldsymbol{w}_{\tau,t,i}\right) = l_{t,ki}^{\mathrm{LoS}}$. The derivation of $\sum_{i \in \mathcal{U}_{\tau,k}} \sum_{t=1}^{F_{\tau,i}} P_{\tau,t,ki}^{\min}$ is

$$\frac{\partial \sum\limits_{i \in \mathcal{U}_{\tau,k}} \sum\limits_{t=1}^{F_{\tau,i}} P_{\tau,t,ki}^{\min}}{\partial x_{\tau,k}} = \sum_{i \in \mathcal{U}_{\tau,k}} \sum_{t=1}^{F_{\tau,i}} \mu_{\mathrm{LoS}}\left(x_{\tau,k} - x_{t,i}\right)\psi_{t,ki}\left(\left(x_{\tau,k} - x_{t,i}\right)^2 + \left(y_{\tau,k} - y_{t,i}\right)^2 + h_{\tau,k}^2\right)^{\frac{\mu_{\mathrm{LoS}}}{2} - 1}$$

$$\approx \sum_{i \in \mathcal{U}_{\tau,k}} \sum_{t=1}^{F_{\tau,i}} \mu_{\mathrm{LoS}}\left(x_{\tau,k} - x_{t,i}\right)\psi_{t,ki} h_{\tau,k}^{\mu_{\mathrm{LoS}}-2} = 0.$$

As a result, $x_{\tau,k} = \frac{\sum_{i \in \mathcal{U}_{\tau,k}} \sum_{t=1}^{F_{\tau,i}} x_{t,i}\psi_{t,ki}}{\sum_{i \in \mathcal{U}_{\tau,k}} \sum_{t=1}^{F_{\tau,i}} \psi_{t,ki}}$ and $y_{\tau,k} = \frac{\sum_{i \in \mathcal{U}_{\tau,k}} \sum_{t=1}^{F_{\tau,i}} y_{t,i}\psi_{t,ki}}{\sum_{i \in \mathcal{U}_{\tau,k}} \sum_{t=1}^{F_{\tau,i}} \psi_{t,ki}}$. This completes the proof. □

As the users association and altitude $h_{\tau,k}$ are determined, one can use Theorem 16.3 to find the optimal locations of the UAVs. However, Theorem 16.3 can be applied for only the two special cases. For more generic cases, it is difficult to use the derivation for finding the optimal UAV locations. This is because the UAV altitude relies on x and y coordinates. Hence a learning algorithm given in [42] and [43] can be used to find a suboptimal locations of UAVs. The conceptor-based ESN learning algorithm can exploit different actions to find the optimal locations of the UAVs.

In the next section, we present some simulation results so as to show the benefits of the deployment of cache-enabled UAVs.

16.2.5 Simulation Results

To assess the performance of benefits of cache-enabled UAVs, we set up a circular C-RAN area. The radius of this C-RAN area is $r = 500$ m. $R = 20$ RRHs and $U = 70$ users are randomly located at this C-RAN area. Table 16.3 lists other simulation parameters. The data used for the training and content request distribution prediction of ESN were collected from the Youku of China network video index. The actual mobility data were collected from the students at the Beijing University of Posts and Telecommunications. Three baseline algorithms: (1) optimal algorithm, (2) ESN algorithm in [2] for the prediction of the mobility pattern and content request distribution, and

Table 16.3 System Parameters

Parameter	Value	Parameter	Value	Parameter	Value
F	1,000	Y	0.13	P_B	30 dBm
X	11.9	N	25	P_R	20 dBm
$\chi_{\sigma_{\text{LoS}}}$	5.3	H	10	$P_{\max}$	20 W
N_{tr}	1,000	d_0	5 m	σ^2	−95 dBm
N_s	12	λ	0.01	$h_{\min}$	100 m
N_x	4	β	2	B	1 MHz
μ_{LoS}	2	μ_{NLoS}	2.4	$\delta_{S_i,n}$	5 Mbit/s
χ	15	ζ_1	0.5	f_c	38 GHz
$\chi_{\sigma_{\text{NLoS}}}$	5.27	η	100	B_v	1 GHz
K	5	C	1	L	1 Mbit
T	120	ζ_2	0.5	N_w	1,000

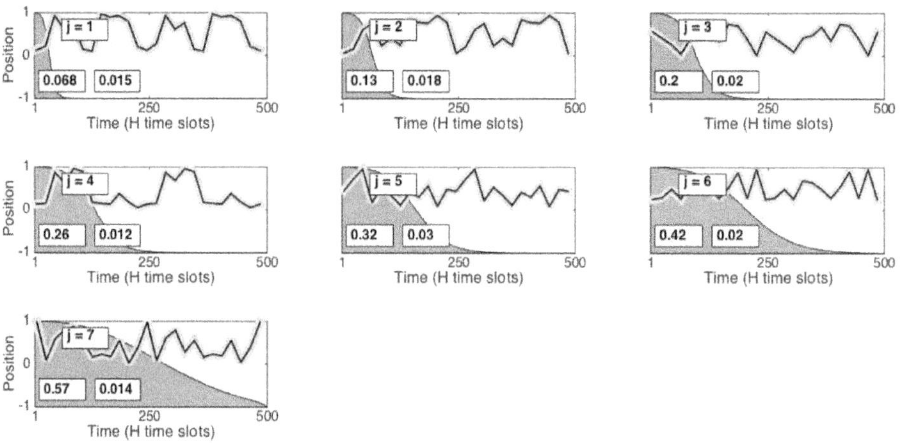

Figure 16.2 Conceptor-based ESN learning algorithm for mobility pattern predictions. Here, the gray curve denotes the mobility pattern predicted by conceptor-based ESN learning algorithm, the black curve represents the actual positions. Top rectangle j represents the mobility pattern index. The legend on the bottom left is the total memory that the ESN has used. The legend on the bottom right represents the NRMSE of each mobility pattern prediction.

(3) random caching with ESN algorithm in [2] for the content request distribution prediction. The accuracy of ESN prediction is measured by normalized root mean square error (NRMSE) [41].

Figure 16.2 shows how the conceptor-based ESN's memory capacity varies as the number of the already learned mobility patterns changes. In this figure, each mobility pattern denotes the trajectory of a given user in one day of a week. The colored region denotes the memory that has been used by the proposed learning algorithm. Figure 16.2 also shows that, as the number of the already learned mobility patterns increases, the ESN memory used for data recording increases. This is because the conceptor-based ESN learning algorithm exploits a limited-capacity memory to learn the mobility patterns. In Figure 16.2, we can also see that, compared to pattern 6, the

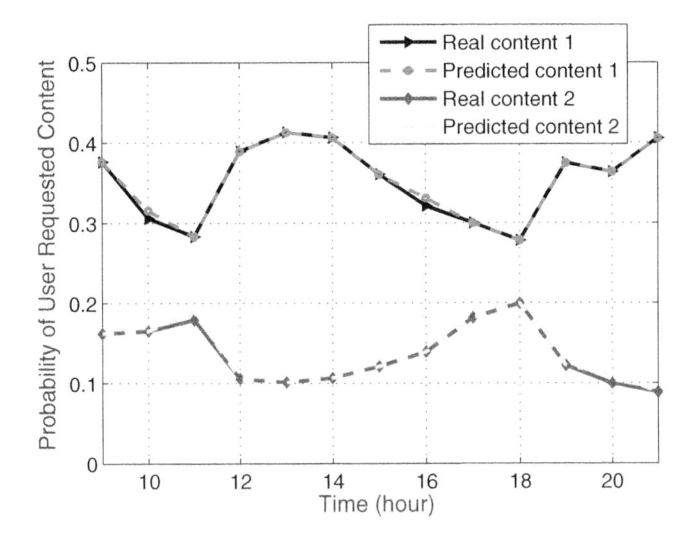

Figure 16.3 Content request probability predictions.

proposed algorithm uses less memory to learn mobility pattern 2. This is due to the fact that mobility pattern 2 is similar to mobility pattern 1. In consequence, the proposed algorithm needs less memory to learn mobility pattern 2 compared to pattern 6. In fact, as the conceptor-based ESN learning algorithm needs to learn a new mobility pattern, it needs only to learn the difference between the mobility patterns that have been learned and the mobility pattern that will be learned.

Figure 16.3 shows how two content request probabilities of a given user change as time elapses. Figure 16.3 shows that the probability of the user requesting content 1 decreases at 9:00–11:00 and 14:00–18:00 and increases at other times. Meanwhile, the probability that this user requests content 2 increases at 9:00–11:00 and 14:00–18:00 and decreases in the rest of the time. This is because content 1 is entertainment content while content 2 is work-related content. From Figure 16.3, we can also see that the sum of the probability of this user requesting content 1 and 2 exceeds 0.5. This corresponds to the fact that a user requests only a small amont of content in each day.

Figure 16.4 shows how the UAVs' transmit power varies when the number of ground users changes. The figure shows that, when the number of ground users increases, the total transmit power of the UAVs increases. This is because, as the number of ground users increases, the number of ground users connected to the RRHs increases. Thus the wireless fronthaul data rate decreases. In consequence, the UAVs must use more transmit power to meet each user's QoE requirement. Figure 16.4 also shows that the proposed algorithm can reduce 33.3% and 20% of the transmit power compared to the conceptor-based ESN learning algorithm without cache and to the conceptor-based ESN learning algorithm without optimization of the locations of the UAVs in a network with 80 users.

In summary, the developed ESN algorithm can be used for the predictions of the content request distributions and mobility patterns. These predictions can be used to determine the optimal location of UAVs and optimal cached contents of UAVs. The results have shown that, given the accurate predictions of users' mobility patterns and

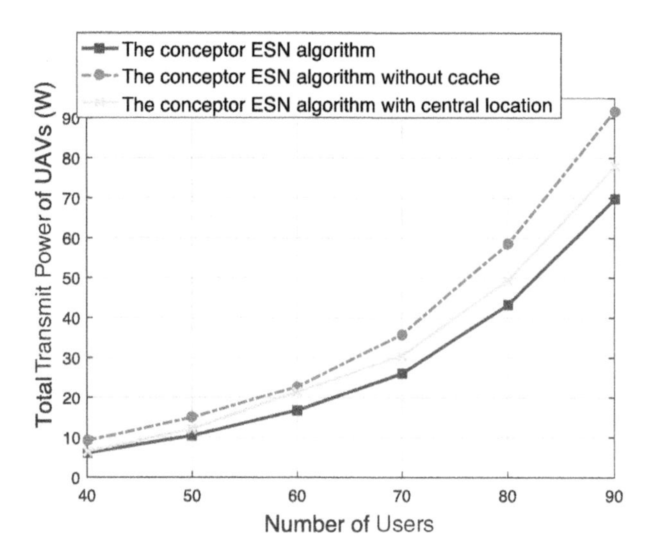

Figure 16.4 Total transmit power as the number of users varies.

content request distributions, the optimal location of UAVs and optimal caching contents can be determined. The results have also shown that the cache-enabled UAVs can efficiently service the users so as to maximize the users' QoE.

16.3 Summary

Caching is envisioned to be a key technique that offloads the traffic over the transmission links between the UAVs and the cloud in cloud radio access networks. As discussed in this chapter, cache-enabled UAVs can cache the most popular contents and directly transmit them to the users without the backhaul transmission so as to maximize the users' QoE.

In this chapter, first, we have developed and analyzed a scenario in which the cache-enabled UAVs are used to service the ground mobile users in a cloud radio access network. We have shown that, in this model, a variety of challenges need to be tackled, such as modeling users' QoE, developing an air-to-ground mmWave channel model and designing a machine learning–based algorithm for predicting of users' mobility patterns and content request distributions. We have discussed how the use of the predictions of users' mobility patterns and content request distributions determine the optimal locations of UAVs and optimal caching contents of each UAV. Through numerical analysis, we have evaluated the performance of predictions for the learning algorithm and the benefit of the use of cache-enabled UAVs.

This chapter has only scratched the surface of the emerging topic of cache-enabled UAVs. While our focus was on the deployment of cache-enabled UAVs and caching content replacement, many important problems remain open. On the one hand, it is of interest to develop new cooperative cache-enabled UAVs models in which all of the cache-enabled UAVs will cooperate with each other to improve the caching hit ratio so as to improve the users' QoE. For example, one cache-enabled UAV that has remaining storage can store the contents that another cache-enabled UAV wants to store but does not have enough storage. On the other hand, it is imperative to jointly consider cache-enabled UAVs with content delivery. In particular, the cache-enabled UAVs can adjust the update rate of caching content according to the content delivery rate. When the content delivery rate increases, the cache-enabled UAVs can increase the update rate of caching content. Certainly, storing contents at the UAV cache will require the information related to the users' behaviors such as users' mobility patterns and content requests. In consequence, there is a need to develop novel learning algorithms to extract and analyze the data related to the users' behaviors so as to predict the users' movements. One potential direction is to apply neural networks such as convolutional neural network to extract the properties of users' behaviors and use recurrent neural networks to predict the users' movements. Finally, in order to deliver emerging wireless service, such as virtual reality services [44], the effect of optimizing metrics such as energy efficiency, delay, and computation must be incorporated into the studied cache-enabled UAV models.

References

[1] Y. Pan, C. Pan, H. Zhu, Q. Z. Ahmed, M. Chen, and J. Wang, "On consideration of content preference and sharing willingness in D2D assisted offloading," *IEEE J. Select. Areas Commun. (JSAC), Special Issue Hum. Loop Mobile Netw.*, vol. 35, no. 4, pp. 978–993, Apr. 2017.

[2] M. Chen, W. Saad, C. Yin, and M. Debbah, "Echo state networks for proactive caching in cloud-based radio access networks with mobile users," *IEEE Trans. Wireless Commun.*, vol. 16, no. 6, pp. 3520–3535, June 2017.

[3] J. Qiao, Y. He, and S. Shen, "Proactive caching for mobile video streaming in millimeter wave 5G networks," *IEEE Trans. Wireless Commun.*, vol. 15, no. 10, pp. 7187–7198, Oct. 2016.

[4] T. X. Tran and D. Pompili, "Octopus: a cooperative hierarchical caching strategy for radio access networks," arXiv.org/abs/1608.00067, July 2016.

[5] Y. Guo, L. Duan, and R. Zhang, "Cooperative local caching under heterogeneous file preferences," *IEEE Trans. Commun.*, vol. 65, no. 1, pp. 444–457, Jan. 2017.

[6] E. Bastug, M. Bennis, M. Kountouris, and M. Debbah, "Cache-enabled small cell networks: modeling and tradeoffs," *EURASIP J. Wireless Commun. Netw., Special Issue Tech. Adv. Design Deployment Future Heterogeneous Netw.*, vol. 2015, no. 1, Feb. 2015.

[7] Z. Ye, C. Pan, H. Zhu, and J. Wang, "Tradeoff caching strategy of outage probability and fronthaul usage in Cloud-RAN," arXiv.org/abs/1611.02660, Nov. 2016.

[8] U. Challita, A. Ferdowsi, M. Chen, and W. Saad, "Artificial intelligence for wireless connectivity and security of cellular-connected uavs," ArXiv, abs/1804.05348, 2018.

[9] X. Xu, Y. Zeng, Y. L. Guan, and R. Zhang, "Overcoming endurance issue: UAV-enabled communications with proactive caching ," arXiv:1712.03542, Dec. 2017.

[10] M. Chen, M. Mozaffari, W. Saad, C. Yin, M. Debbah, and C. S. Hong, "Caching in the sky: proactive deployment of cache-enabled unmanned aerial vehicles for optimized quality-of-experience," *IEEE J. Selected Areas. Commun.*, vol. 35, no. 5, pp. 1046–1061, May 2017.

[11] M. Chen, W. Saad, and C. Yin, "Liquid state machine learning for resource and cache management in LTE-U unmanned aerial vehicle (UAV) networks," *IEEE Trans. Wireless Commun.*, vol. 18, no. 3, pp. 1504–1517, Mar. 2019.

[12] N. Zhao, F. Cheng, F. R. Yu, J. Tang, Y. Chen, G. Gui, and H. Sari, "Caching UAV assisted secure transmission in hyper-dense networks based on interference alignment," *IEEE Trans. Commun.*, vol. 66, no. 5, pp. 2281–2294, May 2018.

[13] M. Mozaffari, W. Saad, M. Bennis, Y. H. Nam, and M. Debbah, "A tutorial on UAVs for wireless networks: applications, challenges, and open problems," arXiv:1803.00680, Mar. 2018.

[14] H. Wang, G. Ding, F. Gao, J. Chen, J. Wang, and L. Wang, "Power control in UAV-supported ultra dense networks: communications, caching, and energy transfer," arXiv:1712.05004, Nov. 2017.

[15] M. Chen, U. Challita, W. Saad, C. Yin, and M. Debbah, "Machine learning for wireless networks with artificial intelligence: a tutorial on neural networks," arXiv.org/abs/1710.02913, Oct. 2017.

[16] M. Agiwal, A. Roy, and N. Saxena, "Next generation 5G wireless networks: a comprehensive survey," *IEEE Commun. Surveys Tutorials*, vol. 18, no. 3, pp. 1617–1655, 2016.

[17] M. Peng, Y. Sun, X. Li, Z. Mao, and C. Wang, "Recent advances in cloud radio access networks: system architectures, key techniques, and open issues," *IEEE Commun. Surveys Tutorials*, vol. 18, no. 3, pp. 2282–2308, 2016.

[18] Z. Zhao, M. Peng, Z. Ding, W. Wang, and H. V. Poor, "Cluster content caching: an energy-efficient approach to improve quality of service in cloud radio access networks," *IEEE J. Selected Areas Commun.*, vol. 34, no. 5, pp. 1207–1221, Mar. 2016.

[19] T. X. Tran, A. Hajisami, and D. Pompili, "Cooperative hierarchical caching in 5G cloud radio access networks (C-RANs)," arXiv.org/abs/1602.02178, Jan. 2016.

[20] K. Hamidouche, W. Saad, M. Debbah, and H. V. Poor, "Mean-field games for distributed caching in ultra-dense small cell networks," in *Proc. 2016 Am. Control Conference (ACC)*, July. 2016.

[21] M. Tao, E. Chen, H. Zhou, and W. Yu, "Content-centric sparse multicast beamforming for cache-enabled cloud RAN," *IEEE Trans. Wireless Commun.*, vol. 15, no. 9, pp. 6118–6131, Sept. 2016.

[22] D. Chen, S. Schedler, and V. Kuehn, "Backhaul traffic balancing and dynamic content-centric clustering for the downlink of fog radio access network," *arXiv.org/abs/1602.05536*, Feb. 2016.

[23] Y. Wang, X. Tao, X. Zhang, and Y. Gu, "Cooperative caching placement in cache-enabled D2D underlaid cellular network," *IEEE Commun. Lett.*, vol. 21, no. 5, pp. 1151–1154, May 2017.

[24] Z. Yang, C. Pan, Y. Pan, Y. Wu, W. Xu, M. Shikh-Bahaei, and M. Chen, "Cache placement in two-tier hetnets with limited storage capacity: cache or buffer?" *IEEE Trans. Commun.*, vol. 66, no. 11, pp. 5415–5429, Nov. 2018.

[25] Y. Sun, Y. Cui, and H. Liu, "Joint pushing and caching for bandwidth utilization maximization in wireless networks," in *Proc. IEEE Global Commun. Conference*, Dec. 2017.

[26] M. Mozaffari, W. Saad, M. Bennis, and M. Debbah, "Unmanned aerial vehicle with underlaid device-to-device communications: performance and tradeoffs," *IEEE Trans. Wireless Commun.*, vol. 15, no. 6, pp. 3949–3963, June 2016.

[27] Y. Zeng, R. Zhang, and T. J. Lim, "Throughput maximization for UAV-enabled mobile relaying systems," *IEEE Trans. Commun.*, vol. 64, no. 12, pp. 4983–4996, Dec. 2016.

[28] M. Mozaffari, W. Saad, M. Bennis, and M. Debbah, "Efficient deployment of multiple unmanned aerial vehicles for optimal wireless coverage," *IEEE Commun. Lett.*, vol. 20, no. 8, pp. 1647–1650, Aug. 2016.
[29] A. Al-Hourani, S. Kandeepan, and A. Jamalipour, "Modeling air-to-ground path loss for low altitude platforms in urban environments," in *Proc. IEEE Global Commun. Conference (GLOBECOM)*, Dec. 2014.
[30] I. Bor-Yaliniz and H. Yanikomeroglu, "The new frontier in RAN heterogeneity: multi-tier drone-cells," *IEEE Commun. Mag.*, vol. 54, no. 11, pp. 48–55, Nov. 2016.
[31] E. Kalantari, H. Yanikomeroglu, and A. Yongacoglu, "On the number and 3D placement of drone base stations in wireless cellular networks," in *Proc. IEEE Vehicular Technol. Conference*, May 2016.
[32] M. Mozaffari, W. Saad, M. Bennis, and M. Debbah, "Drone small cells in the clouds: design, deployment and performance analysis," in *Proc. IEEE Global Commun. Conference (GLOBECOM)*, Dec. 2015.
[33] Z. Yang, C. Pan, M. Shikh-Bahaei, W. Xu, M. Chen, M. Elkashlan, and A. Nallanathan, "Joint altitude, beamwidth, location, and bandwidth optimization for uav-enabled communications," *IEEE Commun. Lett.*, vol. 22, no. 8, pp. 1716–1719, Aug. 2018.
[34] F. Hoppner and F. Klawonn, *Clustering with Size Constraints*. Berlin Heidelberg: Springer, 2008.
[35] T. Yoo and A. Goldsmith, "On the optimality of multiantenna broadcast scheduling using zero-forcing beamforming," *IEEE J. Selected Areas Commun.*, vol. 24, no. 3, pp. 528–541, Mar. 2006.
[36] T. S. Rappaport, *Wireless Communications: Principles and Practice*. Upper Saddle River, NJ: Prentice-Hall, 2002.
[37] T. S. Rappaport, F. Gutierrez, E. Ben-Dor, J. N. Murdock, Y. Qiao, and J. I. Tamir, "Broadband millimeter-wave propagation measurements and models using adaptive-beam antennas for outdoor urban cellular communications," *IEEE Trans. Antennas. Propagation*, vol. 61, no. 4, pp. 1850–1859, Apr. 2013.
[38] O. Somekh, O. Simeone, Y. Bar-Ness, A. M. Haimovich, and S. Shamai, "Cooperative multicell zero-forcing beamforming in cellular downlink channels," *IEEE Trans. Inform. Theory*, vol. 55, no. 7, pp. 3206–3219, June 2009.
[39] K. Mitra, A. Zaslavsky, and C. Ahlund, "Context-aware QoE modelling, measurement and prediction in mobile computing systems," *IEEE Trans. Mobile Comput.*, vol. 14, no. 5, pp. 920–936, Dec. 2015.
[40] M. Lukoševicius, *A Practical Guide to Applying Echo State Networks*. Berlin Heidelberg: Springer, 2012.
[41] H. Jaeger, "Controlling recurrent neural networks by conceptors," arXiv.org/abs/1403.3369, 2014.
[42] M. Bennis, S. Perlaza, P. Blasco, Z. Han, and H. Poor, "Self-organization in small cell networks: a reinforcement learning approach," *IEEE Trans. Wireless Commun.*, vol. 12, no. 7, pp. 3202–3212, June 2013.
[43] M. Chen, W. Saad, and C. Yin, "Echo state networks for self-organizing resource allocation in LTE-U with uplink-downlink decoupling," *IEEE Trans. Wireless Commun.*, vol. 1, no. 1, Jan. 2017.
[44] M. Chen, W. Saad, and C. Yin, "Virtual reality over wireless networks: quality-of-service model and learning-based resource management," *IEEE Trans. Commun.*, vol. 66, no. 11, pp. 5621–5635, Nov. 2018.

17 Physical Layer Security for Edge Caching Wireless Networks

Lin Xiang, Derrick W. K. Ng, Robert Schober, and Vincent W. S. Wong

17.1 Introduction

The wireless cellular networks have been witnessing a rapid growth in video-on-demand (VoD) streaming traffic. Different from conventional voice and data applications, VoD streaming requires both high data rate and low latency during data delivery. Meeting the stringent VoD streaming requirements with off-the-shelf wireless cellular networks has been a great challenge due to capacity limitations in both the radio access and backhaul links [1]. To tackle this problem, a disruptive technique employing caching at the wireless edge has been proposed [2–5]. As has been shown in the previous chapters, by storing a priori the popular video files at edge nodes, e.g., base stations (BSs), access points (APs), and user equipment (UE), caching facilitates not only traffic offloading on the backhaul but also capacity enhancement, latency reduction, and energy savings during radio access. Therefore, caching has been considered as a promising solution to support large-scale VoD streaming for next-generation wireless cellular networks.

On the other hand, due to increasing cyberattacks in communication networks, security and privacy are among the utmost concerns for wireless technologies. For example, edge nodes may be *untrusted*, i.e., they may intercept the cached data. Moreover, wireless transmission is insecure due to its broadcast nature. As a result, the streaming of video data may suffer potential eavesdropping from, e.g., nonpaying subscribers, and the privacy of premium subscribers may not be guaranteed. To tackle the security challenges caused by untrusted and/or eavesdropping nodes, secure caching and VoD streaming schemes that can protect video data and guarantee streaming quality of service (QoS) simultaneously are crucially needed, especially for streaming private and paid video content. Hence in this chapter, we study security issues associated with employing caching in wireless networks.

17.1.1 Literature Survey

The development of security techniques for networks adopting wireless caching has gone through several stages. In the early literature, providing security for cache-enabled networks was considered impossible [6]. This is because, with off-the-shelf hypertext

This work is support in part by Australian Research Council's Discovery Early Career Researcher Award funding scheme (DE170100137), by the Alexander von Humboldt Professorship Program, and by the Natural Sciences and Engineering Research Council of Canada.

transfer protocol (HTTP) and hypertext transfer protocol secure (HTTPS), communication security has been ensured using end-to-end encryption methods [6]. Thereby, video content intended for each streaming UE has to be uniquely encrypted and hence cannot be reused to serve other UEs, which compromises the benefits of content caching. For this reason, security constraint has been usually neglected for video content caching.

In fact, the pessimistic perspective in [6] is overly conservative and can be lifted by security techniques proposed in the recent literature [7–13]. Thereby, reuse of cached content is achieved by either modifying the encryption protocol [7–9] or resorting to non-encryption-based security schemes such as physical-layer security (PLS) techniques [10–13]. These security schemes can be classified into two categories, namely *passive* and *proactive*. For passive security schemes, secure cache placement and data delivery are ensured by performing sophisticated encryption, and caching is not exploited to obtain secrecy benefits. The schemes in [7–9] employed the one-time pad method together with the coded caching scheme proposed in [14, 15]. Specifically, a cache is deployed at each user to store parts of popular video content. To achieve high data rates for serving the UEs, the cached and the delivered contents are encoded intelligently via network coding to facilitate coded multicast delivery [14]. In [7], by encoding the contents together with random keys that are kept secret from the eavesdroppers, the authors propose a secure coded multicast delivery scheme to protect the video data from being deciphered passively over the multicast link. However, due to the shortcomings of the one-time pad method, secrecy can be ensured only when the length of the secret keys is large enough, e.g., at least as large as the content size, which increases the system overhead for secure sharing of the secret keys. In [8], an enhanced secure coded caching employing an advanced key generation and encryption scheme is proposed for device-to-device (D2D) networks. Moreover, considering caching in heterogeneous small cell networks, the authors in [9] design a novel secure delivery scheme to prevent eavesdroppers from obtaining a sufficient number of coded packets needed for deciphering the video files.

To eliminate the overhead of generating and distributing secret keys, secure transmission using PLS has been thoroughly investigated in wireless communications. In particular, PLS exploiting multiple-input multiple-output (MIMO) techniques has demonstrated significant advantages over one-time pad-based methods [16–19]. On the one hand, PLS techniques can enhance communication secrecy by opportunistically exploiting the random wireless channels without adopting secret keys. On the other hand, the abundant degrees of freedom offered by MIMO communication can be exploited to significantly improve the secrecy capacity and/or the power efficiency of the system. Inspired by the MIMO-based PLS techniques, caching schemes facilitating PLS were reported for the first time in [10]. By caching each video content at multiple transmitters, cooperative MIMO transmission of the video content can be enabled among these transmitters, and the resulting large transmit antenna array can be utilized to significantly increase the secrecy capacity. As caching also reduces the data sharing overhead typically needed for cooperative transmission [20], cache-enabled PLS is applicable even when the backhaul links have limited capacity. Cache-enabled PLS has been investigated in scenarios with imperfect [11] and statistical [12] channel state information (CSI) about the eavesdropper(s). In [13], cache-enabled PLS was extended to combat the secrecy challenges caused by untrusted cache helpers. Different from the passive security schemes, by cache-enabled PLS schemes [10–13], caching is exploitable to proactively enhance the security of the system.

The aforementioned works from the literature suggest that cache-enabled PLS techniques provide an advanced mechanism for proactively exploiting caching to improve the security of cache-enabled transmission. In this chapter, we discuss the design and optimization of cache placement and cache-enabled secure cooperative MIMO transmission toward achieving enhanced PLS. The chapter is organized as follows. In Section 17.2, we describe the adopted system model. In Sections 17.3 and 17.4, we formulate and solve the two-stage optimization problem arising in the design of cache-enabled PLS systems, respectively. Section 17.5 presents the numerical results. In Section 17.6, we discuss some research challenges in applying cache-enabled PLS and paths to potential solutions. Section 17.7 concludes the chapter.

17.2 System Model

In this section, we present the system model for cache-enabled secure video streaming in cellular networks. First, we introduce the cellular video streaming system. Then, we discuss how caching is exploited to reduce the latency of video delivery and, at the same time, enable secure video data delivery. We provide a list of key notations in Table 17.1.

17.2.1 Network Topology

We consider a cellular VoD streaming system that includes a video server, M BSs, and K legitimate receivers (LRs); see Figure 17.1. The BSs and the LRs are indexed by sets $\mathcal{M} \triangleq \{1,\dots,M\}$ and $\mathcal{K} \triangleq \{1,\dots,K\}$, respectively. Each BS is equipped with N_t antennas while, for convenience, only a single antenna is deployed at an LR. A library of F video files, indexed by $\mathcal{F} \triangleq \{1,\dots,F\}$, is available at the video server, which is located on the internet edge and provides video streaming services for the LRs with the aid of the BSs. The BSs are connected to the video server, and the LRs via dedicated backhaul links such as digital subscriber lines and wireless links, respectively. When the LRs' requests are received, the BSs are responsible for fetching the files requested by the LRs from the video server and for delivering them to the requesting LRs, which take place over the backhaul links and the wireless links, respectively. The backhaul links are assumed to be secure. However, the wireless data transmission from the BSs to the LRs may be leaked to a passive eavesdropping receiver (ER) within the system. The ER has N_e antennas (the N_e antennas may be either co-located or distributed but connected in performing joint eavesdropping). For ensuring communication security, we assume $MN_t > N_e$.

The considered VoD streaming system is time-slotted and employs HTTP [21]. If requested, file $f \in \mathcal{F}$ will be delivered in L time slots, where $L \gg 1$ usually holds. Thereby, a portion of file f, referred to as subfile (f,l), is delivered in time slot

Table 17.1 Nomenclature Adopted in This Chapter

Operators	Description
$\text{diag}(\mathbf{v})$	Diagonal matrix with the diagonal elements given by $\mathbf{v}$
$(\cdot)^T$, $(\cdot)^H$	Transpose and complex conjugate transpose
$\text{tr}(\cdot)$, $\text{rank}(\cdot)$	Trace and rank of a matrix
$\det(\cdot)$, $\lambda_{\max}(\cdot)$	Determinant and maximum eigenvalue of a matrix
$\sim$, $\Pr(\cdot)$	Distributed as and probability mass operator
$\lvert\mathcal{X}\rvert$, $\mathcal{X} \times \mathcal{Y}$	Cardinality of set $\mathcal{X}$, Cartesian product of sets $\mathcal{X}$ and $\mathcal{Y}$
$\mathbf{A} \succeq \mathbf{0}$ $(\mathbf{A} \succ \mathbf{0})$	Matrix $\mathbf{A}$ is positive semi-definite (definite)
$\nabla_{\mathbf{X}} f(\mathbf{X})$	Complex-valued gradient of $f(\mathbf{X})$ with respect to $\mathbf{X}$
$\lfloor\cdot\rfloor$	Rounding operator
$[x]^+$	$\max(0, x)$
Symbols	**Description**
$\mathbb{R}$ and $\mathbb{C}$	Fields of real and complex numbers
$\mathbf{I}_L$, $\mathbf{1}_L$, and $\mathbf{0}_L$	$L \times L$ identity, all-one, and zero matrices
$\mathcal{K}$, $\mathcal{M}$	Sets of K LRs and M BSs
$\mathcal{M}_{f,l}^{\text{Coop}}$	Set of BSs cooperating in delivery of subfile (f,l)
$\mathcal{F}$, $\mathcal{L}$	Sets of F video files and L subfiles per file
V_f	Size of file $f \in \mathcal{F}$ in bits
$\rho \triangleq (k, f, l)$	Request for subfile (f,l) by LR k
$\mathcal{S}$	Set of user requests
$c_{f,l,m}$ $(c_{f,l})$,	Caching, backhaul loading, and cooperation
$b_{f,l,m}$, $q_{f,l,m}$	decisions for delivering subfile (f,l) at BS m
$\mathbf{w}_\rho$, $\mathbf{w}_{m,\rho}$	Beam-forming coefficients at BS set $\mathcal{M}$ and BS m for LR ρ
$C_m^{\max}$, $B_m^{\max}$	Cache size and backhaul link capacity at BS m
Γ_ρ, R_ρ, R_ρ^{sec}	SINR, achievable data rate, and secrecy rate at LR ρ
$R_{\text{e},\rho}$	Capacity of the ER for eavesdropping LR ρ

$l \in \mathcal{L} \triangleq \{1, \ldots, L\}$. We assume that the subfiles of file f have equal size. The sizes of file $f \in \mathcal{F}$ and subfile (f,l) are denoted as V_f and V_f/L, respectively. Moreover, an LR is assumed to request only one (sub)file at a time. Hence the request of LR k for subfile (f,l), denoted by $\rho \triangleq (k, f, l)$, can be mapped to user k on a one-to-one basis. For this reason, in the following, we may use k and ρ interchangeably to index the LRs. Let $\mathcal{S}$ be the set of user requests. We have $\mathcal{S} \subseteq \mathcal{K} \times \mathcal{F} \times \mathcal{L}$ and $|\mathcal{S}| = K$.

Assume that the fading channel during video transmission is frequency flat. Let $y_\rho \in \mathbb{C}$ and $\mathbf{y}_\text{e} \in \mathbb{C}^{N_\text{e} \times 1}$ be the signals received at LR $\rho \in \mathcal{S}$ and the ER, respectively. The input–output channel model is then given as

$$y_\rho = \mathbf{h}_\rho^H \mathbf{x} + z_\rho, \tag{17.1}$$

$$\mathbf{y}_\text{e} = \mathbf{G}^H \mathbf{x} + \mathbf{z}_\text{e}, \tag{17.2}$$

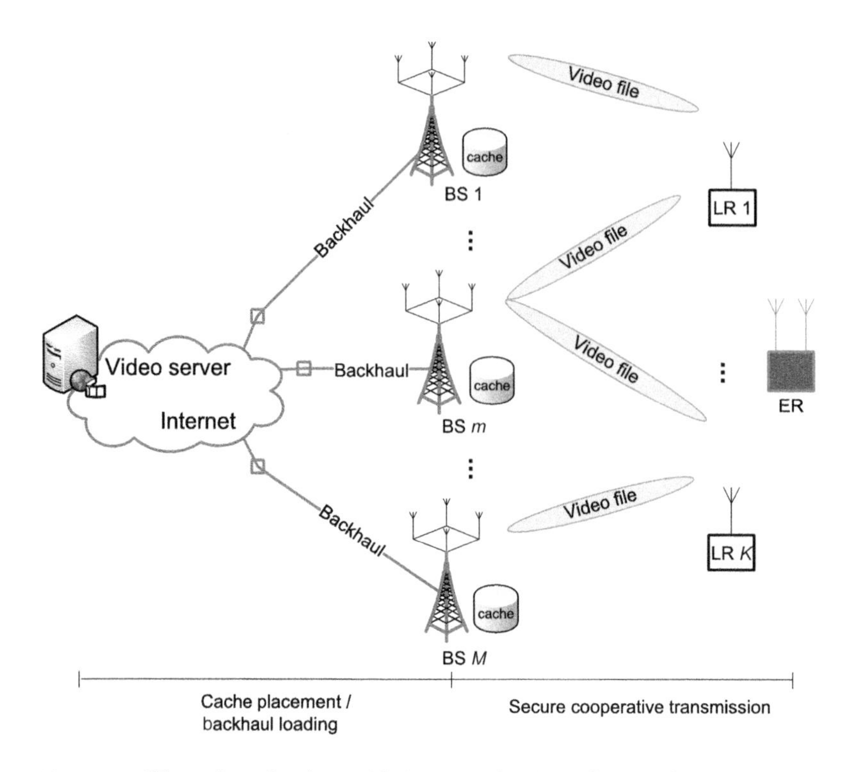

Figure 17.1 Illustration of cache-enabled cooperative beam-forming for secure downlink VoD streaming in multi-cell cellular networks.

where $\mathbf{x} \in \mathbb{C}^{MN_t \times 1}$ is the joint transmit signals of all considered BSs in set $\mathcal{M}$. $\mathbf{h}_\rho = [\mathbf{h}_{1,\rho}^H, \ldots, \mathbf{h}_{M,\rho}^H]^H \in \mathbb{C}^{MN_t \times 1}$ and $\mathbf{G} = [\mathbf{G}_1^H, \ldots, \mathbf{G}_M^H]^H \in \mathbb{C}^{MN_t \times N_e}$ denote the propagation channel vector/matrix from the BSs in set $\mathcal{M}$ and the ER to LR ρ, where $\mathbf{h}_{m,\rho} \in \mathbb{C}^{N_t \times 1}$ and $\mathbf{G}_m \in \mathbb{C}^{N_t \times N_e}$ capture the propagation channels from BS $m \in \mathcal{M}$ and the ER to LR ρ, respectively. Finally, $z_\rho \sim \mathcal{CN}(0, \sigma^2)$ and $\mathbf{z}_e \sim \mathcal{CN}(\mathbf{0}, \sigma_e^2 \mathbf{I}_{N_e})$ are the receiver front-end noises at the LRs and the ER, which follow zero-mean complex Gaussian distributions with variance σ^2 and covariance matrix $\sigma_e^2 \mathbf{I}_{N_e}$, respectively.

17.2.2 Caching and Backhaul Loading

In practice, a backhaul link has to accommodate several types of traffic (e.g., voice, data, multimedia, control signaling) simultaneously, where the backhaul capacity is allocated on a dynamic basis. As a result, VoD streaming may have to utilize an intermittent and limited backhaul capacity [22, 23]. This increases the latency of video data delivery and may even cause congestion on the backhaul links. To tackle this problem, a cache is deployed at each BS. The operation of the cache-enabled system divides into two stages. In the first stage, e.g., during off-peak traffic hours, a portion of the video files is cached at the BSs by utilizing the idle network capacity. In the second stage, i.e., when users issue video file requests during peak traffic hours, the BSs cooperatively serve the requests. Such a two-stage model is often assumed for studying cache-enabled systems. It can also be executed repeatedly over time to adapt the dynamic network traffic and backhaul capacity conditions.

By caching popular video files at the BSs before video streaming starts, the backhaul traffic incurred in the second stage is reduced. This in turn reduces the delivery latency as (portions of) the video data can be fetched from the cache at the BSs in the close proximity of the LRs. In particular, the video files/subfiles can be shared to the BSs either by caching them ahead of time or by loading them via backhaul links instantaneously during delivery. Assume that $c_{f,l,m} \in [0,1]$ and $b_{f,l,m} \in [0,1]$ portions of subfile (f,l) are shared to BS m via caching and backhaul loading, respectively. For modeling simplicity, we assume $c_{f,l,m} = c_{f,m}, \forall l \in \mathcal{L}$. Define binary variable $q_{f,l,m} \in \{0,1\}$ as follows,

$$q_{f,l,m} = \begin{cases} 1, & \text{if } b_{f,l,m} + c_{f,m} = 1, \\ 0, & \text{otherwise.} \end{cases} \tag{17.3}$$

Hence $q_{f,l,m}$ indicates whether subfile (f,l) is available at BS $m \in \mathcal{M}$ during video delivery.

17.2.3 Secure Cooperative MIMO Transmission

On the other hand, caching facilitates cooperative MIMO transmission opportunities for securely delivering the video files to the LRs while mitigating potential information leakage to the ER. In particular, by sharing subfile (f,l) at multiple BSs via both caching and backhaul loading, these BSs can cooperatively deliver the subfile by performing joint transmission. The set of BSs allowed for cooperative transmission of subfile (f,l) is given as $\mathcal{M}_{f,l}^{\text{Coop}} \triangleq \{m \in \mathcal{M} \mid q_{f,l,m} = 1\} \subseteq \mathcal{M}$. Therefore, $q_{f,l,m}$ also indicates whether BS m can cooperate with other BSs in delivering subfile (f,l).

As the system's degrees of freedom increases due to the large virtual antenna array formed by the cooperating BSs, they can be exploited during beam-forming to improve communication security [16–19]. To this end, we design the joint transmit signal of all considered BSs, $\mathbf{x} \in \mathbb{C}^{MN_\text{t} \times 1}$, as

$$\mathbf{x} = \sum_{\rho \in \mathcal{S}} \mathbf{w}_\rho s_\rho, \tag{17.4}$$

where $s_\rho \in \mathbb{C}$ denotes the symbols intended for LR ρ and $s_\rho \sim \mathcal{CN}(0, 1)$ is a complex Gaussian random variable [24, chapter 5]. $\mathbf{w}_\rho \triangleq [\mathbf{w}_{1,\rho}^H, \ldots, \mathbf{w}_{M,\rho}^H]^H \in \mathbb{C}^{MN_\text{t} \times 1}$ is the joint beam-forming vector for serving LR ρ, where beam-forming vector $\mathbf{w}_{m,\rho} \in \mathbb{C}^{N_\text{t} \times 1}$ is adopted at BS $m \in \mathcal{M}$. To adjust the beam-formers, $\mathbf{w}_{m,\rho}$, in adaption to $q_{f,l,m}$ and $\mathcal{M}_{f,l}^{\text{Coop}}$, which further depend on the cache status and the backhaul capacities, cf. (17.3), we impose

$$(1 - q_{f,l,m})\mathbf{w}_{m,\rho} = \mathbf{0}, \ \ \forall m \in \mathcal{M}, \ \forall \rho \in \mathcal{S}. \tag{17.5}$$

Due to (17.5), any BS $m \notin \mathcal{M}_{f,l}^{\text{Coop}}$ can only employ $\mathbf{w}_{m,\rho} = \mathbf{0}$.

Remark 17.1 We note that (17.4) and (17.5) allow to flexibly adjust the BS cooperation topology via controlling $q_{f,l,m}$s (and $\mathcal{M}_{f,l}^{\text{Coop}}$). For example, (17.4) and (17.5) capture both joint transmission with full BS cooperation, by setting $|\mathcal{M}_{f,l}^{\text{Coop}}| = M, \forall f$, and coordinated beam-forming, by setting $|\mathcal{M}_{f,l}^{\text{Coop}}| = 1, \forall f$, respectively. In general, any cooperative set $\mathcal{M}_{f,l}^{\text{Coop}} \subseteq \mathcal{M}$ can be configured using (17.4) and (17.5).

17.3 Problem Formulation

In this chapter, we focus on investigating the secrecy threat from, e.g., a nonpaying video subscriber, where the system has perfect knowledge about the CSIs of all the subscribers. In this case, jamming methods using, e.g., artificial noise cannot improve the PLS for the considered system [25] and thus are not considered herein. The impact of imperfect CSI will be discussed in Section 17.6. To maximize the performance gains of cache-enabled cooperative transmission, we propose a two-stage framework for optimizing the caching and delivery of video contents. As power efficiency is of paramount importance for reducing the operation cost and carbon footprint of future communication systems [26–32], the two-stage optimization problem aims to minimize the total BS transmit power subject to QoS and secrecy constraints. In the first stage, the caching decisions are determined using the statistics or historical records of the user requests. The cache status is updated according to the first-stage (caching) decisions and remains unchanged thereafter. In the second stage when the user requests, the CSI, and the backhaul capacity are known, this knowledge is exploited for optimizing backhaul loading and cooperative transmission at BSs in real time.

17.3.1 Achievable Secrecy Rate

The achievable secrecy rate of LR ρ hinges on not only the channels but also the receiver adopted at the ER. For guaranteeing secure VoD streaming, it is preferable that the secure delivery scheme can ensure communication secrecy even in the worst-case scenario. In particular, when the ER adopts a successive interference cancellation receiver [24], it may attempt to eavesdrop the subfile intended for LR ρ after canceling the interference caused by all other LRs. Consequently, an achievable secrecy rate at LR ρ is given by [16, 25]

$$R_\rho^{\text{sec}} = \left[R_\rho - R_{\text{e},\rho}\right]^+, \quad \rho \in \mathcal{S}. \tag{17.6}$$

Herein, R_ρ denotes the achievable data rate of LR ρ and is given by

$$R_\rho = \log_2\left(1 + \Gamma_\rho\right), \quad \rho \in \mathcal{S}, \tag{17.7}$$

$$\Gamma_\rho = \frac{\frac{1}{\sigma^2}\left|\mathbf{h}_\rho^H \mathbf{w}_\rho\right|^2}{1 + \frac{1}{\sigma^2}\sum\limits_{\rho' \in \mathcal{S}, \rho' \neq \rho}\left|\mathbf{h}_\rho^H \mathbf{w}_{\rho'}\right|^2}, \tag{17.8}$$

where Γ_ρ is the received signal-to-interference-plus-noise ratio (SINR) of LR ρ. Moreover, $R_{e,\rho}$ is the capacity of the ER in decoding the subfile intended for LR ρ and is given by

$$R_{e,\rho} = \log_2 \det\left(\mathbf{I}_{N_e} + \frac{1}{\sigma_e^2}\mathbf{G}\mathbf{G}^H \mathbf{w}_\rho \mathbf{w}_\rho^H\right),\ \rho \in \mathcal{S}. \tag{17.9}$$

17.3.2 Second-Stage Online Delivery Optimization

Assume that the cache status $\{c_{f,m}\}$ and the set of user requests $\mathcal{S}$ are given, e.g., $\{c_{f,m}\}$ is determined in the first stage. The second-stage control is invoked to adjust the BS cooperation strategy $\mathbf{D}_{\text{II}} \triangleq [q_{f,l,m}, b_{f,l,m}, \mathbf{w}_\rho]$. The resulting optimization problem to be solved in the second stage, denoted as problem R0, is then formulated as

$$\begin{aligned}
\text{R0:}\quad & \min_{\mathbf{D}_{\text{II}}}\ f_{\text{II}} \triangleq \sum_{\rho\in\mathcal{S}} \text{tr}(\mathbf{w}_\rho \mathbf{w}_\rho^H) \qquad (17.10)\\
\text{s.t.}\quad & \text{C1: } b_{f,l,m} = (1 - c_{f,m}) q_{f,l,m},\ f \in \mathcal{F}, l \in \mathcal{L}, m \in \mathcal{M}\\
& \text{C2: } q_{f,l,m} \in \{0,1\},\quad b_{f,l,m} \in [0,1], f \in \mathcal{F}, l \in \mathcal{L}, m \in \mathcal{M}\\
& \text{C3: } \sum_{f\in\mathcal{F}} b_{f,l,m} Q_f \le B_m^{\max},\ l \in \mathcal{L}, m \in \mathcal{M}\\
& \text{C4: } \text{tr}\left(\mathbf{\Lambda}_m \mathbf{w}_\rho \mathbf{w}_\rho^H\right) \le q_{f,l,m} P_m^{\max},\ m \in \mathcal{M},\ \rho \in \mathcal{S}\\
& \text{C5: } \text{tr}\left(\sum_{\rho\in\mathcal{S}} \mathbf{\Lambda}_m \mathbf{w}_\rho \mathbf{w}_\rho^H\right) \le P_m^{\max},\ m \in \mathcal{M}\\
& \text{C6: } R_\rho \ge R_\rho^{\text{req}},\ \rho \in \mathcal{S}\\
& \text{C7: } R_{e,\rho} \le R_{e,\rho}^{\text{tol}},\ \rho \in \mathcal{S}.
\end{aligned}$$

Herein, $\mathbf{\Lambda}_m$ is an $MN_t \times MN_t$ diagonal matrix,

$$\mathbf{\Lambda}_m = \text{diag}(\mathbf{0}_{(m-1)N_t\times 1}^T, \mathbf{1}_{N_t\times 1}^T, \mathbf{0}_{(M-m)N_t\times 1}^T), \tag{17.11}$$

whereby $\text{tr}\left(\mathbf{w}_{m,\rho}\mathbf{w}_{m,\rho}^H\right) = \text{tr}\left(\mathbf{\Lambda}_m \mathbf{w}_\rho \mathbf{w}_\rho^H\right)$ holds. Moreover, C3 and C5 limit the maximum backhaul traffic load and the maximum power consumption at BS m to $B_m^{\max}$ and $P_m^{\max}$, respectively. The parameter Q_f (in bps) in C3 denotes the data rate allocated for loading subfile (f, l) via the backhaul links. Assume that each time slot has duration τ. We set $Q_f = V_f/(\tau L)$ or equivalently $c_{f,m} V_f / L + b_{f,l,m} Q_f \tau = V_f / L$. Note that, as (17.3) and (17.5) are difficult to handle due to the "if-else" structure and the bilinear term, they have been reformulated as convex equality and inequality constraints C1 and C4, respectively. The equivalence between C1 and (17.3) can be easily verified. C4 is the big-M formulation [33] for BS cooperation: if $q_{f,l,m} = 0$ or $b_{f,l,m} + c_{f,m} < 1$ hold in C1, it leads to $\text{tr}\left(\mathbf{\Lambda}_m \mathbf{w}_\rho \mathbf{w}_\rho^H\right) = \|\mathbf{w}_{m,\rho}\|_2^2 = 0$, i.e., $\mathbf{w}_{m,\rho} = \mathbf{0}$; on the other hand, if $q_{f,l,m} = 1$ and $b_{f,l,m} + c_{f,m} = 1$, C4 becomes inactive as it is always less restrictive than C5. Thus C4 guarantees $\mathbf{w}_{m,\rho} = \mathbf{0}$ whenever BS $m \notin \mathcal{M}_{f,l}^{\text{Coop}}$ fails to cooperate in delivering subfile (f,l) and hence is equivalent to (17.5). Furthermore, C6 is a streaming QoS constraint, which ensures a minimum data rate, R_{ρ}^{req}, to be provided for LR ρ. C7 provides video data protection, whereby the capacity of the ER is limited within a maximum threshold $R_{\text{e},\rho}^{\text{tol}}$. By imposing C6 and C7, the achievable secrecy rate of LR ρ is guaranteed to at least exceed $R_{\rho}^{\text{sec}} = [R_{\rho}^{\text{req}} - R_{\text{e},\rho}^{\text{tol}}]^{+}$, whenever problem R0 is feasible.

17.3.3 First-Stage Offline Cache Training

Assume that Ω data sets are adopted for training the cache in the first stage. Each data set, indexed by $\omega \in \{1,\ldots,\Omega\}$, is a time series of user requests, CSI, and the available backhaul capacities. The data sets can be obtained either from the system record or generated by Monte Carlo simulations. Moreover, prediction of the users' future requests based on historical user profiles using, e.g., neural networks [34] can be included in the cache training phase to further improve the cache placement. Using these data sets, the first stage controller determines the optimal caching decisions while taking into account their potential impact on the video delivery in the second stage. Thereby, the first-stage optimization space, defined as $\mathbf{C}_{\text{I}} \triangleq [c_{f,m}, \mathbf{D}_{\text{I},\omega}]$ consists of not only caching decisions, $c_{f,m}$, but also auxiliary delivery decisions for training data set ω, denoted as $\mathbf{D}_{\text{I},\omega} \triangleq [q_{f,l,m,\omega}, b_{f,l,m,\omega}, \mathbf{w}_{\rho,\omega}]$. Herein, the feasible delivery set for each training data set ω can be defined as $\mathcal{D}_{\text{I},\omega} \triangleq \{\mathbf{D}_{\text{I},\omega} \mid \text{C1, C2, C4–C7}\}$, similar to problem R0. However, as a slight difference from problem R0, constraints C1, C2, and C4–C7 are defined per training data set. For example, C1 and C2 have to be reformulated by augmenting the data set index

$$\text{C1: } b_{f,l,m,\omega} = (1 - c_{f,m}) q_{f,l,m,\omega} \tag{17.12}$$

$$\text{C2: } c_{f,m}, b_{f,l,m,\omega} \in [0,1], \quad q_{f,l,m,\omega} \in \{0,1\}, \tag{17.13}$$

and C4–C7 can be rewritten in the same manner.

Then, to coordinate the first stage control for power efficiency enhancement, the caching optimization problem is formulated as

$$\begin{aligned}
\text{Q0:} \quad & \min_{\mathbf{C}_{\text{I}}} \ \frac{1}{\Omega} \sum_{\omega=1}^{\Omega} f_{\text{I},\omega} \qquad\qquad (17.14) \\
\text{s.t.} \quad & \mathbf{D}_{\text{I},\omega} \in \mathcal{D}_{\text{I},\omega}, \ \omega \in \{1,\ldots,\Omega\} \\
& \overline{\text{C3}}\text{: } \frac{1}{\Omega} \sum_{\omega=1}^{\Omega} \sum_{f \in \mathcal{F}} b_{f,l,m,\omega} Q_f \le \frac{1}{\Omega} \sum_{\omega=1}^{\Omega} B_{m,\omega}^{\max}, \ m \in \mathcal{M}, l \in \mathcal{L} \\
& \text{C8: } \sum_{f \in \mathcal{F}} c_{f,m} V_f \le C_m^{\max}, \ m \in \mathcal{M},
\end{aligned}$$

which minimizes the empirical average of the transmit powers over all considered data sets. Herein, $f_{\mathrm{I},\omega} \triangleq \sum_{\rho\in\mathcal{S}} \mathrm{tr}(\mathbf{w}_{\rho,\omega}\mathbf{w}_{\rho,\omega}^{H})$ is the instantaneous transmit power for data set ω. $\overline{\mathrm{C3}}$ and C8 are the backhaul capacity and cache capacity constraints, respectively, where $C_m^{\max}$ denotes the cache capacity available at BS m. We note that $\overline{\mathrm{C3}}$ limits only the average backhaul capacity of all considered data sets rather than restricting the backhaul capacity per data set as in $\sum_{f\in\mathcal{F}} b_{f,l,m,\omega} Q_f \le B_{m,\omega}^{\max}$, $\omega \in \{1,\ldots,\Omega\}$. The reasons for adopting $\overline{\mathrm{C3}}$ instead of the latter constraints in the first stage are twofold. On the one hand, in the considered two-stage control, the actual cooperative transmission decisions, which are affected by the backhaul capacity constraints, are deferred to the second stage after the available backhaul capacity is known. Hence $\overline{\mathrm{C3}}$ can avoid a conservative use of the backhaul links when the actual value of the backhaul capacity during delivery is still unknown/uncertain in the first stage. On the other hand, as will be revealed in Section 17.4.3, $\overline{\mathrm{C3}}$ also enables computational convenience in solving problem Q0, whereby a low-complexity cache training can be realized.

17.4 Problem Solution

Problems R0 and Q0 contain nonconvex constraints C6, C7, and binary variables $q_{f,l,m} \in \{0,1\}$ and $q_{f,l,m,\omega} \in \{0,1\}$, for which they belong to nonconvex mixed-integer nonlinear programs (MINLPs) [33]. That is, even if the binary constraints in problems R0 and Q0 are relaxed to continuous convex constraints, the resulting problems remain nonconvex. Moreover, constraint C1 in problem Q0 is bilinear. This type of problem is known to be NP-hard, for which polynomial-time optimization algorithms rarely exist [33]. To balance between performance and computational complexity, we propose two effective suboptimal algorithms for solving problems R0 and Q0, which are based on relaxation techniques and have only polynomial-time computational complexity. Strikingly, we show that the proposed algorithms become optimal when the cache capacity available at the BSs and the number of training data sets become large, respectively.

17.4.1 Optimal Solution of Problem R0 in Large Cache Capacity Regime

We discuss the solution of problem R0 in two steps. First, we show that problem R0 admits a polynomial-time optimal solution when the cache capacity is sufficiently large. Then, based on the derived results, we discuss the general solution of problem R0 in Section 17.4.2.

For given $\mathcal{S}$, define the set of requested files as $\mathcal{F}(\mathcal{S}) \triangleq \{f \mid (\cdot, f, \cdot) \in \mathcal{S}\}$, where $\mathcal{F}(\mathcal{S}) \subseteq \mathcal{F}$. As a file may be requested by multiple LRs, we have $F(\mathcal{S}) \le \min\{|\mathcal{S}|, F\}$, where $F(\mathcal{S}) \triangleq |\mathcal{F}(\mathcal{S})|$. Note that, if $\{c_{f,m}\}$ is given, variables $b_{f,l,m}$ can be eliminated by substituting C1 into C3. As a result, problem R0 can be rewritten as in (17.15), where $Q_{f,m} \triangleq Q_f(1 - c_{f,m})$ is the "effective" data rate required for loading subfile (f,l) via the backhaul link into BS m, after taking into account the cache status of BS m. Note that $\widetilde{\mathrm{C3}}$ and C3 are equivalent. Moreover, the BS cooperation strategy in (17.15) satisfies the monotonicity given in the following lemma.

$$\begin{aligned}
\text{R0:}\quad & \min_{\mathbf{D}_{\text{II}}} \quad f_{\text{II}} && (17.15)\\
& \text{s.t.} \quad \text{C1, C2, C4, C5, C6, C7}\\
& \qquad \widetilde{\text{C3}}: \sum_{f\in\mathcal{F}(\mathcal{S})} q_{f,l,m} Q_{f,m} \leq B_m^{\max},\ m \in \mathcal{M},
\end{aligned}$$

LEMMA 17.1 (Monotonicity of problem R0 [10, 11]). Given two cooperation sets $\mathcal{M}_{f,l}^{\text{Coop},1} \subseteq \mathcal{M}_{f,l}^{\text{Coop},2}$, $\forall (f,l) \in \mathcal{F}(\mathcal{S}) \times \mathcal{L}$, the corresponding optimal transmit powers of the BSs, denoted as f_{II}^1, f_{II}^2, respectively, satisfy $f_{\text{II}}^1 \geq f_{\text{II}}^2$.

Lemma 17.1 implies that fully cooperative transmission strategy with cooperation set $\mathcal{M}_{f,l}^{\text{F-Coop}} = \mathcal{M}$ is optimal if feasible, e.g., in systems with large cache capacity, due to $\mathcal{M}_{f,l}^{\text{Coop}} \subseteq \mathcal{M}_{f,l}^{\text{F-Coop}}, \forall \mathcal{M}_{f,l}^{\text{Coop}}$. In this case, the backhaul constraints C3 or $\widetilde{\text{C3}}$ are inactive since the cache can effectively offload the backhaul traffic. Consequently, by removing C3 and $\widetilde{\text{C3}}$ from R0, the resulting problem is polynomial time solvable without loss of optimality. In general, the optimization problem obtained by fixing the cooperation sets in problem R0, denoted by R0($\mathbf{w}_\rho$) with optimization variable $\mathbf{w}_\rho$, i.e.,

$$\begin{aligned}
\text{R0}(\mathbf{w}_\rho):\quad & \min_{\mathbf{w}_\rho} \quad f_{\text{II}} && (17.16)\\
& \text{s.t.} \quad \text{C4, C5, C6, C7,}
\end{aligned}$$

can be solved to optimality in polynomial time. This is due to a hidden convexity of R0($\mathbf{w}_\rho$). To show this, in the following, we first rewrite the nonconvex constraints C6 and C7 in their equivalent convex forms.

Let $\mathbf{W}_\rho \triangleq \mathbf{w}_\rho \mathbf{w}_\rho^H \succeq \mathbf{0}$ and $\mathbf{H}_\rho \triangleq \mathbf{h}_\rho \mathbf{h}_\rho^H$. Then, C6 can be reformulated as affine constraints,

$$\begin{aligned}
\text{C6} &\iff \Gamma_\rho \geq \kappa_\rho^{\text{req}} \triangleq 2^{R_\rho^{\text{req}}} - 1,\\
&\iff \overline{\text{C6}}: \frac{1}{\kappa_\rho^{\text{req}}} \operatorname{tr}\left(\mathbf{W}_\rho \mathbf{H}_\rho\right) \geq \sigma^2 + \sum_{\rho' \neq \rho} \operatorname{tr}\left(\mathbf{W}_{\rho'} \mathbf{H}_\rho\right). \qquad (17.17)
\end{aligned}$$

C6 and $\overline{\text{C6}}$ are equivalent if and only if $\mathbf{W}_\rho$ further satisfies

$$\text{C9: } \operatorname{rank}(\mathbf{W}_\rho) = 1 \text{ and } \mathbf{W}_\rho \succeq \mathbf{0}.$$

Furthermore, C7 can be rewritten as

$$\begin{aligned}
\text{C7} &\overset{(a)}{\iff} \det\left(\mathbf{I}_{N_\text{e}} + \frac{1}{\sigma_\text{e}^2} \mathbf{G}^H \mathbf{W}_\rho \mathbf{G}\right) \leq 2^{R_{\text{e},\rho}^{\text{tol}}}\\
&\overset{(b)}{\Longrightarrow} \operatorname{tr}\left(\mathbf{G}^H \mathbf{W}_\rho \mathbf{G}\right) \leq \sigma_\text{e}^2 \kappa_\rho^{\text{tol}} \triangleq \sigma_\text{e}^2 \left(2^{R_{\text{e},\rho}^{\text{tol}}} - 1\right)\\
&\overset{(c)}{\Longrightarrow} \lambda_{\max}\left(\mathbf{G}^H \mathbf{W}_\rho \mathbf{G}\right) \leq \sigma_\text{e}^2 \kappa_\rho^{\text{tol}}\\
&\iff \overline{\text{C7}}: \quad \mathbf{G}^H \mathbf{W}_\rho \mathbf{G} \preceq \sigma_\text{e}^2 \kappa_\rho^{\text{tol}} \mathbf{I}, \quad \rho \in \mathcal{S},
\end{aligned}$$

where (a) employs the matrix identity $\det(\mathbf{I}+\mathbf{AB}) = \det(\mathbf{I}+\mathbf{BA})$, (b) is due to $\det(\mathbf{I}+\mathbf{A}) \geq 1+\mathrm{tr}(\mathbf{A})$ provided that $\mathbf{A} \succeq \mathbf{0}$, and (c) follows from the inequality $\lambda_{\max}(\mathbf{A}) \leq \mathrm{tr}(\mathbf{A})$ if $\mathbf{A} \succeq \mathbf{0}$. Moreover, if $\mathrm{rank}(\mathbf{A}) = 1$, equality holds in (b) and (c). Therefore, if $\mathrm{rank}(\mathbf{W}_\rho) = 1$, $\overline{\mathrm{C7}}$ is an equivalent reformulation of C7.

By substituting $\overline{\mathrm{C6}}$, $\overline{\mathrm{C7}}$, and C9, although problem R0($\mathbf{w}_\rho$) is nonconvex due to C9, it can be solved via a relaxation technique. In particular, by removing the rank constraint $\mathrm{rank}(\mathbf{W}_\rho) = 1$ from C9, a convex semi-definite program (SDP) is readily obtained,

$$\begin{aligned} \mathrm{R1}: \quad & \min_{\mathbf{W}_\rho} \ \mathrm{tr}\left(\sum_{\rho\in\mathcal{S}} \mathbf{W}_\rho\right) \\ \text{s.t.} \quad & \overline{\mathrm{C4}}: \mathrm{tr}\left(\mathbf{\Lambda}_m \mathbf{W}_\rho\right) \leq q_{f,l,m} P_m^{\max},\ m\in\mathcal{M}, \\ & \overline{\mathrm{C5}}: \mathrm{tr}\left(\sum_{\rho} \mathbf{\Lambda}_m \mathbf{W}_\rho\right) \leq P_m^{\max},\ m\in\mathcal{M}, \\ & \overline{\mathrm{C6}}, \overline{\mathrm{C7}}, \overline{\mathrm{C9}}: \mathbf{W}_\rho \succeq \mathbf{0}, \mathrm{rank}(\mathbf{W}_\rho) = 1,\ \rho\in\mathcal{S}. \end{aligned} \tag{17.18}$$

which can be efficiently solved using the interior-point method [35] and the standard convex optimization solvers such as CVX [36]. Generally, the optimal value of problem R1 provides a lower bound for that of R0($\mathbf{w}_\rho$). However, for the problem at hand, we can strengthen the result by showing that the optimal solution of problem R1 satisfies $\mathrm{rank}(\mathbf{W}_\rho^*) = 1$, $\rho\in\mathcal{S}$, i.e., $\mathbf{W}_\rho^*$ also defines the optimal beamformer for problem R0($\mathbf{w}_\rho$).

THEOREM 17.2 *Problems R0($\mathbf{w}_\rho$) and R1 are equivalent such that both problems have an identical optimal value. Moreover, the optimal beam-forming matrix $\mathbf{W}_\rho^*$ obtained by solving problem R1 has rank one, i.e., $\mathrm{rank}(\mathbf{W}_\rho^*) = 1$, $\rho\in\mathcal{S}$, and the optimal beam-forming vector $\mathbf{w}_\rho^*$ of problem R0($\mathbf{w}_\rho$) corresponds to the principal eigenvector of $\mathbf{W}_\rho^*$.*

Proof Please refer to Appendix 17.8.1. □

17.4.2 Suboptimal Solution of Problem R0

When the cache capacities available at the BSs are limited, determining the optimal solution of problem R0 has to resort to exponential-time optimization algorithms, such as exhaustive search and branch-and-bound [33]. These algorithms have to enumerate all possible cooperation sets defined by C1, C2, and C3 in the worst case while searching for the optimal solution; hence their overall computational complexity is an exponential function of the number of BSs. To show this in detail, let us define $\overline{T}_m \triangleq \min\left\{\left\lfloor B_m^{\max}/\min_{f\in\mathcal{F}(\mathcal{S})} Q_{f,m}\right\rfloor, F(\mathcal{S})\right\}$, and $\underline{T}_m \triangleq \left\lfloor B_m^{\max}/\max_{f\in\mathcal{F}(\mathcal{S})} Q_{f,m}\right\rfloor$. According to Lemma 17.1, the optimal cooperation formation solutions are contained in the vertices of polyhedral simplexes defined by $\sum_{f\in\mathcal{F}(\mathcal{S})} q_{f,l,m} \leq T_m$ and $q_{f,l,m}\in[0,1]$, where $\underline{T}_m \leq T_m \leq \overline{T}_m, m\in\mathcal{M}$. This implies that, for solving R0, the exhaustive

Algorithm 7 Iterative Suboptimal Algorithm for Problem R0

1: **Initialization**: Set $\mathcal{Q}_0$ by $\prod_{f\in\mathcal{F}(\mathcal{S})}\mathcal{M}_f^{\text{F-Coop}}$, set $k=1$;
2: Solve optimization problem R0($\mathbf{w}_\rho$) for $\mathcal{Q}_0$;
3: **while** $\mathcal{M}_{k-1}^{\text{vio}} \neq \emptyset$ (cf. (17.21)) **do**
4: **for each** $(f,m)\in\mathcal{F}(\mathcal{S})\times\mathcal{M}_k^{\text{vio}}$ **do**
5: Solve optimization problem R0($\mathbf{w}_\rho$) for $\mathcal{Q}_{k-1}\backslash\{(f,m)\}$;
6: **end for**
7: Update $\mathcal{Q}_k$ by $\mathcal{Q}_{k-1}\backslash\left\{(f',m')\right\}$, where (f',m') corresponds to the solution of (17.19);
8: Update k by $k+1$;
9: **end while**

search (branch-and-bound) has to enumerate over $\prod_{m=1}^{M}\binom{T_m}{F(\mathcal{S})}$ choices of the cooperation sets in total (in the worst case).

Due to the overwhelming computational complexity, the enumeration methods are applicable only for small systems but cannot scale to practical large systems. To tackle this problem, we consider a low-complexity iterative algorithm as given in Algorithm 7 to solve R0. The proposed algorithm utilizes the greedy heuristics and the result of Theorem 17.2 that the cooperative beam-forming vectors can be efficiently solved in problem R0($\mathbf{w}_\rho$) for each given cooperation sets.

The iteration of the algorithm is indexed by k. The BS cooperation set obtained at iteration k is given by $\mathcal{Q}_k \triangleq \left\{(f,m)\mid q_{f,l,m}=1, m\in\mathcal{M}, f\in F(\mathcal{S})\right\}$. Algorithm 7 starts with an initial BS cooperation set $\mathcal{Q}_0=\prod_{f\in\mathcal{F}(\mathcal{S})}\mathcal{M}_{f,l}^{\text{F-Coop}}=\prod_{f\in\mathcal{F}(\mathcal{S})}\mathcal{M}$ and then refines the BS cooperation set in an iterative manner. At iteration $k=1,2,\ldots,$ we solve problem R0($\mathbf{w}_\rho$) by optimizing over the cooperative beam-forming vectors while fixing the values of $q_{f,l,m}$ according to $\mathcal{Q}_{k-1}$. The resulting optimal value is denoted as $f_{\text{II}}^*(\mathcal{Q}_{k-1})$. If $\mathcal{Q}_{k-1}$ is feasible, i.e., satisfying C1, C2, and C3, Algorithm 7 terminates and returns the BS cooperation set and beam-formers obtained so far. Otherwise, BS m' is eliminated from the cooperative transmission of subfile (f',l) by setting $q_{f',l,m'}=0$, where $(f',m')\in\mathcal{Q}_{k-1}$ incurs the smallest amount of extra BS transmit power, i.e.,

$$(f',m')\in \underset{(f,m)\in\mathcal{F}(\mathcal{S})\times\mathcal{M}_k^{\text{vio}}}{\arg\min}\left[f_{\text{II}}^*(\mathcal{Q}_{k-1}\backslash\{(f,m)\})-f_{\text{II}}^*(\mathcal{Q}_{k-1})\right], \tag{17.19}$$

$$\mathcal{Q}_k=\mathcal{Q}_{k-1}\backslash\left\{(f',m')\right\}, \tag{17.20}$$

and $\mathcal{M}_{k-1}^{\text{vio}}$ indexes the set of BSs violating constraint C3 by adopting $\mathcal{Q}_{k-1}$, i.e.,

$$\mathcal{M}_{k-1}^{\text{vio}} \triangleq \left\{m\in\mathcal{M}\mid \sum_{(f,m)\in\mathcal{Q}_{k-1}} Q_{f,m} > B_m^{\max}\right\}. \tag{17.21}$$

This process is executed repeatedly until $\mathcal{Q}_k$ becomes feasible, whence Algorithm 7 terminates.

During iteration k, solving (17.19) in Algorithm 7 may incur an enumeration over $F(\mathcal{S})\times\left|\mathcal{M}_{k-1}^{\text{vio}}\right|$ choices of (f,m). In the worst case, the total number of choices is given by $F(\mathcal{S})\times\left|\mathcal{M}_0^{\text{vio}}\right|\times T$, where $\sum_{m\in\mathcal{M}}\underline{T}_m \leq T \leq \sum_{m\in\mathcal{M}}\overline{T}_m$. Since the beam-forming optimization for each choice of (f,m) can be solved in polynomial time, the overall computational complexity of Algorithm 7 is only a polynomial function of M. We note that the proposed algorithm is in general suboptimal. However, if the cache capacity is sufficiently large, Algorithm 7 would terminate without the need to solve (17.19) and, according to Lemma 17.1, the obtained solution is globally optimal.

17.4.3 Solution of Problem Q0

The size of problem Q0 is Ω times larger than that of problem R0. Moreover, constraint C1 of problem Q0 is bilinear in $\{\mathbf{C}_\mathrm{I}\}$. Consequently, neither the enumeration methods nor the greedy method (cf. Algorithm 7) can be conveniently applied for solving Q0. In the following, we solve problem Q0 using the binary relaxation technique. For convenience, let us rewrite constraint C1 as

$$\widetilde{\mathrm{C1}}: c_{f,m} + b_{f,l,m,\omega} \geq q_{f,l,m,\omega}. \tag{17.22}$$

If constraint $\widetilde{\mathrm{C1}}$ is active, we have $b_{f,l,m,\omega} = (1 - c_{f,m})q_{f,l,m,\omega}$ due to $\widetilde{\mathrm{C1}}$, C2, and C8. Thus constraints $\widetilde{\mathrm{C1}}$ and C1 are equivalent.

Next, replace the binary variables in $\widetilde{\mathrm{C1}}$ by $q_{f,l,m,\omega} \in [0,1]$ and denote the resulting constraint as $\widehat{\mathrm{C1}}$. By substituting $\widetilde{\mathrm{C1}}$ with $\widehat{\mathrm{C1}}$ in Q0, we have

$$\begin{aligned} \mathrm{Q1}: \quad & \min_{\mathbf{C}_\mathrm{I}} \ \frac{1}{\Omega}\sum_{\omega=1}^{\Omega} f_{\mathrm{I},\omega} \\ \text{s.t.} \quad & \mathbf{D}_{\mathrm{I},\omega} \in \widehat{\mathcal{D}}_{\mathrm{I},\omega},\ \omega \in \{1,\ldots,\Omega\}, \overline{\mathrm{C3}}, \mathrm{C8}, \end{aligned} \tag{17.23}$$

where $\widehat{\mathcal{D}}_{\mathrm{I},\omega} \triangleq \left\{\mathbf{D}_{\mathrm{I},\omega} \mid \widehat{\mathrm{C1}}, \mathrm{C4–C7}\right\}$. Although problem Q1 is nonconvex due to the auxiliary beam-forming optimization, its hidden convexity can be verified similar to problem R0, cf. Theorem 17.2. As a result, problem Q1 can be efficiently solved.

For its simplicity, the binary relaxation technique has usually been adopted in the literature to obtain low-complexity yet suboptimal solutions for MINLPs. However, for the caching problem at hand, the binary relaxation technique can exploit the problem structure for a high-quality solution as well. In particular, we further show that the solution obtained from the relaxed problem Q1 turns out to be asymptotically optimal when Ω is sufficiently large.

THEOREM 17.3 *When $\Omega \to \infty$, problems Q1 and Q0 becomes equivalent, where both problems have the same optimum objective value and optimal caching decisions.*

Proof Please refer to [11] for a proof. □

17.5 Numerical Examples

In this section, we provide numerical evaluations about the performance of the proposed caching and secure delivery schemes in a cellular network consisting of $M = 7$

Table 17.2 System Parameters

Parameters	Values
Carrier frequency	2 GHz
System bandwidth	10 MHz
Duration of time slot	$\tau = 10$ ms
Number of subfiles	$L = 45 \text{ min}/\tau = 2.7 \times 10^4$
Pathloss	3GPP "Urban Macro NLOS" model [37]
Multi-path fading distribution	Rayleigh
Maximum transmit power at BSs	$P_m^{\max} = 46$ dBm
Noise power spectral density	-172.6 dBm/Hz [49]
Probability distribution of backhaul capacity	$\Pr(B_m^{\max} = 0 \text{ Mbps}) = 0.3$, $\Pr(B_m^{\max} = 3 \text{ Mbps}) = 0.4$, $\Pr(B_m^{\max} = 6 \text{ Mbps}) = 0.3$, $\forall m$
Minimum delivery rate	$R_\rho^{\text{req}} = 1.1 \times R_\rho^{\text{sec}} = 1.65$ Mbps [10]
Maximum leakage rate	$R_{\text{e},\rho}^{\text{tol}} = 0.1 \times R_\rho^{\text{sec}} = 150$ kbps [10]

hexagonal cells. The BSs are located at the center of each cell and the inter-BS distance is 500 m. Each BS is mounted with $N_\text{t} = 4$ antennas and the ER employs $N_\text{e} = 2$ antennas. The VoD streaming system owns $F = 10$ video files to serve $K = 5$ single-antenna LRs, which are randomly placed in the system while respecting a minimum distance of 50 m to the BSs. Each video file has a length of 45 minutes with size $V_f = 500$ MB (Bytes). It can be estimated that, to facilitate secure and uninterrupted VoD streaming, each LR would require a secrecy data rate of $R_\rho^{\text{sec}} = Q_f = 500 \times 8.0 \times 10^6/(45 \times 60) \approx 1.5$ Mbps. The requests at an LR follow probability distribution $\boldsymbol{\theta} = [\theta_1, \ldots, \theta_F]$, where θ_f is the probability of file $f \in \mathcal{F}$ being requested. The requests of the LRs are independent of each other. We choose $\theta_f = \frac{1}{f^\kappa} / \sum_{f \in \mathcal{F}} \frac{1}{f^\kappa}$ with $\kappa = 1.1$, i.e., $\boldsymbol{\theta}$ follows the Zipf distribution as in [38]. Table 17.2 provides a summary of the relevant system parameters. The cache status at the BSs is initialized using the caching solution of problem Q0, where $\Omega = 50$ data sets are generated according to the defined user request, backhaul capacity, and channel models.

17.5.1 Performance Comparisons with Baseline Schemes

For performance comparison, several baseline caching and delivery schemes are also considered:

- Baseline 1 (preference-based caching): Assuming $\boldsymbol{\theta}$ is known, the caching decision solves the following optimization problem

$$\max_{c_{f,m}} \sum_{f,m} \theta_f c_{f,m} V_f$$
$$\text{s.t.} \quad \text{C2}, c_{f,m} \in [0,1], \forall f \in \mathcal{F}, \forall m \in \mathcal{M}.$$

  Thereby, the system will choose the most popular files to be cached at the BSs until the cache capacity is depleted.
- Baseline 2 (identical caching): The system caches an identical amount of each video file at the BSs, independent of the users' preference, i.e.,

$$c_{f,m} V_f = \frac{1}{F} \min \left\{ C_m^{\max}, \sum_{j=1}^{F} V_j \right\}, \ \forall f,m.$$

For a fair comparison with the proposed caching scheme, Baselines 1 and 2 also adopt the proposed delivery scheme, i.e., Algorithm 7.

- Baseline 3 (coordinated beam-forming): The video file requested by an LR is delivered only from the nearest BS that has sufficient backhaul capacity available. We thus have $\sum_{m \in \mathcal{M}} q_{f,l,m} = 1, \forall (f,l) \in \mathcal{F} \times \mathcal{L}$.
- Baseline 4 (full BS cooperation): The backhaul links have unlimited capacity such that all BSs can cooperate to serve each LR. Hence we have $q_{f,l,m} = 1, \forall f,l,m$. For Baselines 3 and 4, the optimal beam-forming vectors are solved via R0($\mathbf{D}_{\text{II},2}$) after fixing the $\{q_{f,l,m}\}$ accordingly.

Figure 17.2 compares the performance of the considered caching and delivery schemes for different cache capacities, where the optimal solution of R0 is obtained

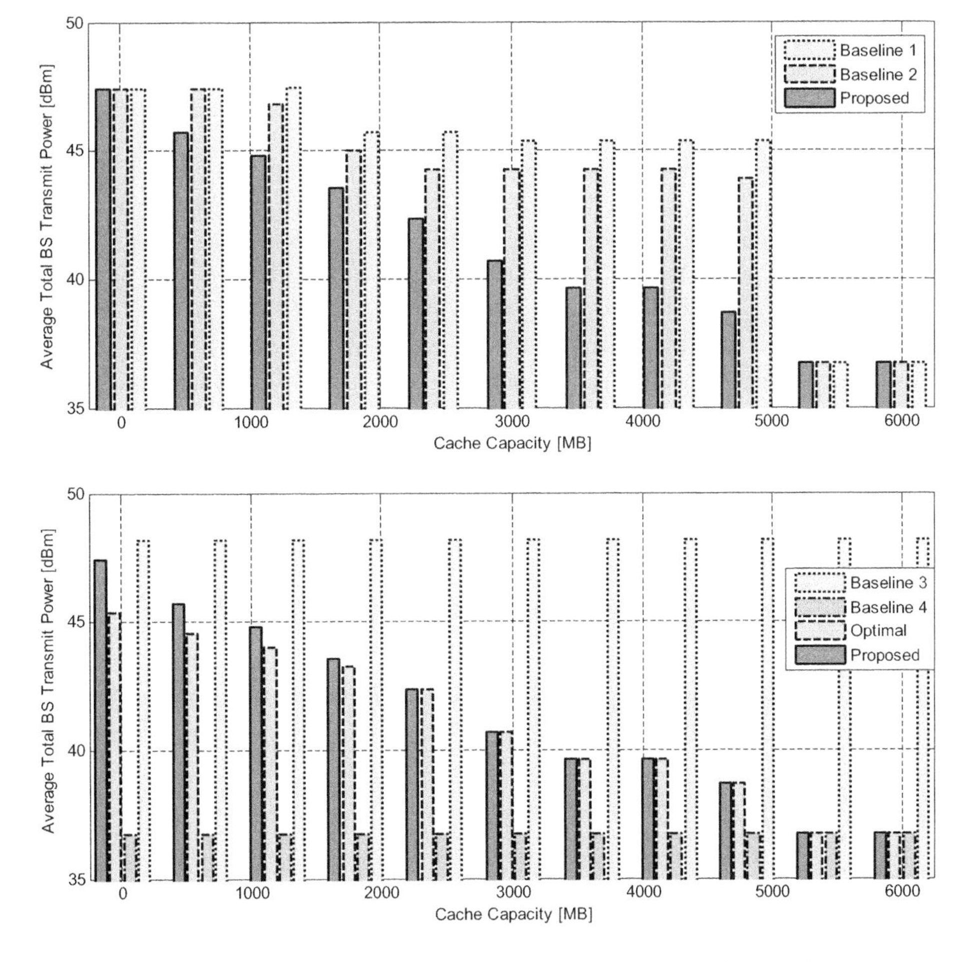

Figure 17.2 Total BS transmit power of considered caching and delivery schemes with respect to cache capacity.

using exhaustive search. Figure 17.2 shows that the total BS transmit powers of the proposed caching and delivery schemes decrease monotonically with the cache capacity. This is because caching enables more BSs to participate in cooperative transmission. As a result, the BSs can utilize a larger virtual antenna array to improve the power efficiency of video delivery. For instance, as the cache capacity increases from $C_m^{\max} = 1000$ MB to $C_m^{\max} = 4000$ MB, the transmit power is reduced by up to 6 dB without degrading the system performance. Comparing the proposed caching schemes with Baselines 1 and 2, only negligible performance gap exists in the small and large cache capacity regimes, as the BS cooperation opportunities are insufficient and saturated, respectively; in contrast, the proposed caching scheme attains much higher power efficiency than Baselines 1 and 2 in the medium cache capacity regime. Figure 17.2 also reveals that the performance of the proposed delivery scheme is bounded by those of Baselines 3 and 4. Moreover, as the cache capacity increases, the performance gap between the proposed delivery scheme and the optimal delivery scheme decreases and even vanishes, e.g., when 40% of the video files have been cached at the BSs. This is because the backhaul traffic is reduced by using large-capacity cache, which further reduces the likelihood of C3 being active.

17.5.2 Impact of Number of Antennas

Figure 17.3 illustrates the secrecy outage probability and the total BS transmit power of the proposed delivery scheme by considering different number of transmit and eavesdropper antennas, respectively. For the problem at hand, secrecy outage arises if constraints C6 and C7 cannot be simultaneously satisfied, whereby problem R0 becomes infeasible. In this case, the secrecy outage probability is given by $p_{\mathrm{out}} = \Pr(R_\rho^{\mathrm{sec}} < [R_\rho^{\mathrm{req}} - R_{\mathrm{e},\rho}^{\mathrm{tol}}]^+)$. Figure 17.3 shows that, for a given cache capacity, increasing N_{t} at the BSs and/or decreasing N_{e} can reduce the likelihood of secrecy outage and the transmit

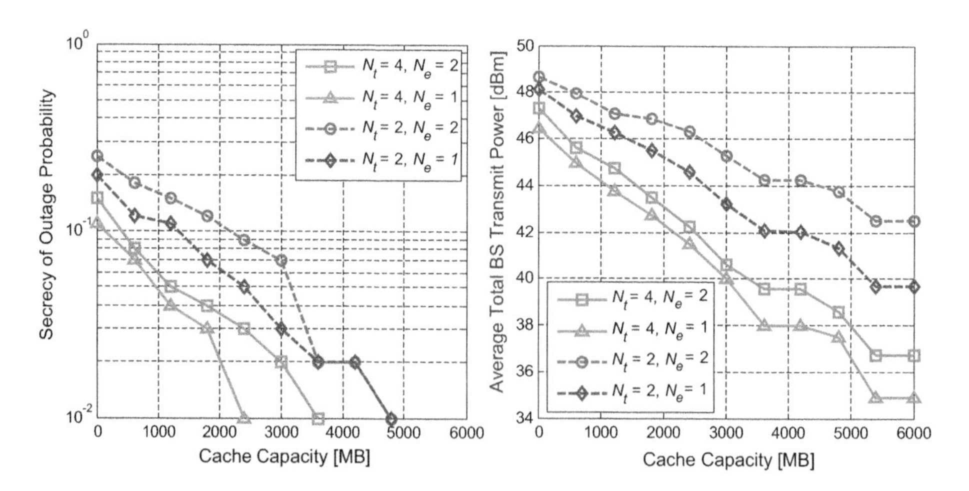

Figure 17.3 Secrecy outage probability and total BS transmit power of proposed delivery scheme with respect to cache capacity for different number of transmit and eavesdropping antennas.

power simultaneously as more secrecy degrees of freedom can be exploited during cooperative beamforming. Moreover, the performance gains can be further enlarged by deploying large-capacity cache at the BSs.

17.6 Research Challenges and Opportunities

In practice, the application of cache-enabled PLS faces several challenges caused by, e.g., other secrecy threats and CSI estimation errors. In this section, we analyze these challenges and point out possible solutions that may inspire future research in this direction.

17.6.1 Trustworthiness of Cache-Enabled Devices

In the aforementioned discussions, the cache devices are assumed to be trusted so that video content can be securely cached and exploited for PLS benefits. However, as caching is being increasingly considered in heterogeneous cellular networks, so are concerns over the trustworthiness of cache devices. In particular, untrusted cache devices may arise in, e.g., small cell BSs compromised to third parties, which may eavesdrop premium video content without paying as well as LRs' private video content. Moreover, a prudent untrusted device may even utilize cached video files for interference mitigation and hence improve the eavesdropping performance [13, 39]. In these cases, caching at the untrusted devices will unfavorably introduce a new secrecy threat, which has to be combated cautiously.

In the presence of untrusted cache devices, an intuitive approach to ensure communication security is to detect these untrusted devices, e.g., via machine learning techniques, and prevent video content from being cached at them. In this case, the proposed cache-enabled PLS schemes are still applicable for trusted devices. However, with this intuitive approach, the cache capacity of untrusted devices cannot be exploited for PLS benefits. Moreover, for imperfect detection of the untrusted devices, caching may compromise the system security.

Another method to tackle untrusted cache devices employs scalable video coding (SVC) based caching [13, 39]. Thereby, each video content is encoded into one base layer and multiple enhancement layers. While the base layer can be decoded independently of the enhancement layers, an enhancement layer can be decoded only after the base layer and all lower enhancement layers have already been decoded [13, 39, 40]. Exploiting this *hierarchical* encoding/decoding structures of SVC, the enhancement layer subfiles can be cached at untrusted devices without degrading system security as long as the eavesdroppers have no access to the baseline subfiles. For example, when a larger number of antennas are equipped at the trusted nodes than at the untrusted nodes, the cache-enabled PLS techniques are applicable to the untrusted devices and the content cached at the untrusted devices can be also exploited for enhancing system security. Inspired by SVC based caching, advanced coding schemes facilitating better exploitation of untrusted cache devices are a promising research topic.

17.6.2 Imperfect, Statistical, and no CSI Knowledge about the Eavesdropper

In this chapter, we have assumed perfect CSI knowledge about the eavesdropper. In practical systems, however, perfect estimation of the transmitter-to-eavesdropper channel is usually infeasible if the eavesdropper remains silent [41]. Due to CSI estimation errors in these systems, cooperative MIMO beam-forming cannot perfectly align the legitimate data signals in the null space of the eavesdropping channels. Consequently, the likelihood of data leakage increases and the secrecy capacity for cooperative MIMO transmission is reduced [42].

One promising approach for ensuring communication security in the presence of CSI estimation errors is to employ artificial noise–based jamming [22, 25, 43, 44]. In particular, as the CSI of the legitimate receivers may be easily estimated with high accuracy, the artificial noise can be designed to proactively interfere the ERs while avoiding interference to the LRs. Applying a deterministic channel estimation error model, robust joint optimization of cache-enabled PLS and jamming was investigated in [11]. The results in [11] suggest that, by employing robust joint optimization of beamforming and jamming, both the power efficiency and the secrecy of the cache-enabled system can be significantly improved.

On the other hand, the CSI of the eavesdropper may also be completely unknown at the transmitter. In this case, cache-enabled joint transmission may not outperform conventional disjoint transmission in terms of PLS. This is because joint transmission may unfavorably enhance the ER's reception. Therefore, there is an interesting trade-off between cache-enabled joint and disjoint transmissions, as investigated in [12].

17.6.3 Active Eavesdropper

So far, we have focused on passive eavesdroppers. Employing cache-enabled PLS to combat active attacks such as spoofing and jamming [45] and pilot contamination attacks [19], is an interesting topic but has not been investigated in the literature.

17.6.4 Other Forms of Cache-Enabled PLS Techniques

Recent works have also exploited caching for developing advanced physical layer techniques including cache-enabled interference cancellation (at receivers), cache-aided non-orthogonal multiple access [46, 47], cache-enabled relaying [48, 49], and cache-aided massive MIMO [49]. Extending these schemes for PLS is feasible. Appealingly, caching may lead to additional secrecy benefits that are infeasible in conventional cellular networks (without caching). For example, a novel PLS technique exploiting caching and superposition coded transmission was proposed in [51, chapter 2]. Thereby, if a video content has been cached at the LRs but is not requested, it can still be transmitted together with the requested video content in superposition to proactively interfere the eavesdropper. Meanwhile, the LRs can cancel this artificially added interference using the cached file, e.g., by re-encoding and modulating the cached data and subtracting its contribution from the received signals [51]. Consequently, the proposed scheme achieves a positive secrecy capacity even if the legitimate channel conditions are strictly worse than the eavesdropping channel. In contrast, for conventional cellular networks (without caching), the secrecy capacity in such a case is strictly zero.

17.7 Summary

While enormous research has demonstrated the benefits of traffic offloading and delivery latency reduction enabled by caching, in this chapter, caching was exploited as a PLS mechanism to improve the secrecy capacity of cellular video streaming. Thereby, caching was utilized as alternative "backhaul links" to facilitate cooperative transmission for large groups of BSs and hence increase the secrecy degrees of freedom in the system. A two-stage nonconvex optimization problem was formulated to determine caching and the corresponding cooperative transmission. To solve the problem efficiently while balancing between complexity and performance, suboptimal algorithms with polynomial-time computational complexity were proposed. Thanks to the special structure of the problem, the proposed algorithms were verified to be asymptotically optimal when the cache capacity available at the BSs and the number of training data sets become large, respectively. Through numerical examples we showed that the proposed schemes can achieve significant gains in PLS and power efficiency. Finally, future research challenges and potential solutions for the application of cache-enabled PLS have been discussed.

17.8 Appendix

17.8.1 Proof of Theorem 17.2

To show the equivalence between problems R0($\mathbf{w}_\rho$) and R1, we need to verify that the beam-forming matrix obtained by solving problem R1 satisfies $\mathrm{rank}(\mathbf{W}_\rho^*) = 1$. In particular, we introduce $\boldsymbol{\alpha} = [\alpha_{m\rho}]$, $\boldsymbol{\beta} = [\beta_m]$, $\boldsymbol{\lambda} = [\lambda_\rho]$, $\boldsymbol{\Phi}_\rho$, and $\boldsymbol{\Theta}_\rho = [\boldsymbol{\Theta}_{1\rho}, \boldsymbol{\Theta}_{2\rho}]$ as the Lagrangian multipliers for $\overline{\text{C4}}$, $\overline{\text{C5}}$, $\overline{\text{C6}}$, $\overline{\text{C7}}$, and $\overline{\text{C9}}$, respectively, with

$$\alpha_{m\rho} \geq 0, \beta_m \geq 0, \lambda_\rho \geq 0, \boldsymbol{\Phi}_\rho \succeq \mathbf{0}, \boldsymbol{\Theta}_{1\rho} \succeq \mathbf{0}, \quad \text{and} \quad \boldsymbol{\Theta}_{2\rho} \succeq \mathbf{0}. \tag{17.24}$$

Then, for problem R1, the Lagrangian is formulated as

$$\mathcal{L}(\mathbf{W}_\rho; \boldsymbol{\alpha}, \boldsymbol{\beta}, \boldsymbol{\lambda}, \boldsymbol{\Phi}_\rho, \boldsymbol{\Theta}_\rho) = \mathrm{tr}\left[\sum_\rho \left(\mathbf{B}_\rho - 2\lambda_\rho \mathbf{H}_\rho - \boldsymbol{\Theta}_{1\rho}\right)\mathbf{W}_\rho\right] + \Delta, \tag{17.25}$$

where Δ includes the constant terms and

$$\mathbf{B}_\rho \triangleq \mathbf{I} + \boldsymbol{\Lambda}_\rho^{\alpha,\beta} + \mathbf{G}\boldsymbol{\Phi}_\rho\mathbf{G}^H + \sum_{\rho\in\mathcal{S}}(1+\kappa_\rho^{\text{req}})\lambda_\rho\mathbf{H}_\rho \succ \mathbf{0}, \tag{17.26}$$

for $\mathbf{\Lambda}_\rho^{\alpha,\beta} \triangleq \sum_{m\in\mathcal{M}}(\alpha_{m\rho} + \beta_m)\mathbf{\Lambda}_m$ and $\mathbf{\Lambda}^\beta \triangleq \sum_{m\in\mathcal{M}} \beta_m \mathbf{\Lambda}_m$. As problem R1 is a convex optimization problem and fulfills the Slater's condition, strong duality holds for R1. According to the duality theory [35], a primal-dual point $(\mathbf{W}_\rho;\ \boldsymbol{\alpha}, \boldsymbol{\beta}, \boldsymbol{\lambda}, \mathbf{\Phi}_\rho, \mathbf{\Theta}_\rho)$ is optimal if and only if it satisfies the Karush–Kuhn–Tucker (KKT) conditions, which are given by

$$\nabla_{\mathbf{W}_\rho}\mathcal{L} = \mathbf{B}_\rho - 2\lambda_\rho \mathbf{H}_\rho - \mathbf{\Theta}_{1\rho} = \mathbf{0}, \tag{17.27}$$

$$\mathbf{W}_\rho \mathbf{\Theta}_{1\rho} = \mathbf{0}, \tag{17.28}$$

$$\mathbf{W}_\rho \succeq \mathbf{0}, \quad \lambda_\rho \geq 0. \tag{17.29}$$

Based on (17.27) and (17.28), we have $\mathbf{W}_\rho \mathbf{B}_\rho = 2\lambda_\rho \mathbf{W}_\rho \mathbf{H}_\rho$. Since $\mathsf{rank}(\mathbf{H}_\rho) \leq 1$, we have the following rank inequalities

$$\mathsf{rank}(\mathbf{W}_\rho) \overset{(a)}{=} \mathsf{rank}(\mathbf{W}_\rho \mathbf{B}_\rho) \overset{(b)}{=} \mathsf{rank}(\lambda_\rho \mathbf{W}_\rho \mathbf{H}_\rho) \tag{17.30}$$

$$\overset{(c)}{\leq} \min\left\{\mathsf{rank}(\lambda_\rho \mathbf{W}_\rho), \mathsf{rank}(\mathbf{H}_\rho)\right\} \leq 1, \tag{17.31}$$

where (a) utilizes the fact that $\mathbf{B}_\rho \succ \mathbf{0}$, (b) follows from (17.27) and (17.28), and (c) is a result of inequality $\mathsf{rank}(\mathbf{AB}) \leq \min\{\mathsf{rank}(\mathbf{A}), \mathsf{rank}(\mathbf{B})\}$. Since problem R1 is feasible only if $\mathbf{W}_\rho \neq \mathbf{0}$, we have $\mathsf{rank}(\mathbf{W}_\rho) = 1$, which completes the proof.

References

[1] R. Knutson, "Video boom forces Verizon to upgrade network," *Wall Street Journal*, Dec. 2013.

[2] V. W. S. Wong, R. Schober, D. W. K. Ng, and L.-C. Wang, *Key Technologies for 5G Wireless Systems*, Cambridge: Cambridge University Press, 2017.

[3] X. Wang, M. Chen, T. Taleb, A. Ksentini, and V. Leung, "Cache in the air: exploiting content caching and delivery techniques for 5G systems," *IEEE Commun. Mag.*, vol. 52, no. 2, pp. 131–139, Feb. 2014.

[4] Q. Li, W. Shi, X. Ge, and Z. Niu, "Cooperative edge caching in software-defined hyper-cellular networks," *IEEE J. Sel. Areas Commun.*, vol. 35, no. 11, pp. 2596–2605, Nov. 2017.

[5] L. Hu, Y. Tian, Y. Jun, L. Xiang, and Y. Hao, "Ready player one: UAV clustering based multi-task offloading for vehicular VR/AR Gaming," *IEEE Netw.*, vol. 33, no. 3, May/June 2019.

[6] G. Paschos, E. Baştuğ, I. Land, G. Caire, and M. Debbah, "Wireless caching: technical misconceptions and business barriers," *IEEE Commun. Mag.*, pp. 16–22, Aug. 2016.

[7] A. Sengupta, R. Tandon, and T. C. Clancy, "Fundamental limits of caching with secure delivery," *IEEE Trans. Inf. Forensics Security*, vol. 10, no. 2, pp. 355–370, Feb. 2015.

[8] Z. H. Awan and A. Sezgin, "Fundamental limits of caching in D2D networks with secure delivery," in *Proc. IEEE Int. Conf. Comm. (ICC)*, June 2015.

[9] F. Gabry, V. Bioglio, and I. Land, "On edging caching with secrecy constraints," in *Proc. IEEE Int. Conf. Comm. (ICC)*, May 2016.

[10] L. Xiang, D. W. K. Ng, R. Schober, and V. W. S. Wong, "Cache-enabled physical-layer security for video streaming in backhaul-limited cellular networks," in *Proc. IEEE Global Commun. Conf. (GLOBECOM)*, Dec. 2016.

[11] L. Xiang, D. W. K. Ng, R. Schober, and V. W. S. Wong, "Cache-enabled physical-layer security for video streaming in backhaul-limited cellular networks," *IEEE Trans. Wireless Commun.*, vol. 17, no. 2, pp. 736–751, Feb. 2018.

[12] T.-X. Zheng, H.-M. Wang, and J. Yuan, "Physical-layer security in cache-enabled cooperative small cell networks against randomly distributed eavesdroppers," *IEEE Trans. Wireless Commun.*, vol. 17, no. 99, pp. 1–14, June 2018.

[13] L. Xiang, D. W. K. Ng, R. Schober, and V. W. S. Wong, "Secure video streaming in heterogeneous small cell networks with untrusted cache helpers," *IEEE Trans. Wireless Commun.*, vol. 17, no. 4, pp. 2645–2661, Apr. 2018.

[14] M. A. Maddah-Ali and U. Niesen, "Fundamental limits of caching," *IEEE Trans. Inf. Theory*, vol. 60, no. 5, pp. 2856–2867, May 2014.

[15] M. A. Maddah-Ali and U. Niesen, "Decentralized coded caching attains order-optimal memory-rate tradeoff," *IEEE/ACM Trans. Netw.*, vol. 23, no. 4, pp. 1029–1040, Aug. 2015.

[16] A. Khisti and G. W. Wornell, "Secure transmission with multiple antennas I: the MISOME wiretap channel," *IEEE Trans. Inf. Theory*, vol. 56, no. 7, pp. 3088–3104, July 2010.

[17] Ashish Khisti and Gregory W. Wornell, "Secure transmission with multiple antennas II: the MIMOME wiretap channel," *IEEE Trans. Inf. Theory*, vol. 56, no. 11, pp. 5515–5532, Nov. 2010.

[18] X. Chen, D. W. K. Ng, W. H. Gerstacker, and H.-H. Chen, "A survey on multiple-antenna techniques for physical layer security," *IEEE Commun. Surveys Tut.*, vol. 19, no. 2, pp. 1027–1053, 2017.

[19] Y. Wu, A. Khisti, C. Xiao, G. Caire, K.-K. Wong, and X. Gao, "A survey of physical layer security techniques for 5G wireless networks and challenges ahead," *IEEE J. Sel. Areas Commun.*, vol. 36, no. 4, pp. 679–695, Apr. 2018.

[20] D. Gesbert, S. Hanly, H. Huang, S. S. Shitz, O. Simeone, and W. Yu, "Multi-cell MIMO cooperative networks: A new look at interference," *IEEE J. Sel. Areas Commun.*, vol. 28, no. 9, pp. 1380–1408, Dec. 2010.

[21] K. J. Ma, R. Bartos, S. Bhatia, and R. Nair, "Mobile video delivery with HTTP," *IEEE Commun. Mag.*, vol. 49, no. 4, pp. 166–175, Apr. 2011.

[22] D. W. K. Ng and R. Schober, "Secure and green SWIPT in distributed antenna networks with limited backhaul capacity," *IEEE Trans. Wireless Commun.*, vol. 14, no. 9, pp. 5082–5097, Sept. 2015.

[23] X. Ge, H. Cheng, M. Guizani, and T. Han, "5G wireless backhaul networks: challenges and research advances," *IEEE Netw.*, vol. 28, no. 6, pp. 6–11, Nov. 2014.

[24] D. Tse and P. Viswanath, *Fundamentals of Wireless Communication*, Cambridge: Cambridge University Press, 2005.

[25] Q. Li and W.-K. Ma, "Optimal and robust transmit designs for MISO channel secrecy by semidefinite programming," *IEEE Trans. Signal Process.*, vol. 59, no. 8, pp. 3799–3812, Aug. 2011.

[26] I. Humar, X. Ge, L. Xiang, M. Jo, M. Chen, and J. Zhang, "Rethinking energy efficiency models of cellular networks with embodied energy," *IEEE Netw.*, vol. 25, no. 2, pp. 40–49, Mar. 2011.

[27] L. Xiang, F. Pantisano, R. Verdone, X. Ge, and M. Chen, "Adaptive traffic load balancing for green cellular networks," in *Proc. IEEE PIMRC*, Sept. 2011.

[28] L. Xiang, X. Ge, C. Wang, F. Y. Li, and F. Reichert, "Energy efficiency evaluation of cellular networks based on spatial distributions of traffic load and power consumption," *IEEE Trans. Wireless Commun.*, vol. 12, no. 3, pp. 961–973, Mar. 2013.

[29] L. Xiang, D. W. K. Ng, W. Lee, and R. Schober, "Optimal storage-aided wind generation integration considering ramping requirements," in *Proc. IEEE SmartGridComm 2013*, Oct. 2013.

[30] W. Lee, L. Xiang, R. Schober, and V. W. S. Wong, "Direct electricity trading in smart grid: A coalitional game analysis," *IEEE J. Sel. Areas Commun.*, vol. 32, no. 7, pp. 1398–1411, July 2014.

[31] W. Lee, L. Xiang, R. Schober, and V. W. S. Wong, "Electric vehicle charging stations with renewable power generators: A Game Theoretical Analysis," *IEEE Trans. Smart Grid*, vol. 6, no. 2, pp. 608–617, Mar. 2015.

[32] J. Zhang, L. Xiang, D. W. K. Ng, M. Jo, and M. Chen, "Energy efficiency evaluation of multi-tier cellular uplink transmission under maximum power constraint," *IEEE Trans. Wireless Commun.*, vol. 16, no. 11, pp. 7092–7107, Nov. 2017.

[33] C. A. Floudas, *Nonlinear and Mixed Integer Optimization: Fundamentals and Applications*, Oxford: Oxford University Press, 1995.

[34] L. Xiang, X. Ge, C. Liu, L. Shu, and C.-X. Wang, "A new hybrid network traffic prediction method," in *Proc. IEEE Global Commun. Conf. (GLOBECOM)*, Dec. 2010.

[35] S. Boyd and L. Vandenberghe, *Convex Optimization*, Cambridge: Cambridge University Press, 2004.

[36] M. Grant and S. Boyd, "CVX: Matlab software for disciplined convex programming, version 2.1," http://cvxr.com/cvx, Dec. 2017.

[37] 3GPP TR 36.814, "Further advancements for E-UTRA physical layer aspects (Release 9)," Mar. 2010.

[38] L. Breslau, P. Cao, L. Fan, G. Phillips, and S. Shenker, "Web caching and Zipf-like distributions: evidence and implications," in *Proc. IEEE INFOCOM*, Mar. 1999.

[39] L. Xiang, D. W. K. Ng, R. Schober, and V. W. S. Wong, "Secure video streaming in heterogeneous small cell networks with untrusted cache helpers," in *Proc. IEEE Global Commun. Conf. (GLOBECOM)*, Dec. 2017.

[40] X. Zhang, T. Lv, and S. Yang, "Near-optimal layer placement for scalable videos in cache-enabled small-cell networks," *IEEE Trans. Veh. Technol.*, vol. 67, no. 99, pp. 1–5, 2018.

[41] Y. Zhong, X. Ge, T. Han, Q. Li, and J. Zhang, "Tradeoff between delay and physical layer security in wireless networks," *IEEE J. Sel. Areas Commun.*, vol. 36, no. 99, pp. 1–13, 2018.

[42] T.-Y. Liu, P. Mukherjee, S. Ulukus, T. Liu, P. Mukherjee, S. Ulukus, S. Lin, and Y.-W. Peter Hong, "Secure degrees of freedom of MIMO Rayleigh block fading wiretap channels with no CSI anywhere," *IEEE Trans. Wireless Commun.*, vol. 14, no. 5, pp. 2655–2669, May 2015.

[43] D. W. K. Ng, L. Xiang, and R. Schober, "Multi-objective beamforming for secure communication in systems with wireless information and power transfer," in *Proc. IEEE Personal Indoor Mobile Radio Commun. (PIMRC)*, Sept. 2013.

[44] Y. Sun, D. W. K. Ng, J. Zhu, and R. Schober, "Multi-objective optimization for robust power efficient and secure full-duplex wireless communication systems," *IEEE Trans. Wireless Commun.*, vol. 15, no. 8, pp. 5511–5526, Aug. 2016.

[45] L. Xiao, C. Xie, T. Chen, H. Dai, and H. V. Poor, "A mobile offloading game against smart attacks," *IEEE Access*, vol. 4, pp. 2281–2291, 2016.

[46] L. Xiang, D. W. K. Ng, X. Ge, Z. Ding, V. W. S. Wong, and R. Schober, "Cache-aided non-orthogonal multiple access," in *Proc. IEEE Int. Conf. Commun. (ICC)*, May 2018.

[47] L. Xiang, D. W. K. Ng, X. Ge, Z. Ding, V. W. S. Wong, and R. Schober, "Cache-aided non-orthogonal multiple access: the two-user case," *IEEE J. Sel. Topics Signal Process.*, vol. 13, no. 3, pp. 436-451, June 2019.

[48] L. Xiang, D. W. K. Ng, T. Islam, R. Schober, and V. W. S. Wong, "Cross-layer optimization of fast video delivery in cache-enabled relaying networks," in *Proc. IEEE Global Commun. Conf. (GLOBECOM)*, Dec. 2015.

[49] L. Xiang, D. W. K. Ng, T. Islam, R. Schober, V. W. S. Wong, and J. Wang, "Cross-layer optimization of fast video delivery in cache- and buffer-enabled relaying networks," *IEEE Trans. Veh. Technol.*, vol. 66, no. 12, pp. 11 366–11 382, Dec. 2017.

[50] X. Wei, L. Xiang, L. Cottatellucci, T. Jiang, and R. Schober, "Cache-aided massive MIMO: linear precoding design and performance analysis," in *Proc. IEEE Int. Conf. Commun. (ICC)*, May 2019.

[51] L. Xiang, *Cache-Enabled Wireless: Physical Layer Design and Optimization*. Erlangen, Germany: University of Erlangen-Nuremberg, 2018.

18 Mobile VR Edge Delivery: Computing, Caching, and Communication Trade-Offs

Jacob Chakareski

We investigate streaming 360° videos to mobile virtual reality (VR) clients in next-generation cooperative 5G cellular systems. The setting under consideration comprises a set of 5G small-cells that are interconnected via backhaul links, aiming to share their resources of computing and caching when serving the clients. We formulate efficient 360° video representations to enable streaming only the contents of remote scenario/scene viewpoint demanded by a user at any time, thus outperforming the current largely inefficient practice of sending a bulky 360° video, that might contain many redundant parts of scene information not browsed by a user. In addition, we formulate an optimization framework that allows the small-cell base station to cooperatively choose streaming/caching/rendering methods that optimize the cumulative reward they obtain while communicating with its users, subject to computing and caching/storage constraints at each small cell base station. We characterize the problem as integer programming, show that it is NP-hard, and formulate an approximation algorithm that shows strong performance guarantees in fully polynomial time. Our framework enables considerable operational efficiency advantage over the existing caching techniques operating in concert with 360° video.

18.1 Introduction

Virtual reality and augmented reality (AR) are two rising technologies with a large impact on our society. While VR brings a virtual or actual location to our perception in an immerse manner (i.e., via *virtual human teleportation* [1]), AR enables embedding (virtual) digital objects into our perceived physical surrounding. Their rapidly growing popularity in the market and the mainstream forebodes future opportunities. In particular, there has been a flurry of related equipment/services/platforms released on the market, developed by startups and large companies [2–12], with increasing investments and acquisitions [13–16]. By 2020, the VR/AR market is expected to reach $150 billion and dramatically shift the existing mobile market revenues at the same time [17].

Supported in part by NSF Awards CNS-1821875, CNS-1836909, ECCS-1711592, and CCF-1528030, and research gifts from Adobe Systems and Tencent Research.

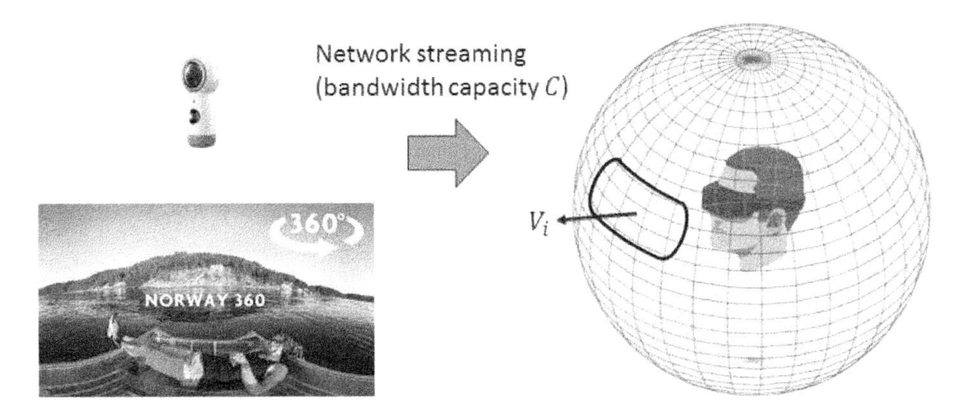

Figure 18.1 Illustration of 360° streaming. Viewport V_i on the 360° sphere.

Networked VR/AR applications are going to play a key role in enriching the world of the internet of things (IoT) [18–22] as well as large ecosystem of 5G tactile internet [23, 24]. Concretely, with the capability of traveling virtually and having a superhuman kind of vision, one can expect that VR/AR might lead to fruitful and diverse changes in technology and society. Still, there are considerable challenges on the road to such a future due to technology limitations and infrastructure costs.

For instance, emerging applications of VR at this stage are constrained to offline configuration and computer-generated content, featuring a 3D scene that is statically 360°-navigable and is constructed for a user with VR device that is externally linked to a powerful desktop computer. In an online scenario, the content will be required to streamed to the user. Examples of this scenario are engineered by internet companies such as YouTube and Facebook [10, 25], providing video-on-demand (VoD) services of 360° video.

The 360° video is a novel video format recorded by an omnidirectional camera that captures incident light rays from every angle. It allows a remote user to have a look around the actual surrounding scenario/scene in 360° using a VR device, with the user being virtually placed at the camera location, as Figure 18.1 illustrates.

However, existing implementations are highly inefficient as they require downloading practically the entire 360° video file in advance, prior to the start of the streaming session. This is required because the data rate of the encoded 360° video panorama is considerably higher than the network bandwidth C available to stream the file. Moreover, the downloaded file features the complete 360° panorama of the scene; however, a user can experience only a small portion of it, at any time, denoted as viewpoint V_i in Figure 18.1. This is necessary to avoid simulator/motion sickness [26] and ideally deliver good quality of experience, as the *intuitive approach of sending only the current V_i required by the user* using traditional server–client delivery architectures, where the server responds to client updates, would preclude application interactivity due to the inherent latency. In reality, these design choices require the user client to first ingest and then manage a huge volume of data, on the order of many hundreds of Gigabytes, which has the following penalizing consequences:

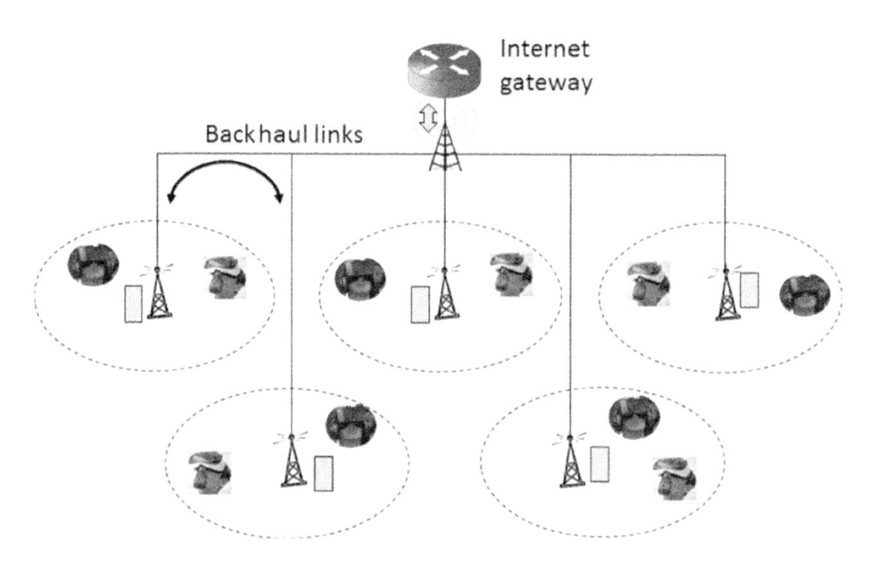

Figure 18.2 Virtual reality communication via small-cell cooperation.

- The sensation of immersion and the interactive nature of the VR application are considerably degraded, as the service is not streaming, but, rather operating in an *offline-like* download and play mode.
- The user can consume only very short low-quality 360° videos, as its client device has limited storage/computing capabilities and its internet access link features a limited data rate.[1]

Both of these consequences considerably impact the quality of immersion experience delivered to the user.

This chapter investigates a future 5G setting, made up of a set of mobile VR clients experiencing streaming 360° videos, as Figure 18.2 illustrates. The VR users are distributed spatially and have network connectivity through a set of small-cells base stations. These small cells are interconnected via backhaul, and provide service to their users in concert area. The base stations also feature limited caching and computing resources they can leverage in serving the mobile VR clients.

In this context, we provide multiple contributions under the setting depicted in Figure 18.2. We first represent the 360° videos that are delivered and containing solely viewpoint of interest of users in the scene. Subsequently, the remaining viewpoints are adaptively delivered to users in navigation. Moreover, the optimization framework we formulate allows the small cell base stations to collaboratively choose caching, edge/fog computing, and streaming methods, thus ultimately aiming to maximize the cumulative benefit/reward during their service. Third, we show that the problem under consideration is NP-hard, therefore a fully polynomial-time approximation algorithm is provided by relying on dynamic programming. The approximation algorithm enables

[1] The 360° video panorama data rate is multiple orders of magnitude bigger than the available network transmission bandwidth C.

us to have a stronger performance guarantees (namely, $(1 - \epsilon)$) compared to existing methods with half the guarantees [27]. Last, we show strong advances in operational efficiency relative to present caching methods operating in concert with emerging 360° video representation techniques.

18.2 Related Work

The 360° video streaming to mobile VR clients in cooperative 5G systems has not been intensively explored before. In this regard, immersive telecollaboration [28, 29], wireless systems with multiple camera and views [30], compression of multi-view video and its communication [31–34], single 360° video streaming through the internet [35–38], and scheduling of video packets subject to rate-distortion optimization [39–45] are some of related works in this regard.

Prior work exploring wireless base station caching includes [46], which studies estimation of content popularity distribution in a small cell and minimization of delay during the content retrieval, characterizing the latter as a knapsack problem [47]. Moreover, [27] studied content delivery delay minimization via caching at small cell base stations (aka wireless helper nodes), featuring low coverage but having large storage capabilities. The decision of node availability therein is made based on the closeness to a node to be served. Similarly, the work in [48] examined optimizing the network parameters under a single small cell cache, while the works in [49, 50] explored cellular networks with backhaul and hierarchical caching. Additionally, hierarchical caching from information-theoretical perspective has been investigated in [51, 52].

18.3 System Models

18.3.1 VR Data Model

The traditional approach: 360° VR video provides a user with a visual immersion of a remote scene that the user can interactively experience and navigate from any perspective. We can illustrate this conceptually via the setting included in Figure 18.1, in which the user is centered in a 3D sphere where the scene video evolving on the inner surface of the sphere. With V_i, we denote the present viewpoint of interest of the user navigating the remote scene. V_i is commonly denoted as the user *viewport* in the literature. The VR device with vertical and horizontal fields of view (namely, a head-mounted display; HMD) will establish the surface area of the 360° video content shown on the device, as delineated in Figure 18.1. For simplicity, the symbol V_i is interchangeably used to denote the direction of viewing of the user associated with the corresponding viewpoint, as the two are uniquely related. In particular, the vector V_i indicating the direction of viewing, represents the 360° surface normal of the respective viewpoint, and its intersection with the 360° sphere is the center of symmetry for the viewpoint V_i.

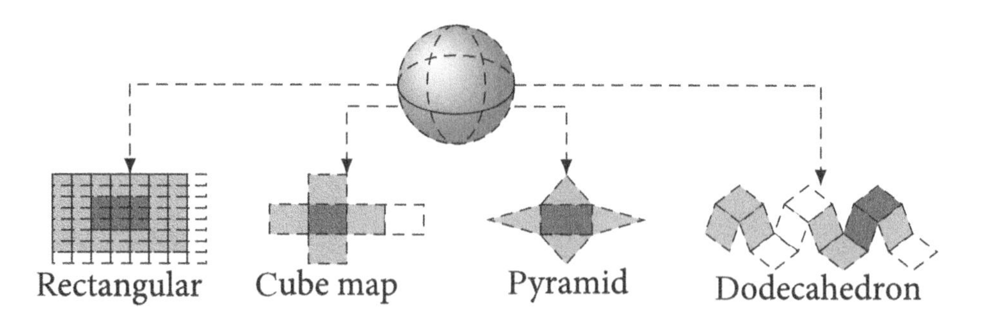

Figure 18.3 The 360° coding. Sphere to planar shape projections.

To leverage the benefits of conventional video compression, which operates on planar surfaces, the 360° video panorama is first projected to a planar shape: equirectangle, pyramid, cube, or dodecahedron, as illustrated in Figure 18.3. The latter three shapes have been studied because an approximately 30% pixel duplication is created in the case of the first projection (equirectangle) [35, 53]. However, they feature their own shortcomings; for example, they introduce projection distortions around the planar shape's edges. The latest modern video codec high-efficiency video coding (HEVC) [54] is then used for encoding the mapped 360° panorama. The rectangular projection is predominantly used in present implementations, and we will consider it here.

Proposed approach: Suppose that the view space panorama linked to the 360° sphere is denoted by $\mathcal{V}$, and a partition therein is represented by $V = \{V_1, \ldots, V_N\}$. We can construct V in multiple ways; for example, we can uniformly divide the complete solid angle of the sphere into N number of equal segments of $4\pi/N$ steradians. Then, the surface normals on the sphere related to each such solid angle segment will represent the respective vectors V_i. Using techniques like optimal 2D (vector) quantization [55], one can also alternatively define V as the set of centroids of the 360° sphere surface, divided into N partitions according to a probability/popularity distribution over the view space $\mathcal{V}$, that is $p_v, v \in \mathcal{V}$. In other words, p_v will represent how often viewpoint v is navigated by a VR user during a session. For sake of simplicity, our intention here is to investigate the first option, where a detailed study on the advantage of second approach is left for future work.

Suppose that the initial 360° sphere surface area related to viewpoint V_i due to the partition V is denoted by S_i^0. We expand S_i^0 in each direction uniformly across the sphere surface to construct a larger area S_i that encompasses views $v \in \mathcal{V}$ adjacent to V_i. The thereby constructed S_i is denoted as the quality emphasized region (QER) for the viewpoint V_i [35]. The 360° video content associated with each viewpoint $V_i \in V$ that is delivered to a VR user is then constructed as follows. The section of the 360° video panorama spanned by S_i is encoded at high quality. The remaining portion of the panorama denoted as $S_i^c = S_{360^\circ} \setminus S_i$ is encoded at low quality, where S_{360° denotes the surface area of the entire panorama and "\" denotes the operator set difference.

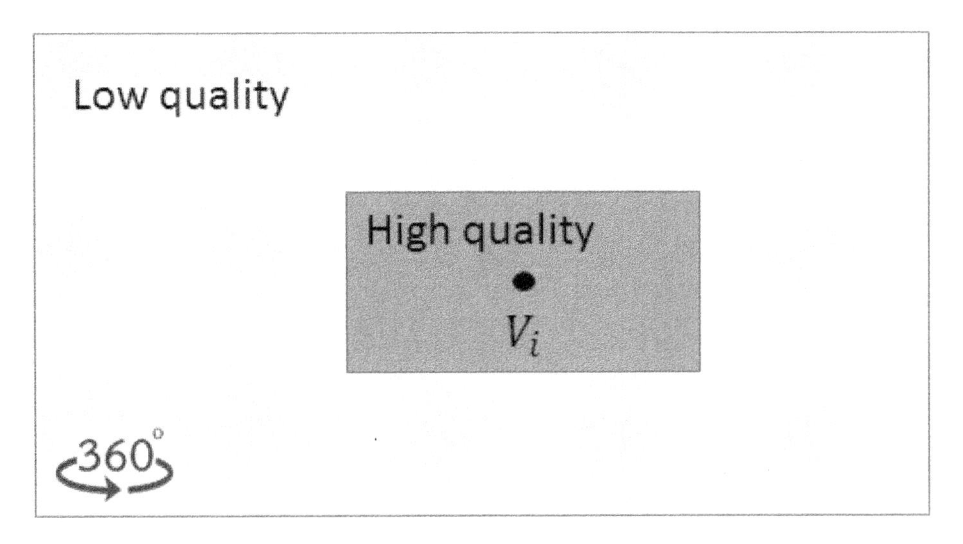

Figure 18.4 Partitioned 360° video panorama. Viewpoint V_i (black dot) is surrounded by a spatial area S_i encoded at high quality (gray). The remaining area of the panorama S_i^c is encoded at low quality (light gray).

S_i is selected to be the maximum possible surface area such that the aggregate encoding rate associated with the 360° video content represented by viewpoint V_i does not exceed the network bandwidth C available to a mobile VR user, i.e., $R(S_i) + R(S_i^c) \leq C$. Figure 18.4 illustrates the partitioned 360° video i associated with viewpoint V_i. Integrating the part of the panorama S_i^c ensures smooth navigation, if the user diverges too far from V_i over time, while encoding it at low quality considerably reduces the data rate required to transmit the resulting 360° video and thus enables streaming 360° video to mobile VR clients. Moreover, expanding S_i^0 to a larger S_i accommodates rendering of minor local variations of V_i and smooth temporal navigation between two adjacent partitions/viewpoints V_i and V_j during which the user will consistently experience high-quality viewports, as ensured by our streaming model, described next. Both of these characteristics will considerably enhance the quality of experience of the user.

18.3.2 The 360° Streaming Model

A 360° video is encoded into temporal segments (groups of consecutive video frames) of duration of several seconds. Each segment can be decoded independently at a client. In our case, at the beginning of a segment, the client informs the server of its present viewpoint v. The server responds accordingly by sending the partitioned 360° video content associated with the viewpoint V_i that is closest to v on the 360° sphere surface. This process is illustrated in Figure 18.5, which captures the streaming model between the server and client that we designed.

Selecting the duration of a segment is important in this context, as the server can only send 360° video content associated with another viewpoint/partition V_i at the

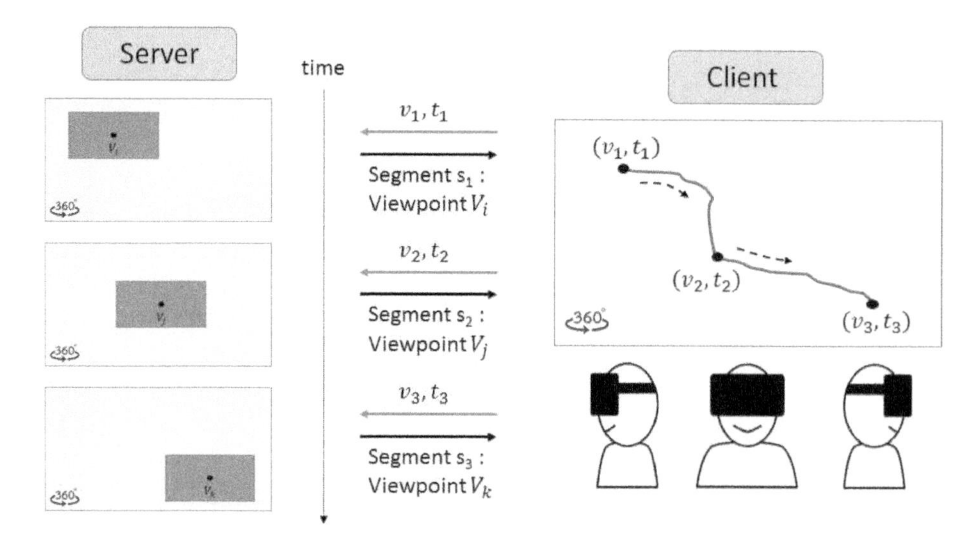

Figure 18.5 A VR client navigates over time from viewpoint v_1 at time t_1 through viewpoint v_2 at time t_2 to viewpoint v_3 at time t_3. The server responds accordingly by streaming segment s_1 at time t_1 with QER for viewpoint V_i, segment s_2 at time t_2 with QER for viewpoint V_j, and segment s_3 at time t_3 with QER for viewpoint V_k. The client experiences consistently high-quality viewports over time and enhanced quality of experience.

beginning of the next segment. Thus the user should be navigating within the QER of the present segment's viewpoint V_i with high likelihood until then, to ensure good quality of experience. Shorter segment lengths increase this probability, however, at the expense of compression and transmission efficiency. We have established that a segment duration of 2 seconds ensures a very high likelihood of retaining the user within the present QER during that time, with minimal impact on compression/transmission efficiency [35].

18.3.3 VR Computing and Data Complexity

Each partitioned 360° video j that is thereby constructed will feature different computational and data complexity characteristics. Let B_j indicate the volume of data generated by video j at encoding. Similarly, let B_j^o denote the computing requirement imposed by video j needed to render a viewport v from its encoded data. In an offline VR system, the rendering of v is enabled by the desktop computer where the virtual reality device is attached. In the case of online streaming, a server is required to implement such a task, by interactively constructing different viewpoints v for user/client (due to navigation) from the corresponding 360° video. Observe that the complexity of the remote scene and dynamics will be proportional to B_j^o, as one can observe from the viewpoint V_j that defines this partitioned 360° video (see Figure 18.4), i.e., for a sophisticated scene one could expect higher value of B_j^o.

18.3.4 Cellular Network Model

In the network setting we are considering, backhaul links interconnect the small cell base stations, providing at the same time internet access to them, as Figure 18.2 illustrates. Note that other types of wireless access networks could be also applied to our setting. We consider N_0 partitioned 360° videos (following the procedures described in Section 18.3.1) and K number of small cell base stations. The small cells can utilize the backhaul links to collaboratively serve their users. We denote a_{ij} as the access rate (or popularity) of video j observed at base station i. This quantity essentially captures how often segments from video j has been requested by users at the ith small cell base station.

Let us suppose that the binary variable X_{ij}^k quantifies the choice to render and serve the VR content representing video j, requested by a user connected to base station i, from the cache of base station k. Hence selecting $X_{ij}^i = 1$ indicates the event of locally rendering/serving the viewpoints of the partitioned 360° video j, such as from the cache base station i. Last, setting $X_{ij}^{K+1} = 1$ denotes the event where VR content j is served via the internet. In our investigation, we assume that each base station i has caching capacity of C_i and computing capability of C_i^o.

18.3.5 Reward Model

The location from where the VR content represented by a demanded 360° video is rendered (and streamed) will determine the network provider's reward/benefit. In particular, this will be reward $R_{i,p}$ (ith base station serving locally) or $R_{i,p}^k$, for $k \neq i$ and $k \leq K$, where $k = K + 1$ denotes streaming from a remote back-end server. We denote with $R_{j,v}$ the reward specifically earned by the provider delivering video j, i.e., the viewpoints v that can be rendered from its QER.

Having diverse reward factors $R_{i,p}$, $R_{i,p}^k$ is motivated by the need to capture the different degrees of benefit the provider will obtain by streaming video j from different places/locations. Concretely, one can intuitively assume that $R_{i,p} > R_{i,p}^k > R_{i,p}^{K+1}$, reflecting the fact that provider can gain better if serving locally and gain less if serving from the most distant location. One can alternatively define $R_{i,p}$ and $R_{i,p}^k$ by allowing them to be inversely proportional to the cost due to delivery from different locations.

Similarly, $R_{j,v}$ can also quantify the different levels of quality of experience aimed to the user, taking into account the degree of details in the scene and level of interactivity performed by the video j. In fact, both factors are usually proportional to the volume of data and complexity of rending video j. Hence one might also associate $R_{j,v}$ with a cost for the service provider. Examples of reward and cost/penalty factors in this context are video quality and interactivity degree for the former (relating to quality of experience) and energy and latency for the latter.

Last, introducing the factors a_{ij} will enable us to capture the base stations' content popularity distributions (probability of user requests), which are typically non-uniform.

18.4 Problem Formulation

Our objective is to maximize the aggregate reward earned by the provider serving the offered pool of N_0 360° videos to users managed by the K base stations. The maximization problem is formulated as follows:

$$\max \sum_i R_{i,p}\Big(\sum_j a_{ij} R_{j,v} X_{ij}^i\Big) + \sum_i \sum_{k \neq i} R_{i,p}^k \Big(\sum_j a_{ij} R_{j,v} X_{ij}^k\Big) + \sum_i R_{i,p}^{K+1}\Big(\sum_j a_{ij} R_{j,v} X_{ij}^{K+1}\Big), \tag{18.1}$$

subject to:

$$\sum_j W_{ij} B_j \leq C_i, \forall i, \qquad \sum_j W_{ij} B_j^c \leq C_i^o, \forall i, \tag{18.2}$$

$$X_{ij}^k \leq W_{kj}, \forall i, j, k, \qquad \sum_{k=1}^{K+1} X_{ij}^k = 1_{\{a_{ij} \geq 0\}}, \forall i, j, \tag{18.3}$$

$$X_{ij}^k \in \{0,1\}, \forall i, j, k, \qquad W_{kj} \in \{0,1\}, \forall k, j. \tag{18.4}$$

In the problem formulation, X_{ij}^k and W_{ij} are the decision variables, where the latter denotes the choice of caching video j at base station i. The first term in the objective function is the reward for caching the requested item at the local base station. The second term captures the reward of having the requested item in nearby base stations. The last term represents the reward for content obtained from the internet. Hence our general goal is to maximize the average reward via a joint design for content placement and routing.

The constraints (18.2) indicate a base station's caching and computing capacity limits. The first condition in (18.3) indicates that an item can be retrieved only from a base station that cached the item. The second condition guarantees no more than one copy of the requested item is delivered to the client. Finally, the constraints (18.4) capture the binary nature of the decision variables X_{ij}^k and W_{ij}.

LEMMA 18.1 *The problem* (18.1)–(18.4) *is NP-complete.*

Proof We can complete the checking of the feasibility of a given solution in polynomial time, by verifying that it meets the constraints (18.2)–(18.4). Hence it remains to show that (18.1) is NP hard.

We carry out a mapping from the well-known 0-1 multiple knapsack problem [47] that is defined as follows. Consider K knapsacks of a set consisting of N items, with $N \geq K$. Denote W_j as the weight of the jth item and P_j as the profit associated to item j. In addition, let Cap_i denote the capacity of the ith knapsack. To proceed, we divide the N items into K disjoint groups and assign a different knapsack to a group. The groups are determined in such a way that the group's total profit is maximized and the total weight does not exceed the corresponding knapsack. The mapping to an instance of our own problem is carried out via the following steps: First of all, each base station's cache is assigned with a knapsack, i.e., $C_i = Cap_i, \forall i$. Next, each content j is linked to the jth item and set $B_j = W_j$ and $\sum_i a_{ij} = P_j, \forall j$. Finally, we set $R_{i,p} = R^k_{i,p} = 1, \forall k \in \{1, \dots, K\}$ and $R^{K+1}_{i,p} = 0$. We note that the necessary and sufficient condition for a value $\mathcal{V}$ to make the 0-1 multiple knapsack problem feasible is that it also makes our problem feasible. As the simplified problem can be executed in polynomial time, our problem of interest is NP complete. □

18.5 Polynomial-Time Approximation

We leverage dynamic programming [56] to characterize a polynomial-time approximation algorithm for solving (18.1)–(18.4). We consider here that $R^k_{i,p}$ is the same for all $i \neq j$ combinations, which fits the problem setting. Thus we denote $\bar{R} = R^k_{i,p}, \forall i \neq k$. Consequently we can rewrite the objective function $O(\{X^k_{ij}\})$ as:

$$\sum_{i,j} R_{i,p} a_{ij} R_{j,v} X^i_{ij} + \sum_{\substack{i,j,k:k \neq i, \\ k \leq K}} \bar{R} a_{ij} R_{j,v} X^k_{ij} + \sum_{i,j} R^{K+1}_{i,p} a_{ij} R_{j,v} X^{K+1}_{ij}.$$

Hence, our optimization problem can be rewritten formally as:

$$\max_{\{X^k_{ij}\}} O(\{X^k_{ij}\}); \quad \text{s.t. (18.2)–(18.4).} \tag{18.5}$$

Our approach is to scale the profit of every cached content at each base station such that they are polynomially bounded in KN. Then, we will be able to solve the scaled instance of the problem using dynamic programming and output an approximate solution to the original problem. Moreover, proper adjustment of the degree of scaling that integrates the approximation factor ϵ is needed to reduce the algorithm complexity to polynomial time in regard to $1/\epsilon$.

Denote $w_{ij} = R_{i,p} a_{ij} R_{j,v} + \sum_{k \neq i} \bar{R} a_{kj} R_{j,v}$ as the maximum profit obtained by prefetching content j at the ith base station i. The scaling factor is chosen as $w^{\max} = \max_{i,j} w_{ij}$. Next, we define the vector product set of all possible combinations of contents and caches as $\mathcal{C} = \{1, \dots, K\} \times \{1, \dots, N\}$ and a precision factor as $p = \lfloor \log(\epsilon w^{\max}/NK) \rfloor$. By definition, the set size is $C = |\mathcal{C}|$ and any member of the set is represented by $c \in \mathcal{C}$.

We point out that the re-indexing enables us to conceptually transform (18.5) into a knapsack problem over $\mathcal{C}$, in which each base station is bounded by multiple knapsack constraints. In particular, if $w^{\max}$ is the maximum reward obtained by storing item $c \in \mathcal{C}$, then $|\mathcal{C}|w^{\max}$ denotes an upper bound on the accrued aggregate profit p. Now, let $S_{i,p}$ be a subset of $\{c_1, \dots, c_i\}$ that takes the smallest caching/computing volume and exhibits the highest aggregate profit, $\forall i \in \{1, \dots, |\mathcal{C}|\}$ and $p \in \{1, \dots, |\mathcal{C}|w^{\max}\}$. We denote with $A(i, p)$ the size of the set $S_{i,p}$. Leveraging the dynamic programming recurrence relation (the Bellman equation [56]), we can formulate the expression:

$$A(i+1, p) = \min\{A(i, p), \text{Size}(c_{i+1}) + A(i, p - \text{Reward}(c_{i+1}))\},$$

Algorithm 8 Fully Polynomial-Time Approximation

1: Initialize $\mathbf{T} = \emptyset, P_0 = 0, \mathbf{Q}_0 = \{(\mathbf{T}, P_0)\}$
2: **for** $\forall c \in \mathcal{C}, l = 1$ to C **do**
3: $(i, j) \leftarrow c$
4: Grow $\mathbf{Q}_l = \mathbf{Q}_{l-1} \cup \{(\mathbf{T} \cup \{c\}, P_{l-1} + w'_{ij}) | \sum_{c' \in \mathbf{T}: i'=i} B_{j'} + B_j \leq C_i, \sum_{c' \in \mathbf{T}: i'=i} B^c_{j'} + B^c_j \leq C^c_i, (\mathbf{T}, P_{l-1}) \in \mathbf{Q}_{l-1}\}$

$$\text{where } w'_{ij} = \begin{cases} \lfloor \frac{R_{i,p} a_{ij} R_{j,v}}{10^p} \rfloor + \sum_{k \neq i} \lfloor \frac{\bar{R} a_{kj} R_{j,v}}{10^p} \rfloor, \\ \quad \text{if} \nexists c' \in \mathbf{T} : j' = j, \\ \lfloor \frac{R_i a_{ij} R_{j,v}}{10^p} \rfloor - \lfloor \frac{\bar{R} a_{ij} R_{j,v}}{10^p} \rfloor, \\ \quad \text{otherwise.} \end{cases}$$

5: **if** $\exists (\mathbf{T}^1, P^1), (\mathbf{T}^2, P^2) \in \mathbf{Q}_l : P^1 = P^2$ and
6: $\Big\{ \Big(\sum_{c \in \mathbf{T}^1} B_j > \sum_{c \in \mathbf{T}^2} B_j$ and $\sum_{c \in \mathbf{T}^1} B^c_j \geq \sum_{c \in \mathbf{T}^2} B^c_j \Big)$ or
7: $\Big(\sum_{c \in \mathbf{T}^1} B_j \geq \sum_{c \in \mathbf{T}^2} B_j$ and $\sum_{c \in \mathbf{T}^1} B^c_j > \sum_{c \in \mathbf{T}^2} B^c_j \Big) \Big\}$ **then**
8: Prune $\mathbf{Q}_l = \mathbf{Q}_l \setminus \{(\mathbf{T}^1, P^1)\}$
9: **end if**
10: **end for**
11: Select $\{(\mathbf{T}^*, P^*)\} \in \mathbf{Q}_C : P^* = \max_P \{(\mathbf{T}, P) \in \mathbf{Q}_C\}$

for the evolution of the optimal caching configuration size, to populate $A(i, p), \forall i, p$. We then seek the optimal caching setup corresponding to a feasible set $S_{i,p}$ that imposes the smallest $A(i, p)$ and simultaneously maximizes p. In the following, we formulate an algorithm that will determine this set $S_{i,p}$ effectively.

In particular, we first introduce an approximated algorithm and subsequently verify its performance in terms of accuracy and fully polynomial-time nature. We formally characterize the details of the proposed algorithm in Algorithm 8. It leverages a data structure $\mathbf{Q}_l$ that tracks the explored state-space via stage l. Specifically, member elements $(\mathbf{T}, P) \in \mathbf{Q}_l$ of the data structure indicate subsets $(\mathbf{T})$ of size $k \leq l$ of the first l entities in $\mathcal{C}$, which represent the maximum earned reward (P) for the predefined cached contents and computing volumes $\big(\sum_{(i,j):c \in \mathbf{T}} B_j$ and $\sum_{(i,j):c \in \mathbf{T}} B^c_j \big)$. For each element l $(c \in \mathcal{C})$, our algorithm features an expansion phase and a pruning phase. In the former, the optimal paths stored in $\mathbf{Q}_{l-1}$ are branched out, taking into consideration the next caching decision variable. While the later keeps only the optimal paths after the expansion. We maintain the computing and caching constraints at each base station during the expansion phase. Last, when stage KN completes, our algorithm completes by selecting the caching configuration $\mathbf{T}^*$ in $\mathbf{Q}_{KN}$, which features the maximum profit P^*.

Given the notation used in (18.5), we then set the optimization variables as follows. If $c \in \mathbf{T}^*$, then $X^i_{ij} = 1$ and $X^{K+1}_{kj} = 0, \forall k$, with $(i, j) \leftarrow c$. In the next step, if $\exists i \leq K$: $X^i_{ij} = 1$ and $\exists k \neq i : X^m_{kj} = 0, m \leq K$, we set $X^i_{kj} = 1$. Finally, $\forall i, j : X^k_{ij} = 0, k \leq K$, we set $X^{K+1}_{ij} = 1$.

THEOREM 18.2 *We denote with OPT the maximum value of the objective in* (18.5). *Moreover, we denote with* $\{X^k_{ij}\}^*$ *the respective solution. Let* $\{X^k_{ij}\}'$ *denote the solution produced by Algorithm 8. It holds* $O(\{X^k_{ij}\}') \geq (1 - \epsilon) \cdot OPT$.

Proof We first divide the corresponding reward of each item pair $c = (i, j)$ by a scaling factor $S = \epsilon w^{\max}/NK$ (with rounding), in Algorithm 8. Hence the resulting reward w'_{ij}, which includes an item c in the scaled problem instance, satisfies $S \cdot w'_{ij} \leq w_{ij}$. Hence the reward obtained by $\{X^k_{ij}\}^*$ can decrease by the largest S for every cached item (according to $\{X^k_{ij}\}^*$), evaluating in the scaled instance. Thus the overall reward decrease can be bounded as:

$$O(\{X^k_{ij}\}^*) - S \cdot O'(\{X^k_{ij}\}^*) \leq CS, \tag{18.6}$$

where O' represents the scaled objective in (18.5).

Now, Algorithm 8 uses dynamic programming to compute $\{X^k_{ij}\}'$. Hence $\{X^k_{ij}\}'$ is the optimal solution to the scaled instance of the problem (18.5). Hence $O'(\{X^k_{ij}\}') \geq O'(\{X^k_{ij}\}^*)$ must hold. We leverage this, to write the following inequalities:

$$O(\{X^k_{ij}\}') \geq S \cdot O'(\{X^k_{ij}\}') \geq S \cdot O'(\{X^k_{ij}\}^*) \tag{18.7}$$

$$\geq O(\{X^k_{ij}\}^*) - CS \tag{18.8}$$

$$= \text{OPT} - \epsilon w^{\max} \tag{18.9}$$

$$\geq (1 - \epsilon) \cdot \text{OPT} \tag{18.10}$$

where (18.9) follows from (18.6) and (18.10) holds as $\text{OPT} \geq w^{\max}$. □

Algorithm 8 features running time that is polynomial in C, as it has to complete a table $A(c, r)$ consisting of $C^2 \lfloor w^{\max}/S \rfloor$ entries ($1 \leq c \leq C$ and $1 \leq r \leq C \lfloor w^{\max}/S \rfloor$. We point out that the scaling we select enables Algorithm 8 to be fully polynomial, i.e., the corresponding running time is also polynomial in $1/\epsilon$, since $w^{\max}/S = C/\epsilon$. A minimization version of Algorithm 8 necessitates reversing constraint in (18.6). In that case, the variables w_{ij} will then capture the expense for storing item j at base station i.

18.6 Experiment Evaluation

We evaluate the performance of our framework via simulation experiments. In the evaluation, we also benchmark our approach (identified henceforth as Opt) to two recent competitive methods. We call the first one LRU, as it implements a policy that evicts the least popular cached item to make room for new fresh data. We call the second reference method NCC, and it implements the framework explored in [46]. To facilitate implementing our framework, here we have implemented the optimization in (18.1), with the max operator being replaced with a min operator. Accordingly, we have replaced the reward factors therein with corresponding cost counterparts. Actual network energy consumption information is readily available in the public domain. Hence the cost factors relate to how much energy is consumed to stream the 360° content to the mobile users. In particular, the consumed energy due to local delivery include only the transmission energy dispensed by a small base station, streaming the content to one of its users. We account for the additional consumed energy, when the content needs to be delivered first from another small base station or a remote back-end server in the internet, as being proportional to the number of network hops that would need to be traversed. We leverage the information from [57], to consider that the latter case will have 15 intermediate network hops. Moreover, we follow the published information from [58, 59], to set the energy consumption of wireless links to 3.5 micro-Joules and the energy consumption of wireline links to 0.5 micro-joule. The number of base stations is set to 10.

The two reference methods leverage video files whose size is randomly chosen from a range of values, with 200 megabytes as the lower end and 800 megabytes as the upper end, in increments of 100. For Opt, we select B_j again at random, however, from a range that features 10 times smaller values, enabled by the efficient viewport-aware representation of the content we formulate. The popularity factors for the video files are generated according to a probability distribution that follows the Zipf law, with a factor of γ, we set to 4/5. This choice has been motivated by earlier work [60]. A base station is assigned a number of mobile users, chosen at random.

We transform the quantities $R_{j,v}$ to their cost equivalents, by inversely normalizing them with a reference value, representing $R_{j,v}$ for the most data voluminous video file. Similarly, we leverage published information from [61], to define κ_E to be the rendering energy per unit of time (second) needed to reconstruct a VR viewport at the temporal frame rate of 60 fps, and set its value to 100 milliwatts. We then characterize B_j^o for any video file j as $\kappa_E B_j / B_{j,\,\min}$, where the last quantity captures the least data volume of a video file. Here, we consider that the energy consumed due to rendering is proportional to the data volume of the video content to be reconstructed.

In the remainder, we evaluate the overall consumption of energy of each system under comparison, against three factors: the storage space at a base station, how many base stations there are in the macro cell, and the computing budget at each base station.

We first examine the impact of the first factor on the system performance of each method. In these experiments, we adopt published information from [62], to set the value of C_i^o to 65 watts, for every base station. The respective results are shown in Figure 18.6. It can be observed that Opt enables enhanced performance, by consuming 3 to 7 times less energy with respect to the competing techniques, for all values of storage space considered. Moreover, its energy consumption drops much faster as more storage space is introduced, as observed from the lower range in the figure. Finally, even when no caching takes place (cache size of zero), enabled by the efficient viewport-aware representation of the content we formulate, Opt still leads to considerable energy savings, as seen from Figure 18.6. To gain additional insights on this aspect, we evaluated how Opt would perform if it streams the entire monolithic 360° panorama to users, identically to LRU and NCC. This approach has been identified as Opt-NVDC in the figure, and it can be seen that it features identical energy cost as the competing methods, when the base stations have no storage space. On the other hand, Opt-NVDC consumes much less energy, when the base stations can store video files, enabled by our framework. We note for example that it leads to 50% energy savings relative to the competing methods, for most of the storage space values examined in Figure 18.6.

Next, we examine the impact of the network size/density, in terms of how many small base stations are deployed in the macro cell. For every base station, the values of C_i

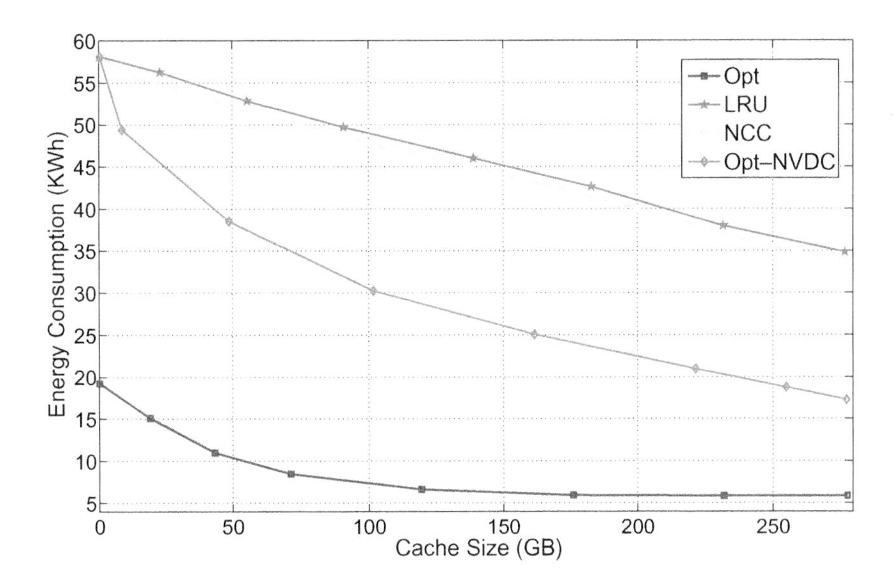

Figure 18.6 Consumed system energy against storage space at a small base station.

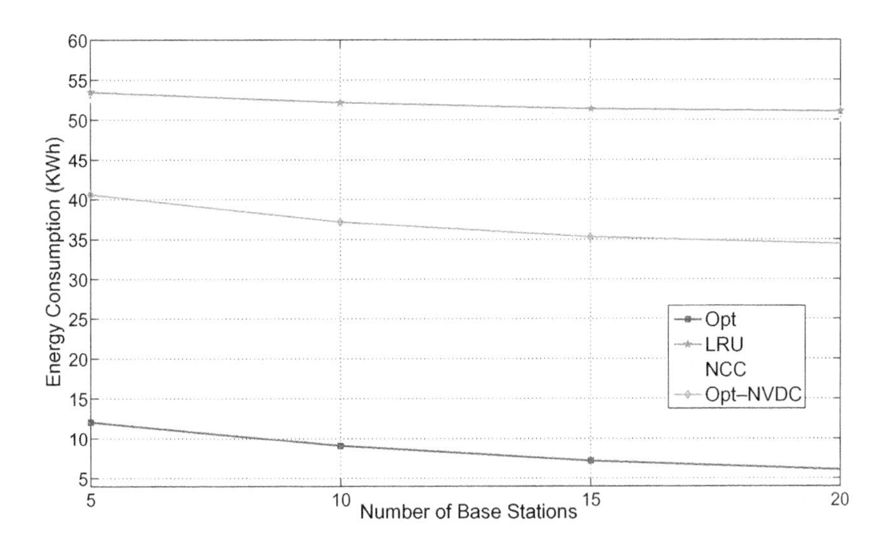

Figure 18.7 Consumed system energy against network size (number of small base stations).

and C_i^o have been fixed to 60 gigabytes and 65 watts, respectively. These results are shown in Figure 18.7. It can be seen that introducing more base stations improves the energy efficiency of all three systems. The savings in energy as the number of base stations increases are only marginal for LRU and NCC, as expected. In contrast, Opt benefits from more small base stations in the system, by more effectively combining and leveraging the higher level of resources available in the system thereby, such that the system efficiency is thus augmented. This is evident from the performance of our approach (and its variant Opt-NVDC) exhibited in Figure 18.7.

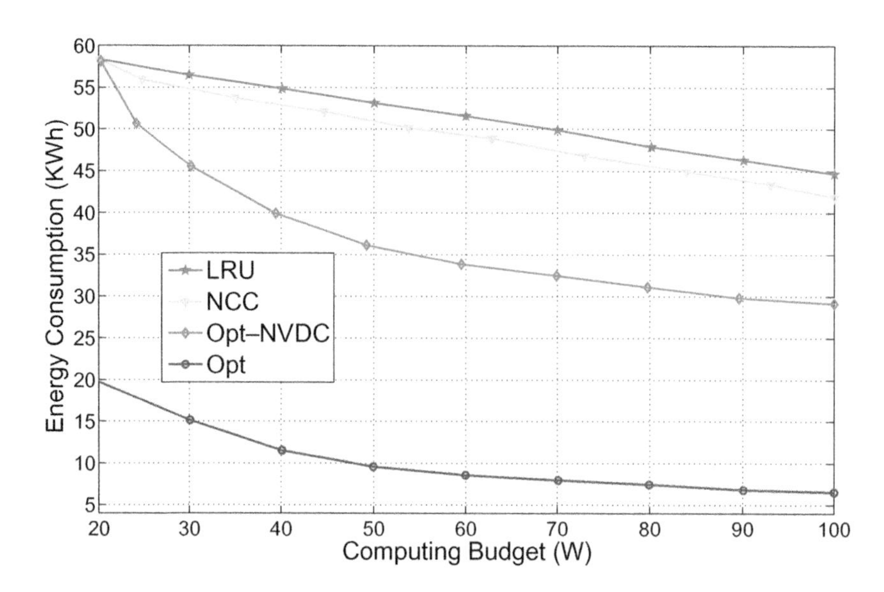

Figure 18.8 Consumed system energy against available computing capabilities at the mobile network edge.

Last, we investigate the impact of available computing resources on energy consumption. We summarize these results in Figure 18.8, considering $C_i = 60$ GB, $\forall i$. Similar trends are observed as those in Figure 18.6. Concretely, for small values of the computing budget C_i^o, all systems under examination consume more energy. This is not surprising, as when insufficient rendering capabilities exist at the edge of the network, the mobile clients need to have their content streamed from the back-end server, which augments the aggregate energy cost of the system. On the other hand, we also observe from Figure 18.8 that once C_i^o starts increasing, the system efficiency in turn starts improving uniformly across all techniques, enabled by the local capabilities for rendering and therefore streaming. It should be noted though that our approach, benefiting from our analytical advances, leverages the available system resources more effectively, to enable lower consumed energy for the system, as evident from Figure 18.8.

18.7 Concluding Remarks

This chapter explored a system framework for streaming 360° content to mobile virtual reality clients in next-generation cooperative 5G systems. Our framework integrates an effective method for representing 360° videos such that they can enable streaming of only the portion of the 360° panorama required by a client at any time. This overcomes a major shortcoming of emerging practices that deliver the entire voluminous panorama, which usually includes information not needed by the respective user. Our framework allows for cooperation among the base stations in selecting joint caching, streaming, and edge computing (rendering) policies, whose objective is to minimize the overall energy consumption of the system, when delivering the requested content to the mobile clients. Constraints that need to be met while meeting this goal is the limited storage and computational resources at each small base station. We formally verify that our problem of interest is NP hard and formulates an efficient solution algorithm that exhibits strong performance approximation guarantees in fully polynomial time. Our experiments demonstrate considerable performance advances over competitive caching methods employed in concert with prevalent 360° video representation practices. These benefits motivate further investigation.

References

[1] J. G. Apostolopoulos, P. A. Chou, B. Culbertson, T. Kalker, M. D. Trott, and S. Wee, "The road to immersive communication," *Proceedings of the IEEE*, vol. 100, no. 4, pp. 974–990, Apr. 2012.

[2] "HTC Vive: This is REAL. Discover VR beyond imagination," www.htcvive.com/us.

[3] "Sony PlayStation VR," www.playstation.com/en-us/explore/playstation-vr.

[4] "Ricoh Theta: 360° degree experience," https://theta360.com/en.

[5] "Samsung Gear 360," www.samsung.com/global/galaxy/gear-360.

[6] "Microsoft holoLens Holographic display," www.microsoft.com/microsoft-hololens/en-us.

[7] "Mixed reality lightfield," Magic Leap, www.magicleap.com.

[8] "Oculus Rift and Gear VR: next-generation virtual reality," www.oculus.com.

[9] "Google Cardboard: experience virtual reality in a simple, fun, and affordable way," https://vr.google.com/cardboard/index.html.

[10] "Facebook 360: a stunning and captivating way to share immersive stories, places and experiences," http://facebook360.fb.com.

[11] "Jump – Google VR technology platform," https://vr.google.com/jump.

[12] B. Darrow, "Project Tango is Google Maps on steroids," *Fortune*, May 2016, http://fortune.com/2016/05/12/google-project-tango.

[13] P. Rubin, "Oculus raises $75 million for the VR goggles of the future," *wired*, Dec. 2013, www.wired.com/2013/12/oculus-vr-funding.

[14] M. Isaac, "Magic Leap, an augmented reality firm, raises $793 million," *New York Times*, Feb. 2016, www.nytimes.com/2016/02/03/business/dealbook/magic-leap-an-augmented-reality-firm-raises-793-million.html?_r=1.

[15] "AR/VR M&A Timeline: Facebook, GoPro, HP, Apple begin to grab startups," CB Insights, July 2016, www.cbinsights.com/blog/top-acquirers-ar-vr-ma-timeline.

[16] "Record $2 billion AR/VR investment in last 12 months," Digi-Capital Research, July 2016, www.digi-capital.com/news/2016/07/record-2-billion-arvr-investment-in-last-12-months/#.V4xVpbAg-M8.

[17] T. Merel, "Augmented and virtual reality to hit $150 billion, disrupting mobile by 2020," Tech Crunch Apr. 2015, https://techcrunch.com/2015/04/06/augmented-and-virtual-reality-to-hit-150-billion-by-2020.

[18] P. F. Drucker, "Internet of things position paper on standardization for IoT technologies," Jan. 2015, www.internet-of-things-research.eu/pdf/IERC_Position_Paper_IoT_Standardization_Final.pdf.

[19] D. Evans, "The internet of things: how the next evolution of the internet is changing everything," Apr. 2011, www.cisco.com/web/about/ac79/docs/innov/IoT_IBSG_0411FINAL.pdf.

[20] M. Kranz, P. Holleis, and A. Schmidt, "Embedded interaction: interacting with the internet of things," *IEEE Internet Computing*, vol. 14, no. 2, pp. 46–53, 2010.

[21] P. Vlacheas, R. Giaffreda, V. Stavroulaki, D. Kelaidonis, V. Foteinos, G. Poulios, P. Demestichas, A. Somov, A. R. Biswas, and K. Moessner, "Enabling smart cities through a cognitive management framework for the internet of things," *IEEE Communications Magazine*, vol. 51, no. 6, pp. 102–111, 2013.

[22] W. Barfield, *Fundamentals of Wearable Computers and Augmented Reality*, Boca Raton, FL: CRC Press, 2015.

[23] G. P. Fettweis, "The tactile internet: applications and challenges," *IEEE Vehicular Technology Magazine*, vol. 9, no. 1, pp. 64–70, Mar. 2014.

[24] "2020: Beyond 4G Radio Evolution for the Gigabit Experience," Nokia Siemens Networks, White Paper, Aug. 2011.

[25] "YouTube: 360° Videos," www.youtube.com.

[26] J. D. Moss and E. R. Muth, "Characteristics of headmounted displays and their effects on simulator sickness," *Human Factors: The Journal of the Human Factors and Ergonomics Society*, vol. 53, no. 3, pp. 308–319, June 2011.

[27] K. Shanmugam, N. Golrezaei, A. Dimakis, A. Molisch, and G. Caire, "Femtocaching: wireless video content delivery through distributed caching helpers," *IEEE Transactions in Information Theory*, vol. 59, no. 12, pp. 8402–8413, 2013.

[28] R. Vasudevan, Z. Zhou, G. Kurillo, E. Lobaton, R. Bajcsy, and K. Nahrstedt, "Real-time stereo-vision system for 3D teleimmersive collaboration," in *Proceedings of the International Conference on Multimedia and Exhibition*, July 2010, pp. 1208–1213.

[29] M. Hosseini and G. Kurillo, "Coordinated bandwidth adaptations for distributed 3D tele-immersive systems," in *Proceedings of the International Workshop Massively Multiuser Virtual Environments*, ACM, Mar. 2015, pp. 13–18.

[30] J. Chakareski, "Uplink scheduling of visual sensors: when view popularity matters," *IEEE Transactions in Communications*, vol. 2, no. 63, pp. 510–519, Feb. 2015.

[31] G. Cheung, A. Ortega, and N.-M. Cheung, "Interactive streaming of stored multiview video using redundant frame structures," *IEEE Transactions in Image Processing*, vol. 20, no. 3, pp. 744–761, Mar. 2011.

[32] J. Chakareski, V. Velisavljević, and V. Stanković, "User-action-driven view and rate scalable multiview video coding," *IEEE Transactions in Image Processing*, vol. 22, no. 9, pp. 3473–3484, Sept. 2013.

[33] J. Chakareski, "Wireless streaming of interactive multi-view video via network compression and path diversity," *IEEE Transactions in Communications*, vol. 62, no. 4, pp. 1350–1357, Apr. 2014.

[34] J. Chakareski, "Transmission policy selection for multi-view content delivery over bandwidth constrained channels," *IEEE Nokia Siemens Networks, in Image Processing*, vol. 23, no. 2, pp. 931–942, Feb. 2014.

[35] X. Corbillon, A. Devlic, G. Simon, and J. Chakareski, "Viewport-adaptive navigable 360-degree video delivery," in *Proceedings of the International Conference on Communications*, IEEE, May 2017.

[36] J. Chakareski, R. Aksu, X. Corbillon, G. Simon, and V. Swaminathan, "Viewport-driven rate-distortion optimized 360° video streaming," in *Proceedings of the International Conference on Communications*, IEEE, May 2018.

[37] X. Corbillon, A. Devlic, G. Simon, and J. Chakareski, "Optimal 360-degree video representation for viewport-adaptive streaming," in *Proceedings of the International Conference on Multimedia*, ACM, Oct. 2017, pp. 934–951.

[38] R. Aksu, J. Chakareski, and V. Swaminathan, "Viewport-driven rate-distortion optimized scalable live 360° video network multicast," in *Proceedings of the ICME International Workshop on Hot Topics in 3D (Hot3D)*, IEEE, July 2018.

[39] J. Chakareski, "Informative state-based video communication," *IEEE Nokia Siemens Networks, Image Processing*, vol. 22, no. 6, pp. 2115–2127, June 2013.

[40] J. Chakareski, J. Apostolopoulos, S. Wee, W.-T. Tan, and B. Girod, "Rate-distortion hint tracks for adaptive video streaming," *IEEE Nokia Siemens Networks, Circuits and Systems for Video Technology*, vol. 15, no. 10, pp. 1257–1269, Oct. 2005.

[41] J. Chakareski and P. Frossard, "Rate-distortion optimized distributed packet scheduling of multiple video streams over shared communication resources," *IEEE Transactions in Multimedia*, vol. 8, no. 2, pp. 207–218, Apr. 2006.

[42] J. Chakareski, S. Han, and B. Girod, "Layered coding vs. multiple descriptions for video streaming over multiple paths," *ACM/Springer-Verlag Multimedia Systems Journal*, vol. 10, no. 1, pp. 275–285, Jan. 2005.

[43] J. Chakareski and P. Chou, "RaDiO Edge: rate-distortion optimized proxy-driven streaming from the network edge," *IEEE/ACM Nokia Siemens Networks, Networking*, vol. 14, no. 6, pp. 1302–1312, Dec. 2006.

[44] J. Chakareski and P. A. Chou, "Application layer error correction coding for rate-distortion optimized streaming to wireless clients," *IEEE Nokia Siemens Networks, Communications*, vol. 52, no. 10, pp. 1675–1687, Oct. 2004.

[45] J. Chakareski, "In-network packet scheduling and rate allocation: a content delivery perspective," *IEEE Nokia Siemens Networks, Multimedia*, vol. 13, no. 5, pp. 1092–1102, Oct. 2011.

[46] P. Blasco and D. Gunduz, "Learning-based optimization of cache content in a small cell base station," in *Proceedings of the International Conference on Communications*, IEEE, June 2014, pp. 1897–1903.

[47] S. Martello and P. Toth, *Knapsack problems*, New York: Wiley, 1990.

[48] E. Baştuğ, M. Bennis, and M. Debbah, "Cache-enabled small cell networks: modeling and tradeoffs," in *Proceedings of the International Symposium on Wireless Communications Systems*, IEEE, Aug. 2014, pp. 649–653.

[49] J. Erman, A. Gerber, M. Hajiaghayi, D. Pei, S. Sen, and O. Spatscheck, "To cache or not to cache: the 3G case," *Internet Computing, IEEE*, vol. 15, no. 2, pp. 27–34, 2011.

[50] H. Ahlehagh and S. Dey, "Video caching in radio access network: impact on delay and capacity," in *Proc. IEEE Wireless Communications and Networking Conference*, April 2012, pp. 2276–2281.

[51] N. Karamchandani, U. Niesen, M. A. Maddah-Ali, and S. Diggavi, "Hierarchical coded caching," in *Proc. IEEE International Symposium on Information Theory*, July 2014.

[52] M. A. Maddah-Ali and U. Niesen, "Fundamental limits of caching," in *Proceedings of the IEEE International Symposium on Information Theory*, July 2013.

[53] M. Yu, H. Lakshman, and B. Girod, "A framework to evaluate omnidirectional video coding schemes," in *Proceedings of the IEEE International Symposium on Information Mixed and Augmented Reality*, Sept. 2015, pp. 31–36.

[54] G. J. Sullivan, J.-R. Ohm, W.-J. Han, and T. Wiegand, "Overview of the high efficiency video coding (HEVC) standard," *IEEE Transaction in Circuits and Systems for Video Technology*, vol. 22, no. 12, pp. 1649–1668, Dec. 2012.

[55] A. Gersho and R. M. Gray, *Vector Quantization and Signal Compression*, New York: Springer, 1991.

[56] R. Bellman, *Dynamic Programming*, Princeton, NJ: Princeton University Press, 1957.

[57] A. Fei, G. Pei, R. Liu, and L. Zhang, "Measurements on delay and hop-count of the internet," in *IEEE GLOBECOM'98-Internet Mini-Conference*, 1998.

[58] J. Huang, F. Qian, A. Gerber, Z. Mao, S. Sen, and O. Spatscheck, "A close examination of performance and power characteristics of 4G LTE networks," in *ACM MobiSys*, 2012.

[59] V. Sivaraman, A. Vishwanath, Z. Zhao, and C. Russell, "Profiling per-packet and per-byte energy consumption in the netfpga gigabit router," in 2011 IEEE Conference on *Computer Communications Workshops (INFOCOM WKSHPS)*, 2011, pp. 331–336.

[60] L. Breslau, P. Cao, L. Fan, G. Phillips, and S. Shenker, "Web caching and zipf-like distributions: Evidence and implications," in *IEEE INFOCOM*, 1999.

[61] B. Johnsson and T. Akenine-Möller, "Measuring per-frame energy consumption of real-time graphics applications," *Journal of Compuer Graphics Techniques*, vol. 3, no. 1, pp. 60–73, 2014.

[62] "CPU power dissipation,"Wikipedia,https://en.wikipedia.org/wiki/CPU_power_dissipation.

19 Economic Ecosystems in Elastic Wireless Edge Caching

George Iosifidis, Jeongho Kwak, and Georgios Paschos

The delivery of content over the internet is a multibillion-dollar business involving different stakeholders: the content providers (CPs) that create and sell content to users; the content distribution networks (CDNs) that manage large-scale content cache servers, and the internet service providers (ISPs) or mobile network operators (MNOs) that are responsible for transferring the content to end request points. The economic interactions of these entities have always been convoluted as their decisions about content pricing, caching, and delivery are inherently intertwined. Moreover, recent advances in network management, including software-defined networking (SDN), network function virtualization (NFV), and in caching architectures, i.e., the deployment of caches at the network edge, further exacerbate these effects, creating new opportunities but also new challenges.

In this chapter, we study the technoeconomic challenges for one of the most promising new caching paradigms, the *elastic wireless edge caching* solution, where third-parties dynamically lease storage resources in a wireless cloud. The main idea is the following: an MNO advertises storage prices for servers placed in proximity to the end users, and various content providers lease on demand capacity to enhance the quality of their services. We describe the main concepts and existing business models for the elastic CDN solution, provide a summary of the related work, and discuss the key differences between in-network and edge caching. We then present a detailed model for this system where the caches reside in cellular base stations. We formulate a problem where cache dimensioning, content caching, and request routing decisions are jointly optimized by a CP in order to reduce content delivery delay subject to a given leasing budget. We design a suite of dynamic solution algorithms, based on the Lyapunov drift-minus-benefit technique, and present numerical experiments that quantify the benefits of elastic over typical static cache deployments. See Table 19.1 for the abbreviations used in this chapter.

19.1 Introduction

The constant mobile data traffic growth [1] and the increasing congestion of internet pipes due to video and other content traffic has spurred many research efforts for the design of content caching infrastructures. However, the economic aspects of these systems have been largely overlooked. Figure 19.1 depicts just how different is the

Table 19.1 List of Abbreviations

Abbreviation	Full Name	Abbreviation	Full Name
CP	Content Provider	CDN	Content Delivery Network
CSP	Cloud Service Provider	ISP	Internet Service Provider
MNO	Mobile Network Operator	Telco CDN	CDN of Telecommunication Operator
BS	Base Station	NFV	Network Function Virtualization
SDN	Software Defined Networking	SBS	Small Base Station
MBS	Macro Base Station	D2D	Device to Device
QoS	Quality of Service	CDNaaS	CDN as a Service
QoE	Quality of Experience	WEC	Wireless Edge Caching

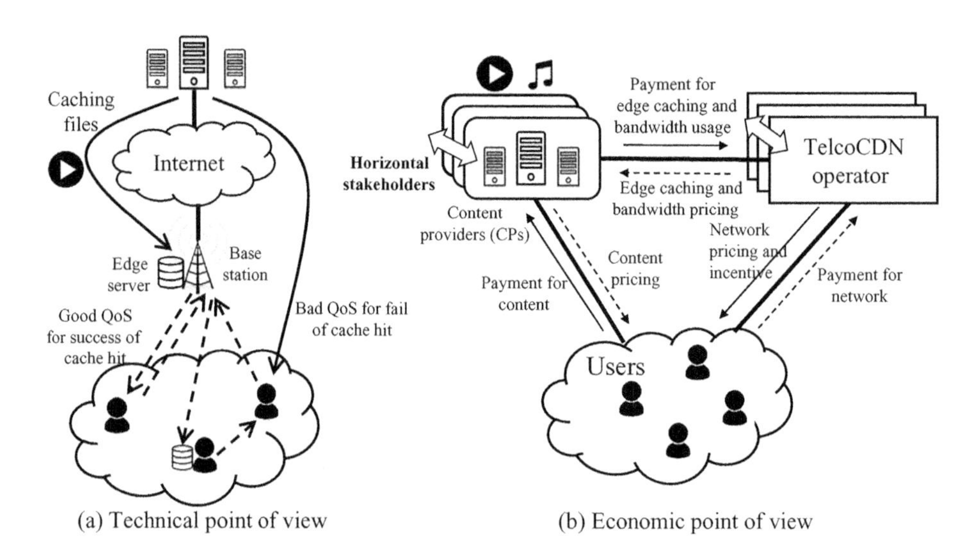

Figure 19.1 Technical and economic perspectives of wireless edge caching. From an economic point of view, vertical stakeholders represent different classes of entities such as users, CPs, ISPs, and CDN providers; and horizontal stakeholders represent different entities within each class. In general, the issues between the vertical stakeholders are pricing for caching or transferring the content and collaboration between them, whereas the issues between the horizontal stakeholders are making service fairness for the same service price.

content delivery process in perspectives of technology and economy. From a technical standpoint, the caching decisions at every edge server are made in order to maximize the caching hit ratio and/or reduce the end-to-end delay. Yet, the key stakeholders involved in this process have most often different goals. This introduces intricate trade-offs about which files to cache, which routes to use, and even how to price the cached content. The focus of this chapter is to highlight these trade-offs, which will be very common in the fast emerging elastic caching ecosystem.

Although memory is typically cheaper than link bandwidth [2], the entire amount of storage capacity in a mobile network can be significant, thus introducing important costs [3]. These costs increase further when it comes to edge storage, which is more expensive than cloud storage in data centers. For example, the cost of the Amazon Web Service (AWS) CloudFront for storing files at edge servers is $0.085/GB, whereas in AWS Simple Storage Service (S3), which uses data center storage, the respective price is four times smaller, i.e., $0.023/GB [4]. On the other hand, caching at the edge reduces backhaul utilization, known to be the performance bottleneck of dense wireless networks. This suggest that leasing caches at wired or wireless networks mandate a careful cost analysis and novel pricing schemes [5].

When it comes to moving massive amounts of video in the internet, a number of stakeholders are involved. For example, CPs such as YouTube, Netflix, and Facebook, produce content; CDNs such as Akamai offer content storage and delivery services; the ISPs or MNOs are responsible for delivering this traffic to the end users who create the demand and pay for this service. In the constantly evolving ecosystem of caching, new roles and new interactions are created among these entities [6]. For instance, companies such as Akamai and Limelite offer CDN as a service (CDNaaS), i.e., they rent their storage infrastructure to third parties. ISPs often choose to build their own CDNs (known as Telco-CDNs), when they decide that in-network storage can reduce their costs. On the other hand, since CPs own the encryption rights of the video content, they often select to install their own cache servers within the ISP or MNO networks. Users may produce their own content (e.g., in YouTube or Facebook) or choose to circulate the videos themselves, e.g., by using peer-to-peer overlays. Caching plays a dominant role in gluing together all these stakeholders and content delivery solutions, leading to many and diverse business models.

In this chapter, we focus on a very promising model, namely the *elastic CDN service*. In eCDNs, there are three key stakeholders: (1) the storage infrastructure owner, a prominent example of which is Amazon AWS ElastiCache [4] or a Telco CDN; (2) a small-size CP that cannot afford a private CDN; and (3) the users that consume the video content. The CP purchases storage on demand and serves the users. The service is called elastic because the storage can be leased dynamically, e.g., on an hourly basis, and to be dimensioned based on the needs at each time instance. This provides unprecedented levels of flexibility and opportunities for budget economization, but complicates the system design and the interactions among these business entities. The idea of elastic CDNs is built upon recent technological advances that enable the flexible control of storage and network resources [7]. For instance, cloud companies offer elastic-*anything*, e.g., AWS provides a variety of options for flexible services such as Amazon EC2 Auto Scaling, Elastic Load Balancing, and Elastic File System [4]. Similarly, Akamai proposed the concept of cloud CDN, where caching capacity can be dynamically dimensioned to host virtual caches [8, 9].

Following this trend, the elastic CDN service provides two significant advantages: it allows us to meet spatiotemporal demand variations by installing caches where and when needed (just-in-time caching), and it enables small-size CPs (such as Pinterest, Snapchat, and Tumbler) to enter the content delivery market with small costs. Traditional CDN pricing reflects the storage usage costs or small-scale traffic fluctuations but is mainly based on a flat-rate price across large geographical regions (e.g., continents) arranged in long-term contracts (e.g., few months to years) based on peak traffic estimations [10]. Therefore, small-size CPs, faced with unpredictable demand, must either invest in building a large private CDN or buy such flat contracts, both of which create a *barrier-to-entry* into the content business. In contrast, elastic CDNs allow these entities to improve content delivery with a flexible pricing scheme. Indeed, the first commercial services, such as Akamai Aura [11] and Huawei uCDN [12], enable us to dynamically scale cache capacity and enable fine-grained payment service [13]. However, due to the complicated nature of caching resource allocation and network environment, the elastic CDN service seems to have very complicated engineering-oriented solutions. In this chapter, we propose a mathematical model toward addressing this challenge.

19.2 Background

Before we delve deeper into the economics of wireless edge caching, we discuss the related work and the main business models for content caching.

Wireless edge caching in heterogeneous networks. The wireless edge caching (WEC) idea, in its current form, has been largely shaped by the femtocaching proposal [14], which optimizes the caching of files in small-cell networks with capacity-limited backhaul links. Many followup works have studied different aspects or different versions of this caching architecture; see [6] for a brief overview. For instance, [15] proposed the capacitated femtocaching model that takes into account the wireless link capacity that is fast drained in massive demand scenarios; [16] proposed the joint design of caching and multicast policies, [17] studied the problem of leasing storage from third parties in wireless networks; [18] proposed the hierarchical file caching at different level of edge clouds; and [19] addressed a joint file caching and user-BS association problem with a consideration of spatial dynamics of content popularity. All these works, however, assume that the caches have a given capacity or that they are dimensioned only one time. Clearly, this is a fundamentally different model than that envisaged elastic CDN solution.

Techno-economical content caching and delivery techniques. Many works on the CDN server placement problem considering the cost of the cache memory [20–23] have been studied. For instance, [20] proposed the joint minimization problem of server deployment, file caching, and routing cost, and solved it exploiting the Benders' decomposition method; while [22] leveraged dynamic programming. On the other hand, [23] formulated a storage budget allocation problem that aims to minimize the content delivery costs in a hierarchical CDN.

Few works have designed mechanisms to enable cooperation among ISPs, CDNs, CPs, and users. For instance, [24] developed an algorithm for the joint design of caching and user association policies in a femtocaching network, where the storage is leased from residential Wi-Fi access points. This idea was further extended in [17], which studied the cooperation between a content provider leasing caches from an MNO, and the impact of the user association policy on the caching performance. Such cooperation mechanisms have been also considered in wired networks, where it has been shown that the coordination of routing and caching decisions between ISPs and

CDNs, respectively, can produce significant cost savings and performance gains; see [25] and references therein.

Business models for content caching.

- *Akamai Intelligent Platform:* Akamai operates CDNs offerring a large amount of internet traffic. They have 0.2 million caching servers distributed over the world, offering 1–10 ms access to content. A recent study [9] proposed the idea of *elastic* CDN, which enables us to change leased cache capacity with a fined-grained time scale. This framework combines storage placement with decisions of caching.
- *Google Global Cache (GGC):* This model consists of caches owned by ISPs. The GGC's aim is reducing network bandwidth costs by locally serving requests for YouTube content [26].
- *Netflix Open Connect:* Netflix partially deploys CDN servers within ISPs [27]. Nevertheless, Netflix faces difficult challenges, because their file library is smaller than YouTube.
- *Amazon AWS:* The Amazon CloudFront is a virtual CDN service that leverages cloud storage to provide content caching services. AWS provides dynamic content caching and bandwidth services where the customers can change their policy every hour.
- *Cedexis and Conviva:* By far, major CPs have contracted with a single CDN, such as Akamai, Level 3 or Amazon CloudFront, or deployed their own CDN, such as Google and Netflix.

19.3 Wireless Edge Caching versus In-Network Caching

Wireless edge caching systems differ from typical in-network caching architectures in many different aspects:

- The population of users reaching a cache of a base station (or, any edge server) in WEC, is significantly smaller than the respective population that creates requests for a core-network cache. This makes the timely collection of statistics about content demand a very challenging task in WEC.
- Contrary to data centers, edge servers have limited storage resources. Hence the cost for leasing storage at the edge is higher than that of cloud storage.
- Finally, WEC systems are inherently dynamic for a variety of reasons, including user mobility, wireless channel state variations, and the overlap of their coverage areas.

This raises the following specific technical challenges in WEC: (1) popularity prediction is difficult, and hence caching efficiency might be compromised; (2) the deployment of storage is of high importance for the system performance, yet is challenging to optimize given the small scope of these caches; (3) the storage investment can reduce the backhaul link utilization; but (4) the cost of leasing edge storage is very high, which in turn calls for dynamic cache dimensioning. In the next section, we study an elastic CDN scenario and propose a solution that tackles these issues.

19.4 Elastic Wireless Cache Lease, Content Caching, and Routing

The elastic CDN model allows small-scale CPs to rent on-demand cache capacity at the various edge servers of ISPs or MNOs, to improve the service quality. We formalize this scenario and develop an optimization framework in a scenario that the CPs have their budget to lease cache capacity with a consideration of spatiotemporally varying file popularity and electricity price for storage. This chapter follows a systematic optimization approach where we jointly consider the decisions about *elastic storage lease*, *content caching*, and *content-user routing*. The goal is to maximize the benefits in terms of average download delay reduction (compared to not using edge caching), conditioned that the CPs do not overspend their budget compared to the time-average storage lease payment. We should note that the delay reduction from edge caching can be directly translated to monetary benefits for the CPs, as the users are extremely sensitive in the content delivery delay, and hence higher prices can be charged to them when the edge caches are employed. We leverage the Lyapunov drift-plus-benefit framework to change an original long-term average problem into online snapshot problems that do not require future content demand and network conditions.

19.4.1 Scenario

The eCDN solution [9] is particularly suitable for small CPs, which can dynamically lease cache capacity. Such CPs operate typically with tight monetary budgets and need to serve fast-changing user demand, and also seek fine-grained control over the leased caches in order to accrue the highest possible benefits. In this scenario, the CPs need to make decisions of leased cache capacity in an edge server attached to a small BS and stored contents in the edge server. Since the content request temporally changes, we need to periodically (every hour) bring them up to date.

This chapter assumes that an MNO has the eCDN, hence the MNO manages both user-small BS association and content caching. In this scenario, the end users subscribe to a content service from a CP and pay a Telco-CDN operator for the network usage. The latter provides both edge caching and data delivery services (from core data centers to the edge servers) to the CP, and mobile internet connectivity to end users. The Telco CDN operator sets the price for the delivery of data and the edge caching service dynamically, based on, e.g., the varying electricity prices.

In our framework, we assume Poisson point process for service area, i.e., each point in the service area receives the requested file from a small BS where the service coverage of small BSs can be overlapped. If the requested file is cached at the routing BS, the retrieving delay can be reduced since the original retrieving delay is that from the original content server to the service point. Moreover, the CP is assumed to let the MNO know its cache rental budget. The aim is at maximizing total service delay

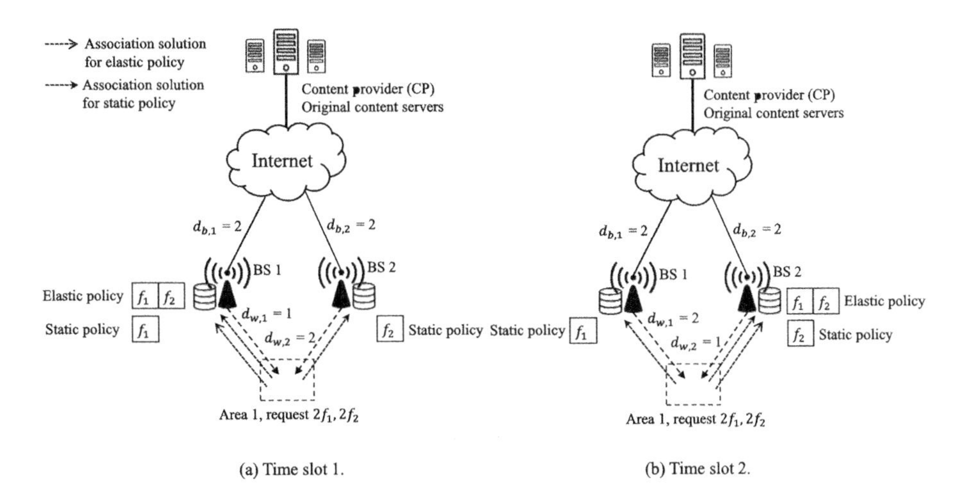

Figure 19.2 An example of the elastic policy and the static policy in a cache rental system. For the simplicity of the model, the only experienced wireless delay is a time-varying parameter and other parameters are static in this model. Note that the association solution for static policy is fair time-sharing between BS 1 and BS 2 for both time slots since time-average delays for both BSs are the same.

benefit, i.e., delay difference between retrieving the corresponding file from the original content server and that from the small BS that caches the corresponding file, constrained by the cache rental budget. The latter is elastic in the sense that the CP can violate it in some rounds and underspend in some others, keeping the average below a given threshold. Clearly, one can consider different business and operational models.

19.4.2 Motivating Example of Elastic Cache Lease

Figure 19.2 shows an example of the elastic cache lease, file caching and area-BS association policy[1], and compares these to a respective static policy in a cache rental system.[2] In this example, a CP provides content service to users in area 1 and a Telco CDN operator has two edge servers attached to each BS that can be leased by the CP. Parameter $d_{b,n}$ denotes the backhaul delay for transferring files between the CP and each BS, and $d_{w,n}$ is the wireless delay when transmitting files from each BS to the users in area 1, respectively where $n \in 1,2$ denotes the BS index. In this example, we consider the following dynamic and static states (or, conditions): *dynamic states* ($d_{w,n}$ changes for all BSs over each time slot) and *static states* ($d_{b,n}$ is constant for all BSs and requests for all files f_1, f_2). We assume that $d_{b,n} = 2$ for both BSs, and that 2 file requests arrive for both files in every time slot. At the first slot, the CP is likely to lease memory space at BS 1 for all files. In this case, all requests would be associated with that BS since the wireless delay from BS2 to area 1 is larger. At time slot 2, the CP is

[1] Here we simply consider BSs instead of small BSs.
[2] To simplify, we do not include the units of all parameters.

likely to lease memory space at BS 2 to store all files; and hence all requests would be associated with that BS, since the wireless delay from BS 2 to area 1 is smaller than that from BS 1 to area 1. However, the CP that adopts the static policy is likely to lease memory space at both BSs (namely, to cache f_1 at BS 1 and f_2 at BS 2) and use fair time-sharing association between the two BSs, since the time-average wireless delays are the same from both BSs to area 1.

We evaluate the elastic policy and the static policy using three metrics: total cache lease costs, total backhaul bandwidth usage costs, and total end-to-end delay (from the closest location that caches the corresponding file to area 1). First, we note that the total cache lease costs for both policies are equal, since the memory space to cache two files is leased by the Telco CDN operator every time slot. Second, the total backhaul bandwidth cost for the elastic policy is zero, and for the static policy it is $4R$, where R is the bandwidth cost per file. The cost for the static policy arises since one request of each file must be transmitted from the original content servers via BS 1 or BS 2 in both time slots. Finally, the end-to-end delay of the elastic policy is 2 (2 file requests for file f_1) and additionally 2 units (2 file requests for file f_2), hence in total equal to 4 at the first slot. Similarly, they are 4 at time slot 2. For the static policy, it is 1 (f_1 from BS 1 caching), plus 3 additional units (f_2 from the CP to area 1 via BS 1), 2 units (f_2 from BS2 caching), and 4 units (f_1 from the CP to area 1 via BS2), which yield a total of 10 units at the first slot. Similarly, we can calculate that for the second slot it is also 10, which bring about an total cost of 20 units. This chapter concludes that in this example the elastic policy offers better QoS with the same memory leasing cost and using less backhaul bandwidth than the static policy.[3]

19.4.3 System Model

The cache rental framework includes a macro BS (MBS) (denoted with s) and multiple small BSs collected in set $\mathcal{J}$. All stations together $\mathcal{J} \cup \{s\}$ cover spatial zone (see Figure 19.3). We partition the area into $\mathcal{I}$ nonoverlapping subareas and use $\mathcal{J}_i \subseteq \mathcal{J}$ to denote the subset of small BSs that are *reachable* by area $i \in \mathcal{I}$. The MBS is reachable from all points in the plane. The Telco-CDN offers storage for leasing at each small BS, that can be used to cache files and facilitate their delivery. Time is slotted in hours $t = 0, 1, \ldots$. For each file f in a catalog $\mathcal{F}$, we denote with $\lambda_{i,f}(t)$ the traffic is requested from area i at time slot t. We assume that $\lambda_{i,f}(t)$ is identically and independently distributed.[4] The traffic demand reflects the spatiotemporal fluctuation of file demand. Hence it is of utmost importance to control file caching across different time slots.

If a location emanates a file request, the corresponding retrieving delay $d_{ij}(t), j \in \mathcal{J}_i \cup \{s\}$ (which is related to the area i where the user is placed) will be generated, and the station $j \in \mathcal{J} \cup \{s\}$ from which the file is retrieved will together determine

[3] In this example, we assume that file popularity can be exactly predicted. There exists many studies that address the prediction of the file popularity using machine learning techniques such as the one in [28] and references therein.

[4] It is possible to extend the model to Markovian arrivals using the framework of [29].

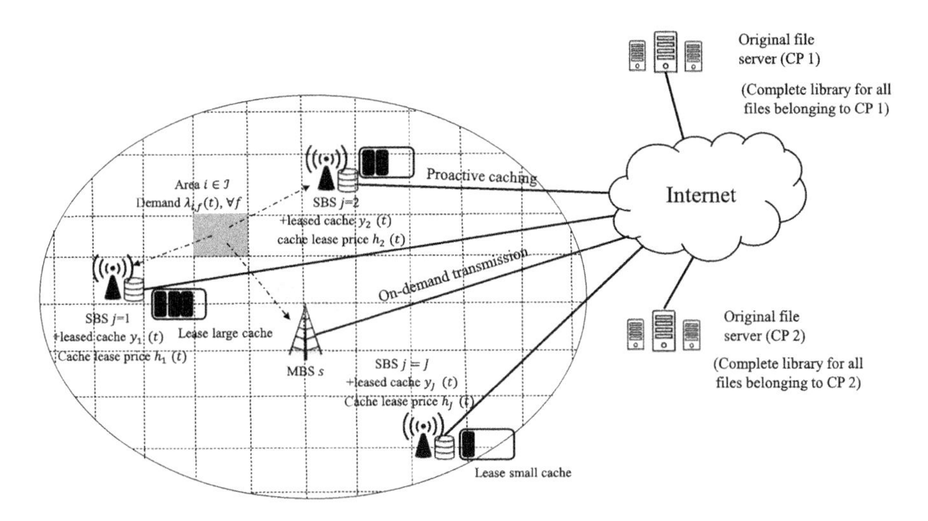

Figure 19.3 Overview of cache rental, file caching, and association in wireless elastic CDNs.

the employed communication path. A possibly different delay is associated with each path for various reasons, such as wireless interference, congestion, and propagation time which are all path specific. When the file is retrieved from the MBS, a remote server is contacted to obtain the file (see Figure 19.3), and the download delay $d_{is}(t)$ is generally large. To improve the QoS offered to end users, the file can be retrieved from a nearby small BS cache instead of the MBS. This improvement can directly lead to revenue increase for the CP, which can, for instance, charge higher prices for the offered service.

The storage is leased at a fluctuating price of $h_j(t)$ dollars per byte, at each small BS j, which is extrinsic to our system. For instance, the price might follow electricity cost fluctuations [30], or may be determined by a spot storage market where idle capacity is dynamically sold in a small time scale. Hence the price is affected by temporal ebbs and flows of traffic and storage demand. We introduce the investment variables $y_j(t)$, which is the leased caching capacity of small BS at time slot t. The investment decisions are subject to an average budget B_{avg}, which must be satisfied over a long horizon. On one hand, cloud service providers, like Amazon AWS, provide storage lease (called S3) and backhaul transmission (called CloudFront) at prices that are adapted every hour, which motivates investment decisions on hourly basis. On the other hand, CPs must meet operational expenditure (OpEx) billing targets only at a larger scale of time, e.g., over a month. Hence, we introduce the time-average budget constraint as follows:

$$\lim_{T \to \infty} \frac{1}{T} \sum_{t=0}^{T-1} \sum_{j \in \mathcal{J}} y_j(t) h_j(t) \leq B_{avg}, \quad \text{(billing constraint)} \tag{19.1}$$

where $\sum_{j \in \mathcal{J}} y_j(t) h_j(t)$ represents the total investment in slot t, the LHS is the time average investment, and B_{avg} is the available *average budget* per hour to be spent on storage.

To determine the average delay experienced within an hour, we must describe carefully how each file request is served. For this, we consider the file caching $z_{j,f}(t) \in \{0,1\}$ and association $x_{ij,f}(t) \in [0,1]$. For example, $z_{j,f}(t) = 1$ if content f is placed at small BS j at time slot t, and $x_{ij}(t) = 1$ if location i requests content f to small BS j at time slot t. Now, we consider the hourly end-to-end delay benefit from file caching for the area i and small BS j as follows:

$$D_{ij}(\boldsymbol{x}(t), \boldsymbol{z}(t); \boldsymbol{\Lambda}(t), \boldsymbol{d}(t)) = (d_{is}(t) - d_{ij}(t)) \sum_{f \in \mathcal{F}} x_{ij,f}(t) z_{j,f}(t) \lambda_{i,f}(t). \tag{19.2}$$

Observe that the delay is a function of the proportion of served traffic to location i ($x_{ij,f}(t)$), existance of file f at small BS j ($z_{j,f}(t)$), and the amount of requests ($\lambda_{i,f}(t)$). Note that the hourly end-to-end delay benefit of each area and each BS is defined as the total reduction of end-to-end delay for that area thanks to edge caching. By improving in this way the QoS, the CP can make more profits from end users. For example, the CP can set the price for the content delivery service in proportion to the delay savings. Then, the unit price multiplied by the sum of end-to-end delay benefit for all areas and small BSs reflects the total revenue of the CP. The total delay benefit in slot t is:

$$g_t(\boldsymbol{x}(t), \boldsymbol{z}(t); \boldsymbol{\Lambda}(t), \boldsymbol{d}(t)) = \sum_{i \in \mathcal{I}} \sum_{j \in \mathcal{J}_i} D_{ij}(\boldsymbol{x}(t), \boldsymbol{z}(t); \boldsymbol{\Lambda}(t), \boldsymbol{d}(t)). \tag{19.3}$$

Below, we will drop $(\boldsymbol{\Lambda}(t), \boldsymbol{d}(t))$ from the argument of g, though the dependence on these parameters remains implied. We note that the total delay and the total delay benefit add up to a constant term (equal to the total delay without caching) and hence minimizing total delay is equivalent to maximizing total delay benefit. We focus on the latter.

A number of constraints must be satisfied at each time slot. Specifically, the entire demand emanating from each area must be served by the small BSs or ultimately the MBS:

$$\sum_{j \in \mathcal{J}_i \cup \{s\}} x_{ij,f}(t) = 1, \quad \forall i, f, t, \quad \text{(service constraint)} \tag{19.4}$$

and the file placement is limited by the available leased storage:

$$\sum_{f \in \mathcal{F}} z_{j,f}(t) \leq y_j(t)/b, \quad \forall\, j, t, \quad \text{(storage space constraint)} \tag{19.5}$$

where b denotes the file size. Even though the size for all files is assumed to be identical for simplicity in this chapter, different file sizes can be modeled if we consider a large file that consists of several chunks with the identical size. Please find a summary of notations in Table 19.2.

19.4.4 Problem Formulation

The system is operated with an *elastic CDN strategy*, which at slot t maps the current state of the system to a decision tuple $\big(x_{ij,f}(t), y_j(t), z_{j,f}(t)\big)$. An elastic CDN strategy is called *feasible* if it satisfies the billing constraint (19.1) and the instantaneous

Table 19.2 Summary of the Notation

Notation	Definition	Notation	Definition
$i \in \mathcal{I}$	Area Index	$d_{is}(t)$	Average Delay for Serving Area i by Remote Servers
$j \in \mathcal{J}$	SBS Index	$d_{ij}(t)$	Average Delay for Serving Area i by SBS j
s	MBS Index	$h_j(t)$	Price to Lease Cache Storage per unit bit
$f \in \mathcal{F}$	File Index	$\lambda_{i,f}(t)$	Demand Profile
B_{avg}	Average Budget Constraint	$y_j(t)$	Leased Cache Space at SBS j
t	Hour Index (time slot)	$z_{j,f}(t)$	File Caching Indicator
$x_{ij,f}(t)$	Association Probability		

constraints of service (19.4) and caching space (19.5) explained earlier. We consider the mobile operator's curiosity: What is the feasible elastic CDN polity that maximizes the average delay benefit? This question can be addressed by the following control problem:

$$\textbf{(P)}: \max_{x,y,z} \lim_{T\to\infty} \frac{1}{T}\sum_{t=0}^{T-1} g_t(\boldsymbol{x}(t),\boldsymbol{z}(t)), \tag{19.6}$$

$$s.t. \lim_{T\to\infty} \frac{1}{T}\sum_{t=0}^{T-1}\sum_{j\in\mathcal{J}} y_j(t)h_j(t) \leq B_{avg},$$

$$\sum_{j\in\mathcal{J}_i} x_{ij,f}(t) = 1, \forall i,f,t, \quad \sum_{f\in\mathcal{F}} z_{j,f}(t) \leq y_j(t)/b, \forall j,t.$$

In other words, the problem **(P)** is to find investment for edge caching, cached files and area-cache association solutions at every time slot to maximize the average delay benefit, i.e., indirectly maximizing the CP's profits, by staying below a predetermined budget.

We should note that the control problem is difficult to solve because

- Crucial factors for the objective such as future traffic demand $\lambda_{i,f}(t)$ and future delay gains $d_{is}(t) - d_{ij}(t)$ are unknown at the time the investment decisions $y_j(\tau)$ are taken ($\tau < t$).
- Due to the time average billing constraint, a large investment $y_j(\tau)$ reduces the available budget in future slots $t > \tau$, which can be problematic in combination with the unknown future costs $h_j(t)$, delays $d_{ij}(t), d_{is}(t)$, and traffic demand $\lambda_{i,f}(t)$.

19.4.5 Lyapunov-Based Elastic CDN Strategy

Since problem **(P)** involves the challenging time-average constraint (19.1), a promising approach is to couple the fate of this constraint with an evolving controllable counter. To this end, let us consider a virtual queue (or counter) as follows.

$$Q_B(t+1) = \left[Q_B(t) + \sum_{j \in \mathcal{J}} y_j(t) h_j(t) - B_{avg} \right]^+. \quad (19.7)$$

Related work [31] proved that if weak stability conditions hold for the virtual queue, i.e.,

$$\lim_{T \to \infty} \frac{1}{T} \sum_{t=0}^{T-1} Q_B(t) < \infty, \quad (19.8)$$

then constraint (19.1) is asymptotically satisfied, in the sense that its residual tends to zero as $T \to \infty$. In other words, satisfying (19.8) implies that the time-average of billing constraint (19.1) is satisfied. Intuitively, the backlog $Q_B(t)$ estimates the total excess budget spent in the previous time slots (instantaneous residual), and therefore $Q_B(t)$ is valuable information for deciding how to invest at slot t. Then, let us focus on slot t. The decision maker is aware of (1) the mean traffic demand profile for the next hour $[\lambda_{i,f}(t)]_{i,f}$, which in practice is achieved by measurements and use of machine learning methods, cf. [28]; (2) the delay profile realizations $[d_{ij}(t)]_{i,j}$, which are readily available by measurements; (3) the prices $[h_j(t)]_j$; and (4) the virtual queue length $Q_B(t)$, while file size b is assumed to be known. Therefore, the elastic CDN strategy is applied on the state $\big([\lambda_{i,f}(t)]_{i,f}, [d_{ij}(t)]_{i,j}, [h_j(t)]_j, Q_B(t)\big)$. To design a strategy that solves **(P)** we leverage a Lyapunov drift-minus-benefit framework in the following.

We first define the quadratic Lyapunov function and arising drift in the following.

$$L(t) \triangleq \frac{1}{2} Q_B(t)^2, \quad (19.9)$$

$$\Delta(t) \triangleq \mathbb{E}\{L(t+1) - L(t) | Q_B(t)\}. \quad (19.10)$$

The meaning of minimizing (19.10) is that we strive to stabilize the virtual queue $Q_B(t)$ or satisfy the billing constraint (19.1). Readers can refer to [29] for more theoretical information on the Lyapunov function and arising drift. In addition, our aim is to maximize the time-average total delay benefit $\lim_{T\to\infty} \frac{1}{T} \sum_{t=0}^{T-1} g_t$ while satisfying the stability of the virtual queue. Therefore, we consider the Lyapunov drift-minus-benefit function (DMB), which balances the drift and the instantaneous obtained delay benefit:

$$DMB(\boldsymbol{x}(t), z(t)) = \Delta(t) - V\mathbb{E}\{g_t(\boldsymbol{x}(t), z(t)) | \boldsymbol{Q}(t)\}. \quad (19.11)$$

Here, V denotes a trade-off parameter between queue stability and maximization of total average delay benefit. In summary, trying to minimize (19.11) in every time slot has the similar meaning with trying to satisfy the original long-term problem **(P)**, i.e., maximizing the average delay benefit and remaining below a predetermined budget on average.

From equation (19.7) and [32], we have the following inequality for $([y_j(t)]_j, [x_{ij,f}(t)]_{ijf}, [z_{j,f}(t)]_{jf})$:

$$\begin{aligned} DMB(\boldsymbol{x}(t), z(t)) &\leq P - V\mathbb{E}\{g_t(\boldsymbol{x}(t), z(t)) | Q_B(t)\} \\ &- \mathbb{E}\Big\{\Big(B_{avg} - \sum_{j \in \mathcal{J}} y_j(t) h_j(t)\Big) Q_B(t) | Q_B(t)\Big\}, \end{aligned} \quad (19.12)$$

where $P = (B_{avg}^2 + |\mathcal{J}|y_{max}^2 h_{max}^2)/2$ is a positive constant, and y_{max} and h_{max} denote the maximum storage that can be leased at any small BS during an hour, and the maximum price respectively. One study [29] showed the optimal solutions of our problem are the same with a minimization of the right-hand-side of (19.12) in every time slot.

We propose the elastic CDN strategy (SBSD), which at slot t takes actions $(\boldsymbol{x}(t), \boldsymbol{y}(t), \boldsymbol{z}(t)) = (\boldsymbol{x}^*, \boldsymbol{y}^*, \boldsymbol{z}^*)$, where

$$(\boldsymbol{x}^*, \boldsymbol{y}^* \boldsymbol{z}^*) \in \arg\max_{\boldsymbol{x},\boldsymbol{y},z} V g_t(\boldsymbol{x}, z) - \sum_{j \in \mathcal{J}} Q_B(t) y_j h_j(t), \tag{19.13}$$

$$s.t. \quad \sum_{j \in \mathcal{J}_i} x_{ij,f} = 1, \forall i, f, t, \quad \sum_{f \in \mathcal{F}} z_{j,f} \leq y_j/b, \forall j, t.$$

The first straightforward result is that SBSD is a feasible elastic CDN policy. First, the instantaneous constraints of service (19.4) and storage space (19.5) are automatically satisfied at each slot by the design of the policy. Then we may observe that SBSD minimizes the RHS of (19.12), therefore using lemma 4.6 in [29], we can show that SBSD also stabilizes $Q_B(t)$, and hence the billing constraint (19.1) is asymptotically satisfied.

Some further remarks are in order:

- As long as the technical requirement "$\boldsymbol{\lambda}(t)$ and $\boldsymbol{d}(t)$ have finite second moments" is satisfied (used in the proof of asymptotic feasibility), SBSD will satisfy the budget constraint.
- Information of the hourly file popularity, delay profile, and instantaneous budget counter make the caching system achieve a close to optimal performance despite absence of future file popularity or future delay profile.

It remains to solve the slot-by-slot problem (19.13) in every time slot. We provide the solutions for two different cases: nonoverlapping and overlapping small BS coverages.

Nonoverlapping small BS coverage. When small BS coverage is nonoverlapping, each area can reach a single small BS cache, which immediately simplifies routing splits $x_{ij,f}(t)$ to $x_{ij,f}(t) = 1$, $\forall t$ if area i can reach small BS j (otherwise, $x_{ij,f}(t) = 0$), for all i, j, f. In essence, each request can be served only by the reachable cache (or the MBS when the file is not cached there). Now, it decouples the small area BS association problem from the file caching and caching capacity dimensioning problems as follows. First, we note that caching file f at small BS j in slot t yields the delay benefit as follows:

$$K_{j,f}(t) \triangleq \sum_i (d_{is}(t) - d_{ij}(t)) x_{ij,f}(t) \lambda_{i,f}(t),$$

which is calculated by $\boldsymbol{d}, \boldsymbol{x}, \boldsymbol{\lambda}$ ($\boldsymbol{x}$ is a parameter here because it is fully determined by the reachability of the cache) and independent of the decisions $\boldsymbol{y}(t), \boldsymbol{z}(t)$. Consequently, the SBSD optimization becomes

$$\max_{\substack{y_j \geq 0 \\ z_{j,f} \in \{0,1\}}} \quad V \sum_{j,f} K_{j,f}(t) z_{j,f} - Q_B(t) \sum_{j \in \mathcal{J}} y_j h_j(t), \tag{19.14}$$

$$s.t. \quad \sum_{f \in \mathcal{F}} z_{j,f} \leq y_j/b, \; \forall j, f.$$

Because y_j is an integer multiplication of the file size b and $z_{j,f}(t)$ has a binary value, the first term of (19.14) can be expressed with a relation of y_j as follows: $V\sum_{f=1}^{\lfloor y_j/b \rfloor} K_{j,\sigma(f)}(t)$, where $\sigma(f)$ denotes file index when we let $K_{j,\sigma(1)}(t) \geq \cdots \geq K_{j,\sigma(|\mathcal{F}|)}(t)$. Then, the problem can be transformed so as to find $y_j^*(t)$ as follows:

$$y_j^*(t) \in \arg\max_{y_j \geq 0} \sum_{f=1}^{\lfloor y_j/b \rfloor} K_{j,\sigma(f)}(t) - \frac{Q_B(t)}{V} h_j(t) y_j. \tag{19.15}$$

This problem can be easily solved by finding the maximum value among $\lfloor y_{max}/b \rfloor$ candidates. Then, the complexity of this algorithm in each small BS becomes $O(\lfloor y_{max}/b \rfloor)$.

Moreover, since we decouple the problem with the area-small BS association, the problem can be decomposed into each small BS's problem. Next, we give the algorithmic steps to find $\boldsymbol{y}_j$ and $\boldsymbol{z}_j$ for all small BS $j \in \mathcal{J}$ optimally in detail:

1. Calculate $K_{j,f}(t) = \sum_i (d_{is}(t) - d_{ij}(t)) x_{ij,f}(t) \lambda_{i,f}(t)$ for all files.
2. Sort $K_{j,f}(t)$ with permutation σ, such that $K_{j,\sigma(1)}(t) \geq \cdots \geq K_{j,\sigma(|\mathcal{F}|)}(t)$.
3. Let $S(e) = \sum_{f=1}^{e} K_{j,\sigma(f)}(t)$, for $e = 1, 2, \ldots$.
4. Find e^* that is the smallest e that ensures $S(e) - S(e-1) < \frac{Q_B(t)}{V} h_j(t)$.
5. Calculate the optimal cache lease capacity: $y_j^*(t) = e^* b$.
6. Calculate file caching:

$$z_{j,\sigma(f)}^*(t) = \begin{cases} 1 & \text{if } f \leq \lfloor y_j^*(t)/b \rfloor, \\ 0 & \text{otherwise.} \end{cases}$$

Then, the algorithm, namely the Optimal algorithm in the nonoverlapping small BS case has the following features:

- Given virtual queue length, storage price, and parameter V, the algorithm finds the amount of storage that if leased it optimizes a weighted sum of delay benefits and budget penalties.
- For the found storage amount that is leased, files are cached at each small BS according to which yields the highest delay benefit, until the available leased storage is completely filled up.

Overlapping small BS coverage. Now, this chapter addresses a scenario that each location can be served from different small BSs. Therefore, we have to consider the area and small BS routing problem whose variables are $x_{ij,f}(t)$ in addition to the lease of cache capacity and file caching, and we may no longer use the trick with $K_{j,f}(t)$, since the file request from user can be addressed from possibly several small BSs and the actual collected delay benefit depends on which small BS is selected. We remind the reader that the SBSD strategy determines the decisions that solve:

$$\max_{\substack{y_j \geq 0 \\ x_{ij,f} \in [0,1] \\ z_{j,f} \in \{0,1\}}} \quad V g_t(\boldsymbol{x}, \boldsymbol{z}) - Q_B(t) \sum_{j \in \mathcal{J}} y_j h_j(t), \tag{19.16}$$

$$\text{s.t.} \quad \sum_{f \in \mathcal{F}} z_{j,f} \leq \frac{y_j}{b}, \forall j,$$

$$\sum_{j \in \mathcal{J}_i} x_{ij,f} = 1, \forall i, f.$$

It is difficult to directly solve (19.16) since $x_{ij,f}$ and $z_{j,f}$ are tightly coupled with each other. Therefore, we remove $z_{j,f}$ from (19.16) and include a constraint $x_{ij,f}(t) \leq z_{j,f}(t)$.

It should be noted that (19.16) as formulated is a mixed integer *nonlinear* program (MINLP) because of $x_{ij,f}(t)z_{j,f}(t)$, which is located inside g_t. Hence we remove $z_{j,f}(t)$ from the objective and add an extra constraint $x_{ij,f}(t) \leq z_{j,f}(t)$. For example, if $z_{j,f}(t) = 1$, $x_{ij,f}(t)$ is not affected by the new constraint, and works as before. Otherwise, $z_{j,f}(t) = 0$, $x_{ij,f}(t)$ becomes 0 due to the new constraint.

Now, the problem is transformed into a mixed integer linear program (MILP). To solve this problem, we can consider two approaches as follows:

- We may solve the linear relaxation of (19.16) and then use a rounding technique to obtain an approximation guarantee, e.g., a possibility is to combine the relaxation with randomized rounding [33]. According to section 4.7 of [34], our approximate solution of (19.16) will provide an elastic CDN strategy with an approximate feasibility and average delay benefit. In turn, the approximate feasibility can lead to a feasible strategy with some extra losses.
- A second approach is to obtain an efficient approximate solution is to apply the idea of "low complexity scheduling" from [35]. This method assigns to the leased cache capacity by smoothly increasing it or decreasing it with small step size. The sign of the change is randomly chosen. Then it resolves our SBSD optimization to get a new average delay benefit, and if these new values outperform previous delay benefits, the random solution is applied.

In this chapter, we take the second method as an example for deriving an algorithm in the general case. In this context, we provide a stability guarantee for the budget queue length $Q_B(t)$, which implies that the produced strategy is asymptotically feasible. The strategy, namely, a *randomized* algorithm is described as follows:

1. For the first time slot, leased cache capacity $y_j^*(1)$ is chosen as $B_{avg}/(|\mathcal{J}|h_{avg})$ for all small BSs.
2. Based on the decided leased storage for each small BS, file caching and user association solutions $(\boldsymbol{x}^*(t), \boldsymbol{z}^*(t))$ are obtained using a greedy file caching (GFC) policy and an optimal user association (OUA) policy for a given file caching solution, which are described in the following.

3. For time slots $t > 1$, leased cache capacity $y'_j(t)$ is chosen as $y^*_j(t-1) + \delta \cdot U_j(t-1)$ where δ denotes small step size and $U_j(t-1)$ is uniformly chosen in $\{-1, 1\}$ for all small BSs.
4. Based on the decided leased storage for each small BS, file caching and user association solutions $(\boldsymbol{x}'(t), z'(t))$ are obtained using a GFC policy with an OUA policy for a given file caching solution.
5. Compare $Vg_t(\boldsymbol{x}'(t), z'(t)) - Q_B(t)\sum_{j\in\mathcal{J}} y'_j(t)h_j(t)$ and $Vg_t(\boldsymbol{x}^*(t-1), z^*(t-1)) - Q_B(t)\sum_{j\in\mathcal{J}} y^*_j(t-1)h_j(t)$ and choose a set of solutions whose objective value is greater as an optimal set of solutions, i.e., $(\boldsymbol{x}^*(t), z^*(t))$.

The GFC policy begins with an empty cache set in all small BSs. Then, this policy iteratively caches files one-by-one in all small BSs where an added file in each step is selected so as to maximize the differential objective value in (19.16). For a given set of cached files, the OUA policy is to choose association variables by

$$x^*_{ij,f}(t) = \arg\max_{x_{ij,f}(t)} (d_{is}(t) - d_{ij}(t))z^*_{j,f}(t)\lambda_{i,f}(1), \quad \forall i, f.$$

This solution is robust due to the comparison mechanism between the solution of the current time slot and that of the previous time slot. Namely, if the budget queue increases due to the excessive investment for cache lease, it reduces the objective value, hence forcing the decision maker to choose the solution of the previous time slot. On the other hand, if the budget queue decreases due to the less investment for cache lease, it increases the objective value, hence forcing the decision maker to choose the solution of the current time slot. This mechanism stabilizes the budget queue.

Moreover, for a given leased cache capacity, a joint file caching and user association problem is shown to be a monotone submodular problem with matroid constraints in respect to cached files in small BSs according to the recent literature e.g., [36]. This implies a greedy-fashioned file caching algorithm in conjunction with the OUA policy (for a given file caching solution) probably achieves a constant factor approximation $(1 - 1/e)$ to the optimal performance.

To quantify the performance improvement of the elastic cache lease and file caching over the static policies, we run simulations under a simple nonoverlapping small BS scenario (10 small BSs, 10 areas, 50 files in each small BS). We assume that each area is associated with the nearest small BS. In this scenario, delay for the corresponding zone by a small BS and delay for the corresponding zone by the original file servers in each time slot are picked from the Gaussian distribution with various parameters and given only positive values. To capture the spatio-temporal diversity of file popularity, the arrival rate of each file is drawn from the Zipf distribution [28] and different Zipf parameters are used for each area and each period of time slots.[5]

We compare the proposed algorithms of optimal and randomized with the static caching and static budget policies. The static caching policy caches the files based on the general content popularity with the static cache investment, i.e., caching the same number of files at all small BSs, whereas the static budget policy uses the static

[5] The sum traffics for all files at each time slot and each area are picked from the Gaussian distribution with various parameters and taken only positive values.

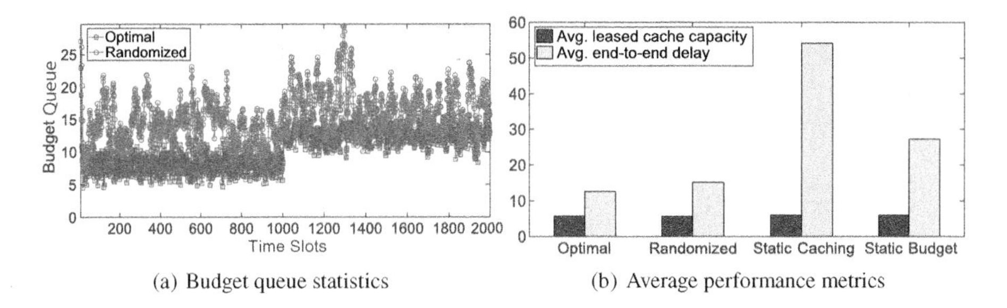

(a) Budget queue statistics (b) Average performance metrics

Figure 19.4 Improvement in performance due to the elastic cache lease and file caching and the elastic behavior in a non-overlapping SBS scenario, with 10 SBSs, 10 areas, and 50 files in each SBS. The average cache lease budget is the same with caching 10 files in each time slot.

cache lease for all time slots but file caching is chosen so as to maximize our objective function in SBSD, i.e., this policy adopts an adaptive file caching for a given cache capacity.

Figure 19.4 shows the budget queue statistics of the elastic algorithms, the end-to-end statistics, and leased cache capacity for all algorithms. The elastic algorithms (i.e., optimal and randomized) opportunistically exploit the dynamics of network delays and file request arrivals with keeping average leased cache capacity, whereas the static algorithms (i.e., static caching and static budget) lease a fixed amount of budget every time slot. For example, when the traffic demand is high and the Zipf parameter is small, then the larger cache capacity is leased, and vice versa. Hence some interesting remarks here are the following: (1) the elastic algorithms perform better than the static algorithms in terms of total end-to-end delay (at least 53% reduction) using even less cache capacity. (2) The randomized policy achieves close to the optimal performance (83% in terms of average end-to-end delay). These results can be found in real spatiotemporal traffic and content popularity variation scenarios where the traffic arrival is high and Zipf parameter is small during the day, while being low and high, respectively, during night hours (temporal diversity); and these statistics depend on the area, e.g., when the CDN simultaneously serves locations in different time zones (spatial diversity). Note that the end-to-end delay indirectly captures the average profits of CPs from end users since the QoS of end users depend on the delay.

Figure 19.5 depicts the end-to-end delay for different average parameters, i.e., average demand, price, and budget. The main differences between the proposed elastic algorithms and the baseline static algorithms for large average demand, price, and budget, arise due to the fact that the proposed algorithms have more degrees of freedom and can exploit a given budget more flexibly than the baseline static algorithms. Therefore, the CP increases its profits from subscribers by offering better QoS. From the perspective of the Telco-CDN operator, e.g., AWS, they are likely to be unhappy since the CP can save money by not using memory at low traffic. However, the real economical benefits of the CDN operator will appear if the remained memory can be covered by other CPs because several CPs have more budget by making more profits from their end users.

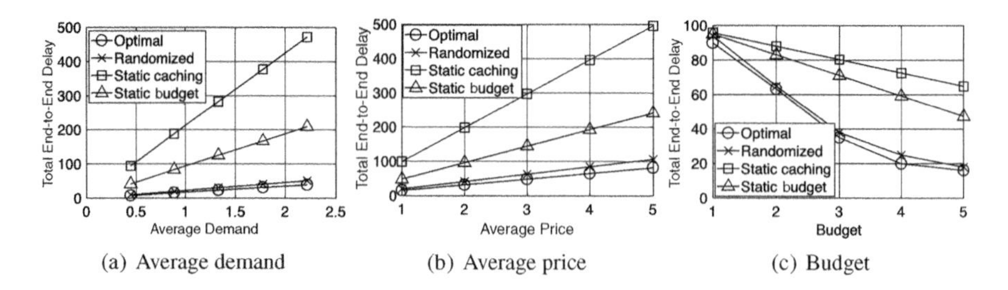

(a) Average demand (b) Average price (c) Budget

Figure 19.5 Total end-to-end delay versus different parameters. The simulation scenario is the same as Figure 19.4 except average demand, average price, and budget. As average demand, price, and budget become larger, the delay benefit of the proposed algorithms get higher. For example, the performance difference between the proposed randomized algorithm and static caching is 1% for budget 1, whereas the performance difference between them is 72.31% for budget 5.

19.5 Open Research Issues

In this chapter, we proposed an elastic edge caching solution for small CPs leasing storage from a wireless Telco-CDN, which relies on a general model and hence captures different technical and economic aspects of this idea. However, this is a novel and quite rich research problem with many challenges that should be carefully addressed. We outline some of the most important challenges in this section.

Fairness mechanism among subscribers: In most of the previous works on caching, including our proposal in this chapter, the objective is to maximize the aggregate performance for all users, e.g., here the total delay savings. However, from an economic point of view, it is often beneficial (sometimes, also necessary) for the CP to offer different classes of services to the users, and each user to subscribe to the one that is suitable for its needs. For instance, some of the services might offer content delivery delay guarantees, while others might not. It is therefore necessary to design a fairness mechanism that will ensure all users within each class enjoy the same QoS and that users in different service classes will be given priority according to their subscription level.

Different objectives of different stakeholders: In this chapter, we analyzed the problem from the perspective of the CP that attempts to optimize the delay savings for a given investment budget. Other stakeholders however, such as CDN providers, MNOs, or even larger CPs (e.g., Amazon AWS, YouTube and AT&T) can have different objectives. For example, AWS would like to select a pricing policy that maximizes its revenue, instead of alleviating the delivery delay of files. Similarly, a Telco might prefer to prioritize the cache-hit ratio, independently of the delay that is induced from the caching policy, in order to save off-network bandwidth. It is thus necessary to analyze the problem from the different perspectives of these stakeholders and, going a further step, understand how these can be aligned.

Incentives: Indeed, the underlying assumption in our model is that the involved entities, i.e., the CP, the Telco-CDN, and the end users will all agree to participate in this cooperative framework. However, this requires a mechanism that will ensure that their incentives are properly aligned. For instance, the coordination between the CP and the Telco-CDN creates additional revenue, both due to bandwidth savings and the improvement of the offered service. It is hence natural to investigate how this profit can be dispersed among these entities so as to incentivize their collaboration.

19.6 Conclusion

The caching economics is one of the less studied areas in the literature of content caching, yet it is quickly increasing thanks to the increasing interests in network and storage softwarization. These technologies revolutionize the business of caching and give rise to new ideas and new types of caching services. And even more opportunities, as well as also challenges, arise when we focus on the inherently volatile wireless/mobile networks. It is clear to us and hopefully to the reader as well at this point that these challenges require a hybrid technoeconomic approach as they cannot be solely addressed with engineering solutions.

In this chapter, we focused on the fast emerging wireless elastic CDN paradigm, which is as promising as is challenging to deploy. We started with a discussion about the current state of the caching ecosystem, presenting the latest business models. We highlighted the difference of this edge caching solution with the standard core in-network caching architecture, both from a technical and from an economic perspective. We then focused on a specific elastic cache lease scheme, and designed a joint policy that dimensions the caches on demand, and decides the content caching and small-request BS routing in order to achieve the minimum possible delay. The decisions are being made by a CP that has a limited time-average (elastic) investment budget. The problem was solved by the design of a dynamic policy that relies on the Lyapunov optimization technique and randomized scheduling to alleviate the complexity for the implementation. Finally, we demonstrated the benefits of the elastic cache leasing over static cache leasing policies and discussed several open issues that must be addressed before this promising idea is widely deployed.

References

[1] "Cisco visual networking index: global mobile data traffic forecast update, 2015–2020," Cisco, www.cisco.com/c/en/us/solutions/collateral/service-provider/visual-networking-index-vni/mobile-white-paper-c11-520862.pdf

[2] S.-E. Elayoubi and J. Roberts, "Performance and cost effectiveness of caching in mobile access networks," in *Proc. ACM ICN*, 2015, pp. 79–88.

[3] M. Leconte, G. Paschos, L. Gkatzikis, M. Draief, S. Vassilaras, and S. Chouvardas, "Placing dynamic content in caches with small population," in *Proc. IEEE INFOCOM*, Apr. 2016, pp. 1–9.

[4] "Amazon web service, https://aws.amazon.com

[5] S. Elayoubi and J. Roberts, "Performance and cost effectiveness of caching in mobile access networks," in *Proc. ACM ICN*, 2015, pp. 79–88.

[6] G. Paschos, G. Iosifidis, M. Tao, D. Towsley, and G. Caire, "The role of caching in future communication systems and networks," *IEEE JSAC*, vol. 36, no. 6, pp. 1111–1125, June 2018.

[7] S. Vassilaras, L. Gkatzikis, N. Liakopoulos, I. N. Stiakogiannakis, M. Qi, L. Shi, L. Liu, M. Debbah, and G. S. Paschos, "The algorithmic aspects of network slicing," *IEEE Commun. Mag.*, vol. 55, no. 8, pp. 112–119, 2017.

[8] Akamai White Paper, "The case for a virtualized CDN(vCDN) for delivering operator OTT video," Akamai, White Paper.

[9] "The elastic CDN solution (akamai-juniper)," www.juniper.net/assets/kr/kr/local/pdf/solutionbriefs/3510532-en.pdf.

[10] M. K. Mukerjee, I. N. Bozkurt, D. Ray, B. M. Maggs, S. Seshan, and H. Zhang, "Redesigning cdn-broker interactions for improved content delivery," in *Proc. ACM CoNEXT*, 2017, pp. 68–80.

[11] "Akamai collaborates with orange on NFV initiative to dynamically scale CDN capacity for large events," www.akamai.com/us/en/about/news/press/2016-press/akamai-collaborates-with-orange-on-nfv-initiative.jsp.

[12] "Huawei uCDN solution," carrier.huawei.com/en/solutions/cloud-powered-digital-services/ucdn.

[13] "Amazon ElastiCache," https://aws.amazon.com/elasticache/

[14] K. Shanmugam, N. Golrezaei, A. Dimakis, A. Molisch, and G. Caire, "Femtocaching: Wireless content delivery through distributed caching helpers," *IEEE Trans. Inform. Theory*, vol. 59, no. 12, pp. 8402–8413, Sept. 2013.

[15] K. Poularakis, G. Iosifidis, and L. Tassiulas, "Approximation algorithms for mobile data caching in small cell networks," *IEEE Trans. Commun.*, vol. 62, no. 10, pp. 3665–3677, 2014.

[16] K. Poularakis, G. Iosifidis, V. Sourlas, and L. Tassiulas, "Exploiting caching and multicast for 5g wireless networks," *IEEE Trans. Wireless Commun.*, vol. 15, no. 4, pp. 2995–3007, 2016.

[17] J. Krolikowski, A. Giovanidis, and M. Renzo, "A decomposition framework for optimal edge-cache leasing," *IEEE JSAC*, vol. 36, no. 6, pp. 1345–1359, June 2018.

[18] J. Kwak, Y. Kim, L. Le, and S. Chong, "Hybrid content caching in 5G wireless networks: cloud versus edge caching," *IEEE Trans. Wireless Communications*, vol. 17, no. 5, pp. 3030–3045, May 2018.

[19] J. Kwak, L. Le, H. Kim, and X. Wang, "Two time-scale edge caching and BS association for power-delay tradeoff in multi-cell networks," *IEEE Trans. Commun.*, vol. 67, no. 8, pp. 1–14, 2019.

[20] T. Bektas, O. Oguz, and I. Ouveysi, "Designing cost-effective content distribution networks," *Comput. Oper. Res.*, vol. 34, no. 8, pp. 2436–2449, 2007.

[21] K. Ho, S. Georgoulas, M. Amin, and G. Pavlou, "Managing traffic demand uncertainty in replica server placement with robust optimization," *Proc. NETWORKING*, pp. 727–739, 2006.

[22] W. Li, E. Chan, Y. Wang, D. Chen, and S. Lu, "Cache placement optimization in hierarchical networks: analysis and performance evaluation," *Proc. NETWORKING*, pp. 385–396, 2008.

[23] N. Laoutaris, V. Zissimopoulos, and I. Stavrakakis, "On the optimization of storage capacity allocation for content distribution," *Comput. Networks*, vol. 47, no. 3, pp. 409–428, 2005.

[24] K. Poularakis, G. Iosifidis, I. Pefkianakis, L. Tassiulas, and M. May, "Mobile data offloading through caching in residential 802.11 wireless networks," *IEEE Trans. Network Service Manage.*, vol. 13, no. 1, pp. 71–84, 2016.

[25] B. Frank, I. Poese, Y. Lin, G. Smaragdakis, A. Feldmann, B. Maggs, J. Rake, S. Uhlig, and R. Weber, "Pushing CDN-ISP collaboration to the limit," *ACM SIGCOMM Comput. Commun. Rev.*, vol. 43, no. 3, pp. 34–44, 2013.

[26] "Google Global Cache (GCC)," https://peering.google.com/#/infrastructure.

[27] A. Berglund, "How Netflix works with ISPs around the globe to deliver a great viewing experience," *Netflix Blog*, 2016.

[28] E. Bastug, M. Bennis, and M. Debbah, "Living on the edge: the role of proactive caching in 5G wireless networks," *IEEE Commun. Mag.*, vol. 52, no. 8, pp. 82–89, Feb. 2014.

[29] M. Neely, "Stochastic network optimization with application to communication and queueing systems," *Synth. Lectures. Commun. Networks*, pp. 1–211, 2010.

[30] "CAISO: California independent system operator," www.caiso.com.

[31] M. Neely, "Energy optimal control for time varying wireless networks," *IEEE Trans. Inform. Theory*, vol. 52, no. 7, pp. 2915–2934, July 2006.

[32] L. Georgiadis, M. Neely, and L. Tassiulas, "Resource allocation and cross-layer control in wireless networks," *Found. Trends Networking*, vol. 1, no. 1, pp. 1–149, 2006.

[33] P. Raghavan and C. D. Tompson, "Randomized rounding: a technique for provably good algorithms and algorithmic proofs," *Combinatorica*, vol. 7, no. 4, pp. 365–374, 1987.

[34] L. Georgiadis, M. J. Neely, L. Tassiulas, L. Georgiadis, M. J. Neely, and L. Tassiulas, "Resource allocation and cross-layer control in wireless networks," *Found. Trends Networking*, vol. 1, no. 1, pp. 1–144, 2006.

[35] L. Tassiulas, "Linear complexity algorithms for maximum throughput in radio networks and input queued switches," in *Proc. IEEE INFOCOM*, Apr. 1998, pp. 533–539.

[36] M. Dehghan, A. Seetharam, B. Jiang, T. He, T. Salonidis, J. Kurose, D. Towsley, and R. Sitaraman, "On the complexity of optimal routing and content caching in heterogeneous networks," in *Proc. IEEE INFOCOM*, Apr. 2015, pp. 936–944.

Index